AUTECOLOGY

Organisms, Interactions and Environmental Dynamics

AUTECOLOGY
Organisms, Interactions and Environmental Dynamics

Gimme H. Walter
School of Biological Sciences
The University of Queensland
Brisbane QLD
Australia

Rob Hengeveld
Faculteit der Aard- en Levenswetenschappen
Vrije Universiteit
HV Amsterdam
The Netherlands

CRC Press
Taylor & Francis Group
Boca Raton London New York

CRC Press is an imprint of the
Taylor & Francis Group, an **informa** business
A SCIENCE PUBLISHERS BOOK

CRC Press
Taylor & Francis Group
6000 Broken Sound Parkway NW, Suite 300
Boca Raton, FL 33487-2742

Cover Picture—The cover picture shows the traces left in the bark of a scribbly gum (*Eucalyptus racemosa*) by the mining of larvae of the scribbly gum moth (genus *Ogmograptis*, family Bucculatricidae). It illustrates the autecological perception of the way in which organisms move haphazardly in the environment - conditions elsewhere are unknown to them.

Printed in the United States of America on acid-free paper

International Standard Book Number: 978-1-4822-1414-7 (Hardback)

Library of Congress Cataloging-in-Publication Data

Walter, G. H. (Gimme Hugh), 1954-
Autecology : organisms, interactions and environmental dynamics / Gimme H. Walter, Rob Hengeveld.
pages cm
Includes bibliographical references and index.
ISBN 978-1-4822-1414-7 (hardback)
1. ECOLOGY--PHILOSOPHY. 2. BIOTIC COMMUNITIES. 3. PALEOECOLOGY. I. HENGEVELD, ROB. II. TITLE.
QE720.W35 2014
577.01--dc23

2014004179

Visit the Taylor & Francis Web site at
http://www.taylorandfrancis.com

CRC Press Web site at
http://www.crcpress.com

Science Publishers Web site at
http://www.scipub.net

Dedication

A few decades ago a story (perhaps apocryphal) was popular among ecologists, probably because it involved a physiologist. Ecology was on the rise and seen to be replacing experimental physiology in the next golden age in biology. So intrigued with the technology of the trade was the physiologist, it was said, that the apparatus constructed for a particular test contained no provision for holding the test organism. Ironically, and in our mirth, we somehow failed to see that in our discipline, ecology, the abstraction and idealism of the theory had similarly failed to provide any place for the organism.

We dedicate this work to those population biologists who persistently reminded us that the theory representing ecology should also be strongly focused on the organism in its environment.

H.G. Andrewartha
W.H. Drury
F. Hüber
H.E.H. Paterson
W.G. Wellington
T.C.R. White

Preface

Biology as a whole is changing. New evidence becomes available daily, from observations that span the entire gamut of possible scales. Recent paleontological insights, for example, have been as influential as those associated with molecular techniques. Accordingly, general perceptions of biological systems have shifted dramatically, and these developments reflect the interplay between theory and new evidence. Advances of this kind represent the most exciting aspect of science, for they radically change the way in which we see the world.

Ecologists, too, have adjusted to new evidence in their primary quest of explaining the distribution, abundance and diversity of species. In general, though, the adjustments to ecological theory have proved rather superficial relative to the implications of the new findings from disciplines not traditionally associated with ecological theory, despite their relevance being widely acknowledged. The paleontological data provide a good example (Walter and Paterson 1994). A more fundamental shift is therefore needed if ecology is to remain in line with the developments in biology as a whole. In ***Autecology***, we develop and justify such realignment to ecological theory. Chapter one explains the basis of what we have done and provides an overview of the structure and place of autecological theory. The points made are expanded through the rest of the book and a glossary with an autecological focus is included.

Autecological research of the nature we recommend has been conducted by many researchers, many of them with applied ecology interests. Clearly, we do not claim novelty in this regard. What we do offer as new is the synthesis of this approach into a general methodology and theory that, together, provide a coherent and comprehensive alternative to the approach and interpretations of demographic ecology (which is represented chiefly by population and community ecology). Such an alternative, with the novel insights it generates, has never been available, not even in the thesis of Andrewartha and Birch (1954), for they could not reconcile their approach with any available evolutionary theory. Although demographic ecology theory has been subject to severe criticism for some time, it has retained its dominance. Its determinism and associated promise of accurate prediction, even with respect to future evolutionary trajectories, has an appeal that seems to be strengthened by want of an alternative.

In passing, we ask the reader to consider the ultimate purpose of ecological theory. We accept it is a generalization, both organizational and interpretive, of the observed patterns of organismal distribution, abundance and association. Consider this in the light of tendencies to divert theory towards the practical aspirations of the observer, whether that is spatial analysis with the aid of satellite-assisted technology or the conservation of organisms. Convenience, although desirable in a theory, should not force the theory to contradict the very nature of the systems it is meant to explain. Drury's (1998) warning, in his *Ecology for Conservationists*, reminds us that most ecologists now probably learn ecological theory before getting the chance to experience organisms in nature to any deep extent. Although this might be an inevitable consequence of our urbanization and institutional existence, the organisms in their own environment should be foremost in our minds when trying to interpret the living world.

The subject matter of ***Autecology*** is designed to penetrate to the heart of ecology and encourage scrutiny of the basic workings of the discipline. It justifies the introspective reconsideration of the basic direction of ecology and therefore poses questions about current practice that are uncompromising to some extent. This synthesis nevertheless represents an open and honest attempt to demonstrate that at least one sound and workable alternative general approach to the present system of ecological investigation and interpretation of distribution, abundance and diversity is available. The alternative we canvas should not be mistaken as an equivalent to the alternative hypotheses commonly presented in ecological writings, where alternative interpretations are advanced to justify a scientific test or explain a set of observations. Instead, we present a fully articulated alternative framework for the whole of ecological theory and practice. The message can therefore be judged only once fully assimilated so that it is understood as a whole. Interim judgments of components of the overall argument are necessary, but the overall argument should not be dismissed simply because existing theory or a new sub-discipline is seen to cope with that particular component or sub-discipline. Such superficial dismissal is likely to be precipitated if too little attention is paid to terminology. Terms and concepts require careful attention, for they are frequently used in subtle ways that imply meaningful differences. Such everyday terms as "species", "population" and "competition" already suffer from such misuse. Unless meanings are specific, precise and used consistently, ecology will not improve as a science.

The theory presented in ***Autecology*** has been designed as a replacement for demographic theory, which was developed to explain the numerical stability of organisms in local situations. This is the view that underpins Darwinian natural selection, to its detriment according to Brady (1982).

For demographic ecology to achieve its original goal, a suite of associated assumptions had to be made as a platform. These premises were generally poorly articulated and thus remained cryptic to most ecologists (Cooper 2001). During the past few decades, various observations have helped expose these assumptions, encouraged their scrutiny and specified what an alternative ecological framework must accommodate (and what it, in turn, assumes).

For example, demographic theory was simply not designed to accommodate the spatio-temporal dynamics of organisms and the "idiosyncrasy" of species. Efforts to overcome these problems have commonly been discussed, ecological theory has been adjusted in various ways and several new sub-disciplines have been developed. Spatial ecology, metapopulation dynamics and macroecology, for example, are intended corrections to problems revealed in the foundations of demographic ecology. Developments like these, although significant and welcome, are still not entirely satisfactory as they are still only *ad hoc* adjustments to relatively peripheral aspects of ecological theory. They do not deal with the problems at the heart of demographic theory and they thus work against the formulation of an integrated theory for the discipline. In other words, they do not do full justice to the problems they should have overcome. Is it sufficient simply to add such *ad hoc* interpretations to old theory? Or is a complete reorganization required to cope with the knowledge that the fundamentals of the discipline are flawed to the extent that research, interpretation and application are all influenced negatively? We justify the latter approach otherwise logical disjunction will inevitably attend interpretation, compromise understanding, and affect the practice of ecology.

Autecological theory has the advantage of generalizing at a level that accommodates the problematic "idiosyncrasy" that species present to demographic ecology and is consistent with the associated "individualistic" spatio-temporal dynamics of species. Among the strengths of this theoretical replacement for demographic ecology is its logical connection to a species concept and model of adaptation that explains the spatio-temporal distribution of species' properties. Neither does it deal with the ecology of organisms primarily in relation to the resource use of other organisms, as in community ecology. All autecological interpretation is specified in relation to the adaptations of individual organisms and their interactions with their immediate environment, and it explains the spatio-temporal adjustment of organisms in relation to environmental dynamics at its various scales. To this end, it does not bisect the environment into its abiotic and biotic components, but deals with the environment as the organisms themselves confront it.

Autecology encourages ecologists to think of ecological processes primarily in terms of individual organisms, their adaptations, and their interactions with the environment and its dynamics. It is here that the foundational assumptions of autecology lie, and they are specified and justified later. The properties of organisms must be considered in relation to their mode of inheritance and the associated process of recombination. They thus have their origins in the species gene pool of their provenance. Ultimately, the origin of such characters relates back to the process of adaptation and this informs how species must be dealt with in ecological theory and interpretation, and has implications for understanding adaptation and speciation (which are also spelled out). Such a readjustment in thought is liable to affect, to some extent at least, all the endeavors of ecologists.

An ancillary aim of ***Autecology*** is to help free biology from an idealism that is increasingly accepted, albeit unconsciously or unintentionally. The imposition of determinism on biological information spans biology at all scales, from the molecular upwards. Although it is relatively easy to demonstrate that deterministic interpretation, and therefore theory, is inappropriate for biology, the development of a well-articulated alternative is not so straightforward, and has proved elusive. Rose (1997) presented an alternative that copes very well with phenomena at the level of the organism and below. His proposal, a return to the multidimensional and interactive nature of development, and the maintenance and reproduction of organisms, is an instructive and refreshing change in emphasis from genetic determinism. Above the organism level, however, we find we are at slight odds with Rose, so ***Autecology*** is our attempt to do for the biology of whole organisms what he has done for the processes that take place within them.

Numerous colleagues and friends have contributed to the development of our ideas and the production of this book. Our collaborations with other scientists and research students have, in particular, been as enriching in this respect as they have been enjoyable. The intellectual stimulation generated by the challenges and insights of all of these people, going back decades, has helped substantially in formulating our interpretation and synthesis, and we wish to express our deep gratitude to them all. Several colleagues have read portions of the manuscript and their comments have been extremely useful in tightening our arguments and smoothing our presentation – we thank Tony Clarke, Greg Daglish, Lindsay Popple, Lauren Kaye, Renee Rossini and Andrew Ridley. Hugh Paterson, Raghu Sathyamurthy, Chris Pavey and James Hereward kindly found time to work through the entire manuscript, to its immense benefit. We are sorry we could not follow all of the suggestions as far as we would have liked. We had to draw a line at some point despite the many interpretations to follow. Many figures had to be redrawn and the perceptive talents of Iman Lissone have worked amazing changes in rendering elegant clarity from complexity. Finally, we thank Anantanarayanan Raman and Phoebe Coulon-McIntosh for their support during this project.

GIMME H. WALTER
ROB HENGEVELD

Contents

Chapter 1

Introduction An Outline and Justification for *Autecology*

At least two general approaches can be taken in designing and conducting research into the questions of distribution, abundance and diversity. This duality is an underlying feature of ecological research and interpretation, but it tends to be downplayed or avoided in most ecological writings. The issue is frequently circumvented with statements about ecology comprising different subject areas, such as population ecology, spatial ecology, behavioral ecology, the explanation of local diversity, applied and theoretical ecology, and so on. Similarly, reference is made to the different aims of ecologists, such as the prediction of population trends, the explanation of local diversity, the generation of general models of ecological processes, the conservation of rare species, or the reduction of pest populations.

Hidden beneath the surface of all ecological endeavors, though, is the awkward fact that fundamentally different approaches can be taken in tackling any specific ecological problem. In Chapter 2 we illustrate this point, as graphically as we can, with reference to an ecological system that has been investigated from alternative perspectives, population (or demographic) ecology and autecology. The reasons for these different approaches are then explored. We thus justify the strengths of what has, for a long time, been called autecology. Despite the existence of the autecological approach and a few strong champions (most notably Andrewartha and Birch (1954)), a theoretical basis has never been developed for it. More significantly, perhaps, it has never been linked successfully to evolutionary theory, as pointed out by Orians (1962), in his criticism of Andrewartha and Birch's (1954) thesis.

Autecology develops a logic and general methodology that integrates ecology more seamlessly than ever before into the general field of biological investigation. Autecological theory is developed directly from an understanding of the properties (or adaptive mechanisms) of organisms, and thus provides a mechanistic basis for linking ecology directly with other biological sub-disciplines. Such a realignment of ecology is necessary because the general theories of population and community ecology do not relate directly to the properties of organisms, so they cannot portray ecological systems accurately and do not link mechanistically to other biological sub-disciplines (see Chapters 7 to 10 for justification). The various offshoots of demographic ecology also fail in this regard, including behavioral, evolutionary, landscape and metapopulation ecology. All of these endeavors are based so strongly on numbers of individuals, together with the optimization of resource use and relative maximization of reproductive output, that only minimal reference to the actual adaptations of organisms is made. Ultimately, the accurate portrayal of ecological systems is subverted.

Demographic ecology sees numbers of organisms driving optimizing selection, through the maximization of fitness. Population density and competition are thus seen to be the fundamental influences in behavioral ecology. They are also seen to drive density relationships, spatial processes and metapopulation dynamics. In line with this demographic approach, species tend to be treated typologically, as adaptive interacting units that make up local communities and so-called regional species pools. Even the field of population dynamics is typological in its focus on the mean, or equilibrium, rather than on population values achieved in the field. The typological treatment of these dynamic systems even goes so far as to designate species, for explanatory purposes, as having "kinds of population dynamics" (Huffaker et al. 1999, Lawton 1999).

The focus of demographic ecology, just outlined, persists despite the presence and abundance of any particular species within a locality reflecting the degree to which the adaptations, requirements and tolerances of individuals of that species match the conditions of their surrounding environment. This ecological intensity, measured in terms of density of conspecific organisms per unit of habitat (among various other measures), reflects the primary phenomenon of organisms matching local environmental conditions. Local density expresses this match and is therefore secondary to it. Density effects undoubtedly do have an ecological influence, but only sporadically and as derived phenomena, which are secondary to the species-specific properties of individuals and their interactions with the environment. Density effects therefore do not provide the best basis for developing robust generalizations and theory. Nevertheless, current theory in population and community ecology is still concerned principally with these secondary processes, and thus overlooks the primary nature of the match between the adaptations of organisms and the environmental conditions to which they are usually exposed.

The physiological, behavioral, and associated structural properties of organisms, and their dependent capabilities, have a profound influence ecologically. Because these properties differ among species, all species differ from one another ecologically, and are said to be "idiosyncratic" or "individualistic" in relation to the expectations of demographic ecology theory. Species have come to differ from one another, often in numerous properties, through their independent adaptive change. How the adaptive process proceeds and is structured cannot be established through the application of equations that do not refer directly to the properties that actually undergo change. This is the major failure of the concept of optimization, which represents the evolutionary arm of demographic ecology. We therefore justify how the primary ecological process of organisms matching the specific environmental conditions they need for their maintenance, survival and reproduction, must provide an explicit basis for ecological theory if it is to be realistic, for these are the aspects central to the life of organisms in nature. From this basis we are able to integrate ecology into the other biological sub-disciplines that deal with the workings of individual organisms.

It is timely to develop autecological theory, because theoreticians have already turned their attention to structured, individual-based population models (e.g., Grimm and Railsback 2005). These models contain parameters representing the properties of individuals, and the structured models allow these properties to be categorized into frequency distributions. Thus, the numbers of individuals are statistically structured with regard to one or more biological properties. Similarly, the environmental conditions that are pertinent to their survival and reproduction can also be represented by frequency distributions, and can be structured statistically in relation to sets of variables. However, environmental conditions are not constant over time, nor are they spatially uniform, and these fluctuations and trends represent the primary driving dynamic in ecology (as expanded in Chapter 4).

Individuals have to cope with this spatio-temporal variability, and have adaptations that allow them to span extensive periods of unfavorable conditions and to move across extensive areas that are unfavorable. Specifically, the complex processes of hibernation, dormancy, diapause, sheltering, movement, and even the ability to "generalize" across resource types, represent species-specific mechanisms to deal with unfavorable aspects of the usual environment of the species concerned. The ecological consequences of these properties, acting together, are best expressed as survival probabilities or mortality risks, and this is how the stochastic elements of the environment are accommodated numerically (Chapter 9). Such risk analyses can be calculated for any particular set of individuals (the "ecological population") in the context of particular environmental conditions. The structure and dynamics of ecological systems are therefore dictated by the closeness of the match between unequal entities, one biological and the other environmental.

Central to the autecological approach are the responses of individual organisms to their immediate environmental circumstances. This is determined by the environmental requirements of the species concerned, and their sensory abilities and responses in relation to their physiological tolerances, buffers and endurances. Individual organisms are thus faced primarily with maintaining themselves within the vagaries of their usual, species-specific environment, and thus survive and reproduce (Haukioja 1982). In so doing, they have no knowledge of the options open to them or to other organisms, nor the options taken by other organisms, the structure of the surrounding population, the species composition of the local area in which they happen to occur, or how to optimize their fitness (Rapport 1991).

Indeed, fitness, population structure and community structure all represent statistical attributes that are constructed retrospectively by scientists. They are not inevitable ecological forces directing the responses of the individual in relation to its environment. That organisms have no ability to predict the environmental circumstances they will face, whether nearby or in the near future, confirms that this is true. They thus cannot anticipate the impact of spatial heterogeneity of environmental conditions at any scale. Although organisms do respond to cyclical environmental events, such as daily and seasonal cycles, the precise timing and magnitude of variation of these cycles cannot be predicted or anticipated by the organisms themselves. Furthermore, the number of factors that can vary, with different degrees of independence of one another, is extensive (as detailed in Chapter 5). Individual organisms really do live in a capricious environment of which they have no overview or perfect knowledge. Neither do they have expectations of future conditions nor of conditions elsewhere.

The only option open to individual organisms is to respond, according to their internal state, to their immediate environmental conditions. If sessile, they either sustain themselves or they fail. Motile organisms (including the motile stages of sessile organisms) have an additional option, for these individuals can move spatially to locate and match suitable conditions elsewhere (Chapter 4). Environmental (or habitat) tracking of this nature results in open aggregations of individuals. Such aggregations form within localities that are suitable for the species under consideration and they have finite turnover rates. In effect, such local ecological populations are temporary statistical phenomena that can be defined only retrospectively in statistical terms. Such concentrations of organisms are traditionally referred to as "populations", a very different use from the original Mendelian (or sexual) population central to population genetics. What this means, in practical terms, is that local populations (and metapopulations) are interpreted in demographic ecology as "units". These, however, have little ecological reality as their limits are specified subjectively by humans. Ultimately such an approach leads to typological interpretation, as when invasiveness is seen as a characteristic of organisms in its own right or species are seen to fit into the structure of an ecological community.

The focus on individual organisms in autecological theory and investigation, by contrast, infers that neither populations nor species can be treated as ecological units that have general properties of their own. That is, such units cannot interact, as a whole, with other such entities within a community, and thus adjust numerically or adaptively on that basis. Populations and species (as gene pools (see Chapter 5)) should not be treated as units that exert an influence on their component individuals or that operate to keep each other in overall numerical balance, as in metapopulation theory. Instead, their numerical representations are statistical effects of the actions and responses of the individuals that comprise them, within a particular environmental setting. Defining the geographical scale of that setting is dealt with later (Chapter 5), but the significant aspect to this statement is that it is related to the responses of the individual organisms of the species of interest.

The practice of autecology is often portrayed as a form of anecdotal natural history, more sophisticated perhaps, but certainly not a scientifically acceptable theoretical approach to ecological interpretation. This book shows, however, that autecology represents a theoretical construct that provides testable alternatives, and is much more than a simplistic and unstructured gathering of relatively easily obtained facts about animals and plants, as implied when the term "natural history" is used pejoratively. We do not imply that the need for an alternative approach to ecology requires an entirely fresh start and a negation of all that has been achieved to date. Much progress has already been made because research of an autecological nature is quite common. In particular, research of an autecological nature is often, from practical necessity, conducted on species that provide resources for humans, have pest or beneficial status, or are conservation priorities. Palynologists and other Quaternary ecologists have also conducted autecological research, for example. A comprehensive theory has, however, never been developed and no species has been investigated comprehensively from this perspective. The reasons for these deficiencies warrant analysis, but is beyond the scope of this book.

Autecology introduces a different way of looking at ecology as a whole. Natural systems are given a new theoretical perspective, principally in terms of individual organisms coping with the vicissitudes of their environment and with reference to our understanding of the statistical consequences of the underlying interactions. Because of this switch in focus for ecologists, the nature and relationships of current ecological theories need to be reconsidered. The complex of ideas relevant to autecology, and its associated models and methodology, need to be understood in isolation of earlier concepts of ecology, and in their entirety.

Autecology does not, however, dwell on the old contentions in ecology, such as those on the intensity of density dependence and competition, population (or metapopulation) stability, or the existence of communities, and so on. We see these as being internal to the current ecological paradigm, called demographic ecology because of the emphasis on demographic factors in explaining ecological patterns. The autecological paradigm provides, instead, an alternative approach in which the issues and concepts addressed in these older debates play a far less central role, to reflect the incidental and variable influences of these demographic processes on the observable patterns in nature. The impact of competition and density dependence, for example, depends ultimately on a plethora of other variables involved in the survival process of individual organisms, and this needs to be reflected more realistically in theory.

That autecology is often seen to serve applied ecology practice may create the impression that ecological theory is in good shape generally and that alternative sets of ideas and approaches are available to ecologists. This perception seems to obstruct, to some degree at least, the argument that a new synthesis is needed for ecology. ***Autecology*** presents the case that these two alternative approaches to explaining patterns in distribution, abundance and diversity have coexisted for some time (Hengeveld and Walter 1999), that only demographic ecology has a widely used theoretical basis, that demographic theory is frequently used inappropriately in interpretation of autecological data, and that such interchange blurs a highly significant distinction. In summary, many autecological ideas have already been expressed and some aspects of a new methodology already exist, but these have to be synthesized and structured into the new perspective that is required.

Autecology continues, in Chapter 2, with an illustration to justify the view that two mutually exclusive approaches to investigating and interpreting ecology are currently available. Two different studies of the same system, lerp-forming psyllids on eucalypts, illustrate graphically how a different starting point, based on different underlying premises of the workings of ecological systems, leads to the collection of different data and the development of different explanations.

From there, the focus is switched strongly to the autecological view of ecological systems. Chapter 3 illustrates the various ways in which individual organisms from an array of taxa, from birds and humans to bacteria and plants, interact with their environment. Significantly, individual organisms do not participate only in these singular interactions; they participate in ecologically relevant interactions throughout their life. This lifetime of specifically timed interactions is conceived in terms of the specifically structured lifeline of the individuals that make up a species. This perspective allows the role of stochastic influences in the life of individual organisms to be considered explicitly, for such influences intrude at each step in the life of organisms. From these perspectives we specify the general aim of autecology and develop a general structure for it through the remaining chapters.

Chapter 4 introduces the primary dynamic in ecology, as perceived from the autecological perspective. The context is spatio-temporal and the driving influence is external to the organisms (rather than being driven internally within populations of organisms). Climatic influences are far more structured and complex than they are portrayed in demographic ecology. The aspect of the dynamic that is emphasized is the current rate at which organisms respond to climatic influences, for that indicates why an alternative to deterministic demographic ecology is required. This view is extended in Chapter 5 to illustrate the scale at which these ecological processes take place. In particular, the spatio-temporal heterogeneity of the environment is influential, and organisms deal with this in relation to their species-specific requirements and tolerances. We can thus specify the relevance of species to autecological understanding. Although autecology focuses on individuals, individual organisms acquire their ecologically relevant features through their derivation from within a species gene pool (sexual organisms) or asexually. This duality between individuals and species introduces a duality to the scale of resolution of ecological investigation (and this, in turn, differs from the hierarchical scale given to ecological systems in demographic ecology theory).

The mechanisms by which individual organisms interact with the environment are detailed more completely in Chapter 6. These are aspects of the organisms' physiology, morphology, behavior, and so on, and this allows the relationship between ecology and the other sub-disciplines in biology to be reconsidered. Significantly, each mechanism of interaction with the environment almost invariably comprises aspects that are genetical, biochemical, physiological, morphological and behavioral; an integrative approach is therefore required to understand them and their consequences for the ecology of the organisms concerned. The mechanistic basis of the ecological interactions of individual organisms is thus generalized in terms of what we refer to as their environmental response systems, and these are strongly species-specific in their composition and operation. A full understanding of the functional basis, geographic distribution, evolutionary stability and ecological consequences of environmental response systems requires an appreciation of their origins, so we also consider the process of adaptation and species formation in this chapter. In turn, autecological principles influence interpretations of these processes.

The understanding of organisms and their ecological context, developed in Chapters 3 to 6, is then considered in relation to the measurement of ecological variables (Chapter 7). In autecology the focus is on survival, reproduction and movement. The context differs from the density related perspective of demographic ecology, and this is discussed in relation to the basic mathematical models that underpin demographic interpretations of ecological systems. We deal with the way in which the environmental response process is structured for modeling purposes. Essentially, autecological modeling is a form of risk analysis, and this is detailed fully in Chapter 7.

Chapter 8 extends perceptions on modeling, but does so from the broad perspective of ecological research as a whole. The initial justification of the ecological pattern (whether general or specific) that demands explanation is a crucial first step but is often ignored. From there, the scientific research required for the reconstruction of ecological processes demands two more steps. The exploration of the variables that could contribute to the observed pattern comes first, and is followed by a reconstruction of the process mathematically. Surprisingly, demographic ecology is still deficient in all three areas, despite its relatively long history of research.

With respect to modeling, in general, the differences between models with a deterministic, chaotic and stochastic emphasis are clarified, before justifying and explaining the use of stochastic models in autecology.

Chapters 9 and 10 are synthetic. The former brings together organisms, their adaptive mechanisms or properties and the way in which they interact with a stochastically influenced, but structured environment. The role of risk analysis in understanding the spatio-temporal geographic dynamics of the individuals that make up a species is expanded, and is illustrated with examples. Each of these exemplifies a particular aspect of such analysis, for no complete example has yet been developed. Chapter 10 takes the synthesis further. It presents a complete summary comparison of demographic ecology with autecology, and this is based on the analysis developed through the earlier chapters. The two approaches can now be compared more comprehensively, as follows.

The primary patterns and associated premises of the two systems of thought are contrasted initially, and this brings together the key aspects related to the way in which each approaches ecological research. The relationship of each ecological paradigm to the other sub-disciplines in biology is drawn. This is a subtle exercise because different sub-disciplines (e.g. behavior and evolution) themselves embody alternative approaches and emphases (one demographic in approach, one autecological) and these have to be disentangled for comparative purposes. Finally, each approach to ecology differs from the other with respect to scientific method, the nature of ecological interpretation and ecological prediction, and these are also contrasted. In short, we end with a strong justification for the logical independence of each paradigm, because they do not represent obverse sides of the same coin, different levels in a hierarchy, or even complementary approaches to the understanding of ecological systems. Autecological theory provides a strong, independent and intellectually satisfying approach to understanding the changing spatio-temporal dynamics of organisms and their local diversity.

Chapter 2

Two Alternative Approaches in Ecology

CHAPTER OUTLINE

1 INTRODUCTION

In this chapter we demonstrate that at least two alternative general approaches to ecological investigation are available. We do so through examples provided by different ecological studies on almost identical ecological systems. Having two such general options available means that, for tackling any specific ecological problem, a decision must be made as to which path to follow. This initial differentiation between possible starting points, despite its importance, is frequently made only tacitly and sometimes, perhaps, unknowingly. More care is needed in this regard because the consequences are serious. The approach taken dictates the type of data collected and that invariably differs from the information collected under the alternative. Inevitably, the interpretations that derive from the data sets also differ from one another. Some overlap is inevitable, for various reasons, but should not detract from the distinctions drawn.

The ecological system used to exemplify these points involves plant-feeding insects called *psyllids* or *jumping plant lice*. These are small sternorrhynchan bugs that suck sap from plants to gather nutrients for their life processes. In terms of ecological study, this is an unusual case because it represents one of the few systems that have been investigated from both of the research traditions that are currently available to ecologists.

The point illustrated in this chapter is that each of the two research traditions in ecology is based upon different premises and these render them mutually exclusive of one another. The premises that underpin scientific investigation in this way are also called basic assumptions, first principles or foundation statements. Ecologists need to consider which set of premises is more appropriate for understanding ecological systems, for that will suggest the most profitable research directions and thus provide the most realistic interpretations. Note that there is no hint here of a continuum in terms of which principles are applied. The two sets of principles represent strict alternatives that cannot be integrated into one another, and they should hold universally. The alternative ecological research traditions have been labeled the "demographic paradigm" and the "autecological paradigm" (Hengeveld and Walter 1999, Walter and Hengeveld 2000). In other words, each of these paradigms dictates a different starting point for any ecological investigation and alternative investigative pathways. Ultimately, interpretations that are mostly mutually exclusive of one another are likely.

2 OUTBREAKS OF AUSTRALIAN PSYLLIDS

Among the Psyllidae of the world, a group of Australian species is unique in that each nymph builds a sugary dome (or lerp) that entirely covers the small insect. This sugar, or honeydew, is a waste product; the nitrogen content of the foliar phloem that they ingest is so low that the bugs take in excessive quantities of water and carbohydrates, simply to obtain sufficient nitrogen for their needs. From the excess carbohydrates, they build their domes. Periods of massive psyllid outbreaks were obvious twice during the previous century in Australia, between 1914 and 1922 and again between 1956 and 1963. Investigations of these outbreaks concentrated on several species in the genus *Cardiaspina*, which feed on various *Eucalyptus* species or gum trees. Generally, the psyllid species are restricted to one species of gum tree, or relatively few species.

What is unusual about these studies was the independent involvement of two ecologists in investigating the cause of the outbreaks, and that they did so from very different perspectives from one another. L.R. Clark was strongly influenced by the dominant research tradition in ecology, which is well developed, widely accepted and covers population and community ecology. Combined, these two aspects of ecology constitute the demographic approach to ecological investigation and interpretation, because of their focus on the influence of population density relative to resource use. The second ecologist, T.C.R. White, was influenced by the revival of the autecological approach by Andrewartha and Birch (1954). Autecology originated among the German plant ecologists of the 19th century (Cittadino 1990), but has still not been fully articulated. Indeed, Andrewartha and Birch (1954, 1984) did not even favor the term autecology, perhaps because of the ambivalence seen in it with respect to whether the focus should be on individual organisms or species. We resolve that ambiguity in Chapter 10 and thus justify retention of the term.

3 THE DEMOGRAPHIC APPROACH

3.1 Analysis and Interpretation

L.R. Clark was based in Canberra and investigated the periodic outbreaks of *C. albitextura* on redgum (*E. blakelyi*). In keeping with demographic ecology principles, he gave prominence to the relative densities of psyllids (i.e. their abundance per unit of vegetation) as the most likely determinants of population numbers and change. In his analysis, summarized by Clark et al. (1967), he estimated psyllid survival rates at various sites, some with low densities and others with high ones. He could thus investigate the relationship between survival and density. The percentage mortality of *C. albitextura* was sometimes correlated with density, which is consistent with density-dependent regulation of animal numbers, but in other cases it was not. In the low-density sites, abundance remained low despite the negative effects of high densities of psyllids having been lifted. Therefore, he explained persistently low densities as "almost certainly due to predation". However, in the outbreak areas, counts showed that psyllid densities did not decrease measurably, despite syrphid (hoverfly) larvae killing up to 35% of the psyllid nymphs in a density-related pattern. Clark et al. (1967) concluded therefore that the syrphid predators could not depress the high densities "because the total percentage mortality due to such predators as birds, ants, and encyrtid parasites decreased with increase in psyllid numbers". Of all these predators, birds were given special prominence, particularly as predators of adult psyllids. This conclusion was derived without the aid of specific evidence and despite a diversity of predators and parasitoids consuming psyllids.

The generation and persistence of outbreaks was attributed "ultimately to the prevailing weather", mainly the occurrence of "unusually cool psyllid seasons", despite all the predators present, and the relatively high levels of predation. Weather was thought, however, to act indirectly, principally through reducing the effectiveness of the parasitoids. The search efficiency of the parasitic wasps would be restricted by bad weather, which was thus considered to disrupt host-parasite synchrony.

In short, the conclusion focuses on explaining why the deterministic density relationships were not functioning in the way in which they were expected to operate. The summary model developed to explain the interpretation is presented in Fig. 2.1. The original legend to the figure is reproduced in full to illustrate the nature of the explanation. It is complex and includes numerous alternative possible influences, inputs and outputs. Generally, it lacks clear direction and does not provide a point around which to "anchor" interpretation and understanding by providing specific expectations about the system. Even discerning what follow-up research is suggested by such a scheme is far from straightforward.

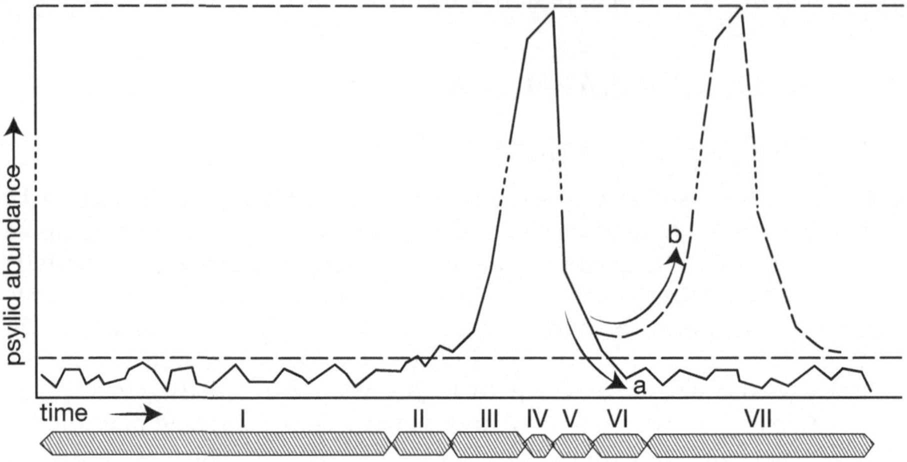

Figure 2.1 Summarized interpretation of the population dynamics of *Cardiaspina albitextura*. The interpretation of the figure, outlined below, follows the numbers along the time axis in the actual figure. Redrawn from Clark et al. (1967), with permission.

I. Psyllid numbers stabilized or almost stabilized because of the combined action of:
- i) predation by birds, ants, encyrtid parasites, etc., on the nymphal stage, the total percentage destroyed being independent of psyllid density,
- ii) the prevailing weather, especially temperature conditions, and
- iii) density-reactive predation by birds on adult psyllids.

II Stabilizing process fails, e.g. because unusually low temperatures reduce percentage parasitism. Psyllid numbers rise to a level at which the hyperparasite *E. psyllae* begins to destroy primary parasites. *Syrphus* sp. also attacks and destroys a few psyllid nymphs.

III Psyllid numbers increase rapidly because of:
- i) decrease in percentage destruction by birds, ants, and encyrtid parasites, and
- ii) failure of *Syrphus* sp. to compensate for waning influence of other predators.

IV Environmental opposition to population growth increases again because of increasing damage to foliage by psyllid nymphs.

V Damage to foliage very severe. Psyllid numbers decrease greatly because of:
- i) density-induced reduction in numbers of offspring per female, and
- ii) number of offspring being excessive for the number of favorable feeding sites available on foliage.

VI Psyllid numbers further reduced to an extent depending on:
- i) intensity of predation by birds, ants, encyrtid parasites, etc., which increases with decrease in psyllid numbers, and
- ii) foliage replacement and associated shedding of infested leaves by the host plant.

VII Two alternative possibilities are indicated on the graph: a) If psyllid numbers are reduced below the level at which the hyperparasite *E. psyllae* operates, they are likely to remain low for some years. b) If not reduced to this extent numbers tend to increase rapidly to the level at which the available amount of foliage again becomes limiting.

Clark's research and analysis excludes virtually all consideration of the adaptations of the psyllids in interpreting their changing abundance. In effect, the concept of adaptation does not constitute part of the science. Apparently, the adaptations of the organisms were not considered central to the way in which the system runs or to the interpretation of the system's dynamics. Thus, the adaptations of the individual organisms are ignored to a large extent, and the primary interpretation hinges around the postulated influences of density (or demography) in setting average abundances. In the explanation, even the stochastic influence of the weather was relegated to a disruptor of the anticipated driving influence of population density. As such, the analysis starts at the wrong end of the chain of events that contributes to psyllid numbers by concentrating throughout on secondary effects of the biological processes involved, notably those involving various mortality factors. Instead, the relevant biological processes include such features as the interaction between the insects' adaptations for feeding and oviposition and the phenology and physiology of the host plant. Psyllid densities and loss to predation or parasitism are thus secondary effects and these are not inevitably expressed in all populations at all times.

In considering only demographic parameters in investigating questions of abundance, unambiguous predictions cannot be made because interpretation is focused on consequences. As a result, the outcomes cannot be tested. This problem is compounded because only a single demographic parameter, mortality, is usually considered. The emphasis on mortality means that reproduction is usually not considered explicitly, except at the population level. This feature is to be found in other demographic population analyses.

3.2 Evaluation

The study by Clark et al. (1967) is not unusual in population ecology, either in approach or in interpretation. The most famous longer-term population studies are similar to it in approach, including those on the winter moth (Varley et al. 1973), cinnabar moth (Dempster 1982), and viburnum whitefly (Hassell et al. 1987, Southwood et al. 1989). These studies have been revealing in terms of documenting how the density of organisms changes locally, as well as identifying the sources of mortality (or unexplained disappearance in some instances). However, their focus on local density dependence presents difficulties in fully understanding what is ultimately driving the system, and thus what process (or processes) are most influential in generating the observed patterns of survival, reproduction and changing abundance. These difficulties are manifest in six specific problems.

1. *The assessment is locality specific, and migration is effectively excluded from consideration.* The winter moth study was confined to five adjacent trees because the apterous females were considered incapable of movement away from or into the site (Varley et al. 1973). However, the hatchling caterpillars drift on air currents whilst hanging onto silken threads

and may thus achieve a rather even redistribution across all trees in orchards (Holliday 1977). Studies focused on localized processes are thus likely to err on this account (Taylor 1986).

2. *The physico-chemical requirements of the various life stages of the species in question are not considered.* Not all environmental influences that impinge on the organisms in question are immediately evident to human observers, even though they may have profound effects on their survival and reproductive rates and may influence the propensity of individuals to migrate. The mortality patterns recorded in life table analyses may well correlate with the pattern of population change, but the interpretation of the pattern remains unsatisfactorily one-sided; it is a pattern of consequences that are locality specific. Interpretations developed on such a basis cannot be extrapolated to other areas with any confidence and the approach does not readily accommodate alternative explanations.
3. *Assessment of the contribution of different mortality factors is not straightforward.* Accurate identification of the particular causes of mortality is clearly important. The recognition and measurement of some causes of mortality (e.g., that inflicted by parasitoids) may seem relatively straightforward, but requires considerable care (Elkinton et al. 1992, Day 1994). For a large part of the mortality, however, the cause often remains rather vague, and some events or processes that relate to a decrease in numbers of organisms may even be missed (Price 1987). The "missing" organisms may, for example, simply have migrated. Consider the large impact of what was referred to as "winter disappearance" in the winter moth study (Varley et al. 1973) in relation to point 1, above. Further, the size of the error in mortality estimates is usually not readily ascertained.
4. *The year-to-year pattern of mortality across the generations that is attributed to each single cause is related to density.* Any mortality factor that acts in a density-related fashion is given special significance in the ecology of the species in question (and is even used in the selection of natural enemies for biological control purposes) as it is taken as proof of density-dependent population regulation. Even if its contribution to overall population change is relatively small it is usually given dominance over other influences in the ecology of the organisms concerned (e.g., Bradshaw 2008).
5. *Environmental conditions, which usually vary within and between generations during the course of the study, are usually not considered.* Alterations to the interaction between herbivore (consumer) and plant (resource) that result from environmental variation are also not taken into account.
6. *Such studies have the potential only to identify a pattern of mortality and, in some studies, recruitment as well, within a single locality.* Since unambiguous predictions cannot be made and tested, causal factors cannot be specified from such a pattern with any certainty.

Clearly, such preliminary results cannot be used for developing effective ecological generalizations about the local system under investigation, and certainly not for ecological systems in general.

4 THE AUTECOLOGICAL APPROACH

4.1 Analysis and Interpretation

The approach of the second study of psyllid outbreaks contrasts strongly with the first and is autecological in nature. White (1970a, b), working from Adelaide, monitored the fate of individual insects, specifically *C. densitexta*, in relation to environmental conditions. He emphasized the role of physiology of both the insects and the plants on which they feed, and he interpreted this interaction in relation to prevailing climatic conditions. The physiological status of their host trees was assessed particularly in relation to the condition of the leaves, for it is on these that the bugs generally remain and feed. In controlled experiments, White estimated the performance of psyllid individuals in terms of their oviposition response to various substrates, reproductive output and survival, and he tested the conclusions with field samples.

Most psyllid species feed and oviposit on flushing growth (e.g., Webb and Moran 1978), but this is not true of the Australian lerp-forming psyllids. *Cardiaspina densitexta* females place their eggs mostly on mature leaves. They avoid young foliage and old leaves, even if they have no alternative (Table 2.1). Leaves are thus accepted for oviposition when photosynthesis reaches its maximum during the few weeks of leaf maturity, just after full expansion of the blade. North-facing leaves receive more sunlight than southern leaves, and so sustain higher levels of photosynthesis and, as a consequence, more nymphs feed on them. Furthermore, nymphs grow faster on leaves on the northern sunny side of the tree, both in winter and in summer. Nymphs continued to feed on leaves that had been placed under dim light in the laboratory, but many of these died, presumably because they could not acquire enough soluble nitrogen, and therefore starved even whilst feeding. Sun and shade thus have differential influences on settling success, as well as on post-settling survival.

Table 2.1 Distribution of psyllid (*Cardiaspina densitexta*) eggs when females were presented with host plant (*Eucalyptus fasciculosa*) leaves of different ages, simultaneously in a choice test and then with each age of leaf presented on its own (no-choice). From White (1970b).

	CHOICE TEST		*NO-CHOICE TEST*	
Leaf age	*Number of eggs*	*All eggs laid* (%)	*Number of eggs*	*All eggs laid* (%)
Old	78	0.4	121	3
Mature	19,526	98.8	3,773	93
Young	155	0.8	163	4

Ultimately, nitrogen availability affects psyllid density and performance within individual trees and across trees. Similarly, it can determine the general level of psyllid abundance in an area. During a succession of wet winters and dry summers, the feeding roots of their eucalypt hosts drown and dry out in the respective seasons. Presumably this stresses the plants and leads to changes in the chemical status of their nitrogen, with more soluble nitrogen becoming accessible to the psyllids. White (1969) postulated that this enhances psyllid survival, and thus increases their density on the affected trees (Fig. 2.2).

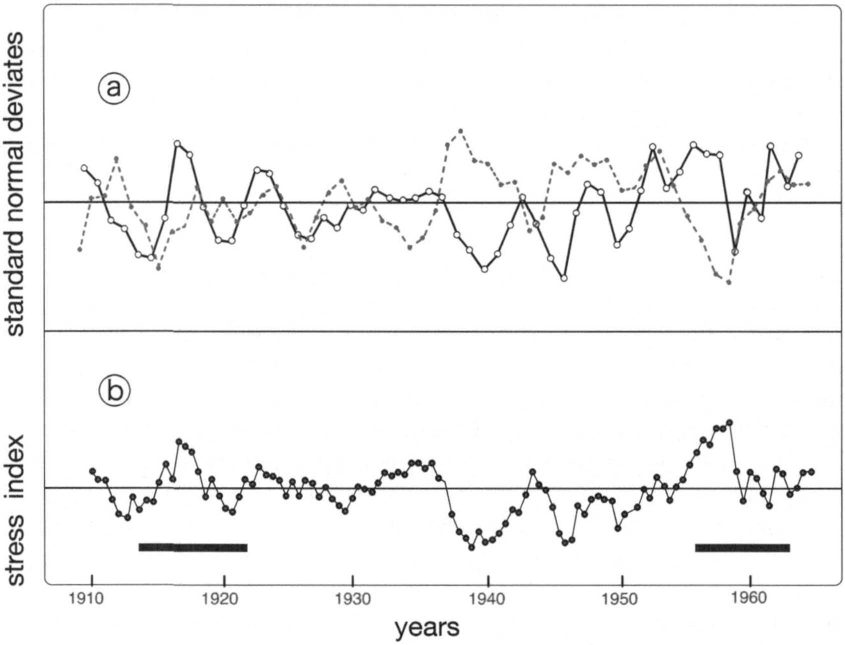

Figure 2.2 Rainfall, stress indices and outbreaks of the psyllid *Cardiaspina densitexta* on *Eucalyptus fasciculosa*. Top graph plots the summer and winter rainfall figures for one locality, Keith, in South Australia. Data points represent three year running means, with the dashed line representing summer (December to February) and the continuous line winter (April to October) rainfall. Both lines are plotted in units of the normal deviate about its mean. Bottom graph plots the stress index, which is simply the algebraic sum of "**b**" (continuous line in top graph) minus "**a**" (dashed line) at each date. The stress index is positive when "**a**" is less than "**b**", which is when the summers were dry relative to the preceding winters. Psyllid outbreaks are indicated by the broad black bands. Redrawn from White (1969), with permission.

4.2 Evaluation

White's (1969, 1970a, b) analysis investigated the primary environmental requirements for the survival and reproduction of psyllid individuals, and did so in relation to the abilities of their host plants to meet those requirements. The extent to which the plants meet the needs of the psyllids evidently determines these insects' densities directly. In other words, when the environment of the psyllids better matches their requirements, psyllid densities increase, both at the level of the individual tree, as well as that of the locality in which they occur. Density thus expresses the match between the biological requirements of the individual insects and the structure of their environment (with the host plant and its nutritional status contributing to their environment and its structure – see following chapters for further explanation). One of the primary processes underlying the existence and reproduction of psyllids on gum trees is their ability to process large volumes of phloem so as to sequester the small quantities of nitrogen present. An altered nitrogen regime, attributable to physiological stress in the host plant (White 1993), is the process that leads to enhanced psyllid survival and reproduction. But still, these changes represent a biological response to a sustained sequence of unusual weather conditions.

The higher population densities observed are thus readily explained in terms of the physiology of the contributing individuals. The physiological processes influence the survival and reproduction rates of individuals, which translates into higher numbers of individuals with longer life spans, and thus the higher densities observed. In other words, population densities are demographic attributes that represent by-products (or effects, *sensu* Williams (1966)) of the individual-environment interaction rather than driving forces that dictate numbers in the following generation, as assumed in demographic interpretations. Of course, extreme densities could feed back on survival and reproductive rates, but this is likely to be sporadic, so their influence is secondary to the primary influences mentioned above. Once the underlying, autecological basis of such a system is understood, the secondary influences of these demographic variables can be tackled. How does predation and parasitism change with psyllid densities? Does it drive the system in density-dependent fashion? Or is the resulting mortality simply density-related, so that higher psyllid densities can sustain more predators and more predation, with little overall effect on the system once it achieves high density? This seems more likely in general (e.g., Smith and Maelzer 1986), as well as for the psyllid situation, as judged from the results and statements presented by Clark et al. (1967) (previous section).

In short, by starting from the ecological requirements and tolerances of the individuals involved, both animals and plants, insights are derived as to the primary processes that drive the survival and reproductive rates of the psyllids. The specific relationship of these Australian psyllids with mature leaves was unanticipated, for psyllids generally feed on flushing growth. Early experience with laboratory rearing indicated what the insects required (White 1993), although field observations might also have revealed these general needs. These expectations can be tested. If observation does not accord with expectation, other factors can readily be considered with reference to expectation. A further strength of the autecological approach to studying psyllids is that the results obtained can be extrapolated to other localities, as testable predictions for example, and they can be generalized at more than one level – that of the psyllid-eucalypt system and also to the level of general autecological models, as covered later (Chapters 7 and 10).

5 EXTENDING EACH OF THE PSYLLID STUDIES

The consideration of each of the studies of psyllids in the previous section helps identify the assumptions on which each research approach is based. This, in turn, helps identify what aspects of living systems need to be taken into account by ecological theory, the topic of this section.

Each psyllid study took, within its general approach, a particular direction. Other specific directions would clearly also be legitimate within each general approach. The range of particular directions available under each approach to ecology would, nevertheless, be consistent with the underlying premises of that approach. These alternative options not only reflect the diversity of factors that influence ecological systems, but also illustrate what alternative explanations are available for testing, even in the autecological approach. Autecology is frequently portrayed as being non-scientific and simply descriptive in not being amenable to test. White's work shows, however, that this perspective is not correct. Temperature could, in theory, also have affected abundance, perhaps through its interactive effects on humidity. That proposition can also be tested. Alternatively, an element or compound other than nitrogen could explain the changing pattern of psyllid densities, and that could also be tested comprehensively. Autecology is, therefore, more than simply "natural history", a term often used pejoratively in this connection. And because autecology has a general direction and a general theory, it is not necessarily anecdotal as is sometimes claimed. The remaining chapters justify this proposition. Note also that autecology does not simply rest on the assertion that density independence is more important ecologically than density dependence, an issue that seems to have undermined discussion previously. Autecology is a lot more than this and focuses on the interaction between the species-specific adaptations of organisms and key elements in their environment, as explained through the rest of this book.

Both studies of the psyllids were locality specific. Nevertheless, information on spatial variation in psyllid abundance was considered in each, but only as a research tool. That is, the individuals in each locality were effectively considered to be members of spatially discrete, stationary populations. Similarly, the time scale was restricted to relatively few generations. Being tied to particular localities and a narrow time frame (both specified by the observer, rather than by the properties and ecological dynamics of the subject organisms) is perhaps understandable, particularly given the small size of each investigative team and the large amount of information that could potentially be collected in such studies. Nevertheless, spatio-temporal scales need to be expanded if a full, realistic ecological interpretation is to be achieved.

The ecologically significant activities of organisms are ultimately dictated by their properties, that is, by their physiological, morphological and behavioral adaptations. In this way, evolutionary considerations enter the picture, and need to be incorporated in studies of distribution and abundance (or, in autecological terms, the spatio-temporal dynamics) of species. Furthermore, no organisms are entirely sedentary, not even sessile ones like plants or barnacles. At some stage of their life cycle they move. Movement, at whatever scale, must therefore also be incorporated, as well as the environmental context within which that movement takes place (Chapters 4 and 9). Thus, we shift focus to an understanding of the species as a whole, and how the spatio-temporal dynamics of its individuals help them to survive and reproduce within a dynamic environment, somewhat in the spirit of Janzen's (1985) "recruitment surface". Here, we are referring to species in an abstract way, as in "*C. albitextura* feeds on redgum". We mean that virtually all individuals of *C. albitextura* are liable, at least when feeding, to be found on redgum. This means, in turn, that our focus is on individuals, but that the concept of species is also relevant in that it indicates how we can generalize about the attributes of the individuals. The focus on individuals and species and on spatial and temporal issues introduces a broader approach to ecological considerations than is evident in either of the psyllid studies. These connections are specified and justified in later chapters.

6 INFERENCES FOR GENERAL ECOLOGICAL INTERPRETATION

The two studies discussed in this chapter were necessarily restricted in scope, as will always be the case in a discipline as diverse and complex as ecology. However, the points we extracted above provide a basis from which we can look more broadly at the ideas and methods associated with each of the two approaches to ecological investigation and interpretation, as well as at their relationships to other areas of biology. We can thus specify the underlying foundational statements of each approach to ecology, and how those relate to the practice, aims and ambitions of each. Each one also has its own relationships with other biological sub-disciplines and, perhaps surprisingly, its own vision for scientific laws and its aims and limits with respect to prediction (Walter 2013), as explained in Chapter 10.

Both studies, taken together, sound a warning to ecologists. The ecological data that anyone collects are not neutral, but are related to the starting point selected initially. Selecting a starting point is frequently not an active process, but derives from the inculcation of premises through prior learning and experience. Textbook influences tend to stabilize such prejudices within disciplines and through time. No starting point is neutral, whatever the branch of science, and neither can a starting point be all encompassing. All such starting points provide a focus that also cuts out particular perspectives and views, the "Bronowski cut", and thus blurs other perspectives (Walter 2003).

For ecological research, the closer the starting point to the primary ecological influences on the study organisms, the more realistic will be the results and the more value they will have. This implies that each study has to make at least one assumption, and this assumption (usually a set of assumptions) is what provides the direction taken in the research. Uncovering the basic assumptions of each approach to ecological research is fundamental to understanding each approach in its own right, and also to understanding why the two approaches cannot be brought together and unified into a single, more comprehensive ecological theory. We return to these aspects in Chapter 10, after developing the structure, workings and aims of autecological theory. This latter task begins in the following chapter, with a closer look at the adaptations with which organisms from various taxa interact with their environment, and how the life cycle of organisms must fit into the seasonal structure (and other dynamics) of their environment. The life cycle, as a concept, has been lost from ecological theory, despite its obvious utility, a point also expanded substantially in the following chapter.

Chapter 3

Basic Ecological Processes and the Autecological Generalization

CHAPTER OUTLINE

- Introduction
- Individual organisms and their ecological interactions
- Examples of specifically structured lifelines
- The Galton board metaphor
- The general aim and structure of autecology

1 INTRODUCTION

In general terms, ecological processes are mediated by interactions that are surprisingly diverse. They are diverse even in relation to the interactions that are emphasized in demographic theory, namely competition and predation. This chapter therefore develops a more inclusive set of generalizations about interactions. It does so by delving first into the diversity of mechanisms, or adaptations, through which organisms interact with their immediate environment, and then relating the operation of these properties of organisms to the structure of the environment in which the organisms live through the different stages of their life.

The detail in the examples covered in this chapter is necessary because interactions mostly tend to be portrayed in terms of their population consequences (rather than their function) and in this respect they are seen simply as positive, negative or neutral. Such an approach to ecological generality even excludes those interactions that involve aspects of the environment itself, like temperature, despite such variables being critical to the survival of all organisms. Interactions also tend to be oversimplified by the common two-way division of the environment into an abiotic component, which is mainly physico-chemical, and a biotic component that is interpreted primarily in relation to the density of the organisms concerned. The mathematical formalization of interactions in relation to the environment deals particularly starkly with interactions. Of all possible interactions in ecological systems, only predation and competition seem to have been modeled generally in this way, mainly through the extension of the logistic equation, and with the primary emphasis on density relationships to represent the biotic component. The term "biotic" is thus used in a far more restrictive sense in demographic ecology than is usual elsewhere (see Glossary).

The following section (Section 2) provides several detailed examples to illustrate how particular adaptations, in a diverse range of organisms inhabiting a diverse range of environments, directly influence the ecology of those organisms. This perspective on the ecology of organisms is then extended, in Section 3, to illustrate how the different adaptations (or interactive mechanisms) operate through the life of organisms, again with a diversity of specific examples. We thus spell out the various ecological consequences of the operations of adaptive traits through the life of organisms and conceptualize the stochastic influences on these interactions metaphorically in Section 4. From this perspective, we develop a summary framework of the goals and approaches of autecology to represent its theoretical basis (Section 5). The various aspects of autecology are then extended and justified through the remaining chapters.

It is tempting to see the organism itself as a life cycle, and life cycles are often portrayed in this way in developmental biology (e.g., Bonner 1993, Arthur 2004). Although the concept of the life cycle and its diagrammatic representation provide a compact illustration, they do have limitations when applied to individual organisms and their ecology. Life cycles do not focus on individuals, but rather the perpetuity of the species. Indeed, the cyclical aspect of these abbreviated generalizations presents an air of deterministic inevitability that misrepresents the capricious nature of the environment inhabited by individual organisms, the responses of the organisms to their immediate circumstances, and the range of fates they may suffer. These aspects are significant autecologically for they ultimately contribute to the dynamic distribution of abundances (or spatio-temporal dynamics) of the species of interest.

"Life cycle" is best left as a generalization to describe the developmental stages through which the individuals pass, and their durations. It is also useful in describing the general problems faced by organisms in different types of environments (e.g., Willmer et al. 2000). For example, freshwater habitats are characterized by their transience and changeability. Organisms therefore need to breed opportunistically in relation to seasonal change, and they should have an associated protective stage in the life cycle. Freshwater organisms can also be expected to experience physiological difficulty in maintaining ion and water balance, especially if they are small and have a high surface area to volume ratio. The continuous downstream flow of streams and rivers has to be countered by behavioral means, and this is especially important for small or immature animals. The adaptive traits that allow survival under such circumstances include a short life cycle with rapid turnover of generations, reduced larval forms and direct development through large, yolky eggs or brood pouches, and persistent larvae that are large or which crawl (as opposed to being pelagic), and so on.

The insights above are basic and provide a good generalized description of the environmental conditions faced by the organisms in question. They also cover the types of adaptations required to face those conditions. Finer detail can be included if the analysis is conducted on a species by species basis. We need, however, to go further if we are to develop a generalized structure of how quantitative analyses should underpin our understanding of the interactions between organisms and their environment, and thus interpret the spatio-temporal dynamic that represents the distribution of abundance of any particular species. The ecology of individual organisms is therefore better conceptualized in terms of its "lifeline", which is the lifetime trajectory of individuals through their development and reproduction and as they move spatially (Rose 1997). This term, whilst drawing from the concept of life cycle, does not only signal a genetic and temporal component to living entities but also the spatial element in the life of individual organisms - their geographical location and their movement. These geographical aspects represent the "lifetime track" of the individual (see Drake et al. 1995, Dingle 1996, Walter 2003). With this distinction between the life cycle of the species, to represent the adaptations carried by the individuals as they develop, and the lifetime track actually followed by any individual in question, we illustrate in the rest of this chapter the following general points about ecological interactions and ecological generalizations.

i) Environmental interactions are structured in terms of the adaptations of organisms and are invariably complex in that they involve numerous steps. Also, their function almost always relates directly to the environment (biotic and abiotic aspects), rather than in response to density itself.

ii) For the survival and reproduction of individuals, the species' life cycle (and thus the lifeline of the individuals) has to "match" the external environment, not least in relation to its seasonal structure (Daubenmire 1974). Note that environmental structure is inadequately represented by the abiotic/biotic dichotomy that currently prevails in ecological interpretation. For each species, a specific subset of environmental variables is relevant to the ecology of the species in question, and different species in a locality are likely to be affected by different subsets of these ecological influences.

iii) The movement of organisms has far-reaching consequences for ecology, which indicates that the mechanisms of movement represent adaptations through which organisms interact with the external environment. Therefore movement, as an interaction, needs to be brought into ecological theory rather than being developed as an adjunct aspect to ecology. Indeed, general theory in ecology has consistently failed because the movement of organisms has not been included centrally in the theory, not even in metapopulation theory, despite the appreciation of all organisms having an "innate tendency toward dispersal" over half a century ago (Andrewartha and Birch 1954, p. 90).

iv) The responses of many organisms to their environment, for a variety of purposes, relies on them obtaining information (or sensory input) from their surrounds, and responding appropriately to that (Dusenbery 1992). Sensory capabilities thus have profound ecological consequences.

v) The mechanistic basis of ecological interactions is invariably intricate, as illustrated by the examples developed in the following two sections (Sections 2 and 3). An understanding of this subtlety, which is overlooked by the optimization theory that underpins evolutionary ecology (see Walter and Donaldson 1994), begs the question of whether selection has rendered these mechanisms effective relative to the heterogeneous and unpredictable environment in which they operate, or whether selection works continuously to improve the postulated efficiency of organisms within a deterministic setting. The former view suggests that effective mechanisms are largely stabilized by selection (as expanded in Chapters 5 and 6). The latter sees organisms under continuous directional selection for enhanced "efficiency", but without a direct link to the mechanistic basis of the interaction ever being considered (Finlay-Doney and Walter 2012a).

vi) Stochastic influences intrude in virtually all ecological processes, usually at each step in the process (as expanded in Chapter 9), and are likely to be misrepresented when portrayed as an average value in deterministic equations like the logistic equation.

vii) Each species is, in at least some ways, ecologically different from even close relatives, because of their differential adaptations to the environment and thus their different life cycles. Such differences account for the differential distribution of species and their ecological idiosyncrasy (or individuality). General theory in ecology must be able to deal with such species-specificity.

The points above are used, in the rest of this chapter, to develop an outline of the most general pattern that is detectable in ecological systems, that of the dynamic and species-specific spatial adjustment of organisms (and thus species) in response to inevitably changing ecological or environmental conditions. Interpretation of these dynamic spatio-temporal patterns is the aim of autecology, and this pattern provides the basis for the developments detailed in the rest of this book.

2 INDIVIDUAL ORGANISMS AND THEIR ECOLOGICAL INTERACTIONS

The most fundamental ecological interactions are those that take place between individual organisms and the environment. How these interactions are conceptualized is an important aspect of autecology, and represents the very heart of the discipline. Individual organisms, without exception, have a range of mechanisms that serve a variety of specific functions with regard to their survival within their usual environmental context and, for those that survive, with respect to the production of offspring. For example, organisms have adaptations to mediate their interactions with various aspects of their surrounds, including the temperature and moisture profiles and resources required to sustain their energy balance, and the mating partners through which they achieve fertilization and syngamy (as illustrated in this section and the next).

Although many of these mechanisms are well known from the research of physiological ecologists (e.g., Willmer et al. 2000, McNab 2002, Chown and Nicolson 2004) and ethologists (e.g., Curio 1976, Manning and Dawkins 2012), their relationship with the central ecological questions of distribution and abundance remain almost entirely unspecified and unclear. The complication for ecological generalization is that the mechanisms carried by individual organisms are species-specific. That is, the individuals that make up each species have a set of such mechanisms that is unique to the species. We know this because species are widely different from one another in their response to climate change (Fig. 3.1) and closely related species invariably differ ecologically from one another, as expressed in their differential habitats or distributions (Fig. 3.2), despite the similarities from their shared evolutionary past.

Naturally, variation across individuals is detected in the operations of the mechanisms that are relevant to their ecological responses, and variation is also evident in the ecological consequences of these operations. Significantly, very different interpretations of such variation are available. The following paragraphs contrast the view of evolutionary ecology with that associated with autecology.

The sub-discipline of evolutionary ecology supports the common view that is associated with the principles of competitive natural selection, where variation among individuals is interpreted in relation to ongoing adaptive change, almost invariably through the process of optimization. Evolutionary ecology is thus based on deterministic reasoning, with random variation serving a deterministic model. The quantification of intraspecific variation and change, however slight, is thus taken as indicative of adaptive evolution in progress through small accretionary changes. Indeed, the diversification of species is explained as an inevitable consequence of such local processes, although the most significant aspects to species, their species-specific (and species-wide) adaptive mechanisms (which are dealt with further in this chapter relative to their ecological consequences and Chapters 5 and 6 in relation to how they evolve), are ignored.

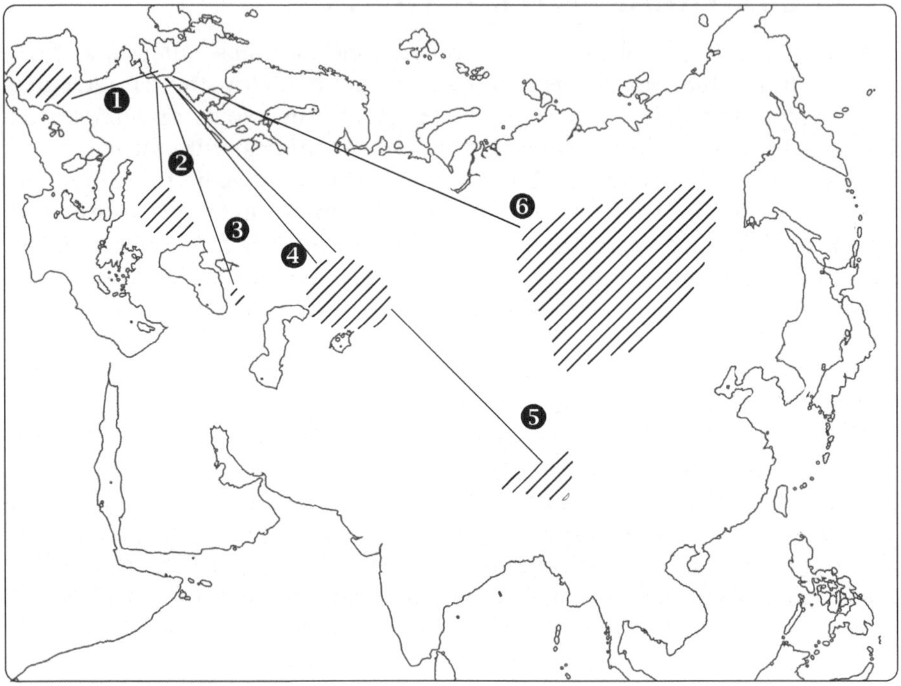

Figure 3.1 Differential change in the distribution of six beetle species in response to climate change. All of these species existed together at the site of present-day Trafalgar Square in London about 100,000 years ago, then shifted independently of one another as conditions changed. The individuals of each species tracked that subset of environmental conditions to which they are primarily adapted. Although we speak of the species shifting distribution, the geographical shift is the result of individuals, in successive generations, moving to areas that provide for their environmental requirements. Redrawn from Taylor and Taylor (1983) by Hengeveld (1990), and reproduced with permission.

Species are thus readily "fragmented" into local entities or populations, each of which is seen to have an independent evolutionary trajectory on the one hand (which is frequently investigated in evolutionary ecology, to verify its validity) and, simultaneously, some link to the species as a whole. The link to species is seldom considered in evolutionary ecology, but is perhaps best exemplified by the multi-regional hypothesis of modern human origins. Here, gene flow across massive distances through "hybridization" has been invoked as the stabilizing influence, as explained in Oppenheimer's (2003, p. 47) diagrammatic summary of this interpretation. The geographic mosaic of coevolution (Thompson 2005) provides another illustration of the general approach. Ultimately, the approach leaves unexplained the way in which species acquire their species-wide (or "fixed") characters, an issue addressed further in Chapters 5 and 6. Suffice to say, this is the reason why species are so frequently downplayed as artificial constructs by evolutionary ecologists (Mallet 1995, Noor 2002). The existence of species-specific adaptations, clear species limits, evolutionary stasis and morphologically identical cryptic species with origins in the distant past stands in the way of Darwin's competitive model of adaptive change, the one that prevails in evolutionary ecology today (Paterson 2005). A general analysis of the expectations of evolutionary ecology concludes with the startling observation that "Understanding why most intensively studied natural populations do not appear to be evolving is an important challenge for evolutionary biology" (Merila et al. 2001). Indeed, evolutionary stasis through time is now more widely acknowledged in ecology than ever (e.g., Eldredge et al. 2005).

The evolutionary ecology approach just outlined suffers attendant methodological difficulties, mainly because case histories are seldom tested from alternative perspectives. For instance, is it certain that no more than one species has been included in the observation? The mistaken inclusion of cryptic species in such studies and interpretations is common and undoubtedly distorts the picture in favor of the expectation of ongoing and differential adaptation across the distribution of species (Paterson 1991, Walter 2003, Popple et al. 2008). This possibility is seldom countenanced in treatments attempting to verify rapid evolution and ecological speciation, for example. In addition, do we know that selection acts on the variation? If so, is it stabilizing selection or directional selection? And if there has been recent slight change (and only slight change has ever been recorded in such studies), does this tell us how diversification takes place at the level of species? In short, the understanding of any reputed intraspecific variation and associated change demands a succession of tests to cover the possibilities outlined above.

Autecology offers an alternative view to the one of locally restricted ecology and local adaptive change outlined above. Although variation and some minor local adaptation should be detectable, especially in organisms that are confined spatially, as on islands (Grant and Grant 2002) or in riverine systems (Endler 1995), the predominant aspect is the spatial adjustment by individuals (Fig. 3.1) (Chapters 4 and 9) and the species-wide nature of the adaptations that mediate ecological interactions (Chapter 5, Section 3). As a first step in developing this logic and formulating a theoretical structure for autecology, we illustrate in the rest of this section the relationship between specific adaptations and some of their ecological consequences. These correlations provide the basis for future tests of the ways in which such mechanisms influence local ecology. The details of how these mechanisms operate also illustrate that it is the

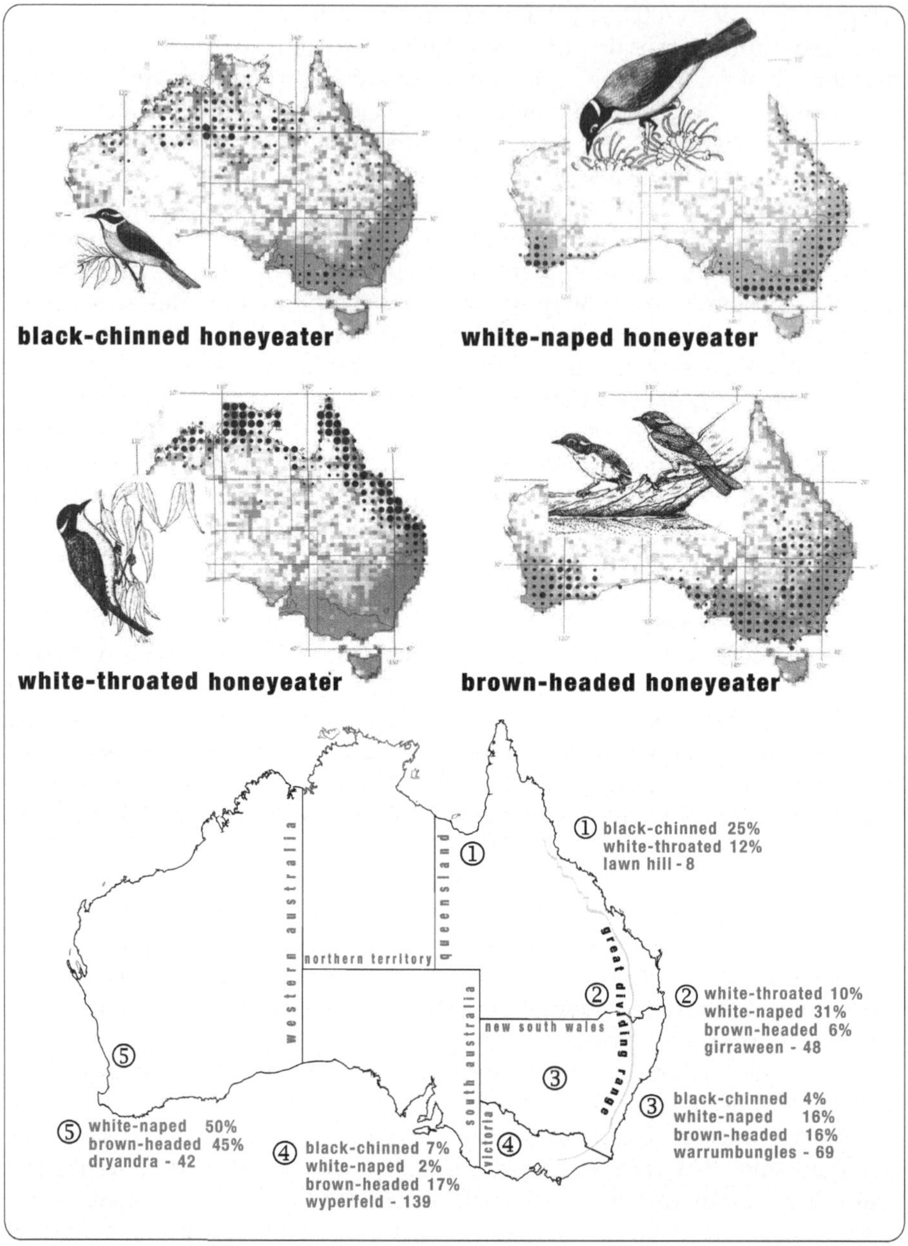

Figure 3.2 A typical example of the differential distribution of closely related species, in this case mainland Australian honeyeaters of the genus *Melithreptes* all of which inhabit rather similar dry sclerophyll vegetation. **Top left:** black-chinned honeyeater *M. gularis*, of which the two subspecies are differentially distributed (*M. g. gularis* ranging south from eastern Queensland, mainly inland of the Great Dividing Range, and *M. g. laetior* in northern Australia, westwards from western Queensland) and which occupy eucalypt and paperbark woodland, acacia scrub

workings of these complex adaptations that are subject to natural selection (rather than the efficiency of the whole per se), with the ultimate criterion being whether they operate effectively or fail. The issue of competitive superiority through enhanced efficiencies, the mainstay of evolutionary ecology, can only ever be a secondary consideration in ecology and evolution (Walter et al. 1984, Walter 1988a, 1995). Naturally, such a shift in perspective has implications for our understanding of how adaptive change does occur and, in particular, interpreting the circumstances under which it takes place, and these considerations are taken up later.

2.1 Resource Acquisition—Foraging in Birds

Foraging for resources is usually portrayed descriptively, sometimes with frequency data across food items, and often in terms of optimization. Although observations may fit expectations generated by optimization theory, one needs to be mindful of spurious correlations because of the unrealistic assumptions that underpin the theory (Pierce and Ollason 1987, Rapport 1991, Walter and Donaldson 1994). Furthermore, no mechanistic connection is discernible between the physical mechanisms with which organisms locate and gather resources, on the one hand, and the idea of efficiency around which optimization theory has been developed (Finlay-Doney and Walter 2012a). To understand how and why organisms forage in particular ways and in particular localities (or habitats), much deeper insights into the physiological and behavioral processes with which individuals of particular species interact with the environment are demanded.

Figure 3.2 *Continued*

and spinifex. **Top right:** White-naped honeyeater *M. lunatus* (with two differentially distributed subspecies, one in Western Australia (*M. l. chloropsis*) and an eastern one (*M. l. lunatus*) that is mainly east of the Great Dividing Range), with both inhabiting eucalypt forest and woodland. **Middle left:** White-throated honeyeater *M. albogularis* is a northern species (which occurs also in the southern lowlands of New Guinea). It is mainly coastal and sub-coastal and inhabits eucalypt and paperbark forest and woodland. **Middle right:** The brown-headed honeyeater *M. brevirostris* has several subspecies of uncertain status and is a bird of drier inland woodland in the southwest and southeast of the continent. It inhabits eucalypt (including mallee) forest and woodland, and occurs on the coast only where heath or other dry environments are present. **Bottom:** The major geographical features mentioned above and lists of the co-occurring *Melithreptes* species in several localities, to emphasize the significance of the question of how each persists in that particular area, given that each presumably has different adaptations to the environment (see text). 1. Northwest Queensland. 2. Coastal border area of Queensland and New South Wales. 3. Inland central New South Wales. 4. Southeast South Australia. 5. Southwest Western Australia. Reproduced and redrawn, with permission, from Barrett et al. (2003). % = proportion of visits (*n* given alongside locality (which is a National Park or State Forest)) in which the species was recorded. Data from Eremaea Birds (2013).

The interaction between birds and their diet is not as straightforward as it may seem, for the adaptive mechanisms of these organisms dictate how they will interact with their environment. The tube-nosed seabirds (Procellariiformes) are pelagic and provide an insightful contrast with the plovers and their allies (Charadriiformes), which are associated with intertidal mudflats and other such open areas. We are only beginning to understand the foraging behavior of these birds and how this is influenced by the sensory capabilities of the organisms themselves. In particular, the differences in foraging behavior and sensory physiology across related species within each of the taxa remain poorly understood, despite their ecological significance.

How do birds locate food on a vast, featureless and dynamically undulating topography? Although the oceans seem uniform, we are discovering just how superficial this assessment has been. For example, a diverse range of oceanic organisms that have almost global distributions are now known to comprise several cryptic species that mistakenly had been "packaged" under a single name. Moreover each of the cryptic species in each species complex has unique ecological specializations associated with an independent geographic distribution. Such mistaken cases include organisms as small as Foraminifera, even those long used in paleoclimatic reconstruction (de Vargas et al. 1999), as common as Portuguese man o' war (Dawson and Jacobs 2001) and as substantial (and seemingly well-known) as killer whales (Pitman and Ensor 2003, Pitman et al. 2007, Leduc et al. 2008). Organisms evidently interact with their oceanic environment in rather specific ways, and we are only now starting to learn how oceanic birds that fly deal with their environment.

The tube-nosed seabirds are all pelagic. They have a well-developed olfactory lobe in their brain, among the largest of any birds (Bang 1966), and this is connected to an elaborate olfactory neuro-epithelium within their nose. Although the birds are difficult to study behaviorally at sea, tests have indicated that relatively large numbers of species are attracted by odors (Verheyden and Jouventin 1994, Nevitt 1999a, b). Birds attracted in this way approach the odor source against the wind and less than one meter from the ocean surface. Seabirds that do not use olfaction made random approaches at greater heights of six meters or more in these studies (Verheyden and Jouventin 1994). Various volatile compounds that are associated with the primary prey of these birds, such as krill and fish, can be mixed with vegetable oil and poured onto the water surface. These scented slicks attract various procellariiform species at up to five times the rate of controls (Nevitt 2000). The species specificity in responses is not yet well resolved but, for example, storm petrels and prions are attracted primarily to dimethyl sulfide (which is released by phytoplankton when krill and other zooplankton eat them (Nevitt 2000, Nevitt and Haberman 2003, Nevitt and Bonadonna 2005)), whereas petrels and some albatrosses respond from a distance to 3-methyl pyrazine (which emanates from macerated krill) (Nevitt et al. 2004). Undoubtedly other attractive compounds have still to be recognized because herring oil and krill extracts attract yet other species (Nevitt et al. 2004). These odors are said to attract at distances of several kilometers and the behavior of the birds, again depending on species, seems to change when they switch from long distance search to localized search and may even differ within species across different habitats, with some fulmars and albatrosses responding to odors in pelagic habitats but not coastal ones near their colonies (Verheyden and Jouventin 1994, Nevitt 2000).

The attractive compounds mentioned above are odorous and are associated with areas of high primary productivity. These, in turn, are associated with bathymetric features such as shelf breaks, seamounts and oceanic frontal zones (Fig. 3.3) (Nevitt and Bonadonna 2005). Birds with a sense of smell can thus pinpoint the localized krill aggregations from which they feed. The advantage of smell in such situations is that it projects the presence of the prey into the air above the water surface, and the oceans have been postulated to present these birds with an olfactory seascape against which they can localize prey. The odor structures that project areas of primary productivity above the ocean surface may persist for weeks and seem to appear seasonally (Nevitt and Bonadonna 2005).

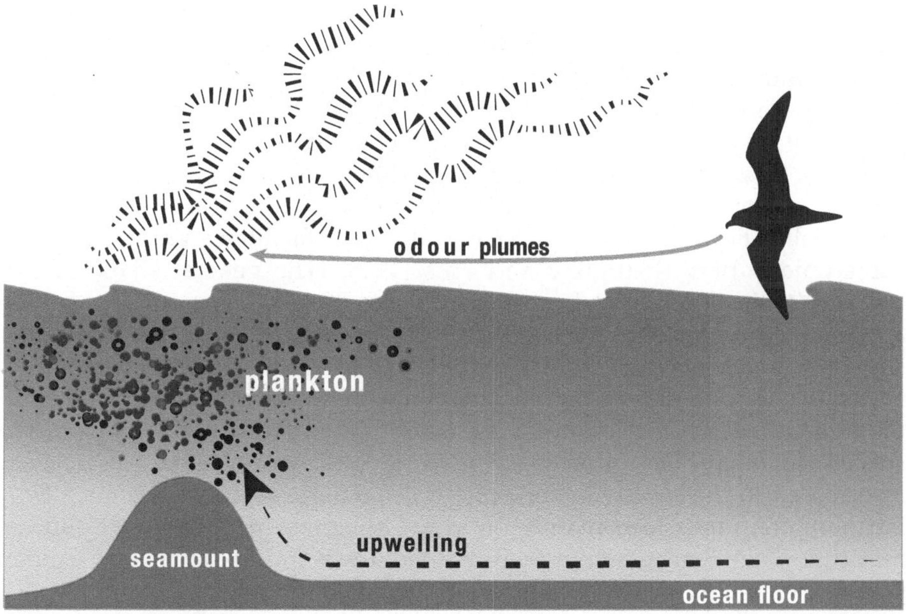

Figure 3.3 Diagrammatic representation of the projection of odors emanating from various planktonic sources in an upwelling associated with an underwater seamount. Procellariiform seabirds detect potentially productive foraging areas through their olfactory abilities. Redrawn from Nevitt (2000).

An appreciation of the sensory physiology of tube-nosed seabirds is clearly informative of the local ecology of different types of these birds. It may also help us to understand the highly localized nesting of many species in this group. A consideration of the Charadriiformes indicates similar associations between their sensory physiology and local feeding ecology. The feeding adaptations of the plovers (Charadriidae) are usefully contrasted with those of the sandpipers (Scolopacidae) to indicate, further, how sensory adaptations dictate where and how different birds can forage. The plovers feed within a diversity of open habitats, from the intertidal to open grassland. They locate prey visually, run to it and capture it with a peck. By contrast, the sandpipers and their close relatives feed mostly on intertidal mudflats, and do so by probing soft substrates with their bill for prey. Their search is driven by tactile information. These foraging differences across the families are underpinned by fundamental structural and functional differences in their brain and sensory structures.

Sandpipers like the dunlin (*Calidris alpina*) and red knot (*C. canutus*) have a high density of sensory structures at the tip of their beak (Fig. 3.4). These are innervated by extension of the trigeminal nerve, and sensory information is fed back to a relatively massive expansion of the forebrain (Fig. 3.5), a feature postulated to form a specialized tactile fovea to enhance sensory resolution and integration of incoming information (Pettigrew and Frost 1985). Plovers have no such expansion (Fig. 3.5). Red knots detect pressure gradients that are generated when the beak is inserted repeatedly into wet sand and the travelling waves suffer interference from inanimate objects such as stones and bivalve prey as they block the flow of pore water (Fig. 3.6). The birds thus use the sensory structures on their beak tip to detect prey before they make contact with it and catch it (Piersma et al. 1998). The bill tip is modified to enable exposure of the internal sensory system without the need for separating the mandible from the maxilla (Fig. 3.7), and presumably this also facilitates grasping of the subterranean prey (Pettigrew and Frost 1985). The general shape of the red knot's bill is different from that of most other *Calidris* species, and is apparently structured to induce evenly distributed pressure, for it is thick, straight and evenly tapered. Their bill structure and sensory capabilities correlate with observations on their foraging behavior in the field, and with the unusual diet of red knots compared with that of most other congeneric species, for knots feed mostly on mollusks and hard-shelled crustaceans, as opposed to the mostly softer prey of congeneric species (Piersma 1996). Furthermore, knots feed mainly on sandy substrata with a surface film of moisture. They tend to avoid well-drained areas, muddy areas and deeper water (Piersma et al. 1998, van de Kam et al. 1999).

These avian examples (see also Hall and Ross (2007) for ecological correlates with avian eye structure) show that the foraging ecology of an organism is associated with a sensory system that works in a particular way. How far different species within each of the genera concerned are different from one another is not yet possible to assess - but autecological theory suggests we should be aware of the possibilities. We know that different scolopacid species do have different habitat requirements for feeding, although several species may feed in an area (Piersma et al. 1998, van de Kam et al. 1999). Clearly, the relationship between diet, usual feeding habitat and sensory specialization requires further investigation if we are to understand more fully the habitat associations of these species and their local ecology.

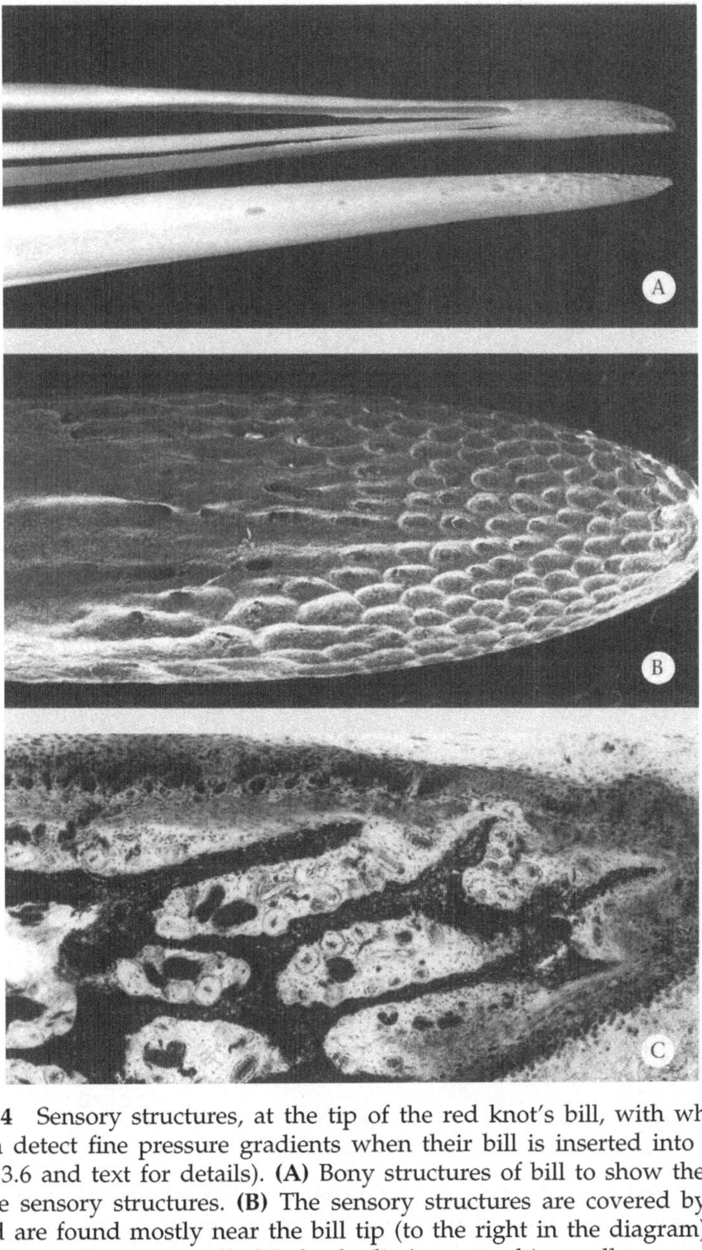

Figure 3.4 Sensory structures, at the tip of the red knot's bill, with which these birds can detect fine pressure gradients when their bill is inserted into wet sand (see Fig. 3.6 and text for details). **(A)** Bony structures of bill to show the pits that house the sensory structures. **(B)** The sensory structures are covered by a horny layer and are found mostly near the bill tip (to the right in the diagram). **(C)** The pits are filled with sensory cells (Herbst bodies) arranged in small groups (10 to 20) around a central nerve (black in photomicrograph). Reproduced, with permission, from van de Kam et al. (1999).

Figure 3.5 Dorsal views of sandpiper and plover brains to illustrate the relatively massive expansion of the sandpiper brain in front (i.e. to the right of the diagram) of the visual area (which is posterior to the vallecular groove indicated by a curved line within the larger part of the brain). The trigeminal expansion (in black) deals with sensory input from the tip of the bill (see Fig. 3.4). Redrawn from Pettigrew and Frost (1985) by van de Kam et al. (1999), and reproduced with permission.

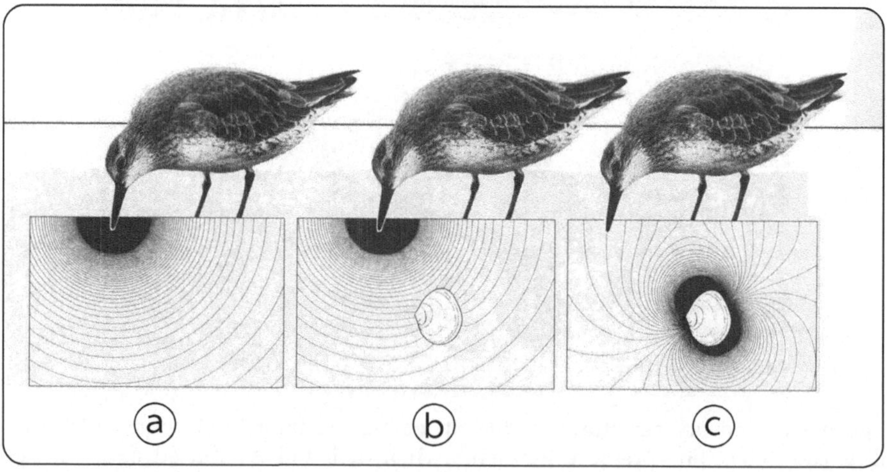

Figure 3.6 Red knots feed by inserting their bills repeatedly into wet sand and generating pressure waves **(a)** that lose their symmetry **(b)** when they encounter a buried obstacle such as a prey item. The resultant asymmetry of the pressure waves **(c)** is detected by the sensory structures in the bill tip that are illustrated in Fig. 3.4. Reproduced from van de Kam et al. (1999) with permission.

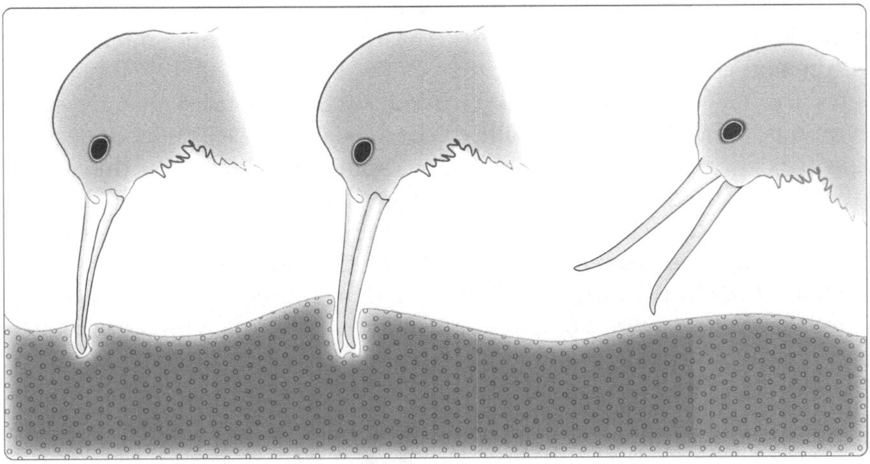

Figure 3.7 Bill of the dunlin to show its resting position **(left)** and the way in which the tip can be opened **(centre)** to expose the tactile sensory structures (Fig. 3.4) within the bill tip, even whilst the mandibles remain largely in apposition. The flexibility of the bill is also illustrated **(right)**. Redrawn from Pettigrew and Frost (1985).

2.2 Resource Acquisition—Feeding in Tropical Fruit Flies (Tephritidae)

Houseflies and their relatives have mouthparts formed into a "sponge", with its functional part at the distal end of a protrusible labellum. Such mechanisms are often considered to function in a generalized way to soak up liquid resources with small particles, and to filter out larger particles incidentally (Smith 1985). The interpretation is, however, not as straightforward as this.

The labellar structure of tephritid fruit flies in the genus *Bactrocera* provides a good example. Until quite recently, few indications were available that the adult flies did anything other than sponge up juices, yeasts and other particles from broken or fermented fruit. Simultaneously, no explanation was available for the unusual behaviors that were associated with the flies' labella, the blowing of bubbles and the regurgitation of fluid droplets onto the substrate and their subsequent uptake (Drew et al. 1983). Speculation, generally unaided by investigation, ranged from temperature regulation through evaporative cooling to osmoregulation by evaporating excess water (see Drew et al. 1983, Vijaysegaran et al. 1997). That other labellate flies do not do the same suggests adult tephritids feed in a different way from them. In addition, experiments have indicated that *Bactrocera* flies are attracted to the volatiles produced by leaf-surface bacteria (Enterobacteriaceae) and feed on them to the extent that sexual maturation is supported, at least in experiments (Drew et al. 1983, Drew and Lloyd 1987).

A microscopic analysis of *Bactrocera* mouthparts returned surprising results in that "sponge" (Oldroyd 1964, p. 202) is a poor metaphor for the function of the labellar structures (Vijaysegaran et al. 1997). Figure 3.8 shows the layers of fine "tubes" (called pseudotracheae) that lie on the operational surface of the labellum. Each such "tube" is made up of a sequence of cuticular rings. Each ring has spine-like projections (microteeth) that project forward to touch the posterior margin of the adjacent ring. The microtooth structure leaves fine holes at regular intervals along the structure (Fig. 3.8) and these lead into the lumen of the pseudotracheae. The holes are small (0.25 to 0.5 μm), much smaller than the yeast cells that the flies were thought to sponge up, and about the diameter of small bacteria. They thus correlate in size with the Enterobacteriaceae upon which the flies have been shown to feed (Fig. 3.8).

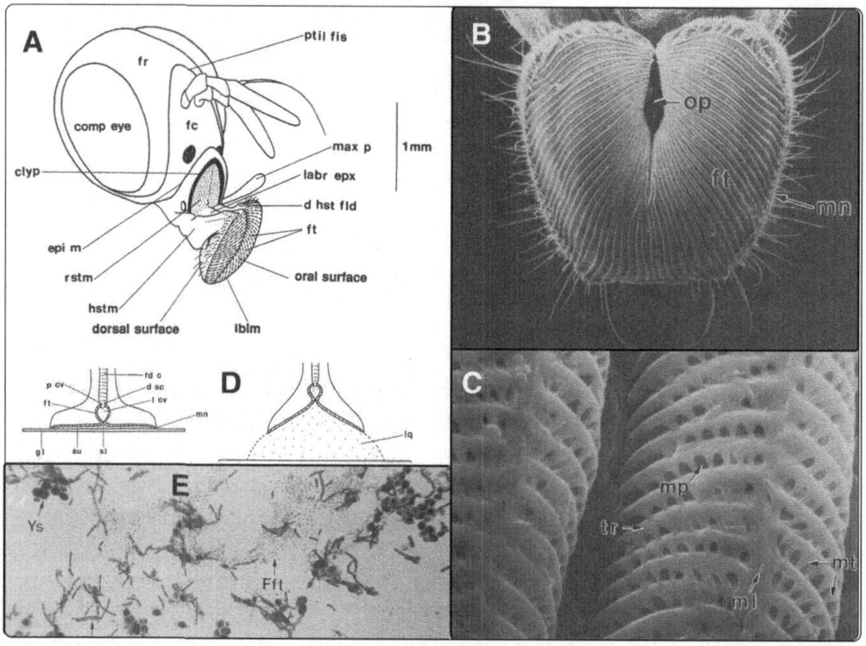

Figure 3.8 The fluid-centered labellar filtering mechanism of adult *Bactrocera* fruit flies (Tephritidae). **(A)** Schematic view of the adult head, to illustrate the mouthparts. Note the extruded labellum (lblm). **(B)** Electron micrograph of the exerted labellum to show the arrangement of the tubular system responsible for feeding, the pseudotracheae or feeding tubes (each thick dark line represents one of these in the area "ft"). **(C)** Electron micrograph of the pseudotracheae to illustrate the ring arrangement of their structure and the series of inlets along their entire length. **(D)** Diagrammatic illustration of the filtering method of feeding, with the left hand image showing how flies filter small particles from a thin layer of moisture (e.g., leachates on a leaf or fruit surface) and from a larger volume of fluid (right). **(E)** Suspension of baker's yeast (Ys), rod shaped bacillus bacteria (Bc) and the much smaller bacteria (Fft) that are selectively taken up through their pseudotracheal mechanism by the fruit flies. Reproduced from Vijaysegaran et al. (1997), with permission.

Additional tests showed that the flies, indeed, do not take up yeast cells, pollen grains or other relatively large particles. They do, however, take up Enterobacteriaceae and dissolved sugars. They evidently take up the bacteria from the leaf surface by suspending them in the liquid they "regurgitate" onto a likely surface and then re-imbibing it together with its newly acquired suspension of bacteria. Much of the feeding ecology of adult tephritids remains to be discovered, but we now know where to look and what to test (Raghu et al. 2002, Raghu 2003).

Understanding the adult feeding of *Bactrocera* will not provide a complete picture of tropical tephritid ecology, for these insects interact with the environment in numerous other ways, from the time they are deposited as eggs (Walter 2003, pp. 195-201). But this knowledge will help to interpret how the life cycle is completed, and how environmental conditions impact on insects in this stage of the life cycle, and thus their local presence and abundance.

2.3 Plant-microbial Interactions in the Soil

The roots of virtually all plants interact with various microbial soil organisms in rather specific ways. Without these various symbiotic interactions the vegetation that surrounds us would undoubtedly look very different from what it does now. Indeed, so important and ancient are the mycorrhizal symbioses, for example, that they may well have had a significant role in plants becoming terrestrial initially (Remy et al. 1994). What is significant ecologically is that these interactions are structured in specific ways, through intricate signal-response systems, to include a very specific subset of microorganisms and exclude the vast majority that could potentially invade the host plant tissues. The structure of the mechanisms that underpin the interactions dictates the specificity of the interactions of the organisms concerned, and thus their ecology. In these interactions we again see specificity at the species level, followed by species-specific ecological consequences.

We focus on the underground interactions that leguminous plants share with rhizobial bacteria. These are the most common associations with nitrogen fixing organisms among a considerable diversity (Gyaneshwar et al. 2011). These interactions tend to be better understood mechanistically than those involving mycorrhizal fungi, although the latter affect the vast majority of plants and have major influences in biogeochemical cycles (Taylor et al. 2009). With regard to rhizobia, even those that are relatively "non-specific" interact with only a small subset of all plants that apparently grow within "reach". The summaries of Lambers et al. (1998), Parniske and Downie (2003) and Limpens and Bisseling (2003) have been adopted below, with some additions.

Seedling plants produce particular flavonoids from their root system, according to their species. Bacteria in the genus *Rhizobium* interact with specific flavonoid compounds in the soil. Further specific interaction results in the initiation of nodule formation, but only if the signals and responses are appropriate, as follows. (Note, in the meantime, that the flavonoid signal structure is not necessarily the absolute arbiter of recognition, because higher doses of some compounds can apparently trigger recognition by the *Rhizobium*, and non-flavonoid signals may also contribute to specificity, so further understanding in this area is required.)

To proceed with nodulation, the specific flavonoid must be recognized by the *Rhizobium* through its binding to a receptor protein and then interact with a specific promoter (or transcriptional regulator, called NodD) in the *Rhizobium* genome. This promoter somehow communicates with the rhizobial nodulation (or Nod) gene complex that synthesizes and secretes the Nod factors. The Nod genes produce about 30 enzymes that together synthesize the lipo-oligosaccharide Nod factor, which contains chitin-like constituents. The specificity of this compound dictates the species specificity in the bacterial-plant interaction and is conferred by its side groups, which include such structures as fatty acids, fucose and sulfates. The lipid component enables the molecule to penetrate membranes. These Nod-factors thus hold the code for penetrating the plant and then inducing nodulation, which is first expressed visibly in an atypical curling of the root hair of the host plant (and which is accompanied by differential cortical cell division).

The process of recognition, by means of specific plant receptors (which are not yet well known), apparently comprises several steps and involves different specific compounds (Limpens and Bisseling 2003, Parniske and Downie 2003). Recognition by the plant of the Nod factor leads to Ca^{+2} efflux, associated hormone changes, root hair curling and infection, as well as mitotic cell divisions in the cortex and pericycle. The root hair curls around the bacteria that activated them, and this cell proliferation becomes the nodule tissue, which is infected via the associated root hair. That is where the bacteria attached and where the plant cell wall degraded to allow bacterial entry. This part of the cell wall also invaginates to form an infection thread, a conduit for the bacteria along the root hair and into the root cortex. The bacteria end up incorporated into organelle-like symbiosomes within the nodule, to facilitate metabolite exchange (Limpens and Bisseling 2003, Limpens et al. 2005). Of added ecological significance is that relatively few root hairs (< 5%) become infected, and only about a fifth of infections result in nodule formation, presumably because plants produce chitinases that break down the chitin components of the Nod factors of "unacceptable" rhizobial species, which is yet another factor in host specificity. Overall, the host plant controls the number of nodules produced, to balance the amount of carbon lost to the bacteria against the amount of nitrogen acquired from them (Limpens and Bisseling 2003).

A consideration of plant-microbial interactions in the soil indicates more than just their specific nature, requirements and ecological consequences. They are variously influenced by the abiotic environment, often substantially so. For instance, the fungal species that interact with roots differ as much in their responses to agricultural disturbances, mineral environment of the soil and seasonal climate, as they differ in their match with host plant species (which, incidentally, is much more specific in nature than was originally presumed) (e.g., Bever et al. 2001, Tagu et al. 2002). Indeed, the abiotic context of these interactions means it is not particularly meaningful to partition the environment into the usual bipartite biotic versus abiotic, for the life of the fungus and the course of the interaction draws requirements from a diversity of influences, both biotic and abiotic, and often both together. Autecological theory resolves this conflict through its focus on the individual organism and the mechanistic basis of its interactions with the environment around it, as explained through the other examples in this chapter and the theoretical implications that are drawn from them.

2.4 Respiratory Mechanisms in Plants

High and low temperature extremes define the geographic limits for the survival and reproductive potential of particular species. Although these responses and limitations may appear to be understood somewhat intuitively (Fig. 3.9), a full understanding entails considerable subtlety and biochemical complexity (Schulze et al. 2005). Even more subtle, perhaps, and far more poorly understood is the way in which the metabolism of the plant must be adapted to deal with temperature change within the range of conditions that span those extremes, the so-called permissive range (Criddle et al. 1994). Such requirements are dictated by the underlying biochemistry of the organism. "Every biological process depends on a series of consecutive and linked reactions, each characterized by its own temperature coefficient. Regulatory controls must adjust to differential effects of temperature on a multitude of complex processes that must be stringently controlled to maintain balanced reaction rates over the range of temperatures encountered" (Criddle et al. 1994).

The Arrhenius temperature coefficient of metabolism of a plant, Criddle et al. (1994) therefore suggest, is related to the plant's growth rate. Increased ambient temperature stimulates the respiration rate of a plant as it increases its metabolic rate. This reflects the increased demand for energy to support the increased rates of biosynthesis, transport, and protein turnover that occur at higher temperatures (Lambers et al. 1998). At a more basic level, concentration gradients of ions and molecules exist across the membranes that form compartments within the cells. These concentration and electrical potential gradients provide the immediate source of energy that drives the synthesis of new molecules, whether they are metabolic or structural. If temperature either slows respiration or renders it too inefficient to provide energy at a rate capable of maintaining the energy gradients in at least a steady state, these gradients will diminish until the structure can no longer be maintained. The system then begins to lose structures within the cell and, if such circumstances persist for too long, repair becomes impossible and cell death follows. Anekonda et al. (1999) thus spell out the link between cellular biochemistry and the interaction of individual organisms with their environment. Ultimately, this influences the spatio-temporal patterns we see in ecology. For a fuller understanding of the diversity of enzymes that underpin respiration and other metabolic processes, and their interrelated roles and requirements for operation, summaries are available in several good texts (Lambers et al. 1998, Raven et al. 1999, Schulze et al. 2005).

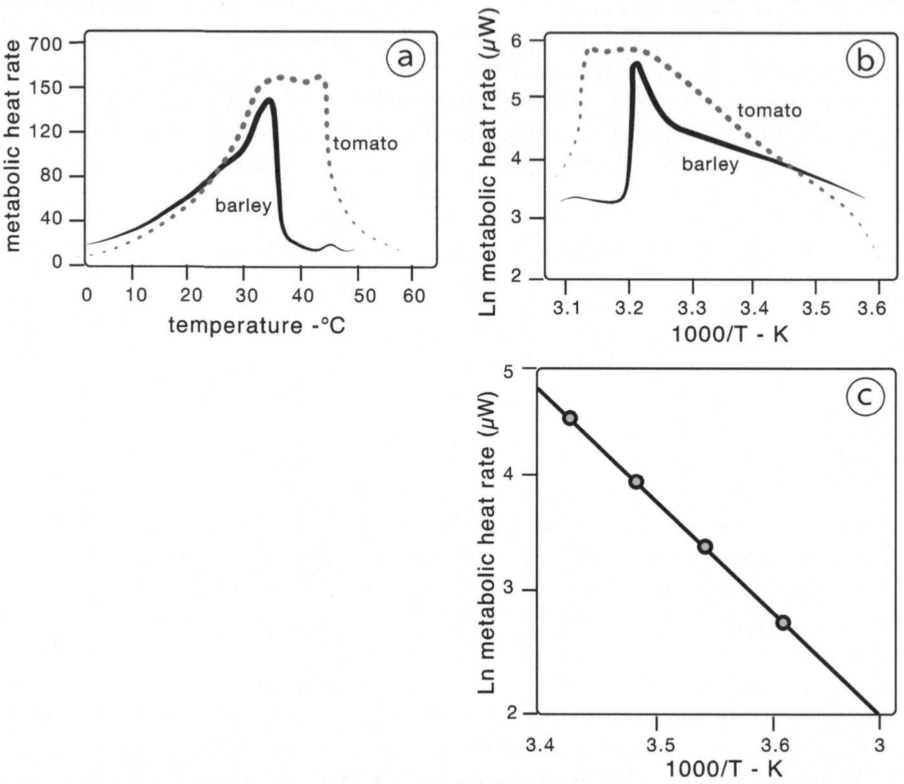

Figure 3.9 The relationship that metabolic heat rate (in μ Watts) holds relative to prevailing temperature in barley (cold adapted) and tomato (warm adapted). (Root tip cells used for the former and leaf cells for the latter.) The response of each plant species to limiting temperatures is consistent with its growth habits. **(a)** Barley metabolism rapidly decreases to near zero above 34°C, whereas the warm-adapted tomato plateaus just below this temperature, at 32°C, and decreases to zero only when the temperature reaches 44°C. The metabolic heat rate accurately reflects the temperature stability of tomato as being 4-8°C greater than that of barley. **(b)** The two species also differ at low temperatures (and here the natural logarithm of metabolic heat rate is plotted against 1000/T (kelvins), which is the Arrhenius plot for these data). Below 11°C (1000/T = 3.51) tomato metabolism decreases abruptly as temperature decreases, but the metabolism of cold-adapted barley remains nearly constant down to near zero degrees. **(c)** Arrhenius plot across the temperature range that returns a linear relationship for the data acquired from tamarack meristem tissue. The slope of the plot estimates μ, the Arrhenius temperature coefficient. Such plots differ markedly across different plant species. Reproduced, with permission, from Criddle et al. (1994).

The complexity just outlined underpins the numerous ways in which subtle differences can evolve in the biochemistry and physiology of plants, and thus influence their ecology. We now use the temperature coefficient, μ (which has units of kelvin), for ecological interpretation, but this is based on the following basic assumption. Values of μ are assumed to be interpretable simply as the consequence of the temperature dependence of some particular controlling process, such as the binding of regulator molecules or kinetics of a particular enzyme (Criddle et al. 1994), but it may be far more complex or subtle than this. Also, photosynthetic metabolism is not included. Again, the story is likely to be much more subtle in ecological terms.

Ecological work in this area of plant physiology is not yet well developed but it does reveal the potential of such studies for understanding the spatio-temporal dynamics of particular species. Research has mostly focused on a relatively extreme aspect of plant geography, the requirements for existence at higher latitudes and altitudes relative to more equable situations. Plants growing in these situations face a much wider range of environmental circumstances than experienced by plants growing elsewhere. How does this reflect differential adaptation across the species concerned, and how do individuals of species that span a breadth of such conditions cope with such variation? The significance of these questions does not relate solely to the adaptation of the plants relative to the temperature extremes or limits that are experienced, which are more straightforward to explain ecologically. The relationships plotted in Fig. 3.9 confirm this general expectation and also illustrate the utility and nature of the Arrhenius plot, which is used extensively in what follows. The seasonal temperature profile within the permissive range also has to be confronted if the spatio-temporal dynamics of plants are to be understood, and findings in this area are expanded in Section 3.1. Note that the results reported in Fig. 3.9 do not span the entire life cycle of the plants concerned, but focus on well-grown plants at different stages of the growing season. Seedlings, in particular, warrant investigation from this perspective.

Woody perennial plant species from colder environments (higher latitudes and altitudes) have temperature coefficients for their non-photosynthetic metabolism that are lower than those for closely related congeneric species from warmer environments at lower latitudes and altitudes (Table 3.1). Notable differences are, however, evident across the genera. The high altitude *Ribes* species, for instance, has a μ value that is higher than those of most low altitude species (Table 3.1), so it is not μ itself that is the primary determinant of the species' ecology and more subtlety is undoubtedly involved in the way in which different species cope with a given environment. Phylogenetic influences seem also to be evident in studies of eucalypt subgenera (Anekonda et al. 1999, Adams 2000, Anekonda et al. 2000), which supports the implication that μ evolves relative to a particular environmental context and, within a species, is relatively stable in an evolutionary sense (despite the variation that can be measured within individuals and across conspecifics). In other words, the variation helps individuals ecologically, as outlined below, but has relatively limited capacity to do so and evidently, within the species, only limited capacity for local adaptive change. This biochemical pathway is not subject to the ongoing optimizing "improvement" expected in evolutionary ecology (but see the results in the paper of Thygerson et al. (2002) on germinating herbaceous seedlings).

Table 3.1 Comparison of Arrhenius temperature coefficients (μ) across closely related species (conspecific varieties in some cases) that grow at different altitudes (from Criddle et al. 1994). High elevation species were all from 3500 m (except *S. sempervirens* at 1400 m) and low elevation species were from 1900 m (except *S. giganteum* at 100 m). Temperature ranges of tests were 5 to 20°C, except for the *Sequoia, Sequoiadendron* (5 to 25°C) and Ribes species (4 to 19°C).

GENUS	*"HIGH" SPECIES*	μ(K) ± *S.E.*	*"LOW" SPECIES*	μ(K) ± *S.E.*
Sequoia/ Sequoiadendron	*sempervirens*	5000 ± 400	*giganteum*	8300 ± 800
Ribes	*montigenum*	8800 ± 1000	*aureum*	11000 ± 200
Prunus	*virginiana* var. *melanocarpa*	4500 ± 2600	*virginiana* var. *melanocarpa*	6100 ± 2400
Salix	*planifolia*	3800 ± 1800	*exigua*	8200 ± 1600
Artemisia	*tridentata* var. *vaseyana*	1400 ± 900	*tridentata* var. *tridentata*	5000 ± 1000
Abies	*lasiocarpa*	4500 ± 500	*concolor*	5500 ± 600

Besides the obvious genotypic variation in μ within a species (Criddle et al. 1994), variation is also evident through the growing season, and μ shifts to lower values as that season progresses. In the combined data from the closely related *Purshia tridentata* and *Cowania stansburiana,* the early northern growing season (May) μ values for plants of various provenance range from 8300 to 11500 K, whereas the equivalent values closer to midsummer (early June) are 5500 to 8500 K (Table 5 of Criddle et al. (1994)). The plants at lower temperatures have relatively higher metabolic rates than those at higher temperatures. In general, the switch to lower μ is later for the plants from higher elevations and latitudes. Biomass accumulation by plants is an exponential function of growth rate (for which μ is used here as a surrogate), so even small differences in μ can have a very large effect on total plant growth (see Criddle et al. 1994). Plants cannot maintain their tissues at an optimum growth temperature as ambient temperature changes, so they grow only in those geographical localities in which the value of μ and its seasonal dynamics match the seasonal temperature profile that the plant experiences (Criddle et al. 1994, Criddle et al. 1996, Criddle et al. 1997, Anekonda et al. 1999).

Species that experience broader temperature ranges during their growth are relatively less sensitive to temperature in having smaller temperature coefficients than species experiencing narrower ranges of temperature (Criddle et al. 1994). Again, this may vary across species, as may the profile of the temperature coefficient with growth stage and season. These changes tend to reflect larger increases in metabolic rate at lower temperatures, so that the plants can develop even whilst temperatures are still low. All plants are not adapted to broad temperature ranges, despite the apparent advantage. Plants that are able to survive over a broad range of temperatures accrue additional energy costs, because energy is required to underpin that particular physiological capability, and the energy thus used is no longer available for growth, reproduction and resource acquisition (Anekonda et al. 1999, Anekonda et al. 2000). Physiological limitation to a narrower ecological range is, however, accompanied by the risk of greatly reduced metabolic efficiency and major damage or death if temperatures extend beyond the narrow tolerance limits (Criddle et al. 1994).

The species investigated so far have been mostly temperate ones, and the findings relate only to woody perennials. Annuals and herbaceous perennials do not show the metabolic relationships detected in the woody perennials and seem to time their growth relative to seasonal temperature profiles (Criddle et al. 1994, and see Section 3 of this chapter) and rainfall (Summers et al. 2009). The metabolism of more tropical species has yet to be investigated in this regard, and the interactive effects of the prevailing temperature, humidity, edaphic and soil water availability profiles across different environments will need to be taken into account if we are to develop a general model of how plant species interact with the environmental factors that make up their permissive ecological range. In other words, we have examined but one environmental axis along which species may be differentially adapted (see Chapters 4 and 5), and the ecological interactions of plants with their environment is far more intricate because other variables impinge on the individuals concerned.

2.5 Localization of Hosts and Mates

Among the most widely appreciated interactive mechanisms of organisms include the host finding adaptations of insect herbivores and parasitoids and the fertilization mechanisms (and sexual communication systems) of bi-parental eukaryotes. These adaptations represent critical components of the life cycle of the organisms concerned. Without accurate function at this level, persistence and reproduction are gravely compromised, if not impossible.

The behavior associated with the localization of their usual host(s) is of clear ecological significance to organisms like insect parasitoids. This behavior dictates the observed host range of the insects concerned. Ectoparasitoid wasps and specialist herbivores clearly illustrate this point. Many species of these organisms can be reared on a wide range of host organisms. With regard to ectoparasitoids, the eggs and developing larvae are situated externally and do not confront the hosts' immune system (Salt 1968). The larvae of a single ectoparasitoid species can therefore feed successfully upon a diversity of host species and host types. Nevertheless,

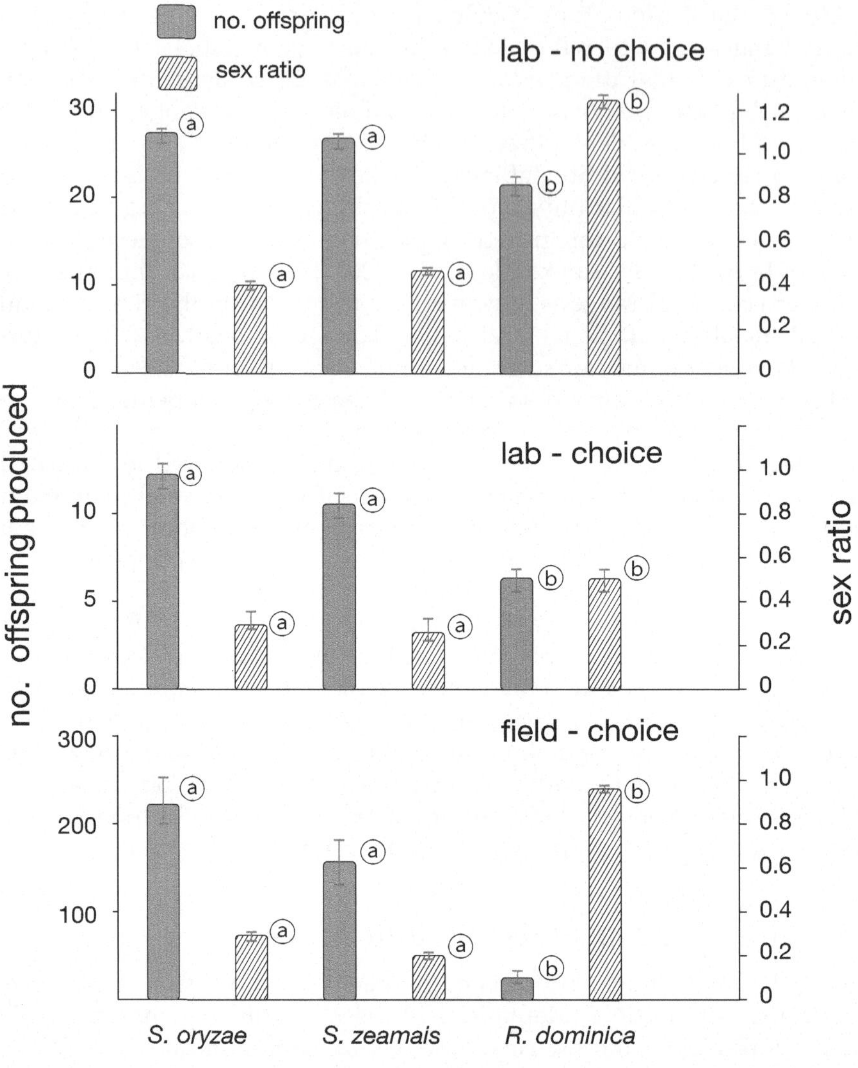

Figure 3.10 Numbers of offspring (and their sex ratios) produced by mated females of the ectoparasitoid wasp *Anisopteromalus calandrae* when exposed to host beetle species (in the genera *Sitophilus* and *Rhyzopertha*) in different combinations, either alone (no choice) in the laboratory or in combination (choice) in the laboratory and field. The laboratory data reflect the mean (±SE) productivity per female ($n = 10$) over three days, and females produced about the same number of offspring across each host species in the no-choice test (albeit significantly lower from *R. dominica*, but not much so) and about the same number from all three hosts together in the choice test (where again production was significantly lower from *R. dominca*, but much more so). These differences in the field were much more extreme (four days exposure to an unknown number of wasps), indicating that the host-searching mechanism is adapted to *S. oryzae*, in particular. The male-biased sex ratio consistently associated with *R. dominica* indicates that this species is treated as an inferior host by ovipositing females. Drawn from data presented by Sayaboc (1994).

ectoparasitic species tend to be much more host restricted when their ecology is quantified in the field, as illustrated in Fig. 3.10, which details the host relationships of a common parasitoid that attacks beetle pests of stored grain. The implication is that the host searching behavior is structured in specific ways, as generalized in Fig. 3.11, and this specific mechanism dictates which host species are attacked in the field, on which plants, and in which vegetation type(s). The spatio-temporal dynamics of the organisms are thus strongly influenced by their sensory and behavioral mechanisms.

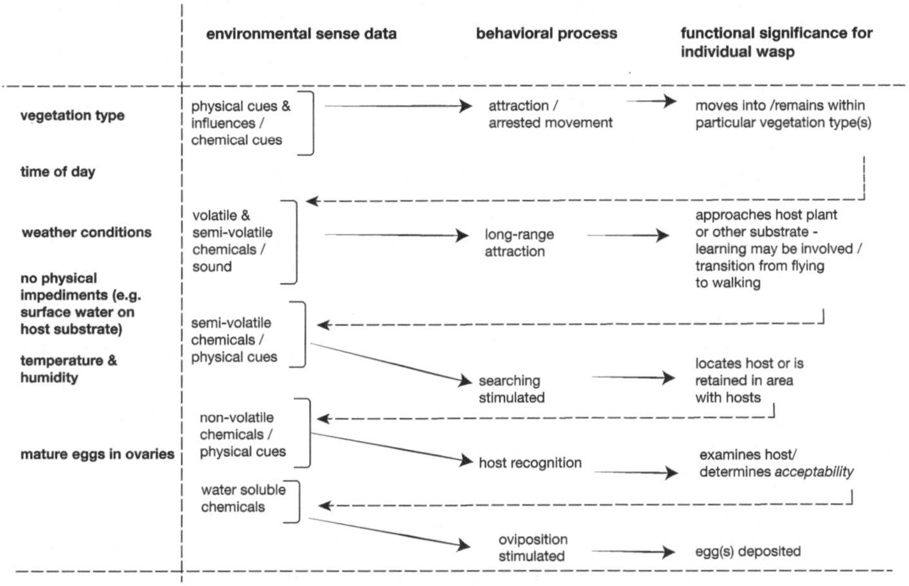

Figure 3.11 General model of the sequential physiological and behavioral processes that result in parasitism of the usual hosts of a parasitoid species. An act of parasitism involves the detection, localization and recognition of an appropriate host within the usual environmental context of the organisms concerned, as well as the deposition of one or more eggs in the appropriate site. Reproduced from Walter (2003), with permission.

Recent research indicates that even species that are seen to be resource generalists have rather specific host searching mechanisms and that their sensory capabilities influence the pattern of their resource use in the field. No such species is well known in this regard, but the following composite picture of the situation reveals the sort of analytical research that is required to understand the organisms in question, and explain how and why the organism-resource interaction is much more specific than commonly portrayed.

i) Many generalists have been found, with appropriately designed research, to be comprised of more than one species, each of which is associated with a subset of the resources (e.g., Paterson 1991, Clarke and Walter 1995, Walter 2003, Bickford et al. 2007, Malausa et al. 2007, Loxdale et al. 2011, Rafter et al. 2013).

ii) Quantified sampling in the field reveals that even resource generalists are strongly focused on relatively few resource types, so an understanding of their ecology requires research to concentrate on the interaction practiced by the majority of individuals, at least initially (Walter 2003).

iii) The basis by which resources are located (by herbivorous insects, for example) or captured (by spiders, for example) are specific, to the extent that most individuals interact with a few particular hosts (e.g., the cotton bollworm *Helicoverpa armigera* (Rajapakse et al. 2006, Rajapakse and Walter 2007)) or capture a small subset of all available prey species (e.g., spiders (Harwood et al. 2001)). See also Walter (2003, Chapter 7) and Finlay-Doney and Walter (2012a).

iv) The nature of the host-searching mechanism of herbivore species that do use multiple host species also influences the relative frequency at which the various host species will be attacked, and thus the local abundance of the herbivores (Manners 2008).

The sexual communication system of bi-parental eukaryotic organisms has elements in common with the host-searching mechanism described above; however, one significant difference between them is the strong co-adaptation that exists between the sexes of every species. There is a level of co-operation between the sexes that does not exist in the host-searching mechanism (cf. the "conflict" between the sexes that some evolutionary ecologists tend to see (Chapman et al. 2003, Tregenza et al. 2006)). From this perspective, the sexual communication system, and the fertilization mechanism of which it forms a part, have no equal when we consider other ecological interactions. Furthermore, recognition is seen simply as that, the unconscious response to signals that match the neural template, much like the recognition of antigens by antibodies in the immune system (Paterson 1981, 1985). We do not see it as a system of decision-making based on costs and benefits (e.g., Sherman et al. 1997), for example, so the postulated recognition of "quality" is not seen as more than incidentally relevant. Organisms are not seen to make a choice, as postulated in sexual selection, but interact on the basis of a system of recognition (Paterson 1982a, 1985, 1993b), as outlined below.

The sexual communication system can be conceptualized diagrammatically as the specific-mate recognition system (SMRS) (Fig. 3.12). The strong co-adaptation between the sexes, the environmental and habitat context of the sensory and behavioral mechanisms and the functional role of the specific behaviors all contribute to the complexity of the mechanism and, ultimately, to the stability of these mechanisms. In other words, individuals that are divergent relative to the functioning of the mechanism within its usual environmental context are more likely not to participate in fertilization (and thus be selected out) than to initiate evolutionary divergence (Paterson 1982b, 1993b). This implies that evolutionary change requires special conditions, and this is covered more extensively in various contexts in Chapters 5 (Section 6), 6 (Subsections 4.3 and 4.4) and 10 (Subsection 5.2.2).

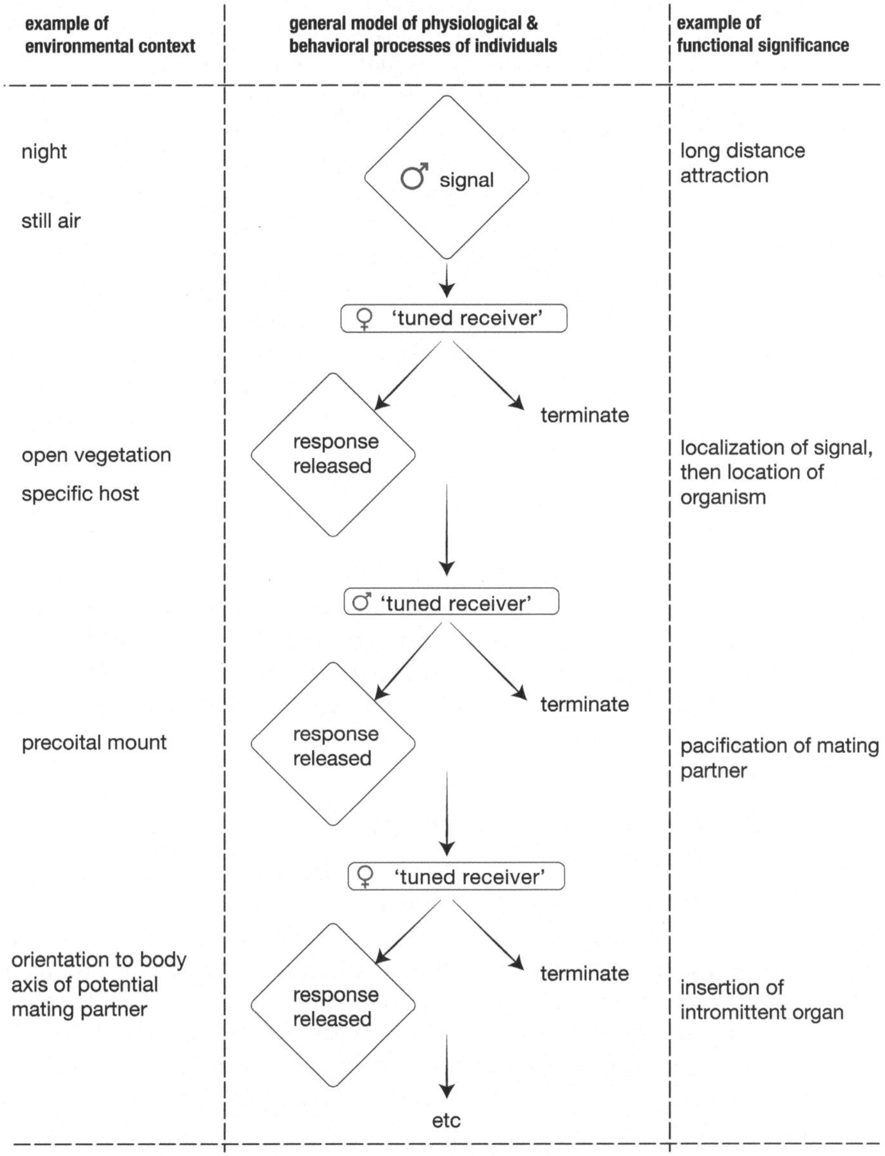

Figure 3.12 Diagrammatic representation (central column) of the sequence of events involved in the Specific-Mate Recognition System of a bi-parental species of animal. Each event should be considered not only in its context in the chain of events that leads to fertilization but also in relation to the nature of the environment (left column) and the specific functional role of the behavioral and physiological processes involved (right column). The number of stages in the sequence is characteristic of the species concerned. Redrawn from Paterson (1985) by Walter (2003) and reproduced with permission.

Ecologically, the particular structure of the SMRS has significant implications, as it is specific with respect to habitat, diel periodicity and seasonality. The environment thus provides the requisite backdrop for the effective operation of the SMRS, and changes to significant aspects of the environment will have ecological consequences for the species concerned (Paterson 1986). Furthermore, the SMRS represents but one functional component of the breeding system as a whole and cannot be readily excised, for the life cycle is an integrated whole. Consider, for example, the Atlantic herring that breed in the western Baltic, off the Kiel coast. Although the fish are categorized under a single species name, or taxon, *Clupea harengus*, two distinct types are present in these waters. They are morphologically distinct from one another and each breeds independently of the other, one in spring (in the Schlei Inlet) and the other in autumn (open sea of the Bay of Kiel), and fish of the two groups school together in winter (Sinclair and Solemdal 1988, p. 194). Apparently, these two populations are independent of one another, in reproductive terms, and may well constitute independent cryptic (or sibling) species, but this needs to be tested directly. If correct, how and why they differentiated warrants close attention, but a preliminary comment can be made about the structure of the SMRS. Maturation of the gonads has to be synchronized across the sexes and with regard to season, and the hormonal control of the emission of sexual signals, and responses to them, has to be synchronized in similar fashion. These aspects are significant because they help to draw the direct relationship between the life cycle adaptations of organisms, their ecological interactions with the environment and the consequent changing abundance of the species locally and geographically.

2.6 Adaptations for Movement

Species are differentially adapted to their requirements for movement and their migration trajectories, as indicated by these features in a group of small birds, leaf warblers in the genus *Phylloscopus*, and in *Homo sapiens*. These extremely different organisms illustrate that the adaptive mechanisms associated with movement are not adjusted simply in terms of an assumed efficiency. They have adjusted in concert with other features of the life cycle to and way-of-life to ensure the whole is appropriate to particular sets of environmental circumstances. The consequences for the ecology of the organisms are profound.

Migration takes place within the context of various other life cycle features; it has to fit in with the seasonal aspects of the life cycle. Phylogenetic analysis indicates that winter molt evolved in the Old World warblers (Sylviidae) several times, independently of one another, and it did so from the ancestral condition of a summer molt (see Fig. 1 of Svensson and Hedenstrom (1999)). Winter molt tends to be associated with species with longer migratory distances. The match, though, is far from perfect (cf. Fig. 2 of Svensson and Hedenstrom (1999)), especially in the genus *Phylloscopus*.

Most *Phylloscopus* species migrate long distances (at least when that is defined arbitrarily as 1000 km or more), but only 10 of 26 species molt during winter (Fig. 3 of Svensson and Hedenstrom (1999)). Even some northern Siberian breeders that overwinter in southern Asia retain the winter molt (*P. inornatus*, *P. proregulus* and *P. fuscatus*). Although "evolutionary inertia" or "constraints" (usually developmental) are invoked in explanation, such interpretations are unsatisfactory for they relate to expectations generated by optimization theory. After all, *P. trochilus*, alone among all the many species of sylviid warblers, has both a summer and winter molt (Svensson and Hedenstrom 1999). Advantages can be envisioned for this, but why then do none of the other species do the same? Are they constrained? Or are we perhaps looking at an unusual feature that has been evolved solely by *P. trochilus*, and that may open the way for novel deployment of this feature should this lineage diversify? If this were to be the case, such a feature would be recognized as a pre-adaptation when looking back from some future time. Nevertheless, some features of organisms are ancestral, but persist (e.g., Wanntorp 1983). These, too, should not be interpreted in relation to perceived constraints.

Seeking ecological correlates with the feature of interest is likely to be more useful than invoking constraints. For example, Svensson and Hedenstrom (1999) suggest breeding duration and habitat variables could be significant. The explanation may, by contrast, be historical and relate to the relative geographical distribution of the breeding and overwintering grounds at the time of speciation. The long-distance migrators that molt in winter may have speciated, and thus acquired their properties, when their migration route was relatively short. Their adaptations have not, if this interpretation is correct, precluded invasion of different areas and longer migrations. Note that southern populations of willow warblers have the same molt pattern as northern ones. That is, this feature is species-wide, so apparently evolved at speciation in a relatively small population (Chapter 5). The adaptation would then have been "carried" to all parts of the current distribution. This all implies that for a more meaningful analysis, additional comparative detail about species' Pleistocene refugia, current geographical distributions and precise timing of winter molt would be required. Also, a better understanding of the species status of the relevant populations is necessary; we note that several *Phylloscopus* species are now recognized to comprise cryptic species complexes, including fairly well known ones like the chiff chaff (*P. collybita* and *P. ibericus*) and Bonelli's warbler (*P. bonelli* and *P. orientalis*) (Helbig et al. 1995, Irwin et al. 2001, Bensch et al. 2002, Salomon et al. 2003).

Migrant birds, across several taxa, have wings that are generally more pointed and convex (Fig. 3.13) than those of non-migratory congeneric species and congeneric species that do not migrate so far (Lockwood et al. 1998, Arizaga et al. 2006, Baldwin et al. 2010). The *Phylloscopus* species that breed across Eurasia have different overwintering areas. Those that migrate furthest, the Arctic warbler (*P. borealis*) for example (Fig. 3.14), tend to have the wingtips relatively more pointed and convex and have larger

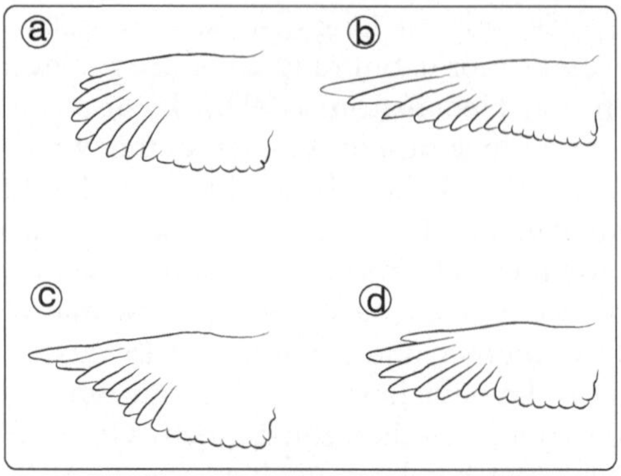

Figure 3.13 Schematic of the wing shape of birds to illustrate wing pointedness and convexity. **(a)** Rounded. **(b)** Pointed (characterized by a shift in the wingtip towards the leading edge). **(c)** Concave. **(d)** Convex (which is a decrease in the acuteness of the wingtip, with the primaries close to the wingtip more extended than those further from the tip). Reproduced from Lockwood et al. (1998), with permission.

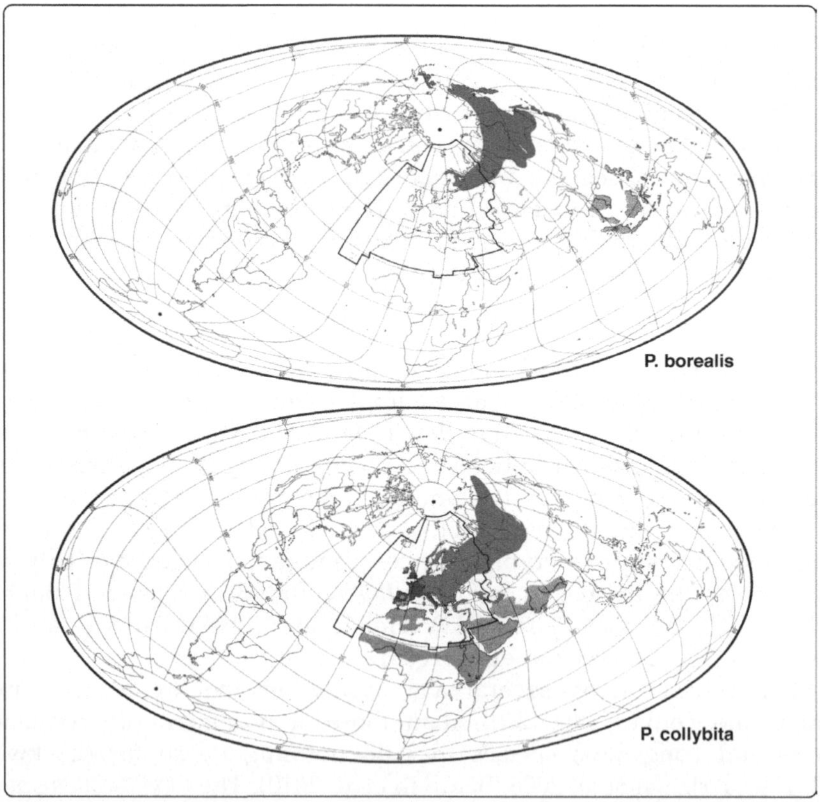

Figure 3.14 *Phylloscopus borealis* and *P. collybita* breeding (northern) and overwintering (southern) distributions to indicate the distances they migrate annually. Reproduced from Cramp (1992), with permission.

aspect ratios than species that do not migrate as far, including the chiff chaff (*P. collybita*) (Fig. 3.14) (Tiainen and Hanski 1985, Marchetti et al. 1995).

Although the differences in wing shape outlined above seem slight, they do not seem to be trivial ecologically. Also, they may well be of long standing, as illustrated by comparison across the two subspecies of willow warbler in Scandinavia. These are distributed on opposite sides of a geographical divide, and have independent migratory trajectories. Differences across the divide also include body size, plumage color, responses to stress and starting date of molt, all of which suggests that the division represents the secondary contact of two populations that underwent allopatric adaptation (Bensch et al. 1999). *Phylloscopus trochilus acredula* migrates in spring and early summer, to breed in northern and east Europe. South of that line in central Sweden is the breeding distribution of *P. t. trochilus*, with overlap of about 300 km in which hybridization is said to occur readily (which is explicable in there being no measurable or functional differences in song structure discovered to date). Individuals of the subspecies *trochilus* migrate, in autumn, in a southwesterly direction whereas *acredula* individuals fly southeast. Hybrids are said to fly intermediate routes but data are relatively few. The significance in this example is twofold: (i) the line along which the two subspecies abut in their distribution has persisted for at least 150 years, perhaps as many as several thousand years, and (ii) the migratory responses of each subspecies appear to be subspecies wide. They thus acquired their differences independently of one another, possibly during the last Pleistocene glaciation (although the age of the species might indicate otherwise (Price et al. 2000)), and the migratory responses of each work in conjunction with the other relevant adaptations of each subspecies.

The environmental influence on seasonal migration in birds is perhaps readily appreciated as an ecological adaptation, but all adaptations for movement carry ecological consequences. Even the capacity for locomotion of our own species, *Homo sapiens*, has significant implications for the ways in which we could originally interact with the environment, and the way we have done so since, although technological innovation relating to our thermal relations, energetics and ability to do work has changed that somewhat.

Bipedal locomotion is unusual, especially among mammals, and its evolution has demanded an extensive range of concomitant adjustments. Simultaneously, its full expression has had profound ecological consequences, even for hominids. Our usual walking gait is so "normal" to us that we tend to forget that we habitually and unconsciously balance our bodies upright on the relatively small feet of our hind limbs, with our knees fully extended, and all without a tail for fine adjustments (Spoor et al. 1996). We consequently do not readily perceive how the dramatic adaptive changes our species underwent were driven by the original life, death and reproductive implications of the selective environment of the time, or how the changes simultaneously transformed our subsequent ecology (Lovejoy 2005b).

Modern humans and some of their ancestors, alone among the primates, habitually use a striding gait (Lewin 2005). Bipedalism in hominids clearly had several ecological advantages, but which subset of them was the prime driver in the evolution of bipedalism is "one of, if not the biggest problem confronting human paleontology" (Arsuaga and Martinez 2006). The long, stable stride is fast over short distances, yields great stamina in running or walking, keeps the head in relatively cool strata and attracts comparatively less solar radiation during the hottest times, and frees the hands and arms for manipulation and carrying (Arsuaga and Martinez 2006). The last mentioned might have been particularly relevant to provisioning altricial young with huge resource demands for development. The ecological consequences were clearly substantial. Within this context, it is only the technological and economic activities achieved by humans, and not the adaptive modification just discussed, that can be interpreted in terms of optimization and efficiency.

Bipedalism did not evolve gradually through small refinements that accumulated additively over protracted time periods. Pre-adaptation was co-opted and associated with a relatively rapid shift that produced an organism, *Australopithecus*, that was fully bipedal and which had probably left the trees except to take refuge (Lovejoy 2007, Lovejoy and McCollum 2010). The pre-adaptive beginnings were acquired a long time before bipedalism was achieved, and these early changes were evolved without any connection to bipedalism itself. For example, the postcranial anatomy of humans was pre-adapted to bipedalism through its original adaptation to a suspensory life in trees, a way of life still practiced by the other higher primates (Lovejoy 2005a). Elements of these changes were present in *Proconsul* as long as 22 million years ago (Lewin 2005, p. 105). In particular, the shortening of the lower back and its associated rigidity, which is still present in the extant great apes, is part of the human bipedal heritage. Lower back rigidity was essential for the hips and femora to adjust to upright walking, to the point that the lumbar spine was the focal point for the earliest adjustments to bipedalism (Lovejoy 2005a). Part of the process was a freeing of the lumbar spine, which is mobile compared with that of the great apes and also is somewhat longer. These changes to its obvious and otherwise "great ape" morphology allowed, together with changes to the pelvic bones, the forward curvature (or lordosis) of the spine to adjust to the new center of gravity required for effective bipedalism (Lovejoy 2005a). Muscle function, size and attachment all changed to keep the organism upright. These changes were accompanied by changes to the head's shape and its position of articulation with the backbone and by changes to the shape and operation of the foot. The latter became a balancing platform for transferring thrust rather than being essentially a grasping organ (Lewin 2005).

Bipedal locomotion, although fully developed through the processes outlined above, was further refined through subsequent adaptive events. Adjustments to the sensory system were significant in this regard – the plane of orientation of the semicircular canals in the labyrinthine canals of the inner ear is crucial to perceiving angular head motion (Spoor et al. 1994, 1996). These sensors are thus involved in the unconscious coordination of body movement in the vertical plane through the vestibular reflexes. This system changed in structure to provide more sensitivity for the coordination of upright bipedal behaviors, like jumping and swerving. The anterior and posterior semicircular ducts are oriented vertically and therefore are presumably more sensitive in their vertical plane within the head. They are larger in *H. erectus* and modern humans than in earlier hominids and the great apes, which implies that some refinements in bipedal capabilities took place at different times, presumably during particular speciation events. By contrast, the horizontal lateral semicircular duct is relatively smaller than in the apes (Spoor et al. 1994, 1996). These adjustments have been interpreted as an adaptation to increase mechanical sensitivity and the improved ability to resolve the small changes in fast vertical head rotations that characterize the agility associated with modern human bipedal running and jumping on irregular substrates. They thus help maintain body coordination (Spoor et al. 2003).

Intriguingly, Neanderthals differ somewhat from the *H. erectus* and *H. sapiens* configuration, having relatively smaller vertical canals, which correlates with these hominids running less and being less agile than modern humans and, by implication, *H. erectus*. This interpretation is supported by the relatively heavier mass and build of Neanderthals, and their shorter limbs. Faunal analyses of prey remains suggest that Neanderthals mostly targeted large, gregarious herbivores of open environments and that they were mostly resident, although they did shift as horse, bison and reindeer moved locally (Vaquero et al. 2001, Hockett and Haws 2003). Evidence indicates they hunted more by ambush. They evidently jumped onto large prey, for their skeletal injuries reflect the injury pattern suffered by modern rodeo riders (Berger and Trinkaus 1995), and this also correlates with their being walkers rather than runners. The features of the Neanderthal inner ear seem, therefore, to be derived (Spoor et al. 2003). Their way of life was thus acquired together with their adaptations for locomotion, and together such features determined their subsequent ecology, which differed substantially from that of the highly mobile modern humans (Wynn and Coolidge 2012).

Modern humans are, however, not just bipeds. They are bipeds with relatively enormous brains, so successful parturition is an additional and powerful selective force. Indeed, birth canal size is said to have provided a continual anatomical crisis throughout the Pleistocene, and the rate and scale of human brain enlargement was remarkable: "Modern human infants are excessively altricial and must be jettisoned 'before their time' so as not to endanger their mother's life" (Lovejoy 2005b, p. 122). The fossil record shows that bipedalism was well advanced before the extremes of encephalization evolved in the genus *Homo* (and some of their immediate ancestors). However, the femur of modern humans (and *H. erectus*) is disproportionately long compared to that of the ancestral types, and it seems not to have been associated with any obvious environmental change or requirements (Lovejoy 2005b). Perhaps increased predatory pressures or an environmental requirement for covering distance rapidly would have selected for enhanced velocity, and this would have been acquired against a background of negative selection from increased injury rates associated with having longer limbs (Lovejoy 2005b).

The demands of bipedalism and the birth process seem to have interacted developmentally, usually in somewhat negative ways. For example, the requirements of bipedalism permitted only a marginally satisfactory birth canal to evolve, when this would otherwise have been under positive selection rather than increasingly altricial birth being selected (because smaller infants suffer increased mortality rates). The various pelvic modifications evidently impeded some adjustments that would have been beneficial to bipedalism (Lovejoy 2005a). Nevertheless, these adjustments were associated with an increase in limb length. Besides the reasons mentioned above, pleiotropy could have been involved (Lovejoy 2005b). Evidence for the latter is only indirect, with the mechanical difficulty during birth being as much as 14 times higher in relatively short females, and birth inlet shape being most strongly correlated with femoral length, rather than any of the other measures included in an exhaustive survey. Selection may, therefore, have been on stature, not directly on limb length (Lovejoy 2005b). When evolutionary change is considered mechanistically in this way, one gets a good idea of how unrealistic is the vague notion of efficiency in interpretations of evolution in terms of optimizing selection.

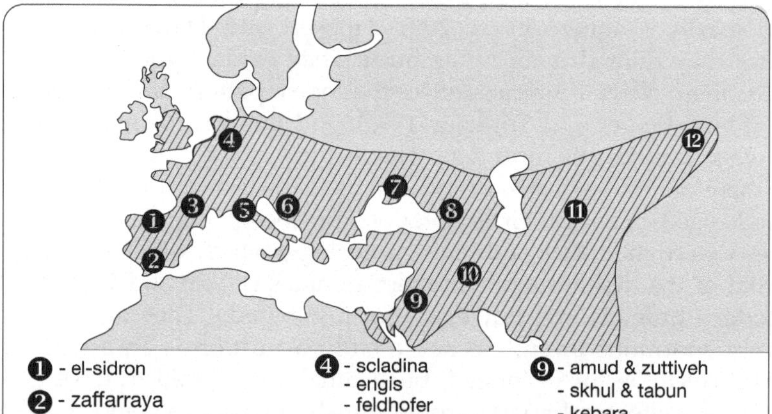

Figure 3.15 Representative sites at which Neanderthal fossils have been found, to illustrate that the known geographical distribution of the species was limited to western Eurasia. The northern limits correlate roughly with the last Pleistocene glacial maximum, so any evidence of their existence that might have been north of this line is likely to have been obliterated. Redrawn from McKie (2000) and Balter (2009).

Bipedalism in humans is only one mechanism by which the individuals concerned interact with their environment. Other mechanisms are also at play, and species-specific differences are evident when what is known about the ecology of *H. neanderthalensis* is compared with that of *H. sapiens* of the time (for they were contemporaneous for 20 to 30,000 years, but with vastly different geographical distributions (Fig. 3.15)). The cold adaptations of Neanderthals are well known, they may also have been adapted to rougher terrain (Higgins and Ruff 2011), and may well have had very different population densities and social systems based on different foraging methods from those of nearby *H. sapiens* (e.g., Vaquero et al. 2001, Wynn and Coolidge 2012, Ready 2013).

In other words, the mechanism of human locomotion has influenced the ecology of the species from the outset, but its population consequences cannot be isolated entirely from the other features that influence human ecology. The same holds for the mechanisms of all the other organisms considered above. In general terms, ecological understanding and theory needs to incorporate the totality of interactions in which individuals of any particular species interact with their surrounding environment.

2.7 Conclusions—Environmental Interactions and Evolutionary Change

The mechanisms that underpin the environmental interactions of an organism are the props that support the persistence of that organism at particular times in its life and thus, acting sequentially, the trajectory that is its lifeline. Autecological theory therefore focuses on the direct relationship between the individuals of the species of interest and the environmental factors that impinge on them, rather than working indirectly via density and competition. The autecological perspective thus specifies the way in which organisms need to be treated with respect to theory in ecology, as follows.

Autecology focuses on the requirements for an organism to survive through each stage of its life and to reproduce. That is, each of the interactive mechanisms that are crucial to each given life stage of that organism and set of environmental circumstances inhabited by it must work effectively to achieve particular ends that are ecologically significant to it. These are the functional ecological operations that carry the organism along its lifeline. As organisms develop and grow, enter different physiological states, or morph into different ways of life, their mechanisms of interaction may change in their functional configurations and different mechanisms may become operative. Such changes are manifest ecologically in the different requirements and tolerances that are now evident in that selfsame individual. Furthermore, different species respond independently of one another to any given environment, through the differential actions of the adaptations carried by the individuals that make up each particular species.

Of particular significance for understanding ecology is that the lifeline of the organisms that make up each species is, in sum, adapted to a particular sequence of environmental circumstances, seasonal for example, or different habitats, as organisms migrate or metamorphose. All organisms move to a lesser or greater extent during their lifetime and the specific set of changing requirements and tolerances of the organism's lifeline have to be accommodated in any new locality (or environment) in which the organism finds itself. Ecological generalization needs to accommodate this aspect of the interaction of organisms with their environment, as well as the diversity of ways in which this is achieved by different species, which are, again, individualistic with respect to one another in this regard. In the rest of this chapter, the ecology of particular species in a diversity of taxa is illustrated with respect to season, metamorphosis, and movement, to emphasize the magnitude of this task.

The synopsis above explains why, in autecology, density-related, numerical influences are considered to be secondary factors in nature (Andrewartha and Birch 1984). Without the primary (or functional) interactions working effectively, density effects cannot even come into play. Demographic ecology, by contrast, stresses density dependence as being of primary significance through assuming that the environment is, on average, of equivalent influence across individuals and across species, which explains the ongoing search for patterns that transcend species and the differences between them. This "all other things being equal" argument is, however, soundly and consistently defeated by the idiosyncrasy of species and the stochastic nature of the influences in the life of organisms and therefore in the dynamics of ecological systems. The autecological view of organisms also has significant implications for interpreting evolutionary processes. These points are dealt with in subsequent chapters.

3 EXAMPLES OF SPECIFICALLY STRUCTURED LIFELINES

3.1 Plants

Consider Fig. 3.16. The correlation between the yield of peas and the seasonal temperature range where they grew in the USA provides an indication of what this species requires, thermally, from germination to seed set if it is to produce viable seeds. Seasonal temperature range is the difference between the mean temperature at the planting stage and that for the harvesting stage. In short, these plants are heat sensitive, not only in terms of temperature level but also in relation to the time sequence in which the heat they need is provided, with an optimal combination of a relatively warm spring and cool summer (Fig. 3.17) (Wang 1960, 1962). In other words, the individual's lifeline has to match the particular sequence of thermal conditions that prevail in a locality if the plants are going to thrive and reproduce. Not surprisingly, pea production is centered on areas that fulfill these requirements best, those with relatively low T_D values (12-15) (Fig. 3.18). Although the map in this figure contains temperature information for only one entire growing season, it does demonstrate that outside the central area of pea production the thermal requirements of the plants are less likely to be met.

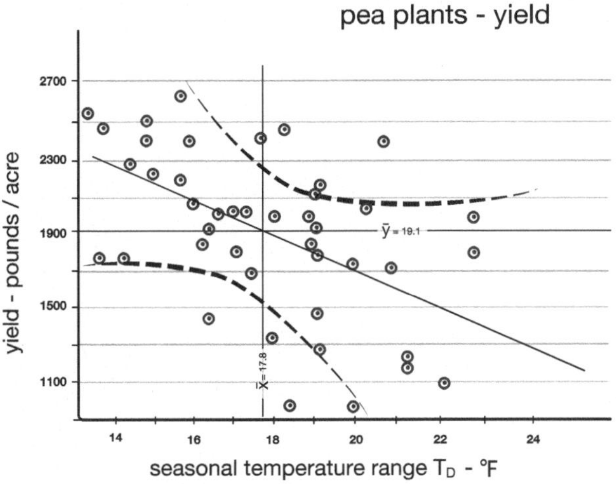

Figure 3.16 Yields of peas over 43 years (1918 - 1960) in Wisconsin plotted against the seasonal temperature range (T_D), which is the difference between the monthly mean temperature during harvesting and the monthly mean temperature during planting. The numbers represent particular years, the x-axis the mean yield and the y-axis the mean T_D value. Most data points are concentrated in the middle of the regression line, between T_D values 16 and 20°F. Those points outside the envelope (dashed lines) probably represent damage from insects, disease and drought. Redrawn from Wang (1962).

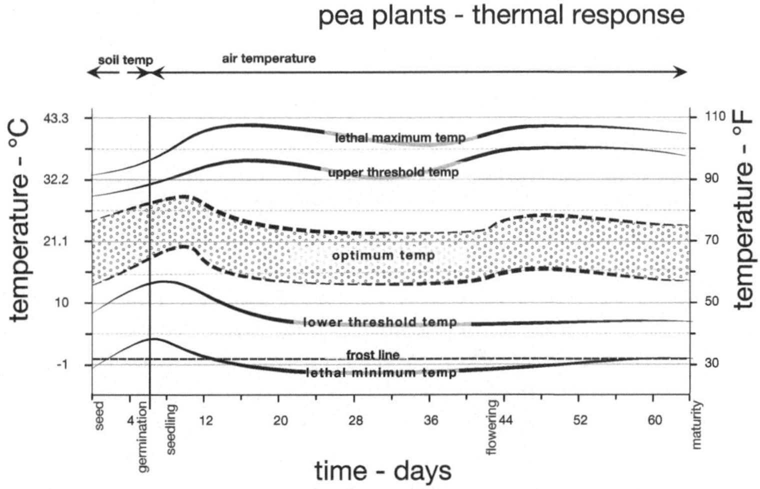

Figure 3.17 Thermal responses of pea plants. This image collates the changing thermal requirements of pea plants as they pass through their various life stages and their changing tolerances to extreme temperatures. The environmental temperature (soil temperature for germination and air temperature thereafter) is the effective temperature for periods of 6 to 12 hours. Redrawn from Wang (1960).

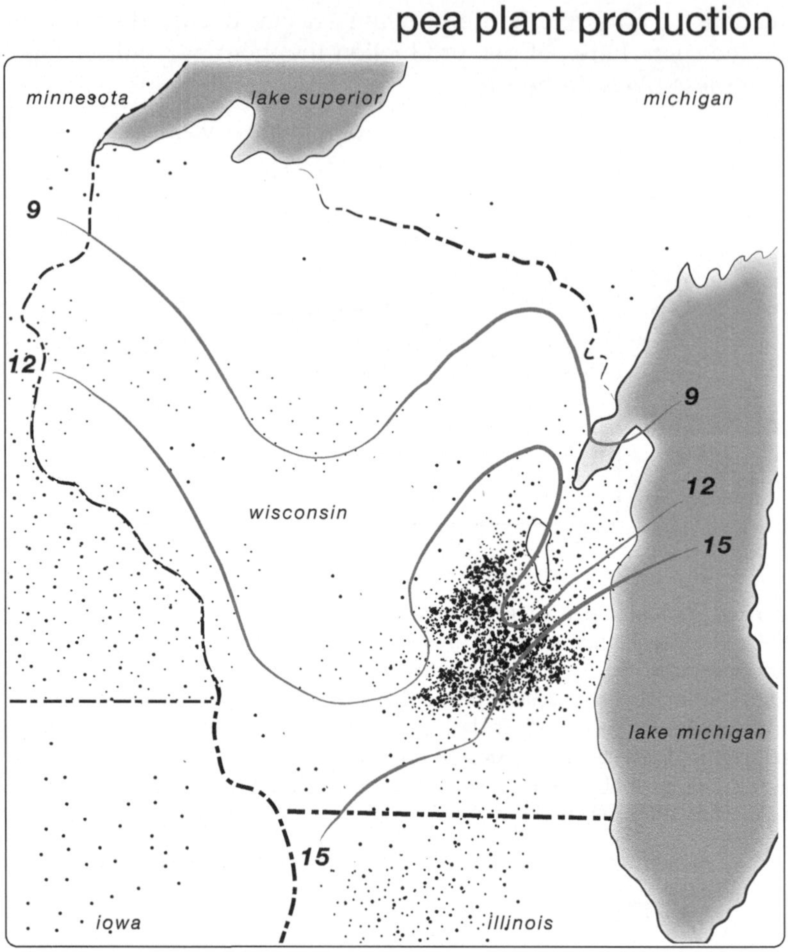

Figure 3.18 The distribution of pea farms (dots) in Wisconsin and surrounding states in relation to the T_D isolines relevant to understanding pea yield (see Fig. 3.16 for definition of T_D). Areas with T_D values of 12 to 15 give maximum yield (see Fig. 3.16). Redrawn from Wang (1962).

A similar data set to that for peas is available for the thermal requirements and tolerances of tomatoes (Fig. 3.19) (Wang 1963). The two species differ from one another in the seasonal pattern of their thermal requirements and tolerances, and also differ from other annual species, including sweet corn (Fig. 3.20). These differences across species, even in their interaction with only one environmental variable, help explain the ecological idiosyncrasy of species. The tomato data, for example, also show that average temperatures relevant to one developmental stage are not fully informative of the thermal interactions of organisms. When yield is plotted against the proportion of days, during the blossoming period, with maximum temperatures of 92°F (= 33.3°C) (which is close to the upper threshold temperature) or above, the relationship is an inverse one (Fig. 3.21). The hotter the days, the lower the yield. Furthermore, sequences of such days have an even greater impact, as indicated by a steepening of the slope for longer hot spells (Fig. 3.21). Notice, too, that thermal requirements are differentiated from thermal tolerance. Heat is required as a resource, although too much or too little is inimical.

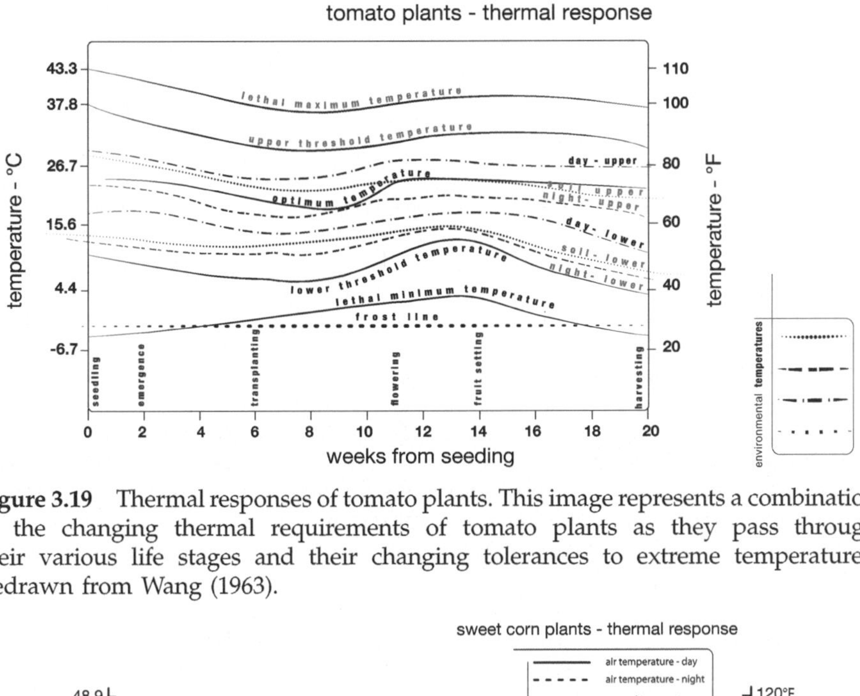

Figure 3.19 Thermal responses of tomato plants. This image represents a combination of the changing thermal requirements of tomato plants as they pass through their various life stages and their changing tolerances to extreme temperatures. Redrawn from Wang (1963).

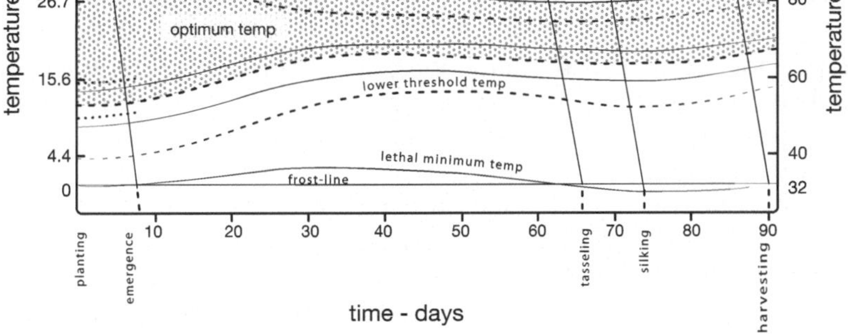

Figure 3.20 Thermal responses of sweet corn plants. This image represents a combination of the changing thermal requirements of sweet corn plants as they pass through their various life stages and their changing tolerances to extreme temperatures. Redrawn from Wang (1972).

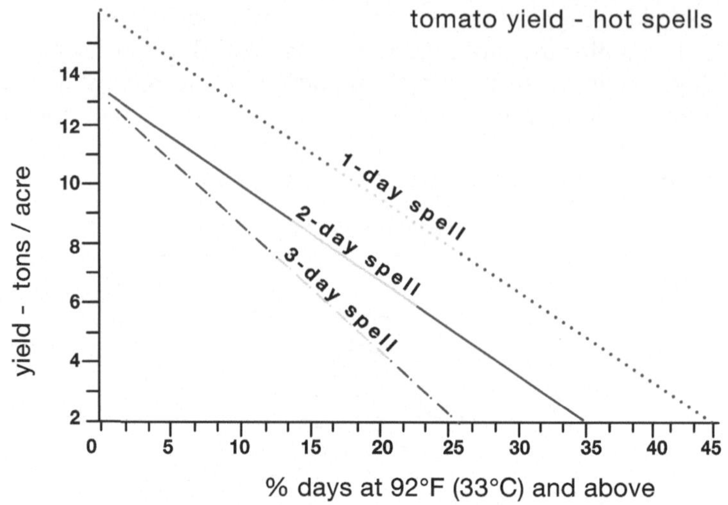

Figure 3.21 Tomato yield plotted against the percentage of days with temperatures equal to or above 92°F (= 33.3°C) during blossoming. As the proportion of time at these higher temperatures increases, so yield declines. The data have been analyzed in three ways, for all days that have temperatures at or above 92°F, for periods during which two days in a row achieve those temperatures, and for those periods during which three days in a row achieve those temperatures. When higher temperatures are present for more extended periods (i.e., more than one day at a time), less time spent at higher temperatures has as great an impact on yield as do longer overall periods at higher temperature that come in shorter bursts. Redrawn from Wang (1963).

For tomatoes, information is also available for rainfall requirements, but just for the blossoming period (Fig. 3.22) (Wang 1963). A more complete picture of the species' requirements can thus be derived, as illustrated through what is best for the plants (defined here by yield) with regard to the combination of rainfall and temperature (Fig. 3.23). Moreover, the degree of the match varies stochastically across seasons. The other environmental influences on plants should not be forgotten, including humidity, soil type, drainage, insolation, frost free days at critical periods, windless days at critical periods, mycorrhizal interactions, and so on. The entire life cycle has to be at least adequately provisioned in all of these respects. The complexity of the species-specific organism-environment interaction can be appreciated from this perspective.

Higher vertebrates and some invertebrates have a lifecycle that is continuous, like that of plants. The structure of their life cycle also has to match the temporal pattern of variation of environmental conditions in which they are located. Unlike plants, though, many animals can undertake seasonal return migrations (and even sporadic non-return migrations) to ensure the various requirements during their life cycle match the spatio-temporal variation in environmental conditions, as illustrated in the following subsection.

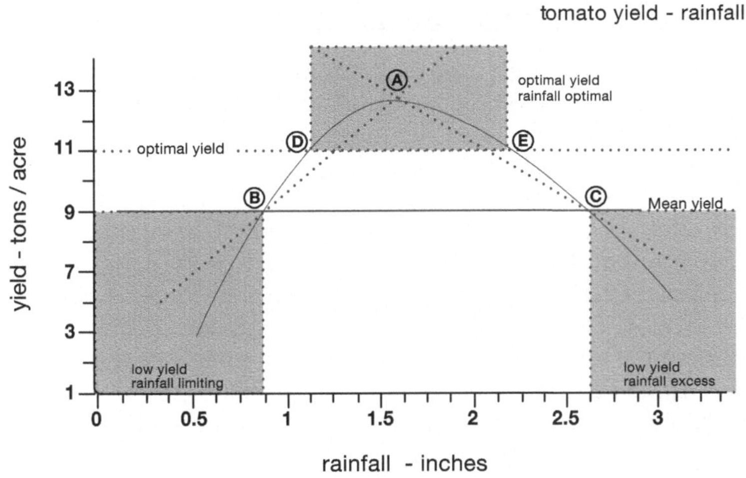

Figure 3.22 Tomato yield plotted against rainfall. The latter is the relative maximum rainfall, a reflection of the wettest period relative to the entire growing season (and is the four weeks of blossoming in tomatoes) (Wang 1963). There are two trends, with the left linear regression line representing the effects of low rainfall and the right hand one the effects of high rainfall. The intersection points **B** and **C** describe the lower and upper limits in the rainfall requirements of tomatoes, whereas **D** and **E** represent the upper and lower limits for optimal yield (mean yield + half standard deviation). **A** represents optimal yield and the parabola defined by these points of intersection represents the relationship between yield and relative maximum rainfall. Redrawn from Wang (1963).

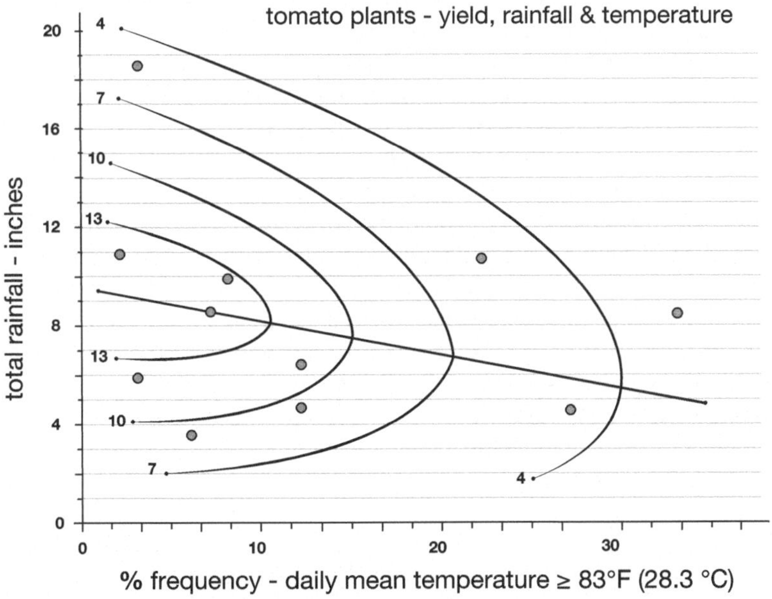

Figure 3.23 Tomato yield isolines (tons/acre) plotted relative to rainfall and temperature data. Dots represent the available data points. Redrawn from Wang (1963).

3.2 Adélie Penguins—Birds of the Pack Ice

The adélie penguin is highly unusual among penguins in being associated with pack ice and in breeding on the Antarctic mainland. A comparison of its life cycle and seasonality with these features of the other species in the genus *Pygoscelis*, the chinstrap and gentoo penguins, is revealing because these three are ecologically very similar. The close phylogenetic history of the species, with the adélie and gentoo penguins being sister species (Giannini and Bertelli 2004), reveals how adélies have adapted to survive in a rather different environment from that inhabited by the other *Pygoscelis* species. The adaptive differences across the species are reflected in the differential distribution of the three species and their differential population trends in localities where they occur together. These points, and what follows, are taken from the insightful and readable summary of Ainley (2002), but with a few supplements.

The adélie penguin is an obligate migrant. Seasonally, the birds move north in the southern winter to escape the fast ice that forms around Antarctica. Because fast ice is thick and caps vast areas of sea surface, it provides the penguins no access to the water and therefore their source of food. The birds therefore track areas where drift ice or ice floes cover 15-80% of the sea surface, which means they track north as the pack ice is pushed that way. This pack ice habitat provides them with easy access to the water and is the environment to which the adélies are primarily adapted. Specifically, adélie penguins hold their breath longer than do chinstraps, and their muscles evidently have superior oxygen holding capacities. Their diving is not for depth, but for spending time searching out the crystal krill that inhabits the small cavities and cracks under ice floes, in the photic zone where the algal food of the krill can photosynthesize. Where there is relatively more water than ice, the birds tend to take more fish. Adélies do not have to achieve great depth in diving, especially in winter when light levels are low, for they hunt by sight (but see Ainley and Ballard (2012)). In these adaptations they differ from chinstraps.

Chinstrap and gentoo penguins have a more northern distribution than the congeneric adélies. They occur from the northern extremities of the Antarctic Peninsula to the sub-Antarctic islands, out of the pack ice zone. Adélies do overlap with them to a small degree, and at a few breeding sites adélies are present with one or both of the other species. The nesting of chinstrap penguins is negatively affected when the sea ice and pack ice persists into their usual breeding season around their island breeding sites, for they are not adapted to these conditions at all. Chinstraps hunt through large volumes of open water and are faster swimmers than adélies and need ice-free conditions to hunt. Further, they are only partial migrants, and then only to get away from ice when it intrudes. Adélies that breed alongside chinstraps in the same area, by contrast, feed well under such ice conditions (Rombola et al. 2003), and their breeding is not negatively affected.

For breeding, adélies need hummocks or raised ground that exposes their nests to the wind. The nests are thus protected from snowdrifts and are raised above any floodwater (Fig. 3.24; and see photograph on p. 146 of Martínez (1992)), essential protection even in summer at the far southern latitudes at which this species breeds. The birds also need stones of a suitable size to build a slightly elevated nest. Readily accessible sites are rather limited around the Antarctic coastline, and most of those not associated with the Antarctic Peninsula are associated with polynya (large openings maintained in the ice by winds or up-wellings) (Ainley 2002). Where fast ice is protected from damage, by headlands for example, and persists at widths of 2 to 3 km between potential breeding beaches and the pack ice, adélie penguins tend not to establish breeding sites. Chinstrap penguins, by contrast, breed on steep slopes, and gentoos nest on low flat ground; both of these situations are more benign environments in terms of snow and melting ice.

Figure 3.24 These adélie penguins nested, atypically, on low ground and this later became a stream of melt water from the glacier in the background. Nesting success was low as a consequence and the group dwindled, with only those pairs with nests on slightly higher ground remaining. Photograph Gimme Walter.

Adélies are adapted to cope with a short summer. After breeding they swim north to their over-wintering sites well away from their nesting grounds. Here, they undergo their annual molt, at sea, on the winter pack ice. This association with the floating ice whilst they cannot swim keeps them and non-breeding birds in contact with their pack ice habitat. After molting they can swim again and feed. Their pre-migratory hyperphagia, or increased feeding and storage of body fat, sustains the birds while they cross the extensive stretches of sea ice to reach the beaches where they nest. Until the sea ice shelves (which cover the sea from the beaches out) break up, to form floating pack ice, the nesting birds have little access to foraging habitat.

By comparison, chinstrap and gentoo penguins cannot fast and do not migrate such distances. In any case, they must wait to molt at the breeding colony as they cannot molt at sea. In other words, their physiological capacities relative to feeding, fasting and molting confine them to areas with a longer summer (at least compared with that experienced by adélies) and limited pack ice. If sea ice does form around their breeding sites, and does not break up and dissipate soon enough in summer, chinstraps and gentoos are inclined to abandon their eggs if they have laid, as they have to feed.

Adélies arrive early at the breeding colony. They show relatively little mate fidelity across seasons compared with chinstraps (which helps ensure fertile eggs are laid early) and have a short courtship relative to the other *Pygoscelis* species. Egg laying of adélies is a month earlier than for chinstraps, despite them mostly nesting much further south, and two weeks earlier than gentoos. Further, adélie egg size is the largest of the *Pygoscelis* species, their newly hatched chicks weigh more than those of the others, the adults stop guarding the chicks earlier, and the chick stage as a whole is shorter than in the other species. Adélie penguin chicks aggregate in "crèches" earlier, fledge earlier than in the other two species and weigh the least relative to adult size at fledging. These aspects all relate to a shorter season and getting the young to the stage when they can find food in suitable ice habitat. Chinstraps and gentoos have a relatively longer season and the presence of ice around their breeding grounds plays only a negative role.

The reliance of adélie penguins on ice of a particular structure is best illustrated when their wintering is considered, again in comparison to that in the other pygoscelids. Adélies spend winter associated with pack ice. These birds need ice floes in loose concentrations (15 to 80%) so they have easy access to the sea. This keeps them relatively far south compared with chinstraps, for example (Fig. 3.25). The adélies keep south of the Antarctic Circumpolar Current, which defines the unstable or dynamic zone in Fig. 3.25, an area in which changes in wind direction can suddenly freeze vast areas of the sea surface, conditions that neither adélies nor chinstraps can survive. In winter, adélies are presumed not to range south of the Antarctic Circle, for their hunting requires light (see Ainley and Ballard (2012)).

Overall, the adélie penguin is one of the best-studied bird species on earth. This focus undoubtedly derives from an unusual confluence of factors, only some of which are biological. This species is abundant, breeds in summer (cf. emperor penguin (Williams 1995)), and their colonies are accessible. This accessibility is provided, indirectly, through the intense political interest in Antarctica. Further, the absence of closely related species

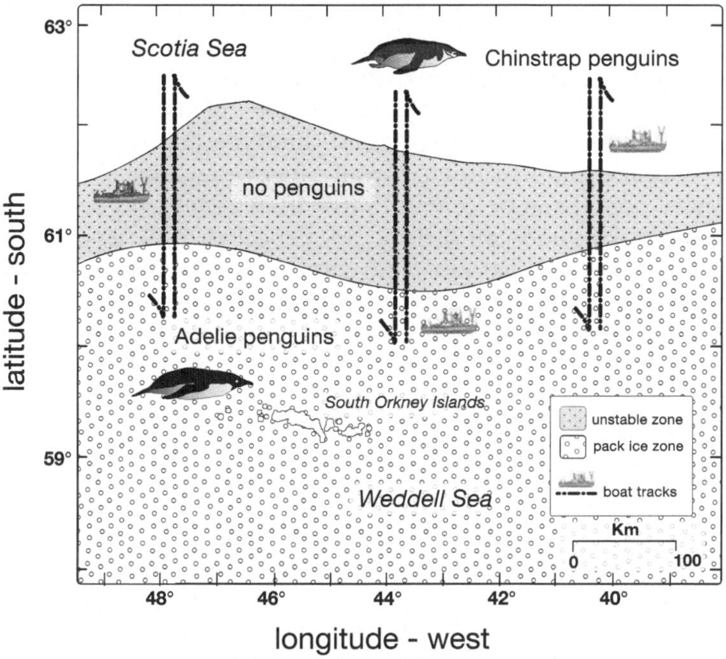

Figure 3.25 The dashed vertical lines represent the tracks of a survey ship that sailed north - south transects across the confluence of the Scotia and Weddell Seas. It was run during winter when adélie penguins have to move north of the fast ice to feed after breeding. Adélie penguins were observed only in the pack ice (or persistent) zone south of the so-called unstable (or dynamic) zone. The latter can freeze over rapidly, which prevents feeding by penguins. By contrast, chinstrap penguins were found only north of the unstable zone, away from all ice. Redrawn from Ainley (2002).

at most of their breeding sites has undoubtedly helped turn the focus from competition studies to a much fuller consideration of the lifeline and annual cycles of the birds, and the adaptations by which they complete these cycles. The interpretation of their lifeline, above, demonstrates just how significant are the adaptations of this species to understanding their distribution and abundance (i.e. their ecology). The adaptations studied so far, such as breeding site, nest construction and feeding ecology, are species wide (Ainley 2002). In other words, these traits are fixed across the species gene pool and this undoubtedly occurred at speciation (see Chapter 5, Section 6). The only detectable change since then is in their mitochondrial gene sequences, which demarcate two lineages within the species. These diverged about 75,000 ybp (Ritchie et al. 2004) but are similar to one another in the adaptive characters mentioned above. The consequence, ecologically, is that the key elements of the breeding biology of each of the *Pygoscelis* species "largely reflect each species' adaptations to conditions in the center or larger part of its respective range" (Ainley 2002, p. 192). This is a pattern that is typical in ecology. Further, the diversification of penguins and their ecologies appear to have been set down at some period when Gondwana was breaking apart (Kooyman 2002). The principle entailed in this example, that the distribution and abundance of organisms are best understood relative to the adaptations of the species, the annual cycles and life cycles of the species and the dynamics of the environment to which they are adapted holds for all other organisms. It does not hold just for those organisms perceived to live in extreme environments, where density dependence and competition are expected to be less important. In any case, any effects of density and competition can be interpreted only in relation to a good understanding of the organism-environment dynamic and its ecological consequences.

3.3 Large Terrestrial Mammals

The ecology of large terrestrial vertebrates is, likewise, dictated by the adaptive requirements of their lifeline and by their life cycle matching environmental circumstances, both seasonally and throughout their life. Critically, their reproductive requirements must fit into the dynamic environmental framework that prevails. For this, their mating system needs to be coordinated, by environmental cues setting their key diurnal and seasonal clocks through their actions on the intricacies of the endocrine system of each sex (Young 1975). Mating must thus be initiated at the appropriate time of year for the gestation period to deliver young at a time when environmental conditions are likely to be best for their survival and development (e.g., Vors and Boyce 2009). Again this match is subject to stochastic interruption through environmental change or unpredictability (e.g., Ginnett and Young 2000), as well as genetic or phenotypic variation in the gestation duration (all from Kiltie (1984)). In other words, survival of mother and young is subject to stochastic influence, against which adaptations of the life cycle work to maximize the probability of success. A changing environment that shortens periods favorable for successful parturition and development of the young, or renders these conditions less regular or predictable, creates problems of life cycle synchrony.

Relatively few herbivorous mammal species living today have an average gestation longer than 12 months, at least without their having delayed fertilization or implantation. All of them are large, at least about tapir size (~200 kg, although mammal weights vary considerably with age, sex, condition and subspecies). They include the elephants, rhinoceroses, camels (but not llamas and relatives), giraffe, okapi, tapirs and some African *Equus* species (ass (*E. asinus*), Grevy's zebra (*E. grevyi*) and Burchell's or plains zebra (*E. burchelli*)). The combined Recent (i.e. the current geological epoch) world distribution of these species is given in Fig. 3.26, which shows that they occupy only those regions characterized by seasonal cycles in rainfall and fluctuations in temperature that are not extreme (including absence of snow cover). In short, and with only the exception of the Bactrian camel (Fig. 3.26), they are restricted to the tropics and subtropics. These areas have longer seasons favorable for birth and all of the species with a gestation longer than 12 months are not as seasonal in their breeding as are extra-tropical species with gestations less than 12 months (Kiltie 1984).

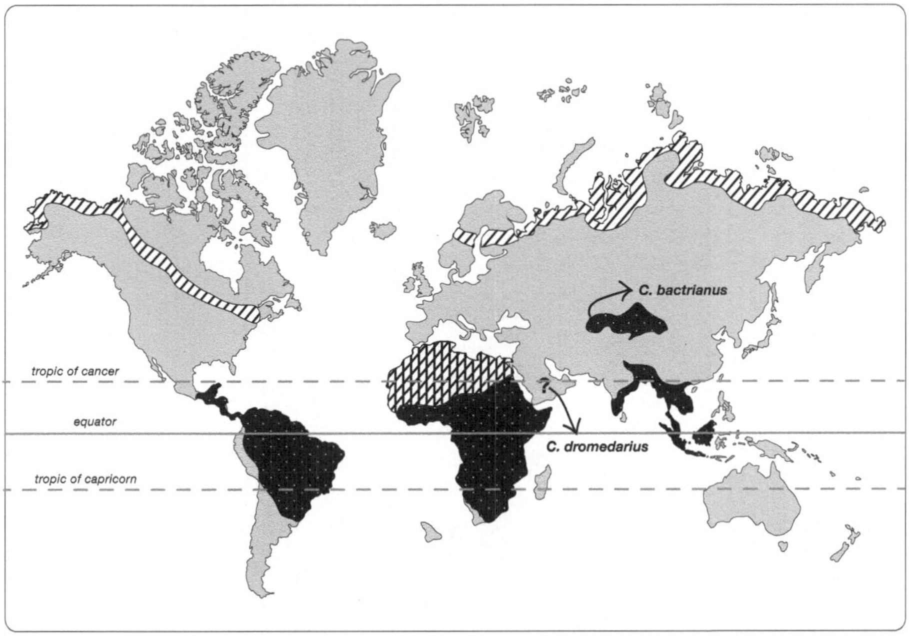

Figure 3.26 Recent cumulative distribution of large mammals (tapir size and larger) that have a gestation period that exceeds 12 months. These are distributed in the areas shaded black within the tropics, with the exceptions indicated (Bactrian camel (*Camelus bactrianus* - arrow) and *C. dromedarius*, which is extinct in the wild but the question mark indicates its suspected original provenance). The Sahara (cross hatched) precludes the current persistence of large mammals, although species with a supra-annual gestation did exist there after the mega-faunal extinctions in extra-tropical regions. The northernmost Pleistocene distribution of large mammals (woolly mammoths) outside this zone (single hatching) is also indicated. See text. Redrawn from Kiltie (1984).

The only exception, the Bactrian camel, is endemic to the arid regions of central Asia and presumably does have a gestation longer than 12 months (Kiltie 1984). Bactrian camels are now rare in their original environment (Tulgat and Schaller 1992, Hare 1997, Reading et al. 1999, Luzhang et al. 2005) and the species seems not to have been investigated further in this regard. Besides this camel, all large ungulates that have survived until Recent times outside the tropics have gestations of less than 12 months, and they give birth within a time frame of two months or less. Among the larger ungulates, including moose, this is achieved by giving birth to less mature young after a shorter gestation than would be normal for their size as indicated by the comparisons detailed

Table 3.2 Comparison of female body weight against mating time, gestation, birth dates and offspring mass for tropical and subtropical ungulate species and temperate species of similar body weights. Moose and bison stand out as unusually large temperate species that give birth to relatively small young after a short gestation for their size. These properties ensure the young are born during the brief seasonal interval when conditions are most suitable for them. Note that size varies substantially in several of these species, especially those comprising different subspecies, and it does so with sex, season, physiological condition and stage in the reproductive cycle.

SPECIES	*FEMALE MASS* (kg)	*MATING PERIOD (peak for 1st heat)*	*GESTATION* (dd) *(ave/ range)*	*BIRTH TIME*	*OFFSPRING MASS* (kg) *[% female mass]*	*NUMBER OFFSPRING*	*SOURCE*
LOW LATITUDE SPECIES							
Plains zebra *Equus burchellii*	302 ($\bar{x}$)		360-390	Oct - Mar (85%)	30-35 [9.9-11.6%]	1 (always)	(Smithers 1983)
Cape buffalo *Syncerus caffer*	350-620		330-346	Dec - Feb (peak, but across 120dd)	45 [7.3-12.9%]	1	(Sinclair 1977, Smithers 1983, Groves and Leslie 2011)
Kudu *Tragelaphus strepsiceros*	200		210	Jan - Mar (most)	16 [8%]	1 (always)	(Smithers 1983)
Eland *Taurotragus oryx*	390-600		271	Aug - Oct (peak)	32-36 [7.0-7.8%]	1 (mostly)	(Smithers 1983, Groves and Leslie 2011)
HIGH LATITUDE SPECIES (data from northernmost subspecies where relevant and available)							
White-tailed deer *Odocoileus virginianus*	67-112	3 Nov-7 Dec	200-210	24 May-14 June	2-4 [3.7%]	1-3	(Miller et al. 2003, State of New Jersey, Division of Fish and Wildlife 1996-2005)
Barren-ground caribou *Rangifer tarandus* ssp.	82-92 (autumn)	Oct-early Nov (high synchrony)	225-235	Late May-early June (80-90% in 10 day period)	6 [6.5-7.3%]	1 (twins rare)	(Cameron et al. 1993, Miller 2003)

Muskox *Ovibos moschatus*	210-220	Aug	242	mid Apr-early June (94% by 8 May)	10 [4.6-4.8%]	1	(Gunn and Adamczewski 2003)
Wood bison *Bison bison athabascae*	493-567 ($\bar{x}$ = 540)	Late Jul-mid Aug peak		Late Apr-early Jun		1 (twins rare)	(Reynolds et al. 2003)
Plains bison *Bison bison bison*	417-454 ($\bar{x}$ = 440)	July-mid Aug peak	262-272 & 277-293	Median 2-8 May (last wk Apr & 1st 2 wks May)	14-18 [2.96%]	1 (twins rare)	(Rutberg 1984, Berger 1992, Reynolds et al. 2003)
Alaskan or giant moose *Alces alces gigas*	297-510 ($\bar{x}$ ~ 427)	28 Sept-12 Oct	231 (87% between 225 & 236)*	Last wk May (95% in 14-20 day period)*	$\bar{x}$ = 16 [3.7%] (twins 13.5) [3.16%]	Mostly one, but also twins.	(Bowyer et al. 1998, Keech et al. 2000, Bowyer et al. 2003)

* To ensure enough time at high latitude to provision young, and 2nd estrus females (failed during first one) even shorten duration of gestation (see Bowyer et al. 2003, p. 937).

in Table 3.2. In particular, compare female mass, gestation and birth mass of the moose with that of the plains zebra and white-tailed deer. Parturition of moose takes place in specific environments and within narrow temporal limits, to the extent that births appear to be synchronous (Schwartz and Hundertmark 1993, Bowyer et al. 1998, Bowyer et al. 1999), again signaling the sensitivity of this stage of the life cycle and the importance of seasonal synchrony to take maximum advantage of the short favorable period within the extreme seasonal environment for a mammal of this size (see also Rutberg 1984).

The reproductive properties outlined above for large mammals with a long gestation are, by inference, characteristic of the mega-fauna that went extinct at the end of the Pleistocene, including woolly mammoths, rhinoceroses, tapirs and larger camels and horses. Kiltie's (1984) analysis suggests strongly that the less equable climatic conditions in extra-tropical regions at the end of the Pleistocene contributed to the decline of these large species. Not only did the unique dry grass steppe vegetation (the mammoth steppe), to which these species were primarily adapted, change to less nutritious shrub steppe (Guthrie 1990, Guthrie 2006), but the time favorable for production of offspring also became shorter and less predictable. In essence, the larger herbivorous mammals with longer gestation periods, and mating times that were cued by environmental conditions not linked directly to the favorable period for birth, were likely to have been disadvantaged by producing their young at seasonally unfavorable times (Kiltie 1984). This climate-based explanation presents a clear alternative to the currently popular Pleistocene overkill model, which attributes the extinctions to hunting by humans (Koch and Barnosky 2006, Rule et al. 2012). The climatic interpretation is supported by evidence showing extinctions occurred before humans arrived in some areas and coexistence of humans and megafauna in others, as well as correlations with climatic and vegetation changes and acknowledges that different species need ecological analyses in their own right to determine why they went extinct (e.g., Trueman et al. 2005, Guthrie 2006, Grayson 2007, Campos et al. 2010, Nikolskiy et al. 2011, Price 2013).

One of the problems for these large mammals is that the intraspecific variation in gestation period increases with an increase in mean gestation time (Fig. 3.27). Reproductive failure is inevitably more common for them. Another difficulty for large species with a gestation longer than 12 months is the time they may lose for breeding. Figure 3.28 illustrates the case schematically. With an 11-month gestation and conception following shortly after birth (as seen in some equids), a large species can breed annually. By contrast, a species with a 13-month gestation loses alternative seasons. In reality, the situation is not so straightforward, being influenced by the particular adaptations of species and various ecological influences, but it does illustrate the point.

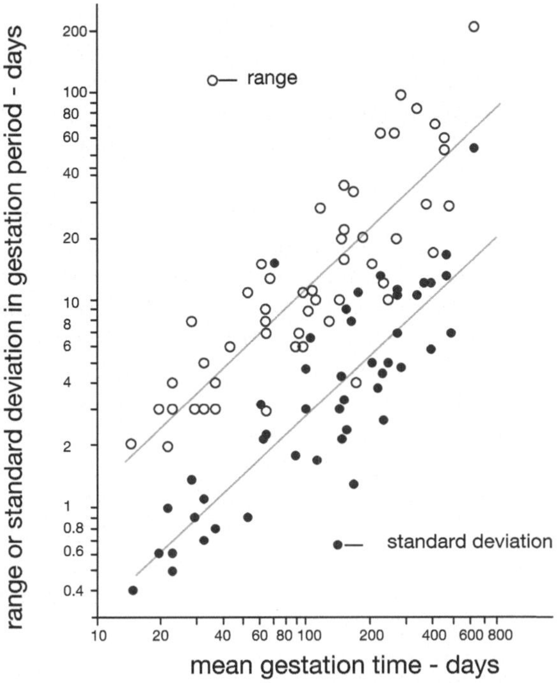

Figure 3.27 Intraspecific variation in mammalian gestation. Log plot of two measures of variance (standard deviation and range) in gestation period against mean gestation period. Reproduced from Kiltie (1984), with permission.

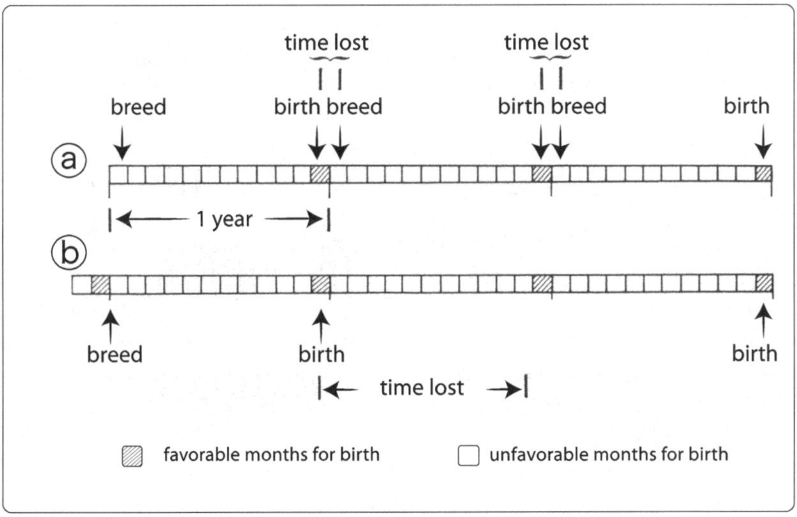

Figure 3.28 Schematic diagram to compare breeding frequency in two species that have only a one month period favorable for birth each year, one with a minimum interval between litters of 11 months **(a)** and one with a 13 month interval **(b)** The former can breed annually, whereas the latter must skip every second year. Reproduced from Kiltie (1984), with permission.

Different species of large mammals often live under similar environmental conditions. Even though they may be relatively closely related to one another, they are likely to be differentially adapted to the prevailing conditions. The different species consequently respond independently to environmental variables and this affects where they occur locally and how they must respond spatially to change. Large kangaroo species in the genus *Macropus* provide one of the better-worked examples, from the few available, as summarized by Tyndale-Biscoe (1973, 2005).

Both the red kangaroo and euro are arid adapted macropod species that co-occur over vast areas of inland Australia, where rainfall is sparse and erratic. Both have good thermoregulation and water balance at high temperatures (40-44°C) and moderate relative humidity (roughly 49%), but when humidity is increased (to about 65%) they show signs of thermal stress. They pant more, but cannot evaporate as much water, their pulse rate doubles and rectal temperature increases as much as two degrees. This all means that if water is freely available, these kangaroos dissipate heat successfully when temperatures are high and humidity is low. But water is often unavailable and relative humidity low where they live and water balance becomes a problem. Under such circumstances, the only two options left include the avoidance of high temperatures and the efficient concentration of urine.

Euros avoid high temperatures by shading among rocky outcrops and in caves, which provide excellent thermal buffering, and they also shelter in excavations they make under dense stands of mulga trees (Fig. 3.29). They thus reduce the need for evaporative cooling and so conserve body water. These animals can survive without free water if such shelters are available. Free water releases them from their heat refuges, even when temperatures are relatively high. Their abilities in this regard become more apparent when it is appreciated that they lose 20% of their body weight in seven days if held in enclosures without rock shelters and free water. We can thus understand why euros are sedentary and relatively solitary.

Red kangaroos differ ecologically from euros in not habitually associating with rocks or other heat shelters (Fig. 3.29). They are more mobile than euros, more gregarious, and live in open country. When it is hot, they reduce activity to a minimum and lie on their sides in as much shade as can be found under low bushes. Their fur is dense and reflective, especially in the area of the pale hip. (Euros, by contrast, are dark and non-reflective.) Their heat load is thus ameliorated substantially even though the radiant temperature may be as much as 30° above body temperature and the air temperature about the same as their body temperature, for they effectively reduce solar radiation to as little as 20% of the radiation that strikes open ground. Air flow in these open sites helps with evaporative cooling through panting and licking the extremities of their limbs, but especially by panting because its cooling effect helps reduce the deep body temperature.

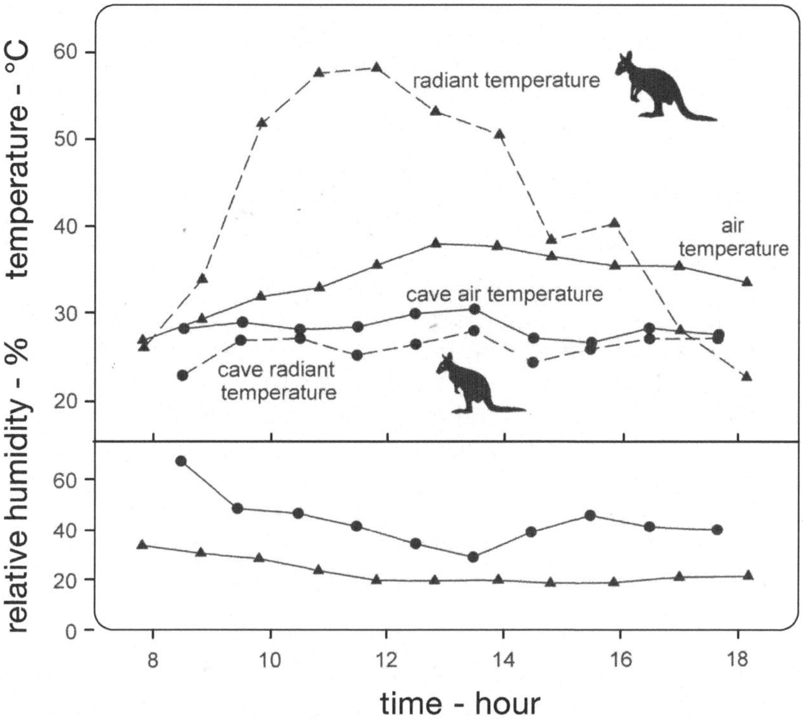

Figure 3.29 Daily variation in the air temperature, solar radiation and humidity of the resting sites of red kangaroos (triangles) relative to these variables recorded in the cave site of euros. Reproduced, with permission, from Tyndale-Biscoe (2005).

The water that red kangaroos lose through evaporative cooling is replenished by selecting better quality food than that used by euros. Since they cannot use the low grade food eaten by euros around their rocky refuges, they track newly sprouted grasses and ephemeral dicotyledons across a vast landscape for their relatively high protein diet, and retreat to the few persistent sources of more nutritious food when rainfall is scarce. Their stomach contents therefore tend to be green even when the surrounding vegetation is brown. Euros, by contrast, are adapted to use low grade feed efficiently, by conserving nitrogen. They remain in nitrogen balance by recycling more urea than red kangaroos, so the concentration of their urine (and especially the concentration of urea) tends to be lower than that of the other species. This, however, also reflects their poorer diet when the environment is dry. Their wellbeing and thus abundance is thus influenced by nutrition rather than water.

In summary, these two widely distributed species of kangaroos "have emphasized different aspects of their common macropod heritage in adapting to the desert" (Tyndale-Biscoe 1973, p. 127). The euro keeps near caves and conserves water, but must be sedentary and is adapted to the nearby less nutritious diet (through urea recycling and a low nitrogen and basal metabolism). Red kangaroos retain mobility, but at the cost of a more prodigal use of water for thermal regulation, so that they can travel to exploit nutritious, succulent plants. Although they depend on water more than euros do, they conserve water better.

3.4 Invertebrates with Discrete Phases to the Life Cycle

Many invertebrates, both aquatic and terrestrial, undergo dramatic metamorphic change at particular stages of their life cycle, as do some vertebrates such as frogs. The holometabolous insects, which have an immobile pupal stage for developmental transformation of the growing larval phase to the sexual adult, perhaps best illustrate the different habitat requirements, adaptive mechanisms and environmental interactions that invariably underpin the various stages that make up such complex life cycles. The ecological consequences of such a divergence are momentous, as are the requirements for seasonal synchrony of the different life stages and the localization of specific environments to support transitions and sustain the developmental stage that follows.

Aspects of the processes involved in these "discrete phase" life cycles have been quite intensively investigated, including some components of host searching behavior in parasitoid wasps and phytophagous insects (see Subsection 2.5), the environmental triggers that drive pupation and ecdysis of adult insects (e.g., Highnam and Hill 1977), and so on. Nevertheless, no organisms have been comprehensively investigated with regard to the way in which the life cycle of individuals of a species matches the environments they habitually occupy, or how these adaptations preclude the species from other environments. Clues as to how organisms achieve this match are available in the literature and a composite from such information is developed below, with particular reference to holometabolous insects and sea urchins.

We begin by pointing out the critical points in the life cycle of each type of organism. In insects, the specific site in which eggs are deposited is crucial, as are the behaviors of neonate larvae upon hatching, the larval feeding site (i.e., the specific resources available and their quality), the time and site of pupation, the timing of adult eclosion and the achievement of fertilization. Quiescence or diapause is frequently part of one or other of these processes, and serves not only to span climatically unfavorable periods, but also to synchronize the life cycle with the seasonal environmental cycle in which it evolved. These features translate, in turn, into the geographical distribution of the species concerned, as revealed also in the previous examples covered in this section.

The details of holometabolous insect and marine bentho-planktonic life cycles vary endlessly according to species, and some examples are outlined below to illustrate the principles involved. The key points outlined above are affected by the species-specific behaviors of the individuals concerned (and which, in turn, are directed by their underlying physiological processes), environmental conditions and their seasonal structure, environmental extremes, and environmental signals. Each of these key aspects can, as hinted at in the examples already covered in this section, be distorted by stochastic influences, but these are dealt with conceptually and in more detail in Chapters 8 and 9. Here we deal with the specific ways in which some terrestrial and marine invertebrates match their life cycles to the dynamic structure of the environment.

Edith's checkerspot (*Euphydryas editha*), a nymphalid butterfly, exists as several subspecies that are distributed from northern Mexico to southern Canada and between the coast and the Rocky Mountains (Murphy et al. 2004). Ecologically it is one of the most intensely studied insects and aspects of the life cycle relative to the environmental circumstances that prevail in different parts of the species' geographical distribution have been synthesized by Parmesan (2003). The uncertainty that persists as to the exact status of the different "types" or subspecies of *E. editha* is a hindrance in developing interpretations of their adaptations and ecology, as described below, but our understanding of these organisms and their ecology nevertheless provides crucial autecological insight.

Despite the different subspecies associating with different vegetation types and host plant species, and being distinctly different in size and color in many cases, interpretations of their ecology and evolution are integrated as that of a single species (Fig. 3.30). Their ecology is thus seen as showing "complex patterns of host and habitat use" to the extent that host-plant preferences are seen to vary across space, through time and with habitat and from flight season to flight season, although "distinct populations of individuals that look similar and fly in ecologically similar habitats nonetheless exhibit local adaptations to resource quality and availability" (Murphy et al. 2004). Despite no direct evidence being available on the species limits within this complex, or strong scientific tests having been conducted in this sphere (see Ehrlich et al. 2004, Murphy et al. 2004, Saccheri et al. 2004), the variation is seen to cross taxonomic and population boundaries to the extent that "genetic variation is continuous" (Hanski et al. 2004, p. 254). This conclusion is consistent with the basic premise of evolutionary ecology, that ongoing adaptation is continuous in local populations across a species' distribution, even though this has not been subject to strong test in *E. editha*. Viewed from such a perspective, the arrival of a second differentially adapted subspecies that uses different host plants in an area occupied initially by another subspecies (that has since declined) would appear to represent a host shift that had evolved *in situ*. Instead, such cases would normally suggest that the populations represent true species because subspecies, by definition, cannot persist together in sympatry for conspecifics mate at random (e.g., Mayr 1963, Ford 1974). Properly constructed tests of these aspects are warranted.

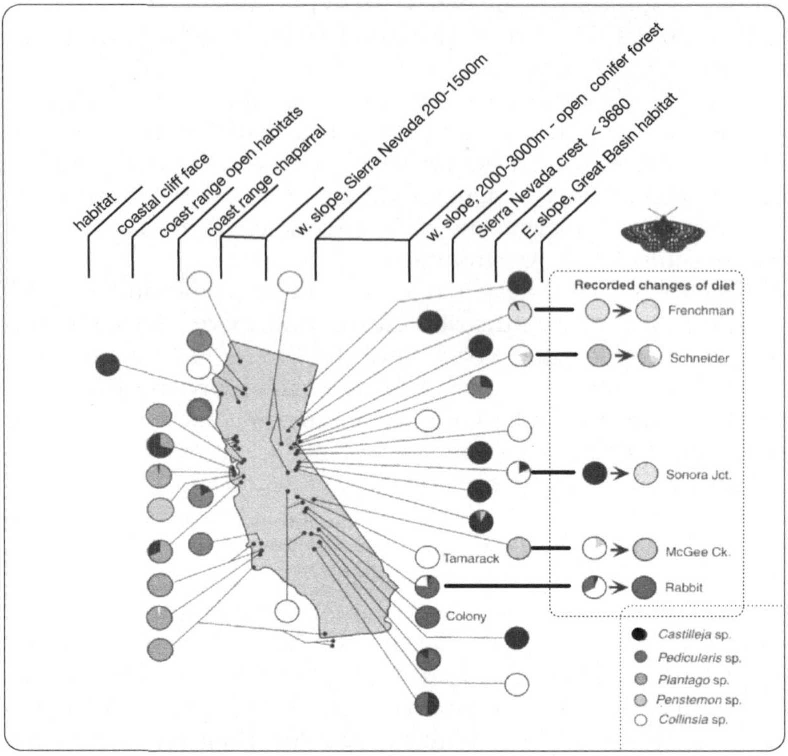

Figure 3.30 Proportion of eggs and/or pre-diapause larvae of Edith's checkerspot butterfly found on each host plant genus at different sites in California. Arrows connect observations made in different years at the same site. Reproduced from Hanski et al. (2004), with permission.

The following outline of the *E. editha* life cycle (Parmesan 2003, Murphy et al. 2004) therefore concentrates on the Bay checkerspot, the subspecies distributed in the San Francisco Bay area. Usually it is referred to as *E. e. bayensis*, although its correct name is *E. e. editha* (Hellmann et al. 2004). The annual cycle is illustrated in Fig. 3.31. Adults fly in March and April (early northern spring) and egg masses (20 to 350 eggs each) are deposited on or near *Plantago erecta*, the primary host for the larvae. Young larvae often feed communally in a silken tent. The major threat to their survival is summer drought, which comes inevitably with the Mediterranean climate of the area, for their host plants senesce in late spring. To survive, larvae must reach the 3rd instar and enter diapause before their food source becomes unavailable. Mortality through host senescence at the egg and larval stage is commonly more than 99%, but larvae can enhance their chances by moving to alternative hosts like *Castilleja densiflora* or *C. exserta*, which senesce later (but which also germinate later, so are not available to most ovipositing females). Late autumn rain germinates the primary host plants. The larvae, now solitary, complete feeding through winter and pupate about February.

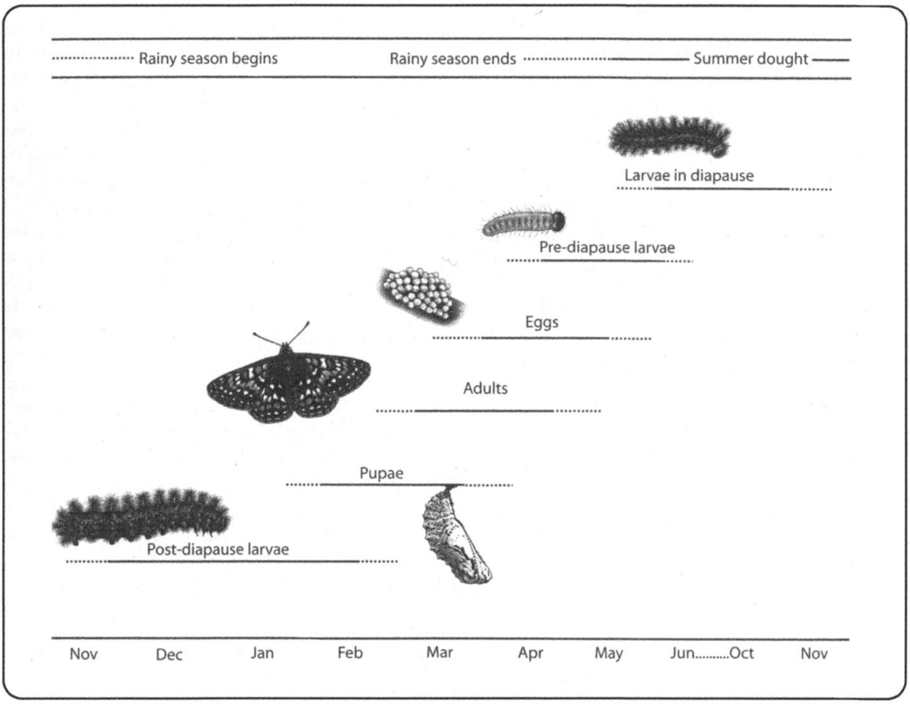

Figure 3.31 Schematic of the Bay checkerspot butterfly *Euphydryas editha editha* life cycle. Adult females lay eggs in the spring and these hatch into pre-diapause larvae that feed until the late spring and early summer. Larvae then diapause through the summer drought and emerge with autumn and winter rains as post-diapause larvae. Reproduced, with permission, from Murphy et al. (2004).

Completion of the life cycle depends on the interactions between various adaptive mechanisms of the organisms and the prevailing environment. The adaptive mechanisms include adult eclosion at the appropriate seasonal stage, after nectar plants become available. Early oviposition is thus ensured and is followed by rapid early development of larvae, larval summer diapause and post-diapause winter-feeding. The life cycle, as structured by these features, fits into the prevailing structure of the environment provided by the serpentine soils area of coastal California. Debate as to whether the species was better suited to environments of the area before they had been modified for widespread agriculture, when other soils supported a diversity of native plants, has not been resolved (Murphy et al. 2004). Climatic influences drive the whole dynamic between the Bay checkerspot and its local environment, through its influence on the timing of the butterfly's life-cycle relative to that of the larval host plants and the nectar plants used by the adults (Parmesan 2003).

The dynamic of the interaction can, furthermore, be modified locally in various ways. In particular, conditions that slow the plants' phenology relative to that of the insects increase the chances of insect survival. For example, dry spring periods impact more on larvae on south facing slopes, which are more sunny and relatively warmer than north facing slopes, for the host plants senesce too soon for the larvae to develop sufficiently (Weiss et al. 1987, Weiss et al. 1988). By contrast, during moderate spring times in such sites, the larvae survive and are present after the winter diapause. Oviposition cannot be geared to avoid such influences, for they are stochastic by nature, and this must presumably dampen any effects of directional selection on these organisms. Also, host plants growing in deeper soils (as modified by gophers for example) do not senesce as quickly, so larvae survive better there, and sites with alternative host species also help provide continuity of the life cycle. Again, stochastic influences modify the local ecology of a species.

The indirect influences that ramify as a consequence of such local impacts can be far reaching. Following the 1977-78 drought one population did well, in a locality with a diversity of slope aspects. This diversity evidently worked through females that emerged early on a south-facing slope being able to take advantage of suitable hosts that had by then become available on a north-facing slope. On southern slopes, post-diapause larvae grew up to 31% faster and pupal development was up to 56% faster compared with that on north facing slopes. Individual larvae on south facing slopes attained an average of 3.2°C higher body temperatures and pupae were an average of 12°C hotter than individuals on north-facing slopes. This incidental compensation would take place only in the event of a single bad year, because no butterflies would emerge early on the south-facing slopes in the following year, with adverse consequences for the synchrony of plants and insects on the remaining north-facing slopes. Other bad years caused local extinctions of *E. editha* populations, showing just how critical is an appropriate match between the life cycle and its adaptive mechanisms and the structure of the seasonal environment. Even overly wet years have been associated with population crashes (all summarized by Singer and Ehrlich 1979, Parmesan 2003). The differential adaptations of the different taxa within the *E. editha* complex have yet to be spelled out, for these details tend to be blurred by the species being treated typologically even while its constituent populations are treated as fluid entities.

Although the codling moth *Cydia pomonella* is associated with an extensive list of host plant species, it is primarily adapted to apple species with relatively large fruits of the *Malus sylvestris* type and to the climate of their endemic distribution in southern and central Eurasia (Shel'deshova 1967). The lower temperature threshold for pre-imaginal development is about 10°C and to complete growth the temperature sum above this lower level should be about 610 to 625 degree-days, and the availability of this amount of heat determines the northern limits of the species' distribution (Fig. 3.32), except where winter temperatures fall below minus 32°C. Survival through winter is sustained by larval diapause, but most freeze to death at temperatures lower than minus 20°C, with only some surviving down to minus 31°C. No geographic variation has been detected in this cold resistance of the species. In areas that have temperatures below minus 32°C coupled with heavy snowfall, the diapausing larvae may survive because the snow layer insulates them, as happens in Siberia and Kazakhstan (all from Shel'deshova 1967).

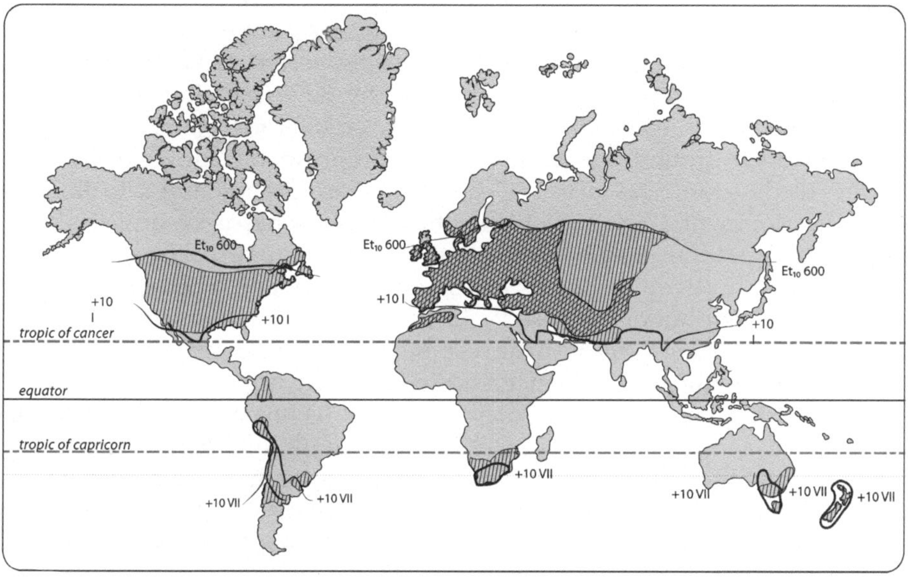

Figure 3.32 The original geographical distribution of the codling moth (cross hatching), based on the distribution of large-fruited wild apples. The other shading represents areas invaded with the development of agriculture, up to the mid 1960's. Et_{10} 600 is the isoline of temperature sum 600 day degrees above 10°C. +10 I is the winter +10°C isotherm in the northern hemisphere (January) and +10 VII is that for the south (July). Redrawn from Shel'deshova (1967).

The onset and cessation of diapause in the codling moth is mediated principally by day-length, with reduced day-length in autumn inducing diapause and increased day-length in spring bringing them out, and it is this mechanism that adjusts the life cycle to different day-lengths at different latitudes. The resumption of development in spring depends on the larvae having experienced prolonged exposure to cold below the developmental threshold. Thus, the seasonal alternation of temperature conditions over the course of the year is essential to the existence of the codling moth in any area, for without a cold enough winter, the life cycle is not reset to the season and to the phenology of its wild hosts. This is the feature that dictates the southern limit of the species in the northern hemisphere and keeps its northern limits south of the subtropics in the southern hemisphere (Fig. 3.32). This seasonal synchrony through the imposition of a minimum temperature is also a primary feature in the ecology of the wild apple species with which the codling moth is associated (all from Shel'deshova 1967).

Many cold-adapted insect species are univoltine. The tipulid crane fly *Tipula subnodicornis* has larvae with an unusually flat response to rising temperature (Fig. 3.33), at least within limits. The functional aspect of such a growth response seems to be its ensuring an annual life cycle over a relatively wide range of climates (Butterfield and Coulson 1997). A mean temperature rise from 10 to 20°C would have no real impact on the duration of the larval growth period in *T. subnodicornis*. By contrast, such a temperature change would shorten the period of larval growth in the cabbage white butterfly *Pieris brassicae* by 80% (see also Lokkers and Jones 1999). *Pieris* caterpillars develop on cruciferous plants, and these grow on open ground and so are typically ephemeral, growing fast and setting seeds soon. The larvae of *P. brassicae* are thus adapted to finalize their development (Fig. 3.33) before the plants senesce or desiccate, for higher temperatures usually hasten both of these latter processes. Typically, *P. brassicae* is multivoltine and its temperature relations are consistent across the species' distribution (Lokkers and Jones 1999). Undoubtedly, different multivoltine species differ in their response to the environment, with more than one environmental influence exerting an influence on its local ecology. The hesperiid butterfly *Atalopedes campestris* is expanding northwards as seasonal temperatures warm. Although minimum temperatures are increasing more rapidly than maxima, this species has not spread to all areas that have a higher winter minimum than minus 4°C, because summer temperature also plays a role and must be high enough for the completion of sufficient generations (Crozier 2004). The more generations, the more likely that enough individuals will be produced in summer that sufficient survive the winter cold into the following summer.

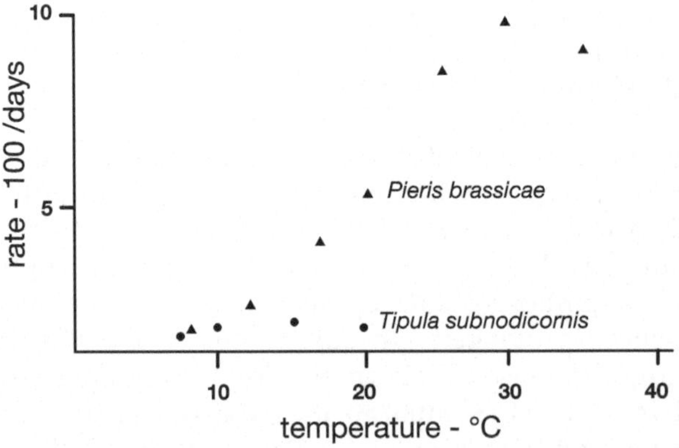

Figure 3.33 Comparison of the growth rate response to temperature in the univoltine crane fly *Tipula subnodicornis* and the cabbage white butterfly *Pieris brassicae*, which usually has at least two generations per year and often more. Reproduced from Butterfield and Coulson (1997) with permission.

In the sea urchin life cycle, the critical aspects are the time and site of larval metamorphosis, spawning time, and adult feeding site. Sea urchin adults provide a strong contrast to holometabolous insect adults, as they are relatively sessile. Their planktonic larvae, by contrast, constitute the motile phase and they spend hours to many months in the water column, depending on the taxon. The relatively sedentary adults cannot actively track down a mating partner and rely on spatial concentration of individuals and tight synchrony of gamete release to achieve fertilization, as shed gametes are evidently diluted very rapidly (Pennington 1985). Larval movement depends on water currents within a complex hydrodynamic environment, but is helped to some extent by the ability of larvae to steer in response to specific environmental signals (see Steinberg et al. 2002). These combined adaptations help ensure settlement as well as metamorphosis from the planktonic to the benthic form, in a suitable site for further development of the organism and its reproduction. The surfaces colonized usually require the presence of other living organisms, from whence the signals to settle come, with non-organismic or "inanimate" surfaces probably rarely used (Steinberg et al. 2002).

Research on the Australian sea urchin *Holopneustes purpurascens* illustrates more specifically how life cycles are structured for survival by bentho-planktonic marine organisms. The seasonal reproductive cycle of the species is somewhat unusual among echinoids in not only having the usual strong seasonal component, with a spawning peak from late winter to early summer, but also several smaller sporadic spawnings (Williamson and Steinberg 2002). These do not correlate with lunar cycles and may be keyed locally, perhaps by photoperiod, temperature, diet or density, and synchronized with pheromones as proposed for other urchins (Pennington 1985, Williamson and Steinberg 2002). This flexibility in the timing of reproduction may be a result of the patchy distribution of the species within localized kelp beds in the Sydney region of eastern Australia (Williamson and Steinberg 2002). Sedentary urchins in small populations or at low density face dilution of their gametes and reduced fertilization rates. Even being separated by one meter reduces the chances of successful fertilization to less than 10% (Pennington 1985). Spawning in response to local cues could help *H. purpurascens* reproductive rates in the smaller, isolated populations (Williamson and Steinberg 2002), although this has not been specifically investigated. Unfortunately, detailed information on the nature of the environmental cues, the distribution of those cues and their influence on the ecology of the organisms is not yet available for the various systems that synchronize the life cycle of sea urchins, although a considerable amount of information is building up about the specific cues that mediate settlement and metamorphosis (Steinberg et al. 2002).

The life cycle of *H. purpurascens* is complex and apparently highly derived relative to that of other urchins. The planktonic apluteal larva undergoes metamorphosis into the benthic stage, which lives within the canopy of sub-tidal algae (Williamson and Steinberg 2002). Chemical cues from macro-algae strongly influence where the larvae ultimately settle for metamorphosis. Compounds from the red alga *Delisea pulchra* influence larval behavior strongly and they ultimately settle and metamorphose mostly on this species (Swanson et al. 2004, 2006, Williamson et al. 2004). In nature, most *H. pupurascens* individuals, and all of the large size class animals, are found not on red algae, but on kelp (*Ecklonia radiata*). Despite the risk of predation, most individuals move ultimately to kelp, on which they achieve much better growth and presumably reproductive output (Williamson et al. 2004, Williamson and Steinberg 2012). For some reason, though, kelp does not stimulate settlement or metamorphosis and individuals in the smallest size classes are never found on kelp. The reason for the functional disjunction across these two algal species with respect to the urchin life cycle has yet to be clarified, as has the influence of their relative distribution locally on survival rates of moving individuals.

3.5 Conclusions—Conceptual Developments

The examples in this section provide a guide as to what is possible in the development of ecological understanding by giving pre-eminence to the strong linkage of organisms, through their life cycle and their species-specific adaptations, to the dynamic environmental circumstances that surround them. The next step is to conceptualize this organism-environment interaction and develop robust general theory around that. Our aim is to provide not only a theoretical framework for the way in which the adaptive mechanisms of a species "match" the organisms to the spatio-temporal dynamic structure of their environment, but also to illustrate the rudimentary nature of the understanding we have in this area, and that that understanding warrants further scientific test and development. Even the interpretations offered in each of the examples covered earlier in the chapter are hypothetical to a large extent, and each warrants further test to extend understanding of them. The advantage of this approach to ecology is its opening up of the way in which we see organisms in their environment, rather than focusing on epiphenomena like population densities and community structure, and this is as true of human ecology as that of any other species (Walter 2008).

In the following chapter we extend the general interpretation we have outlined here. Before that can be done, we briefly conceptualize the ecological aspects of the lifeline of an organism through the standard statistical metaphor of a ball falling down a board and colliding with pins deployed in a regular pattern. Francis Galton developed this device to illustrate the effects of stochasticity on probabilities. This diversion into stochasticity allows a reconsideration of the general aim of autecology, as a discipline, and the structure of that theory.

4 THE GALTON BOARD METAPHOR

The lifeline of an individual, as developed through the examples covered above, can be considered conceptually in terms of each adaptive mechanism having to operate effectively to keep the organism alive within its environmental surrounds. That operation can be negatively affected in numerous ways, some of them intrinsic. For example, extremes of genetic variation are likely to result in failure, which can be treated statistically in stochastic terms because of the random nature of mutation and recombination. Extrinsic factors can be similarly dealt with, because the mortality associated with movement or migration, predation, climatic influences, and so on, is likely to be random because it is influenced stochastically. These processes can be represented metaphorically and schematically, for a single individual, by a ball being released from a funnel to run down a vertical board with pins (ecological hazards) inserted in a regular pattern (Fig. 3.34). Each specific process during the ball's downward trajectory, whether that is the fall of the ball or its bounce against pin or wall, can be interpreted mechanistically in purely deterministic terms. Nevertheless, the path followed by the ball cannot be predicted with any certainty at any stage of its run. It can be documented only after the run has been completed. Stochastic influences, determined initially by the original state and location of the ball, and enhanced during each "interaction" with an ecological hazard, thus affect the outcome too much for there to be any certainty (as illustrated with reference to the motion of billiard balls in Fig. 3.35 and the strong influence of even slight differences in starting position (Fig. 3.36)), despite the fact that each bounce is, in itself, entirely deterministic.

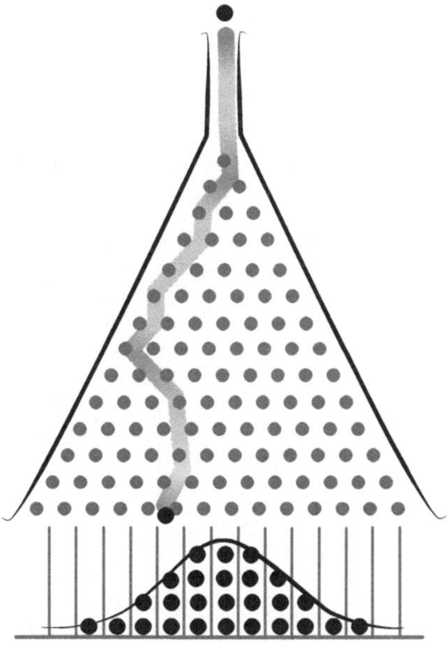

Figure 3.34 Galton board model (also known as a quincunx) to illustrate how stochastic influences alter the trajectory of equivalent balls dropped from a funnel. The typical distribution of fallen balls is presented at the bottom of the board. You can see the operation of this apparatus at MathsIsFun.com (2010).

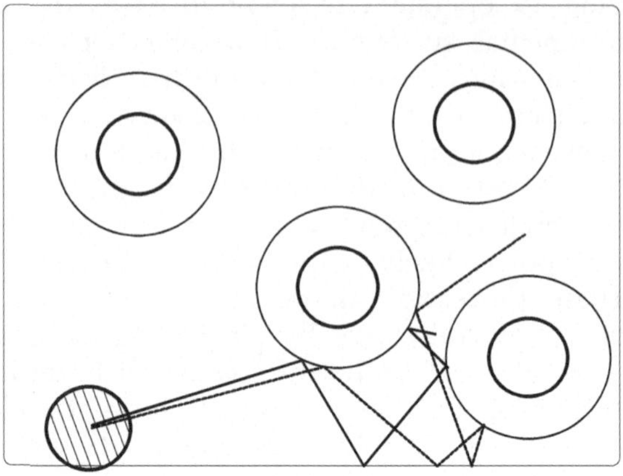

Figure 3.35 Each interaction of ball and pin (Fig. 3.34) can be described deterministically. This point is illustrated by a billiard ball (hatched) interacting with various convex obstacles. The ball starts in the bottom left corner, and the subsequent trajectory of its center is marked (continuous line). An imaginary ball has the same starting position as the other ball, but is started in a slightly different direction (dashed line). After a few collisions, the two trajectories are so different as to belie their common starting point. This illustrates that the exact passage of the falling ball in Fig. 3.34 is unpredictable because of the stochastic nature of the process as a whole (as illustrated further in Fig. 3.36). Reproduced, with permission, from Ruelle (1991).

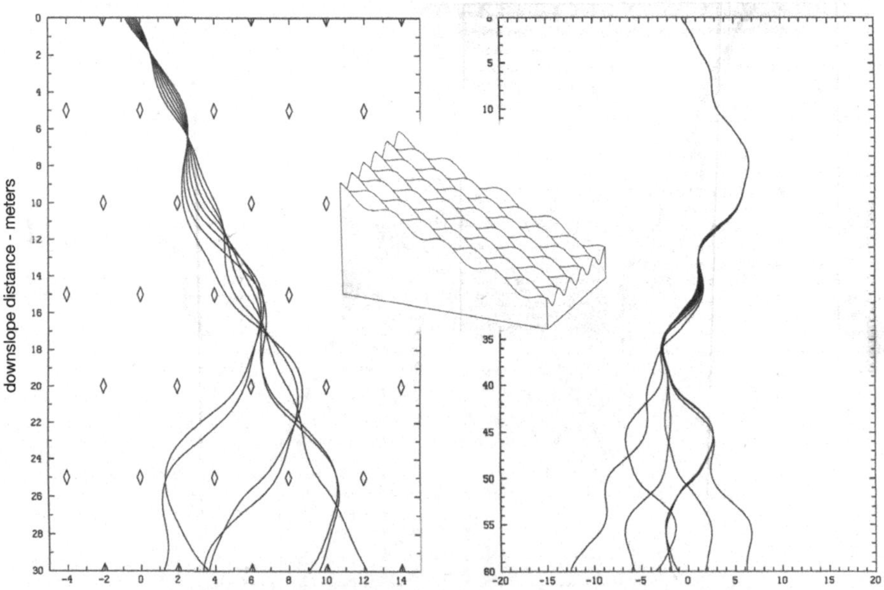

Fig. 3.36. Trajectory of seven boards that start down a ski slope that contains many moguls (inset image) within a short distance of one another. In the left hand diagram the boards start 10cm apart from one another, in the right hand one they are only 1mm apart at the start. Reproduced, with permission, from Lorenz (1993).

If many balls are dropped into the funnel simultaneously, their individual paths are similarly subject to the same stochastic influences of their initial states. Thus, the balls become distributed in the return boxes at the bottom of the pin board in a stochastically predictable way.

To render the metaphor more realistic relative to the lifeline of organisms within their dynamic environment, the stable and rigid Galton board can be made more heterogeneous and dynamic in both space and time. The hazards are made to move and to vary in density. The vertical board could be made spatially variable in terms of the ease with which the ball falls down the surface, to stimulate the spatial variation in the quality of the organism's environment. To this end, areas of the board could be made rough or sticky, to slow the falling ball. Slow movement of the ball across such areas could equate to periods of habitat occupancy interspersed with movement across unsuitable areas. These areas could be made to shift with time, and to "dissolve" and reform, as observed with environmental circumstances in nature (as explained in Chapter 4).

If each ball is considered to represent an individual organism, its survival could be assessed by assigning holes to certain parts of the board, as designators of mortality, and the containers at the bottom could indicate the level of fecundity the individual falling into it has achieved. Further stochasticity could be generated by varying the state of each hole and box relative to its message of doom or designation of fecundity, respectively. The different life stages of species with a discontinuous life cycle (e.g., larvae and adult in metamorphic animals or haploid and diploid stages in plants) differ physiologically and ecologically from one another and so can be thought of in terms of their running down independent pin boards. We thus derive a measure of the reproductive output of individuals, which is then returned to the funnel for the next generation to start.

This outline provides a metaphor for what needs to be understood in ecology and what needs to be achieved in terms of the development of general mathematical models, topics to which we return in Chapters 8 and 9.

5 THE GENERAL AIM AND STRUCTURE OF AUTECOLOGY

Organisms occupy a spatially dynamic, stochastically varying environmental setting that is structured in terms of various temporal scales of climatic and biogeochemical variation and change. Because organisms cannot predict where or when suitable conditions will be found, their life is influenced by this stochastically influenced dynamic variation. Moreover, the organisms themselves are not all the same. Besides differing across species, they also vary in terms of their physiological status, morphology, size, age, and so on, and stochastic influences play a role here, too. They will thus respond to the surrounding environmental circumstances differently from one another and usually independently of one another. In many ways, their lifeline is well represented by the Galton board metaphor of Fig. 3.34.

The key to understanding the dynamics of such a system rests with the specification of the methodology of autecological research and the isolation of those features that contribute to the observed dynamic pattern in nature. The focus in autecology is functional, in that it is built in terms of the biological mechanisms through which ecological interaction takes place, and it is prospective in that it looks ahead as to what options are open to organisms within particular environmental settings and developments. The primary pattern, concepts and methods associated with such an approach are outlined in Table 3.3.

Table 3.3 Tabular presentation of the primary pattern in ecology, from the autecological perspective, with details of the processes involved and ways in which the system is described and modeled. This flow chart represents an epistemological or conceptual view of autecology. The practice of autecology is dealt with in subsequent chapters.

IN GENERAL	*IN PARTICULAR*
PRIMARY PATTERN IN ECOLOGY	* Dynamic spatio-temporal distribution of species (species-specific)
PROCESS THAT SPECIFIES THE PATTERN	* Interactions between the adaptive mechanisms of organisms and their structured but dynamic environment (biotic and abiotic features)
PRIMARY MEASURES OF OUTCOME OF INTERACTION	* Individual survival rates * Net reproductive rate (R_0) (see Glossary)
PROCESS DRIVING THE R_0 DYNAMIC	* Climatic, biogeochemical and vegetation dynamics and variability
OBSERVABLE CONSEQUENCES	* Stochastic influences on the survival of individuals * Spatial and physiological adjustment of individuals * Shifting spatio-temporal distribution of species' ecological intensity (see Glossary)
STATISTICAL CHARACTERIZATION OF THE PROCESS	* Mechanistic models * Stochastic measures * Survival probabilities * Risk distributions

The most consistent pattern that is discernible in ecology, the primary pattern to be explained, is the species-specific dynamic spatio-temporal pattern of distribution of species. Each species is independently adapted to a particular subset of environmental influences, and even if species share a phylogenetic heritage, they do differ from one another ecologically (as in the *Pygoscelis* penguins, *Macropus* kangaroos and *Melithreptes* honeyeaters covered earlier in this chapter). The process responsible for the observed pattern is the interaction between the individual organism and the environment, as mediated by the species-specific adaptive mechanisms carried by the organism. Each species thus responds to a subset of the structured and dynamic environmental axes that make up the total environment in any area (Chapters 4 to 6). The responses of the organisms can be measured in terms of their survival and reproduction rates under different environmental circumstances, as influenced by stochastic intrusions and the spatial and physiological adjustments of the organisms (Chapter 7). The changing distribution of local abundances and the shifting distribution of the species can thus be characterized statistically and understood mechanistically with the aid of research designed to explore those ecological processes and reconstruct them mathematically with the aid of the appropriate mechanistic models (Chapters 8 and 9). Such a reconsideration of what is significant in the ecology of organisms thus allows development of a general theory for autecology that stands in strong contrast to demographic theory (Chapter 10).

Ecologists have long searched for demographic patterns that would unify ecology. These have been sought so as to eliminate the obvious differences across species and the variation induced by climatic and other variables. This would have, had it worked, been extremely convenient for ecological generalization and prediction. The classical pattern in demographic ecology is the one Darwin used in the construction of the logic of natural selection. He envisaged populations to be numerically stable (Brady 1982), a view derived from the "balance of nature" metaphor that had been developed and promulgated by the natural theologians (Worster 1977). Regular and recognizable patterns of resource use among species (and within communities) are expected, but such patterns in particular have proved elusive (Kikkawa 1977, Greig-Smith 1986). Frustration with the failure to find the expected patterns of "community structure" has driven the search to ever greater extents, with regional scales of resource partitioning being sought (Ricklefs 1987) and macro-ecological patterns being canvassed as a potential solution (Lawton 1999, Blackburn and Gaston 2001, 2006). No consistent or useful general pattern of this nature has been uncovered and autecological understanding suggests that none exists.

To uncover a consistent ecological pattern among species, the search parameters need to be modified. To this end, the following guidelines derive from (1) a consideration of the failure to find consistent numerical patterns at the local population and community level, (2) an appreciation of Darwin's ecological perspective, in which he focused on the local rather than the global, (3) the fact that the underlying principles of demographic ecology derive from the local perspective, even when extended to larger spatial scales, (4) the knowledge that species behave in an ecologically idiosyncratic way, and (5) the empirical evidence that all species track climate change on local and geographical spatial scales, appreciating that island populations are different in this regard, although altitudinal options (Hengeveld 1985b) and island hopping opportunities (Carson 1987) may be open to them.

The only consistent ecological pattern across all species, then, is the dynamic spatio-temporal adjustment of species in response to altered environmental conditions, and the fact that this adjustment is species-specific (Table 3.3). Such a pattern is very different in principle from that which is typically sought by ecologists, whose preference in this regard usually extends to something more akin to the Periodic Table of the elements, into which species can be inserted according to their ecological properties (e.g., Southwood 1977, Berry 1989, Keddy 1994). The chemical elements have regular and constant structures and differ from one another in a systematic way, so their interactions are predictable. By contrast, adaptation and speciation do not occur in any such systematic fashion (see Chapter 5), which implies that no such pattern exists except perhaps fortuitously on occasion.

The perception of autecology outlined in Table 3.3 starts from the dynamic and stochastic nature of ecological systems, as opposed to the deterministic equilibrium state from which ecological systems are scrutinized in demographic ecology. Moreover, it is opposed, just as much, to the notion of "non-equilibrium" because the latter is relevant only if an equilibrium state can legitimately be expected (Walter 2008). Understanding the dynamic according to the methodology outlined in Table 3.3 is the focus of autecology. That is, the only way in which to understand any particular ecological system mechanistically is through investigation and quantification of the survival and reproduction probabilities of individual organisms within their usual environment.

The mathematical modeling of such patterns and dynamics is dealt with later. In principle, though, the primary processes that generate the pattern just described are modeled. The model is generalized in that the basic structure holds for all species, and the specificity of species is represented by the particular variables that are included and their specific values, as seen in autecological invasion models (van den Bosch et al. 1992, Hengeveld and van den Bosch 1997) (Chapter 8, Section 6) and those for dynamic energy budgets (Kooijman 2010). In the invasion models, the values of the environmentally significant variables may differ geographically within a species, because of climatic, biogeochemical and related differences, and this translates into geographical differences in invasion rates.

Organisms adjust geographically to environmental input as a consequence of their interactions with their immediate environment. The processes that ultimately drive the dynamics of ecological systems are climate, biogeochemistry and vegetation structure, with the overall dynamic being stochastically variable, as explained more fully in Chapter 4. The relevant dynamics include the energy and mass flow dynamics within organisms, their survival and reproduction rates, their spatial dynamics and, ultimately, their evolutionary dynamics. This explains why autecology is concerned so intimately with the way in which organisms match their surrounding environment (Chapters 5 and 6) and stochasticity, which has to be incorporated into ecological models and interpretation at a very basic level (Chapters 8 and 9). Ultimately, autecology focuses on mechanistic understanding and models of the processes that influence their survival probabilities and risk distributions, as it is these processes that help to keep a prospective focus in ecology.

Chapter 4

Spatio-temporal Dynamics Climatic Variables and the Response Rates of Organisms

CHAPTER OUTLINE

1 INTRODUCTION

Ecological systems are dynamic. No one questions this. What is at stake in ecological theory, though, is which aspects are dynamic, what drives the dynamic and how that dynamic can be conceptualized most realistically. Demographic ecology views systems as tending toward stability, driven primarily by density-related processes. The real dynamic, from this perspective, is the on-going adaptive change in local populations as they evolve in relation to demographically generated competition. Climatic and other abiotic influences are seen as relatively constant but with occasional deviations that disrupt equilibrium, with these diversions then tending to be corrected through density compensation. Populations are thus seen to be balanced or, at least, tending towards a state of balance, as explained by Cooper (2001). That is, local populations should increase and decrease around a state of equilibrium. The concept of community is related to the ideas just outlined and stems from the notion of a local balance being achieved across its constituent species, and evolutionary ecology explains how adaptive adjustment for greater efficiency is achieved within this competitive context (e.g., Ghiselin 1974, Mayhew 2006), often through on-going coevolution (e.g., Thompson 2005, 2009a).

Although ecologists may now avoid terms related to balance, equilibrium and even competition, perhaps in efforts to circumvent these hotly debated concepts, populations and communities are legitimate units of study only if they are considered to be structured relative to perceived equilibria. Even if they are claimed to be "non-equilibrium", a perspective that is increasingly accepted, such a state is specifiable only in relation to the expectations of an equilibrium of some sort (Walter 2008). The dynamics emphasized in demographic ecology are consequently those that restore equilibrium conditions, or those that stabilize ecological systems (Pimm 1991, Cooper 2001, 2003). This direction is strongly evident in the questions that drive so much ecological research, about population regulation, density effects, metapopulation dynamics, coexistence, community structure, macro-ecological scales, and so forth. Within such systems, the organisms themselves are considered to adjust locally through optimizing adaptive processes driven by fitness benefits that are selected relative to the lesser performances of others. In macroecology, processes that operate uniformly at even greater scales are sought. Verification of this perspective appears to be the dominant activity in current evolutionary ecology and is notably evident in investigations of local adaptation and ecological speciation (Walter 1999, 2003).

Autecological theory views ecological systems in a perspective different from that outlined above, because "equilibrium" is an idealized misconception of the state of ecological systems (Worster 1977, Simberloff 1980) and equilibrium conditions are not characteristic of such systems (Rohde 2005, 2013). The primary driving influences in the dynamics of ecological systems are not related to such postulates as population density, equilibria, ecological stability or efficiency through adaptation relative to competitive ability. Rather, organisms respond directly to the driving forces imposed by the environment, because changes in the environment provide the primary dynamic in ecological systems. The organisms respond to this dynamic through the suite of complex adaptations (and their associated requirements and tolerances) that they carry sequentially through their life cycle (e.g., host-searching mechanisms, temperature requirements and motile abilities (Chapter 3)). Although the organisms may be responding to biotic changes locally, such as an increase or decrease in host availability, the primary dynamic is driven by climatic and biogeochemical influences. Hence the idea of "abiotic forcing" in ecological systems and evolutionary interpretations (Hengeveld and Walter 1999, Walter and Hengeveld 2000, Schmidt et al. 2004, van Dam et al. 2006), as opposed to the usual emphasis on competition and competitive selection. Since climatic and biogeochemical influences vary at all temporal scales, ecological systems cannot be characterized in terms of equilibria, and the proposed equilibrial processes cannot, in turn, impose any regular pattern on the variation generated externally in this way. In brief, species are stable and adjust biochemically, physiologically and behaviorally to a spatio-temporally dynamic environment. Adaptive change is not an inevitable and continuous process, but requires particular ecological circumstances (which are covered in Chapters 5, 6 and 10).

This chapter justifies the autecological perspective further, as follows. Section 2 puts the climate into ecological perspective. This is necessary because ecologists have tended to treat climate and its influences on organisms in an over-generalized and rather static way, although this has changed positively with the appreciation of anthropogenic climate change and its impact (Kingsolver et al. 2011, Radchuk et al. 2013). In consequence, the average measurements that humans use to characterize the weather over a particular period, and to generalize about it, have tended also to be used in relation to the ecology of organisms, especially when ecological systems are simply partitioned into biotic and abiotic components. In other words, the climatic variation that is perceptible at numerous spatial and temporal scales of resolution (see following section) is simply reduced to a single average measure of variation to represent "temperature", "humidity" and "rainfall", although the emphasis is commonly placed on extremes. Biogeochemical change is treated similarly, especially when organisms are the agents of the change. The influence of organisms on the nature of the environment is often ecologically consequential for other organisms, and sometimes even crucially so, in providing living place, altering vegetation, changing substrate structure, and so on. The dynamics of such influences are impacted by numerous variables, including abundance of the agents of change, local conditions (physical, climatic, geological), and the like. Particular ecological questions may require that dynamic to be understood. In dealing with such circumstances, ecologists have referred to organisms that change the nature of the local environment significantly as "engineer species", but this metaphor portrays organisms and their ecology as deterministically as it treats them typologically, and so is not particularly useful.

Section 3 provides an historical perspective of ecological dynamics to demonstrate how climate has influenced ecological systems through time, to illustrate how climatic and biogeochemical influences should be considered in terms of spatial and temporal variation, and how the diverse range of such variables affect the individuals of different species differentially. It shows what organisms have to face, often from one moment to the next, in terms of the capricious nature of climatic influences. Section 4 extends this latter perspective to consider ecological dynamics in terms of the rate at which individual organisms respond to climate and its vagaries at scales relevant to studies of contemporary ecological systems. This section emphasizes the fine scale at which climatic variables impose upon organisms. From this perspective, we start to extract generalizations that are relevant to developing autecological theory, as explained in the discussion section (Section 5).

2 CLIMATE IN ECOLOGICAL PERSPECTIVE

2.1 The Diversity of Climatic Influences

How does climate vary? The ways in which organisms interact with environmental variables provide the context for discussing climate from an ecological perspective. What is significant is the scale at which one examines climate, and both the temporal and spatial scale are relevant ecologically, and average values are almost inevitably used at each scale. Although means and other measures of central tendency are useful, they usually provide only a rough indication of the nature of the environment as experienced by organisms of different species (Helmuth et al. 2005). They do not, for example, contain information on the amplitude of variation or about its regularity, both of which are likely to be meaningful ecologically. Averages also circumvent information about variation across different time frames. This is problematic in that such variation structures the environment and influences organisms at different scales, perhaps most evidently in arid systems (e.g., Morton et al. 2011). To use a relatively obvious example, temperature profiles across the seasons have a decidedly different ecological impact on organisms than do diurnal temperature profiles, and if such distinctions are not strictly drawn, their ecological relevance is easily lost.

More detailed, specific information is required if we are to understand the nature of climatic influences on organisms. Of substantial help in this regard is a consideration of the temporal and spatial structure of climate, for this helps us to consider how species respond differentially to climate (Fig. 3.1). This differential leaves ecologists with no option than to identify the significant climatic variables in a specific way for each species if ecological systems are to be understood. The same holds for other aspects of the environment, especially biogeochemical processes, vegetation structure, and their influences. A single set of data from a locality thus cannot provide a reliable indication of the ecological suitability of that locality for a range of species; any coincidence is likely to be superficial, although in some cases such coincidence may reflect common ancestry (e.g., Huntley et al. 1989).

2.2 An Analysis of Climate Variability in Europe

Figure 4.1 presents mean surface air temperatures for the Northern Hemisphere over the century 1880 to 1980 (Lough et al. 1983). Within this period climatic changes are apparent, with some periods warmer than others, but nested complexity is evident within this pattern. Across seasons of the year the differential between the 20-year warm period and the 20-year cold period identified in the figure is statistically significant. In winter and autumn, the differential was significantly greater than that in summer and spring (Table 4.1). The same table also reveals a subtlety; the extent of the difference varies spatially, for the differential experienced in the Arctic was very different in extent and seasonal profile from the average for the whole of the Northern Hemisphere. Regional differences in the extent of temperature variation are evident as well, with most Arctic seasons showing more extreme variation than seen in the hemisphere as a whole.

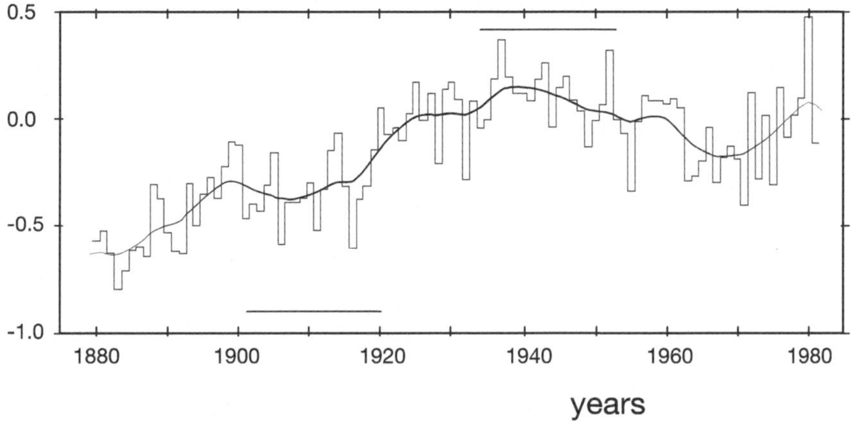

Figure 4.1 Variations in mean surface air temperature in the Northern Hemisphere over the century 1880 to 1980. The curve represents 20-year filtered values. Also indicated are the 20-year warm and cool periods used for comparative (Table 4.1) and modeling purposes. Redrawn from Lough et al. (1983), where the 'filtering' to obtain the curve is also described.

Table 4.1 The difference, in degrees Celsius, between the warmest 20-year period (1934–1953) and the coldest (1901–1920) (as demarcated in Fig. 4.1). The average change in temperature is not spread evenly across all seasons of the year. All values are significantly different at the 1% level. From Lough et al. (1983). The comparison is made independently for the Arctic and the Northern Hemisphere as a whole, which shows, too, the differential across geographical space.

WINTER	*SPRING*	*SUMMER*	*AUTUMN*	*ANNUAL*
ARCTIC				
2.1	1.1	0.5	1.3	1.3
NORTHERN HEMISPHERE				
0.5	0.4	0.4	0.5	0.4

Lough et al. (1983) used the data of Fig. 4.1 and the analysis of Table 4.1 to select periods with statistically different temperature regimes as a basis for modeling the consequences of climatic warming in Europe. To represent spatial variation as realistically as possible, geographical data for Europe on sea level pressure, temperature and precipitation were also included (see Lough et al. (1983) for sources). Should the European temperature rise by about 1°C, on average, this increase would not be evenly spread across the entire year. Winter, for example, is likely to be different from the other seasons, as a belt of negative values shows in Fig. 4.2, with this spanning about 40-60 degrees north. This belt represents a decrease in temperature despite an overall warming environment, and is a consequence

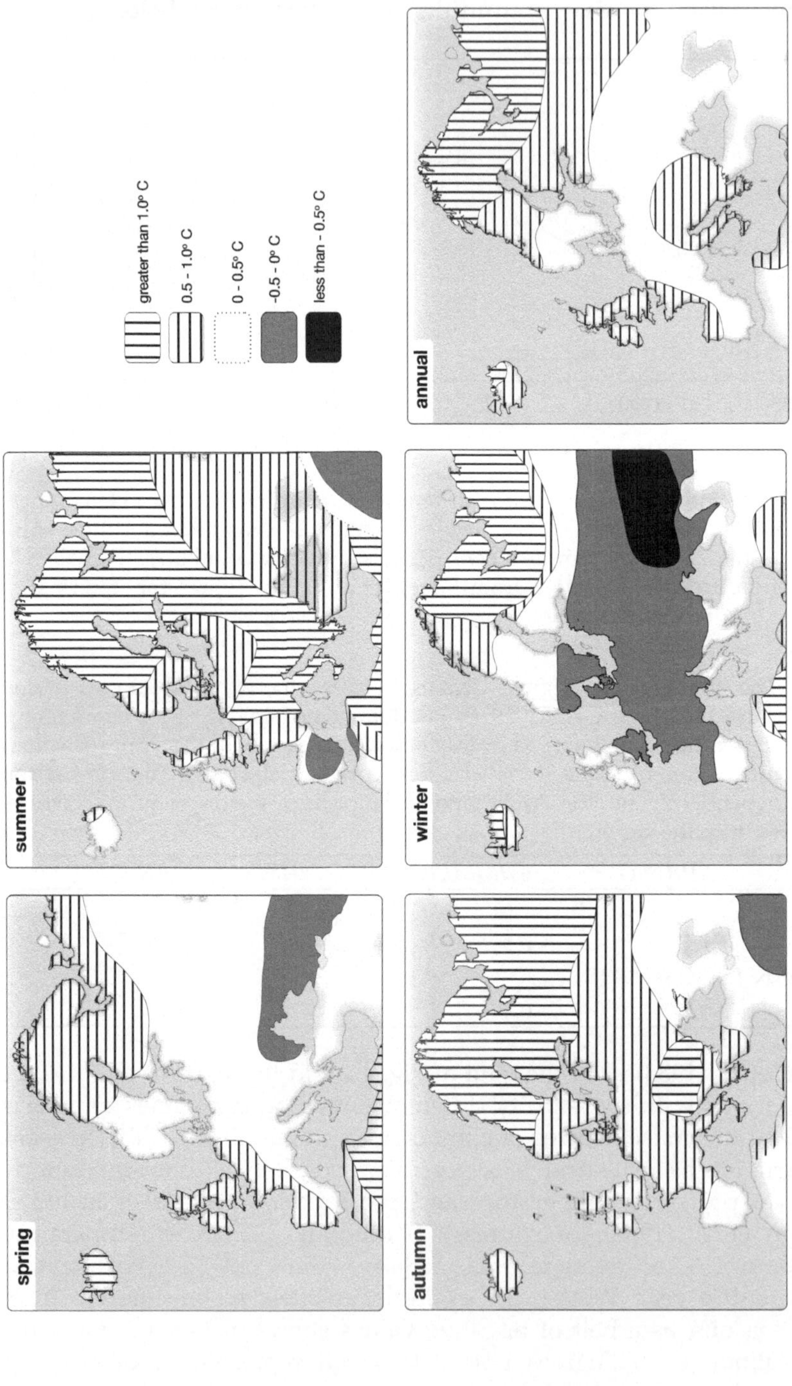

Figure 4.2 Distribution of European temperatures after an average temperature rise of 1°C. The information is presented as the difference between the warmer and the original, colder periods. Seasonal data presented separately, and overall, annual changes are also given. Redrawn from Lough et al. (1983).

of changes in the distribution of pressure (Lough et al. 1983) (Fig. 4.3a). Such localized spatial variation with regard to altered climatic conditions has been measured during the course of this century (Gribbin 1990, p. 12).

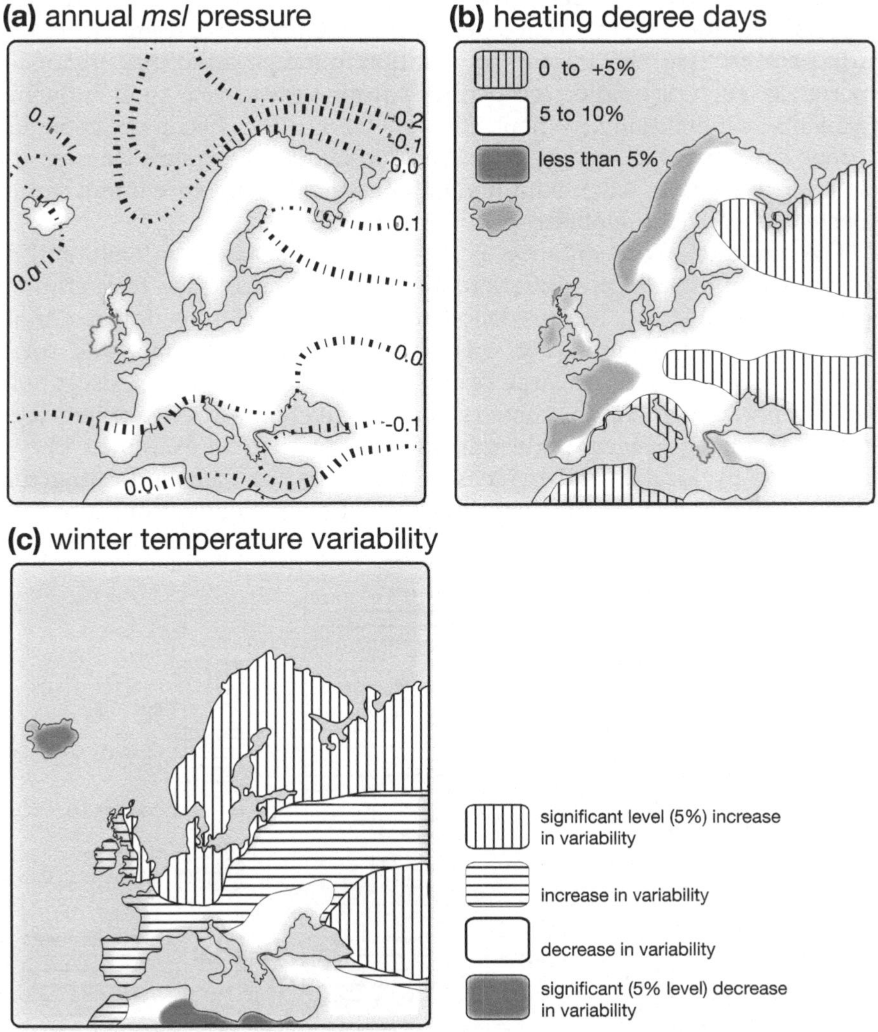

Figure 4.3 Same scenario as illustrated in Fig. 4.2, but representing the mean sea-level pressures (MSL) (in mb) **(a)** and the difference involving the heating degree-days **(b)** and the variability in winter temperature **(c)** These diagrams indicate the counterintuitive conclusion that in a generally warmer environment, summers may be shorter (b) and winters more variable, with some being more severe than present (c). Redrawn from Lough et al. (1983).

Other ecologically significant thermal features are likely to change as well, as the following two examples show. One, the overall duration of warm and cold periods is likely to be differentially significant to the ecology of species, as the former represent the periods for growth and reproduction, and the latter the time for migration or winter maintenance through harsh conditions. Figure 4.3b shows that despite an increase in average temperature, the warmer periods were shorter than before. But again, there is spatial variation in this outcome, with some areas experiencing longer warm periods. Two, the amount of variation in temperature that organisms experience, such as winter temperature from year-to-year, will influence organisms differentially. When climate in Europe becomes generally warmer, an extensive area of increased variability is expected across the north and west (Fig. 4.3c). So a higher frequency of severe winters can be expected, despite the global rise in temperature.

The amount of precipitation is anticipated to change as well, and so too the variability in precipitation. In general, the higher temperatures represented in Fig. 4.1 are predicted to be associated with lower annual rainfall over most regions (Fig. 4.4), such as in northern Italy and south central France where fairly large decreases are expected. Again, exceptions to this trend are evident, for some regions are likely to experience increased rainfall. The geographical pattern in rainfall can also be examined in two other ways, by locality and by season, with each of them indicating that at finer scales of spatial and temporal resolution ecological conditions are variable.

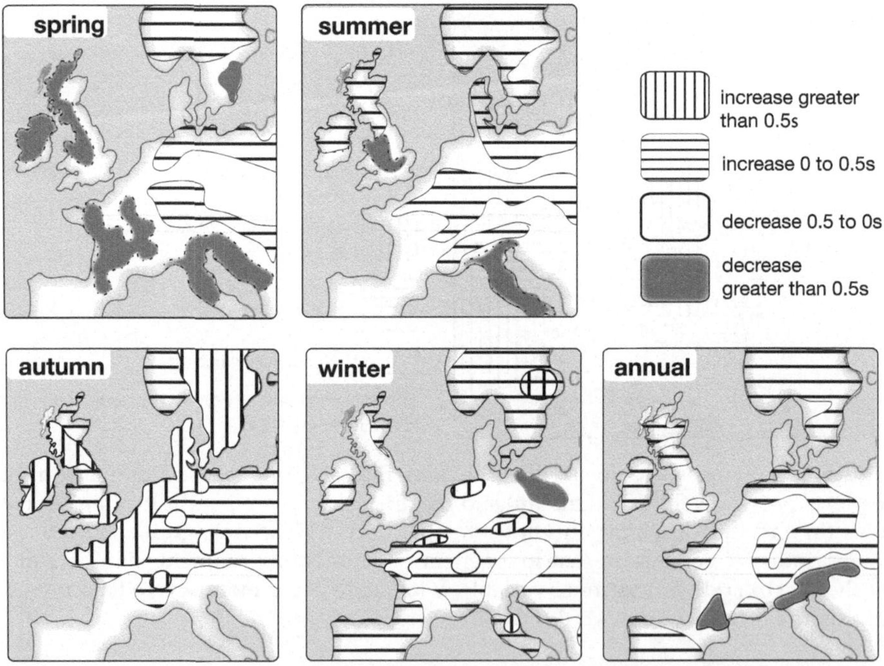

Figure 4.4 Same scenario as in Fig. 4.2, but here the distribution of differences between warmer period rainfall and current rainfall is given for the seasons and overall (annual). Redrawn from Lough et al. (1983). *s* = standard deviations.

The above analysis, which is not unrealistic, has important consequences for how we treat climate relative to ecological systems. The annual trend breaks down spatially, to the extent that localities will differ from one another in the seasonal pattern in which their rainfall comes, even if they should have suffered the same annual change in their rainfall (Table 4.2). A higher global temperature would result in a greater seasonal variability in rainfall with drier springs for most of Europe, and wetter autumns and winters. The variation in precipitation between years is superimposed on these seasonal patterns (Fig. 4.5) with statistically significant increases in most seasons and in different areas. This implies that over the years sporadic precipitation will affect geographical regions differentially.

Table 4.2 The seasonal change in rainfall (warm period rainfall minus colder period rainfall) across two areas (southern England and central France) that are subject to the same overall (or annual) change in rainfall. The values represent multiples of the standard deviation. Extracted from Fig. 4 of Lough et al. (1983).

SEASON	*SOUTHERN ENGLAND*	*CENTRAL FRANCE*
Spring	Decrease: 0.5–0	Decrease: > 0.5
Summer	Decrease: 0.5–0	Increase: 0–0.5
Autumn	Increase: > 0.5	Increase: 0–0.5
Winter	Decrease: 0.5–0	Increase: 0–0.5

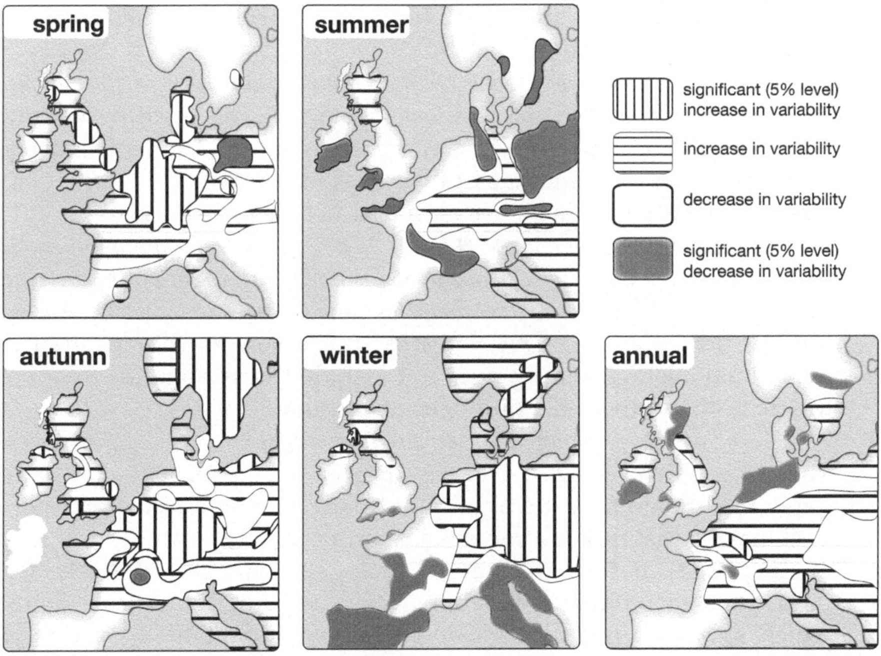

Figure 4.5 Same scenario as in Fig. 4.4, but here the geographical distribution of change to the variance in precipitation is given for the seasons and overall (annual). Redrawn from Lough et al. (1983).

In short, climatic variation cannot be represented in an ecologically meaningful way by taking averages across extensive areas or periods. Organisms simply do not experience climatic variables in this way.

2.3 Climate Variability and Its Ecological Consequences

The examination of climatic variation over geographical space, outlined above, reveals several points relevant to understanding the ecology of species and the way in which organisms respond to climatic influences at the various ecological scales delimited above.

1. *A change in climate is not imposed uniformly across a landmass.* The common picture of shifts in temperature, with latitude for example, followed by equivalent shifts in the geographical distribution of groups of species is inappropriate, as indicated by the data on Pleistocene beetles (Fig. 3.1) and other organisms (see Section 3). Although concordance has been revealed across species in some studies, the responses of the species concerned are not entirely uniform (Parmesan 2003, Parmesan and Yohe 2003).
2. *Regional variation is the norm, and this is true even when climate is considered to be "stable". Consequently, regions that are climatically similar in one respect (or similar in one dimension of climate) are likely to differ from one another in other respects.* Since species respond to particular combinations of climatic variables, the response to climatic variation and change becomes unpredictable across species, for all evidence suggests it will be individualistic.
3. *A change in one variable in one of the seasons is not necessarily matched by equivalent changes in other seasons, whether in time or space.* Warmer summers do not necessarily mean warmer winters. Therefore, species that differ from one another in their seasonality will be differentially affected by climatic change, within and between regions.
4. *Not only do the seasonal maps (Figs. 4.2–4.5) differ from one another, the annual map (with the yearly average) for each of the variables has little in common with the equivalent map for any of the four seasons. Annual averages are therefore not representative of any season.* This means that the annual average simply does not include the seasonal fluctuations to which the individuals of a species respond.
5. *The extent of this variation is likely to affect different species differentially.* The probability of two or more species having exactly the same requirements and tolerances across all spatial and temporal dimensions of climatic variability is remote (as illustrated in the examples detailed in Chapter 3). Besides, additional ecological influences (mostly also influenced by climatic variations) add to the dimensions across which species differ ecologically (Chapter 5, Section 2). Although not inevitable, some similarity in response is most likely in species that share recent ancestry, but even then there are almost invariably regional differences in distribution (Figs. 3.2 and 4.6), despite any similarities in habitat use to one another.

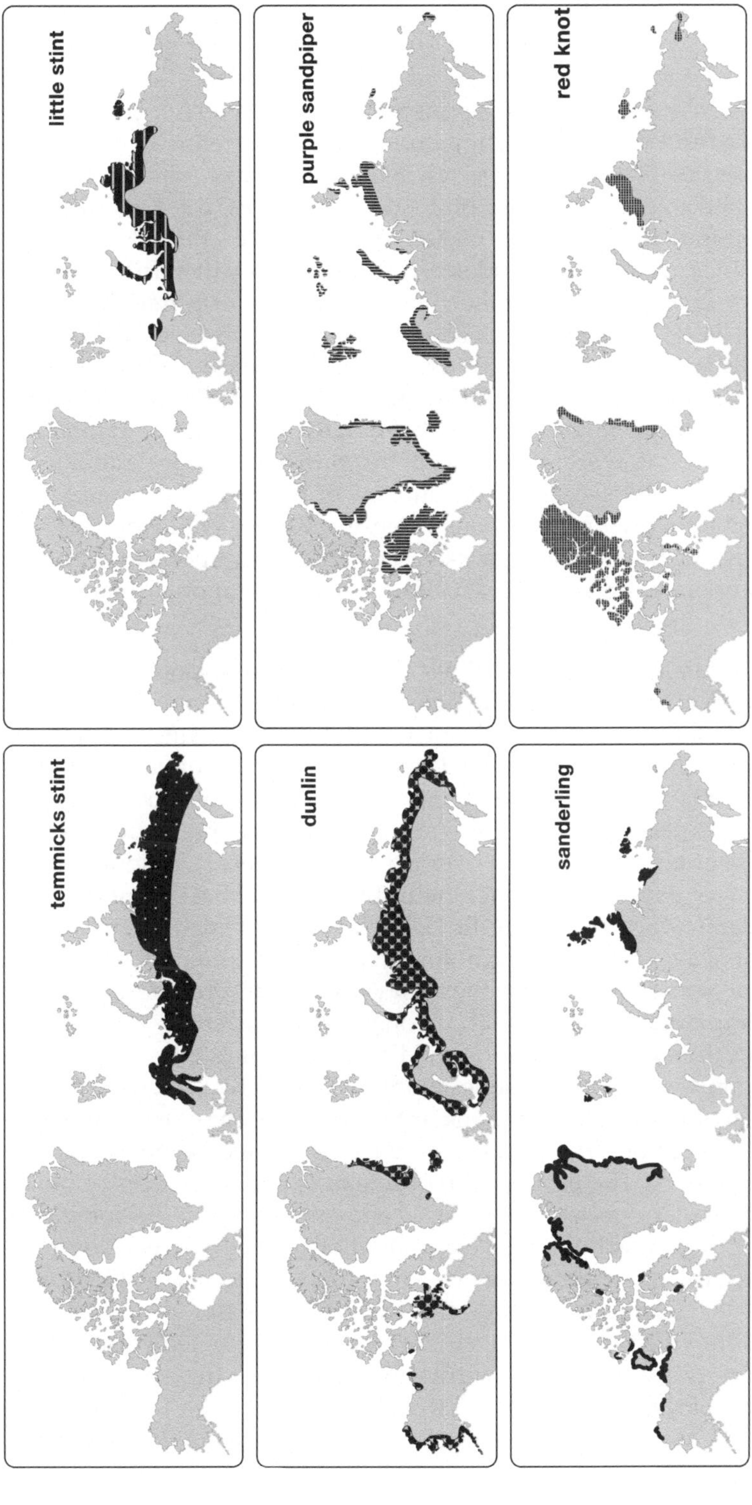

Figure 4.6 Breeding distribution of various Western Eurasian (roughly westward from the Taymar Peninsula) wading birds of the genus *Calidris*, to indicate the general similarity of their distributions coupled with the significant variation. Redrawn from Voous (1962).

6. *Examination of the data from which the two twenty-year periods were extracted for analysis (Fig. 4.1) indicates that neither of these periods was stable. They were both variable, but at different levels. For organisms, this has ecologically significant consequences.* That is, different sequences of years differ from each other. They thus impact differentially on the specific properties of different species. Even at a broad scale, perennials are affected in terms of their build up of reserves, for example, whereas the population densities of annual plants are affected more directly. For an annual to grow and set seed successfully therefore requires only one year of suitable climatic conditions, but for a tree the conditions have to be suitable across several years, and even to set flowers and viable seeds trees may need a particular sequence of conditions. Mulga (*Acacia aneura*) in central Australia, for example, sets viable seeds only in those few years that have sufficient summer rainfall, which stimulates flowering, followed by sufficient winter rainfall, which supports seed set. And germination of those seeds requires, in turn, specific follow-up conditions (Preece 1971a, b, Winkworth 1973). This again illustrates that organisms with different life history parameters will operate dynamically at different spatio-temporal dimensions from one another.

The points above illustrate how the ecological responses of organisms should be seen in relation to the many variables that characterize the spatio-temporal dynamics of climate. Further, variation is evident in the intensity of each variable that can be measured, such as maximum daytime temperature or average maximum summer temperature. Even the duration for which each variable persists is likely to vary. For example, high summer maxima may be experienced often and over a period of weeks, or sporadically and for only a few days in a season. Finally, the seasonal background into which a particular feature of climate fits is also likely to vary in an ecologically significant way. Unusually high summer temperatures, for example, may endure for some weeks, but they may follow or precede a winter with higher temperatures than usual, or one that is extremely cold. Ecological effects can thus become compounded or be ameliorated.

In summary, climatic variables manifest themselves upon organisms in a diversity of ways, through maxima, minima, durations of exposure, levels of variation, and so on. In effect, these variables change kaleidoscopically and at high frequency. The ecological requirements and tolerances of organisms of any particular species are diverse, and the central issue for individual organisms is survival and then reproduction if they successfully maintain themselves to the appropriate age and time (Haukioja 1982, Rose 1997, Walter 2008). Furthermore, the various life processes of organisms may require different environmental circumstances, such as temperature regimes or moisture conditions, even within individuals as they pass through the various developmental phases that constitute their life cycle (Chapter 3). Life has to be maintained within the ever-changing environment that surrounds the organisms and which provides their living conditions. The significant question in ecology therefore relates to the way in which organisms, with their strict developmental sequence and changing diversity of requirements, actually cope with an environment that imposes itself so diversely, heterogeneously and stochastically. Note that any emphasis on notions of competition and fitness are acceptable only if it is granted that the underlying life processes are achieved readily and without external challenge, premises that are patently not appropriate.

3 ECOLOGICAL DYNAMICS

Organisms have particular requirements in terms of the climatic variables that impose on them. Although they require certain conditions in terms of these variables, they cannot tolerate extremes of these same variables. Their specific requirements and tolerances mean that in a varying environment, they must track conditions that are appropriate to their maintenance. Environmental conditions will inevitably "move", in directions that change periodically and unpredictably, at ever-changing rates, and over distances that also vary. Such shifts may even be discontinuous, with a suitable area of environment becoming unsuitable and another area becoming suitable elsewhere. Individuals track suitable conditions as best they can, and we see that species shift individualistically or idiosyncratically (Fig. 3.1). This process of habitat tracking or environmental tracking therefore entails movement over larger or smaller distances, depending on the locality of the organism and the kaleidoscopically changing dispersion of environmental conditions. Significantly, a given climatic change has different consequences for different species, even though some consistency may be detected across a suite of species at times (see Parmesan 2003). More usually, one species is more sensitive to temperature conditions in a locality, whereas another is more sensitive to rainfall there. Similarly, "spring" species are likely to respond to a particular change in a different way from that in which "autumn" species will respond, and so on. Such different consequences are likely even if the organisms are responding to ecological influences at the same scale of variation.

Sessile organisms also shift in the way just outlined, but with a delay that is represented by the lifespan and generation time of the sessile stage that itself remains in the now unsuitable area. Trees, like the lime tree (*Tilia cordata*) in Lancaster (England) and possibly the Southern or Antarctic Beech (*Nothofagus antarctica*) in Queensland (Australia) (Read and Hill 1985, Taylor et al. 2005), that once produced viable seeds may persist within a locality for thousands of years after the climate has shifted to being unfavorable for seed production in those species (Fig. 4.7). Their persistence may be aided by asexual reproduction (Pigott 1992). The geographical distribution of mature plants may thus, at times, be more extensive than the area in which sexual reproduction takes place regularly.

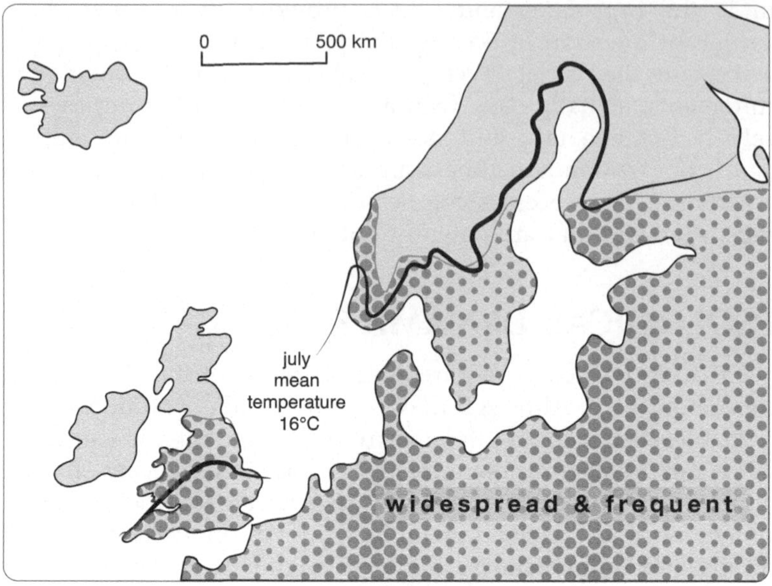

Figure 4.7 Distribution of lime trees, *Tilia cordata,* in northwestern Europe, in relation to the isotherm for mean temperature of 16°C in July, not corrected to sea level. Reproduced from Pigott (1975) and Ford (1982), with permission.

3.1 The Quaternary Perspective—Beetles

Findings by Quaternary ecologists have forced ecologists to confront the geological scale at which they investigate and interpret contemporary ecological phenomena. The spatio-temporal focus in Quaternary ecology extends to the continental level and usually spans either the last 10,000 (Holocene), or the last 100,000 years (late Pleistocene) of the two million years of the Quaternary (e.g., Pielou 1991). The results obtained by Quaternary ecologists at these relatively large spatial and temporal dimensions have demonstrated unambiguously that the issues relevant in ecology – where individuals of particular species occur and their changing abundance – are influenced substantially by processes that take place well outside of the local study area and at time frames far more extensive than previously considered. Furthermore, they have raised serious questions about the rate of adaptive change of organisms and, by implication, about the ecological preconditions that are likely to lead to adaptive change (Walter and Paterson 1994, and see Chapters 5, 6 and 10).

Because fossils are essentially morphological traces, the stability of the physiological and behavioral mechanisms of species has to be inferred. Two arguments support such an inference. One, the different species track subsets of environmental conditions that are appropriate to their survival and reproduction (Coope 1979, Hengeveld 1979, 1990, Huntley 1991), which suggests that they have not undergone physiological change

(Coope 1987a, Huntley 1991). Two, the physiological and behavioral mechanisms of species, which are genetically coded, remain constant across their current geographical range, except for relatively minor genetically-based variations, as in clutch size (Lack 1954), or in age at first flowering, for example (Gray 1997). The spatial stability in these characters is the ecological measure of species stability (see Walter et al. 1984, Walter and Paterson 1994, Rajapakse and Walter 2007). This point is expanded and justified further in Chapter 5, Section 3.

Paleontologists reconstruct past distributional shifts from fossil and sub-fossil material and they can even reconstruct past environments. Sub-fossil insects and plants are preserved in peat and date back to about 120,000 years ago within the Pleistocene (e.g. Coope 1979, Elias 1994). They have proved particularly valuable for paleontological reconstruction (Atkinson et al. 1987), with beetles proving most significant (Elias 1994, Porch and Elias 2000). The hard coleopterous exoskeleton ensures the material retains details of their external structure. Sub-fossils of this nature can also be taken apart relatively easily, to reveal the hard parts of their internal structure.

An unusual feature of beetles is that the hard cuticular structures of the insect penis, or aedeagus, are generally species specific, often having elaborate shapes and projections. The identification, to species level, of fossil insects to this unerring degree is unusual, but it is critical for reconstructing the past accurately (Paterson 1986, Atkinson et al. 1987, Turner and Paterson 1991). For example, representatives of species that were living together in a particular locality in present-day Trafalgar Square (London) about 100,000 years ago can be traced to their present distribution thousands of kilometers away (Fig. 3.1). That each of these species now occurs independently, with each separated from the others by thousands of kilometers, demands answers about how they could have co-occurred in the past, for example. Information of this nature tears at the heart of general ecological models that confine themselves to local density-related demographic processes and issues of local community structure. It suggests several major adjustments to the basic aims in ecology, because these species:

1. remained unaltered, at least morphologically, for periods in excess of 100,000 generations, and
2. shifted their distribution in response to altered conditions, and they did so independently of one another, which implies that the species responded differentially to climate change.

The material covered so far in this chapter puts us in a position to explain why the "Trafalgar Square" beetles once overlapped one another in distribution at one period and then, when climate changed, they "dissociated" and moved independently to remotely disjunct areas. The historical conjunction of the species in England implies that the climatic requirements of each species, as well as their habitat and resource requirements, were once satisfied in that area, and their tolerances were not exceeded. Each species, in other words, was there for different reasons, with each being adapted to, or requiring, a subset of all of the environmental conditions prevailing at that time. The beetles that were once at "Trafalgar Square" were therefore influenced differentially when the environment responded to climatic change, to the extent they followed different pathways and ended up as currently distributed, independently of one another (see also Davis (1981b, 1986b), Graham (1986) and Huntley and Birks (1983), for example). Such a proposition is testable, and a good start has been made with the mutual climatic range method, which determines the common elements in temperature requirements of extant species represented in the fossil record (see Elias (1994, pp. 62-63 and 74ff)).

The equivalent of the situation above can be seen today where the geographical distribution of each of several species falls together in one area (see Fig. 3.2). Because we observe the individuals of a particular species more or less continuously in an area where their required conditions are provided, and because our scale of measurement of climatic variables in such an area tends to exclude influences that to these species would be extreme, we render the influence of climate invisible to ourselves. We are therefore puzzled by the dissociation of the various species when climate changes, because we have imposed the same set of average environmental measurements upon all of them. We thus fail to see the differences in their requirements, but the fact that one species (e.g., white-throated honeyeater) overlaps with particular species in the south of its distribution (white-naped and brown-headed honeyeaters (Fig. 3.2 (2)) and different ones in the north (black-chinned honeyeater (Fig. 3.2 (1)) means we should not be surprised in this regard. Perhaps one species, if we return to the beetles of Fig. 3.1, requires temperature minima that do not go below a certain level if its eggs or first instar larvae are to survive, a second species needs sufficient day degrees between early spring and midsummer to ensure the critical stages of the life cycle are completed within the season, a third persists only if winter maxima are not too low for successful diapause, and the larvae of a fourth require pools of cold meltwater (e.g., *Helophorus lapponicus* (Angus 1983)), a species previously on the British Isles but now evidently restricted to parts of Scandinavia and the Iberian peninsula in western Europe. Furthermore, climate may influence other requirements of the species. For example, the distribution of the specific host species may shift in response to climate change, or vegetation may shift or change in composition or texture, and thus influence reliant species (Hengeveld 1990, Walter and Paterson 1994). Climatic and biogeochemical influences that relate directly to the physico-chemical requirements and tolerances of the species concerned, or indirectly to its biotic needs, are thus of direct ecological importance.

As outlined above, paleontological evidence tells us a great deal about what we need to incorporate into general ecological models. However, it does not specify in any way how to proceed from there, because every species is ecologically idiosyncratic, as the following contemporary examples show.

3.2 Dutch Ground Beetles

Carabid beetles are highly mobile pedestrian predators. The adults of some species run down their prey on the soil surface, whereas others capture prey within cracks and crevices in the soil. Dutch carabids have been examined from two perspectives, each of which indicates the immediate nature of their interaction with the dynamics of their environment. Land that was newly exposed when the IJsselmeer Polders were drained between 1945 and 1968 presented an opportunity to study the invasion dynamics of ground beetles, the distribution of which is relatively well known outside of polder areas (e.g., Lindroth 1949, Turin et al. 1977). From such work, the significance of movement in the ecology of these insects has become clear, as did the primary nature of the relationship between the various species and their environmental requirements. Because the carabid fauna of the Netherlands is so well known, and is relatively well represented in collections, the changing abundance of almost all species represented in the Netherlands could also be tracked, decade by decade for nine decades, starting about 1890. Thus, the immediacy of the ecological impact of even "gentle" climatic changes was demonstrated.

3.2.1 Invasion of New Polders

Drainage of the IJsselmeer Polders exposed a large area (thousands of square kilometers) to invasion by terrestrial organisms and the development of vegetation and a local fauna. Which species established and which did not, and what characteristics did they have? And at what rate did they invade? Invasion precedes establishment, but the latter happened so soon after drainage that rates of invasion for any of the numerous species that did establish could not be measured precisely. Also, from very early on, the kind of immigrant species and the specificity of their environmental (or habitat) requirements and biotope relationships indicated an acute spatial adjustment by mobile individuals at a broad, geographical scale. We know, therefore, that aerial movement of the beetles is a persistent process, at least at the dimension of tens to hundreds of kilometers. There is no reason to believe this movement is directional. Rather, it must be random with respect to the locality of newly favorable localities, and it occurs with high frequency, as indicated by the massive mortality of insects that fly over the ocean (Elias 2010, p. 46). What proportion of all individuals that move in this way fails to locate a suitable place to live because of movement in inappropriate directions or being displaced too far away would be difficult to establish, but does add to the list of criticisms of evolutionary optimization (Pierce and Ollason 1987, Rapport 1991, Walter and Donaldson 1994).

The characteristics of the species that established could be ascertained, at least in relation to the characteristics of all potential invaders (Hengeveld and Hogeweg 1979), a class now called the "regional pool of species" (Ricklefs 1987, Gaston 2003, Mora et al. 2003). Comparison of the Dutch distribution of those species that did successfully invade the polder (Fig. 4.8a) with those species that did not (Fig. 4.8b) shows that the invaders are widely distributed in the Netherlands. In contrast, those that did not establish are distributed just to the south of the polder, on an east-west axis, with some species very close to it, and with some also in the southern extension of the country (Limburg). Distances in these cases could not have inhibited invasion and establishment, because source populations were separated from new habitats by only tens of kilometers. It seems evident that the species' environmental requirements were responsible.

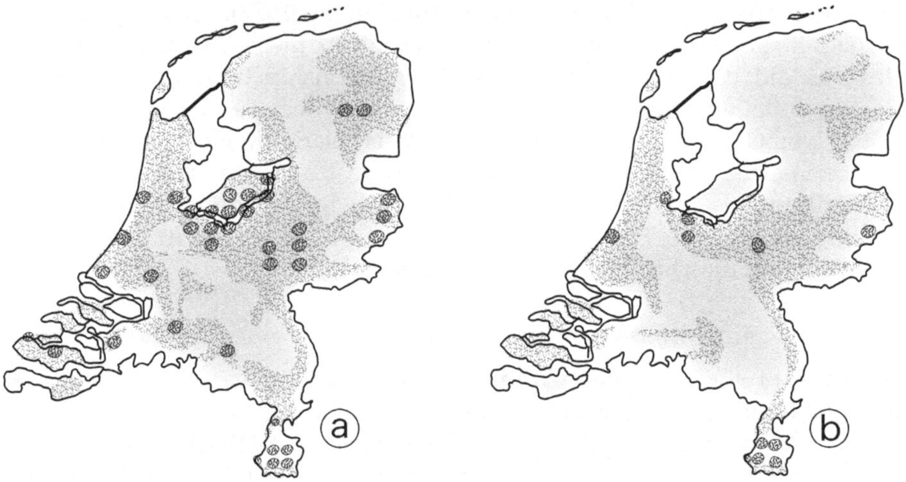

Figure 4.8 Density-prints of the Dutch carabid species that invaded the newly established IJsselmeer Polder area **(a)** and those that did not invade it **(b)**. The invaders did not necessarily come from areas close to the new polder. The darker the shading, the more species that are represented. Redrawn from Hengeveld and Hogeweg (1979).

A cluster analysis of all Dutch species revealed the existence of 10 groups of species, each characterized by particular habitat associations (Hengeveld and Hogeweg 1979). Those that invaded the polder came principally from two groups of species, groups 3 and 4 (Fig. 4.9). Such density prints represent the number of species of a particular cluster per grid square; the darker the mark, the more species of that group that are present. Group 3 comprises species that are widespread across the Netherlands and are associated with arable land. At the time of sampling they were restricted to the northern parts of the new polder, which had already been drained and cultivated. The species in group 4 are associated with marshes, and are distributed mainly in the major river deltas on the western coast of the Netherlands. On the polder they were associated with the still undrained swamps of the southern polder. (Note that in the original publication, group 4 is incorrectly said to comprise dune-inhabiting species, but group 8 represents the dune species, and these never invaded the polder area.) The soil of the polders is mainly clay, which explains the absence of species from other groups, especially those of pasture (2), diluvial sand (6), riverbanks (7), dunes (8), and loess/limestone (9) (Fig. 4.9).

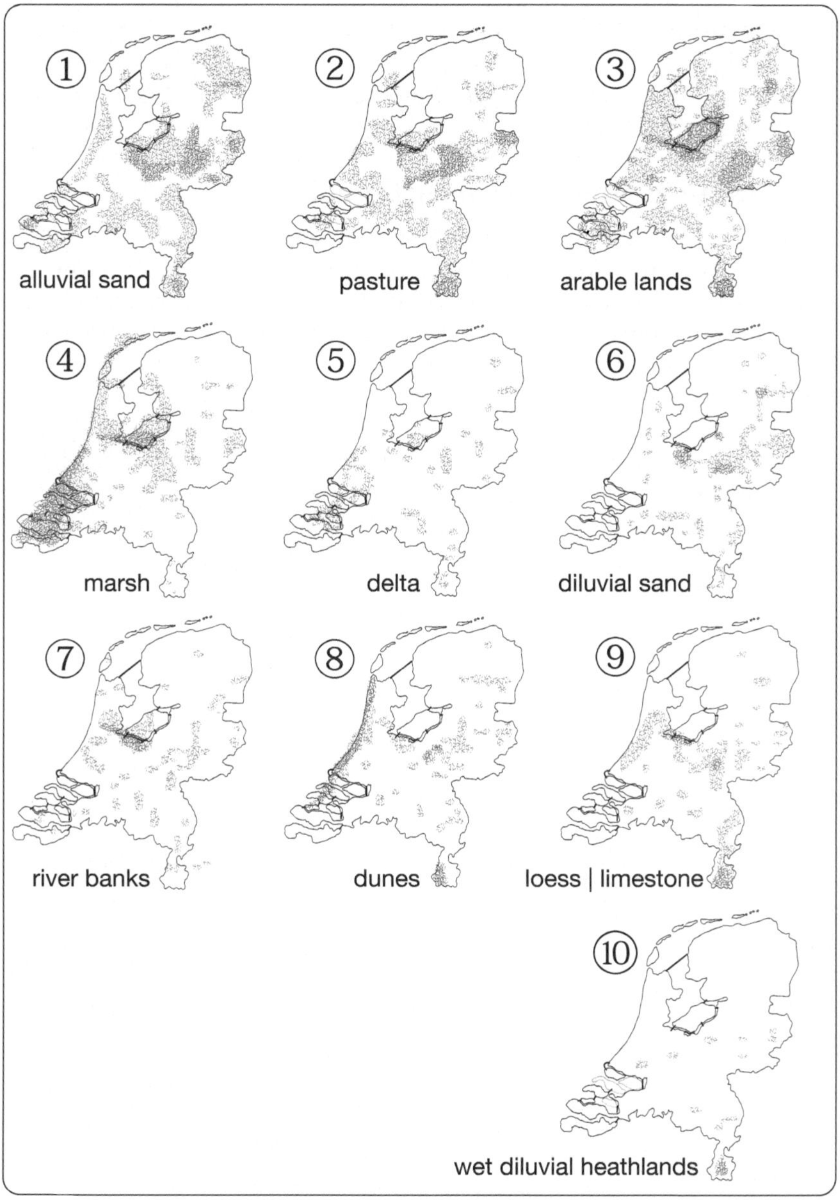

Figure 4.9 Density-prints of the distribution of the ten clusters of Dutch carabid species, to indicate that only those associated with arable lands (Group 3) and marshes (Group 4) invaded the newly established IJsselmeer polders. These clusters are based upon the similarity of the distribution patterns of carabid species within the Netherlands. Redrawn from Hengeveld and Hogeweg (1979).

In summary, edaphic factors were important determinants of which winged carabid species invaded the new polder land. By contrast, accessibility of the new habitat played only a minor role, if any, at the spatial dimensions investigated. For species without wings, the polders as a whole were inaccessible, except where they were directly connected to the mainland. Furthermore, density prints indicated that the beetles do not experience the Netherlands as a homogeneous area (Fig. 4.9), despite its highly cultivated nature.

3.2.2 Changing Abundance over Decades

The abundances of the 365 species that comprise the Dutch ground beetle fauna have fluctuated markedly over the relatively short time frame of the nine decades of 1890 to 1975. No statistical structure was detected among species in their changing abundance pattern across the study period. But when species were grouped according to various of their environmental requirements, statistical patterns became evident, specifically in relation to the species' moisture relationships (Fig. 4.10), usual biotope (Fig. 4.11), and their European distribution (Fig. 4.12) (Hengeveld 1985a). Note in these figures that Axis I has not been considered because it simply accounts for variation in abundance across species, which is expected.

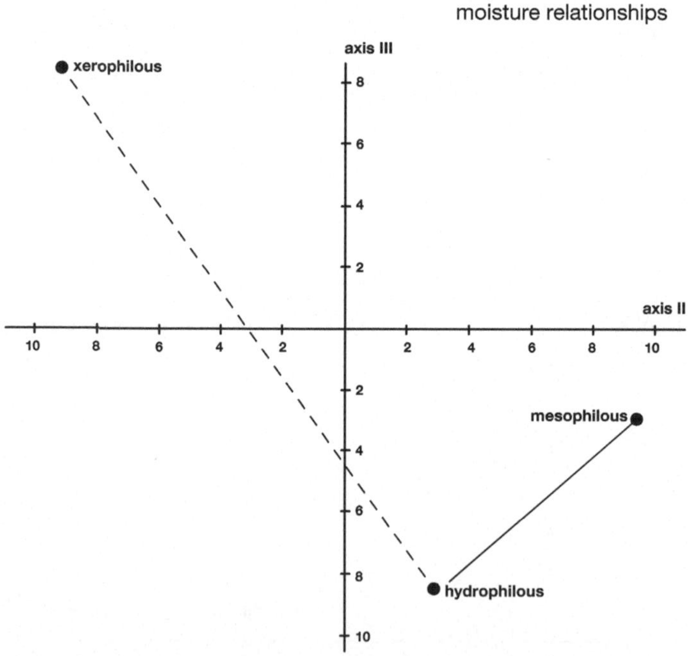

Figure 4.10 Relationship between the moisture preference of Dutch carabid species (xerophilous, mesophilous and hydrophilous) and their position on Axes II and III of a principal components analysis of their changing abundance over the nine decades of 1890 to 1975. The solid line indicates $p \approx 0.25$, and the dashed line $p \approx 0.10$. Redrawn from Hengeveld (1985a).

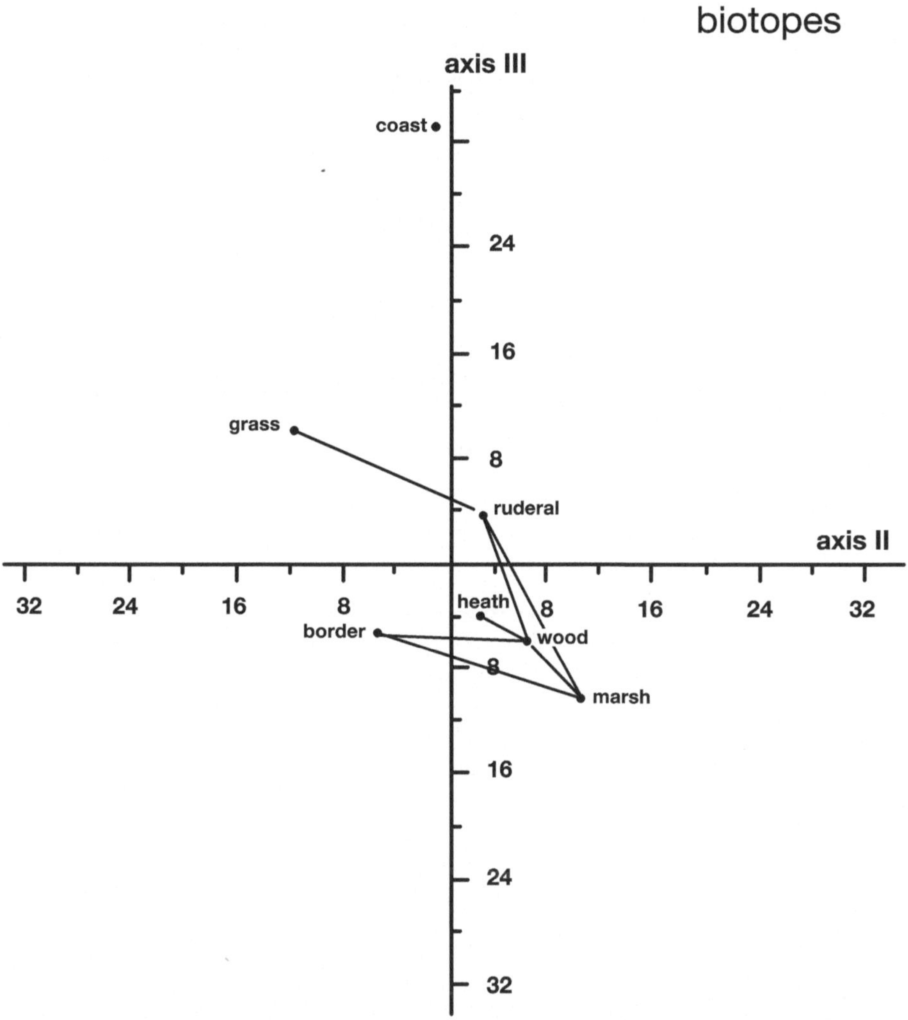

Figure 4.11 Relationship between the biotope usually occupied by Dutch carabid species (including coast or dunes, grass, ruderals, heathland, woodland, marshes and border or ecotonal areas) and their position on Axes II and III of a principal components analysis of their changing abundance over the nine decades of 1890 to 1975. The lines indicate non-significant differences between groups, with $p > 0.05$. Redrawn from Hengeveld (1985a).

The traits represented in Figs. 4.10 to 4.12 are statistically associated with one another, and biologically they are likely to be related through the species' physiological mechanisms and tolerances (Hengeveld 1985a). The various ecological groupings of species showed different patterns to their changing abundances over the study period, which are summarized in Fig. 4.13. This figure proposes that Axis II represents moisture levels varying over time, with relative drought to the left and moist conditions to the right, and that Axis III represents temperature variation, with relatively warm conditions in the decades at the top and cool ones in those below. The groups of species, their ecological characterization and patterns of population change are superimposed.

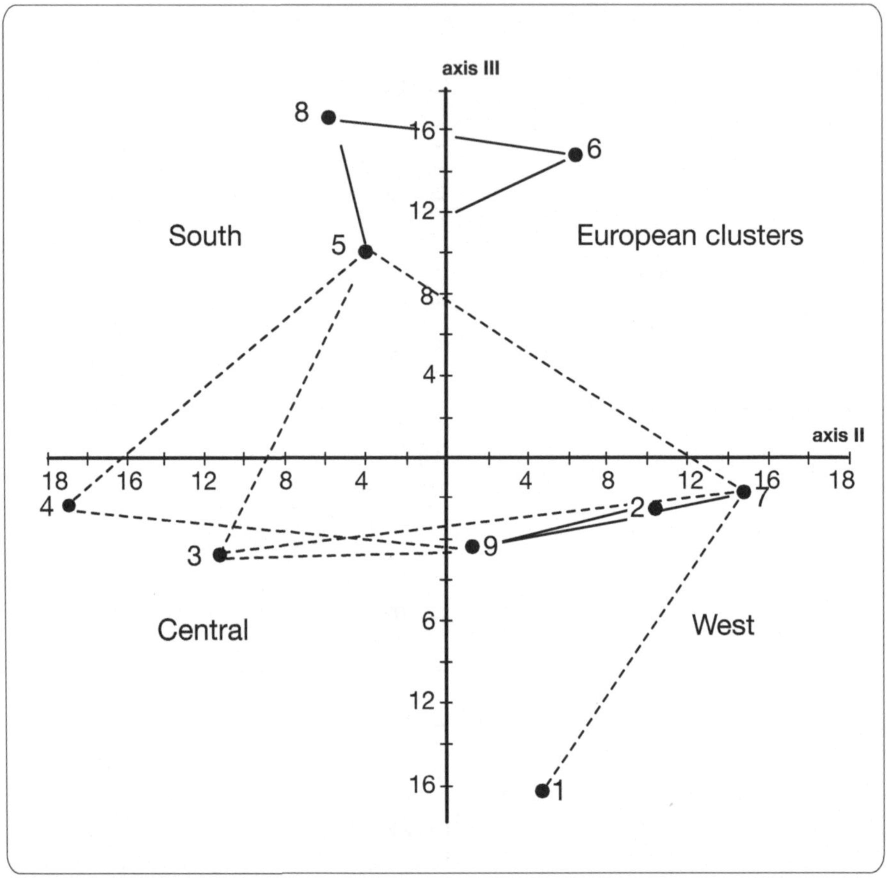

Figure 4.12 Relationship between the overall European geographical distribution of Dutch carabid species (nine clusters emerged) and their position on Axes II and III of a principal components analysis of their changing abundance over the nine decades of 1890 to 1975. The solid lines indicate $p > 0.25$, and the dashed lines $0.25 > p > 0.05$. Redrawn from Hengeveld (1985a).

Those species on the left of Axis II (Fig. 4.13) were most abundant during the first couple of decades; those on the right hand side of Axis II were most abundant during the last few decades of the investigation. Species at the top of Axis III predominated during the 1930's, 1940's and 1950's. By contrast, those at the bottom of Axis III were most scarce between 1930 and 1960. Finally, some species remained at relatively constant levels, and fall at the intersection of the axes. Ecologically, species at the top of Axis III are xerophilous and tend to inhabit open and dry biotopes in southern Europe. By contrast, those at the bottom of the figure are hydrophilous, northern European species of open, wet biotopes. On the left hand side is a group of mainly central European species that inhabit drier grassland and ecotonal areas. Those on the right hand side are mesophilous, inhabit more closed and moist biotopes, and are typically northern European.

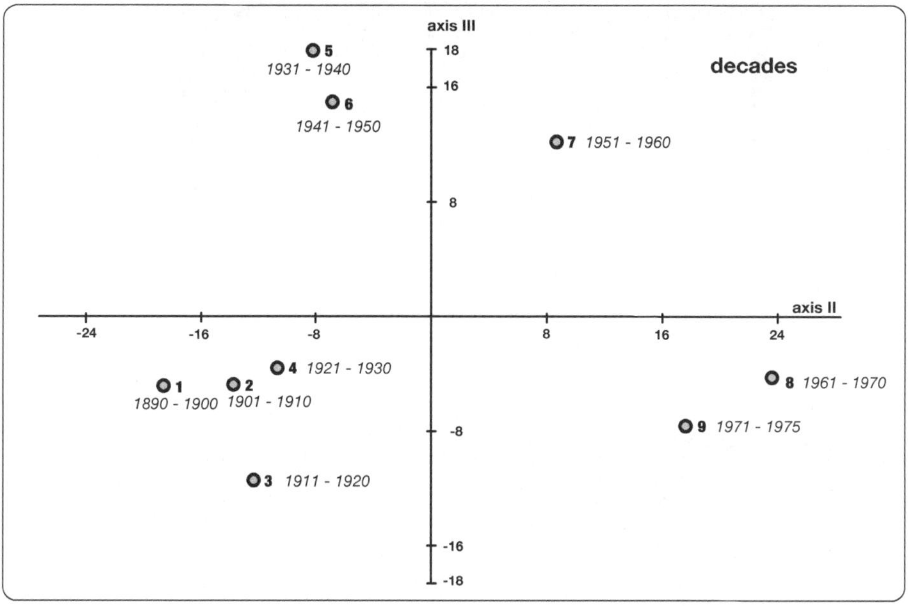

Figure 4.13 Summary diagram to indicate the relationship between the ecological characteristics of Dutch carabid species (Figs. 4.10 to 4.12) and the position of their scatter across Axes II and III. Their position on the axes indicates the pattern of their dynamics across the nine decades of 1890 to 1975 covered by the study. Redrawn from Hengeveld (1985a).

The above categorizations, developed from an understanding of the known ecology of the species, together with the pattern of the changing abundance of each grouping, matches the recorded changes in temperature and rainfall during the period under investigation. Figure 4.14 shows a gradual rise in spring and summer temperatures between 1890 and 1975, with a low temperature trough in the 1960's. Rainfall varied less than mean temperature (Fig. 4.15), but values tended to be higher in the second half of the investigation period, suggesting a trend toward wetter weather. The precise variable responsible for any observed change in abundance cannot be extracted from the data, and even variables like the duration of the growing season across years could influence abundance of species. The growing season varies in duration across the study period, but was generally longer than average from the 1930's to the 1960's (Fig. 4.16).

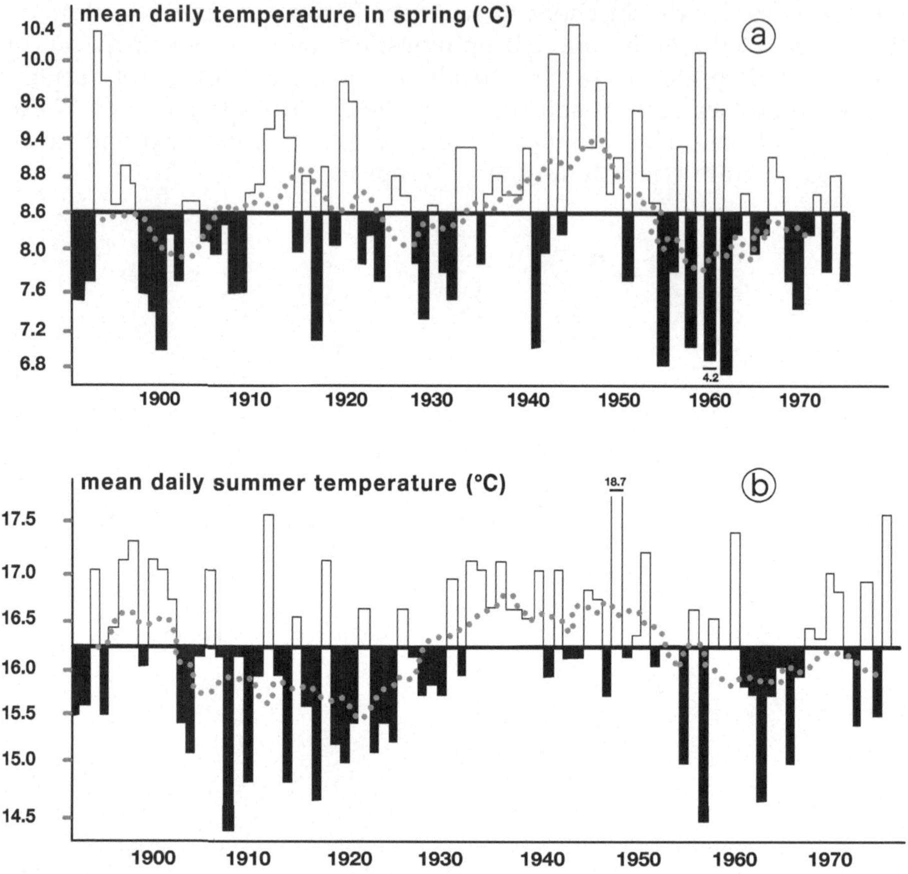

Figure 4.14 Mean daily temperatures in spring **(a)** and in summer **(b)** in each year for the period 1890 to 1975, and which pertain to the results presented in Figs. 4.10 to 4.13. The white columns represent above average values, and the black ones below average values. The dots represent values of the 10-year moving average. Reproduced, with permission, from Hengeveld (1985a).

The trends identified in the spatio-temporal dynamics of the Dutch carabids have not been fully resolved. Indeed, this cannot be done with any certainty for large numbers of species, because each would have to be investigated in its own right. However, the similarities that are evident in the dynamics of the species in the various groups suggest that shifts in climate have had a primary influence in the rise and decline of the various species. More recent compilations undoubtedly show the effect of climate change in inducing shifts in species distributions, but these have sometimes excluded habitat effects to target climate unambiguously (Parmesan 2003).

Changes to the environment by humans have been suggested as an alternative to climate as the primary driver of the changes in distribution of the Dutch beetles (Turin and den Boer 1988, Desender et al. 1994). Although anthropogenic alterations to the landscape are likely to have

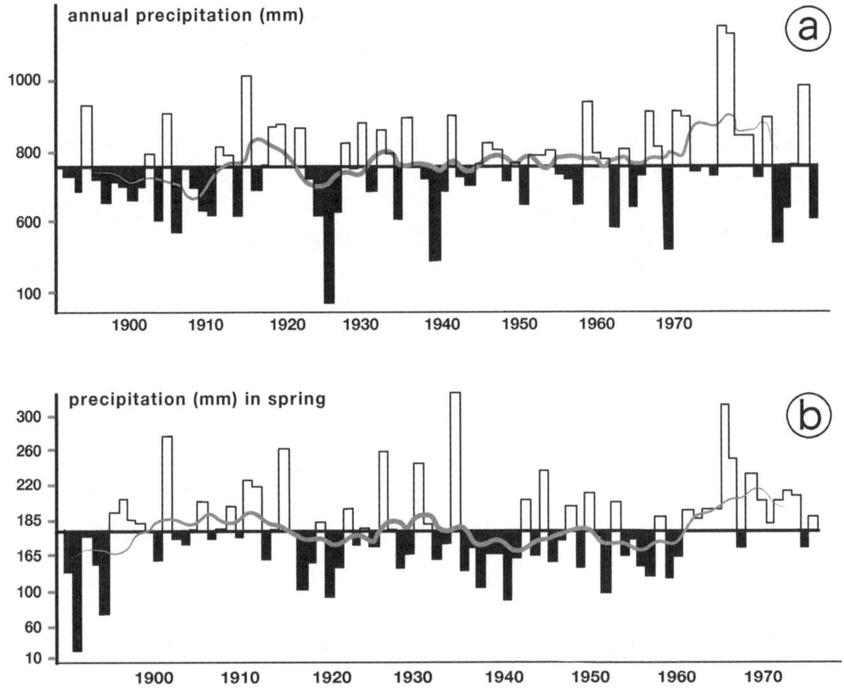

Figure 4.15 Total precipitation (mm) for the entire year **(a)** and in spring **(b)** in each year for the period 1890 to 1975, which pertains to the results presented in Figs. 4.10 to 4.13. The white columns represent above average values, and the black ones below average values. The dots represent values of the 10-year moving average. Reproduced, with permission, from Hengeveld (1985a).

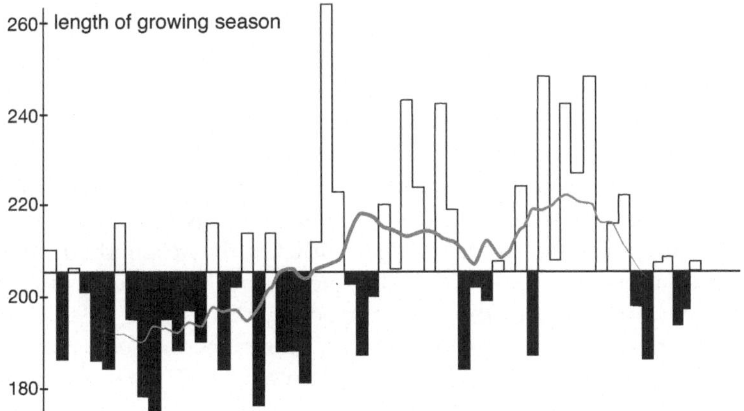

Figure 4.16 Duration of the growing season in each year (as indicated by data from Scotland), from 1914 to 1970, and which pertain to the results presented in Figs 4.10 to 4.13. The white columns represent above average values, and the black ones below average values. The dots represent values of the 10-year moving average. Reproduced, with permission, from Hengeveld (1985a).

influenced beetle distribution and dynamics, they cannot account entirely for the patterns evident in the recorded dynamics (Fig. 4.13). In particular, the fauna associated with the 1960's and 1970's, a generally cooler and wetter period, was different from the fauna before that period. And the return of the "dry, hot" fauna with reversal of conditions after the 1970's demonstrates that climate is more important than anthropogenic factors, because the latter have caused a unidirectional decline in environmental conditions. Note also the rapidity of the response of the total fauna to a change of only five years' duration, from 1970 to 1975. Furthermore, climate change has been the driving process in massive alterations to the distribution and faunal composition of beetles in sites across the world (Fig. 4.17), with transition times between cold and warm adapted faunas being only a few centuries in duration and possibly just decades (Coope 1987a, b, Ashworth 1997).

In relation to the principal argument of this section, no stable point can be specified for the species considered. There is no equilibrium that can be identified, and change in environmental conditions and the spatio-temporal dynamics of species is the norm.

3.3 Beech Trees in the Holarctic

Beech trees in North America (*Fagus grandifolia*) and Europe (*F. sylvatica*) have been separated from one another for a long time, something between 25 and 10 My BP (Huntley et al. 1989). The continental distribution of these two species has been tracked over the past 12,000 years on both continents. Both have shifted as Holocene climate has changed. Several lines of evidence show clearly that climate has played a primary forcing role in dictating the continental level distribution and abundance of the species. These include isopoll maps (Fig. 4.18), which are related to known climates (Huntley and Webb 1989), and surface response analysis (Huntley et al. 1989), which relates distribution and representation (as a measure of relative abundance) to climatic variables (Fig. 4.19).

Of direct relevance to this section is the rate of change of the distribution of the species. Independent studies of these two *Fagus* species, as well as a third, the Japanese beech, *F. japonica,* indicate that their distribution has shifted at an average rate of 150 to 300 meters per year. Such rates are not high by comparison with other plant species, for average rates of as high as 1500 m/yr, and even 2000 m/yr, have been estimated (Huntley and Webb 1989, Huntley 1991). However, Bennett (1997) has suggested an alternative explanation. The chance of finding pollen grains from trees in small, isolated pockets is very small, so the absence of pollen grains does not exclude the possibility that the species is present, but is sporadically distributed and at low density. Should a widespread, but rare, species become more abundant, following northward "improvements" in climate, the pollen record would indicate the rate at which our detection of the

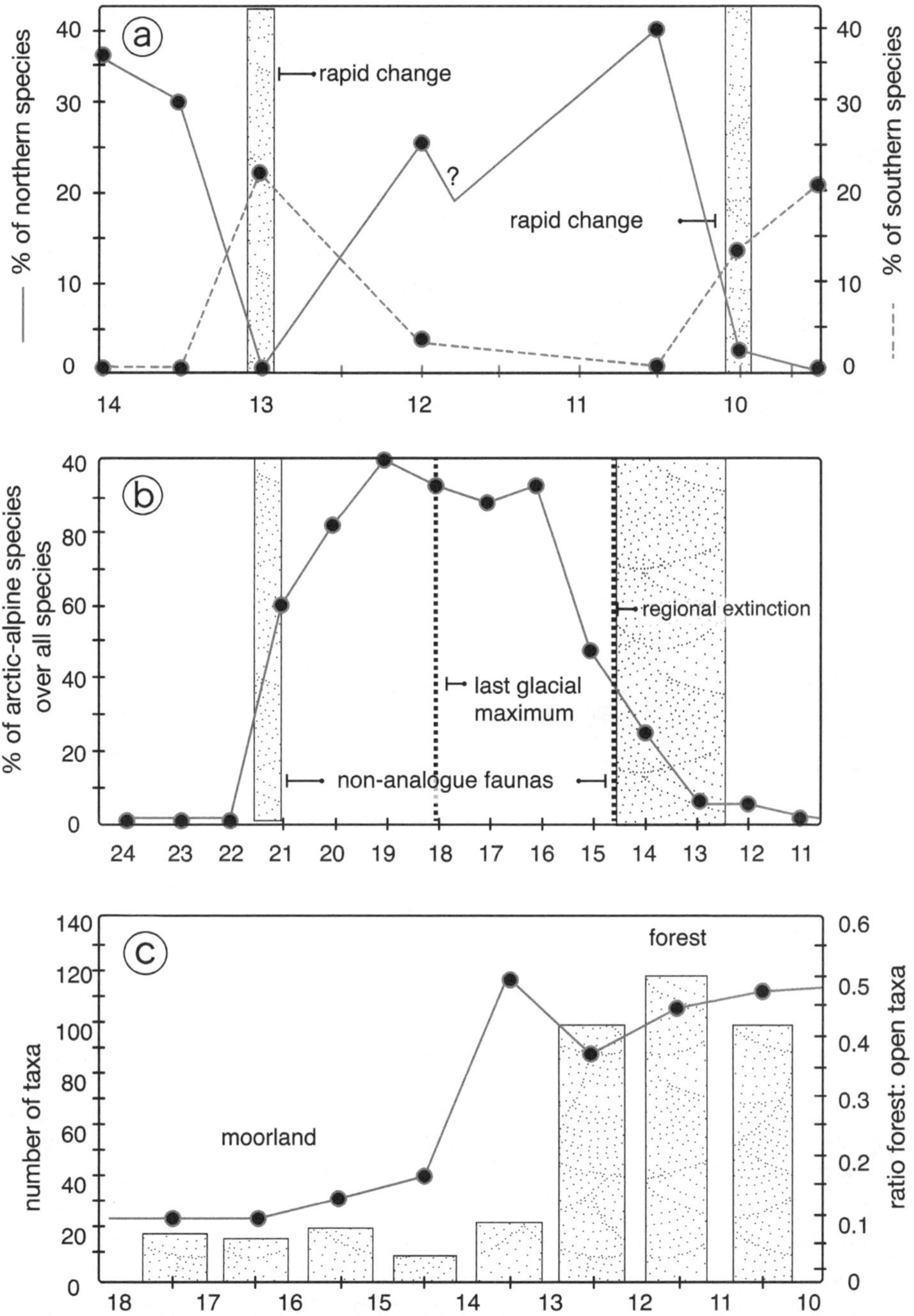

Figure 4.17 Response of beetles to climate change at the end of the last glaciation, as measured by the percentage of species or number of taxa. **(a)** British Isles. **(b)** Midwest of North America. **(c)** Chilean Lake Region. The x-axis represents thousands of years before present. Reproduced, with permission, from Ashworth (1997).

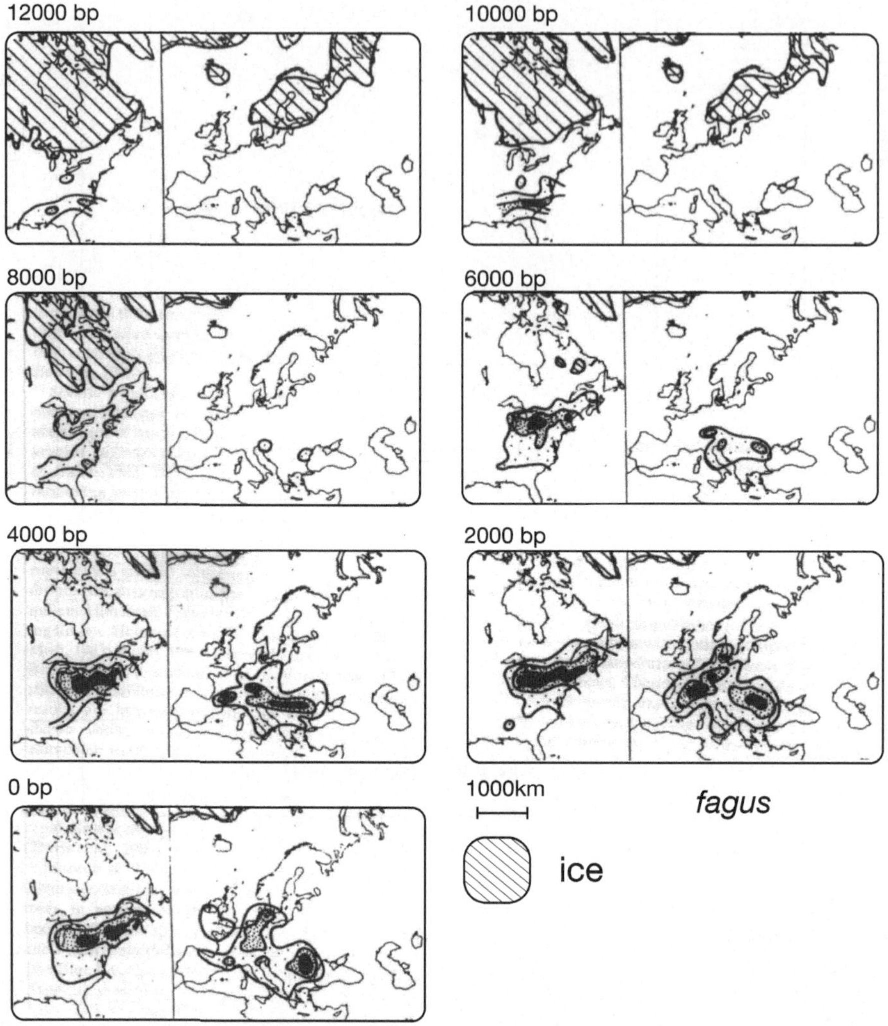

Figure 4.18 Isopoll maps that illustrate the changing distribution and abundance of North American (maps on left in each column) and European (maps on right in each column) beech trees, *Fagus grandifolia* and *F. sylvatica*, respectively. The isopolls are drawn at 1, 5 and 10% (of sum of all major terrestrial pollen types) for North America, and 2, 10 and 20% for Europe. Reproduced, with permission, from Huntley et al. (1989).

species shifts with climate, whereas its own rate of spread, from the isolated pockets, would be much faster. In either case, we are dealing with systems that are not at equilibrium in relation to the density of conspecific individuals or in relation to species composition (community structure).

The spatial dimension to which the data on distributional change has been resolved is relatively large, namely the continental level. The temporal dimension of resolution is at thousands to tens of thousands of years, rather than at the successional level of vegetation or soil development, which varies at a dimension of hundreds to thousands of years. Thus, processes operating at the successional level should not affect the pattern of migration extracted from the data (Huntley and Webb 1989). However, other factors do play an important local role in the ecology of the organisms, including vegetation dynamics, presence and abundance of other species, and physical disturbance, but these effects do not show up over the larger areas sampled by palynologists (Huntley et al. 1989).

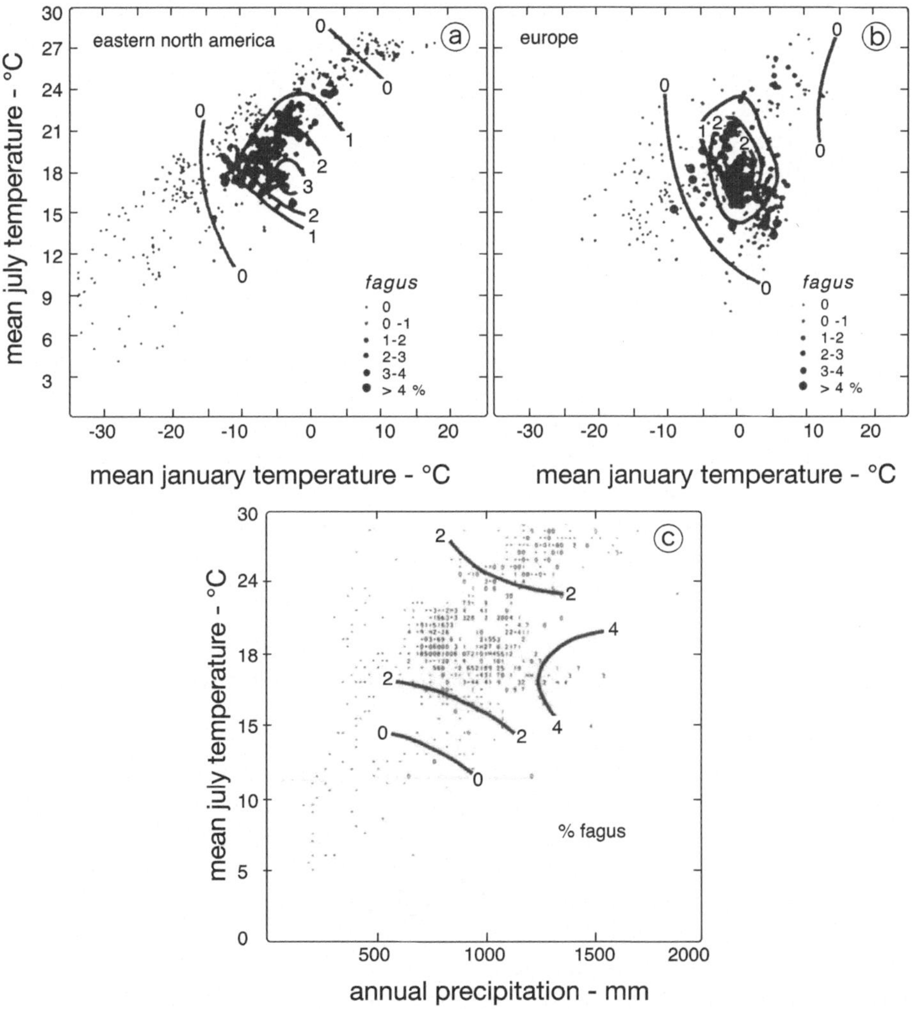

Figure 4.19 Various response surfaces of beech (*Fagus*) pollen distribution and abundance in relation to environmental variables, mean July temperature, mean January temperature and annual precipitation. **(a)** and **(b)** plot data for North America and Europe, respectively, and deal with temperature data alone (reproduced, with permission, from Huntley et al. (1989)). **(c)** plots North American data and deals with summer temperature and rainfall (Reproduced from Bartlein et al. (1986), with permission).

That the palynological data do not carry evidence of the various non-climatic, local influences could be taken to imply that climatic influences vary at such a large dimension that they can be ignored in studies of local ecology. Such an argument has been used to justify that local studies are equilibrium studies in which all else can be taken as constant or equal (e.g., Pimm 1991). Such an inference is mistaken, for even the epicenter of the distribution of species has shifted as climate has shifted. Furthermore, each species is likely to have several centers of distribution (Huntley and Webb 1989), which each rise and decline and shift relative to one another (Fig. 4.18). Such "fragmentation" resolves at smaller dimensions than the continent-wide one, but carries important implications for considering local ecology and even the future of local conservation efforts (e.g., Ford 1982).

Over the short term, such shifts as those described above are, to us, often virtually imperceptible, but clearly the physiological processes of individuals have been affected directly. For individual plants and their propagules, even those as large and immobile as trees, such influences are clearly substantial. Survival, reproductive output, and effective dispersal distance are all likely to be affected. The local dynamic must therefore be much faster than the rates implied by the broad-scale study, and also faster than anticipated by equilibrium models. Demographic models in ecology were never designed to cope with the sources of variation outlined in this chapter. The *ad hoc* adjustments that have been made to the theory (Chapters 7 and 8) cannot, therefore, do justice to dynamics of this nature. Again, the autecological emphasis on individuals and their species-specific adaptations and the spatio-temporal dynamics of the environment has been designed specifically to deal with the idiosyncrasy or individuality of species and their dynamics.

3.4 Spatio-temporal Perspective on Breeding Bird Densities

Bird densities have been counted across numerous breeding seasons in various parts of the world. To get an accurate impression of what these data sets reveal, the information should be examined across species and localities. The only generalization possible with regard to such data sets is that densities of the various species change independently within localities, and they do so across localities as well. Selecting a single species to demonstrate local stability is not an accurate reflection of what happens species-wide in the ecology of organisms. Therefore, reference to the song thrush data reproduced in Fig. 4.20 as representative of stability (Pimm 1991) is misleading. We present several bodies of data to indicate why the local ecology of birds and, by inference, other organisms, cannot be treated in terms of equilibrium and independently of physico-chemical variables. In other words, all else cannot be treated as equal in ecology, a major flaw in demographic models.

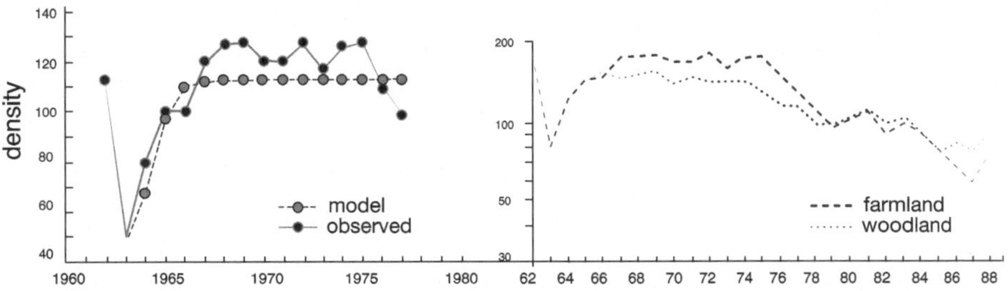

Figure 4.20 Abundance of the song thrush (*Turdus philomelos*) from British counts. **(Left)** represents the time series available to Pimm (1991), which suggested the "population" was stabilizing. **(Right)** is the longer time series that became available subsequently, and which indicates, instead, a pattern of "long-term decline" (Marchant et al. 1990), a trend that has continued since (British Trust for Ornithology (BTO) 2012). Reproduced with permission, thanks to BTO.

First, additional information on the abundance of song thrushes in Britain is available, and the longer time series presented by Marchant et al. (1990) (Fig. 4.20) shows a somewhat different picture from the one Pimm (1991) originally presented. The species has been in long-term decline, although the picture is complicated by anthropogenic influences (Peach et al. 2004). Furthermore, the image obtained from such graphs lacks any spatial element, because numbers derived from various localities have been combined. Inevitably, different localities will show different patterns of changing abundance (e.g., see the extension of the data set for this species to 2012, which shows that numbers have remained low despite a slight increase recently, and the different trends detected in different areas (British Trust for Ornithology 2012)). Spatial separation of abundance data for the wren in Britain (Fig. 4.21) illustrates the point graphically, although the effects of mortality cannot be separated from those generated by movement (Cawthorne and Marchant 1980).

Greater spatial resolution would add to the perspective of the changing intensity of local occurrence, and simultaneously across localities changes in ecological intensity may well take place in opposite directions. In other words, one cannot automatically consider that any particular study site is within the center of abundance of the species studied, where density-related processes are often assumed to be more likely to be manifest than elsewhere. The distribution of a species is most realistically considered in terms of a shifting statistical distribution that comprises many centers of abundance, and with their status relative to one another changing continually (e.g., Fig. 4.18).

So far, we have dealt with single species, but to complete the argument we draw on data from numerous bird species. Many species show changes in distribution and abundance, as illustrated by two data collections that are summarized in Tables 4.3 and 4.4. The first table shows that in Britain, some species have been in general, gradual decline, whereas others have been increasing gradually. Yet other species showed a sudden recent

Table 4.3 Summary of the pattern of abundance of breeding British birds. Each species was assigned to a general category on the basis of the summary statement made about it by Marchant et al. (1990). Such variation in trends across species is evident also in more recent data sets (British Trust for Ornithology 2012).

ABUNDANCE OVER TIME	*NO. OF SPECIES*	*PERCENTAGE*
General Stability	17	16.2
General Increase	18	17.1
General Decline	13	12.4
Ongoing Variation	15	14.3
Recent Increase	12	11.4
Recent Decline	25	23.8
No Reliable Information	5	4.8
TOTAL	**105**	**100**

Table 4.4 The number of breeding bird species, specified as typically Northern or Temperate in occurrence, that showed a change in distribution coincident with climate change (specified as warming or cooling). The number of species is also given by time period, 1900 to 1950, and since 1950. Some species appear in both time classes, so the total of these two columns is always greater than the number of species involved. The direction of the distributional change is specified, and is classified as an advance or retreat relative to the Northern/Temperate status of the species. Data summarized from Appendices 2.1 to 2.8 in Burton (1995).

TEMPERATE OR NORTHERN	*SHIFT*	*CLIMATE CHANGE*	*NO. OF SPECIES*	*NO. OF SPECIES BY TIME*	
				1900-1950	**1950-1995**
Temperate	Advance west	Warming	64	44	51
Temperate	Retreat south	Warming	32	31	15
Temperate	Retreat east	Warming	7	7	5
Temperate	Retreat south	Cooling	24	2	24
Northern	Retreat north	Warming	52	51	7
Northern	Advance south	Warming	9	9	7
Northern	Advance south	Cooling	55	6	55

decline whilst others simultaneously increased. Only a few can be said to have a pattern of abundance that indicates stability. Such summaries can be taken only as indicative of there being no average situation that can be generalized across all, or even many species. Allotting a species in this respect to a particular category is not an easy thing to do, as the combined presentation (Fig. 4.22) of the abundance data for several species and the published description of that data shows. Table 4.4 presents similar data, but on a geographical scale, showing that many species have shifted substantially in their distribution (Fig. 4.23). Despite the species-specificity of such responses by organisms, there remains a tendency to generalize in a single dimension. When consistency is seen across many species (as in

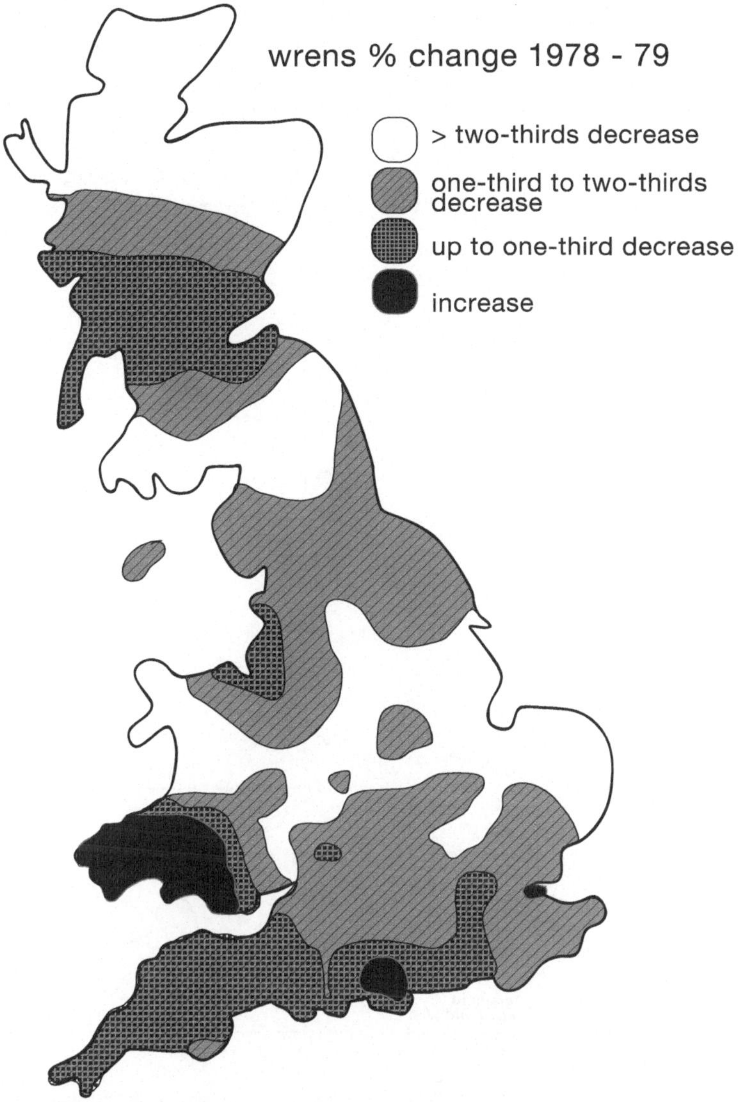

Figure 4.21 Plot of percentage changes in numbers of wrens (*Troglodytes troglodytes*) across England, Wales and Scotland during the 1978/1979 winter. Reproduced, with permission, from Cawthorne and Marchant (1980).

the lepidopteran and other studies reviewed by Parmesan and colleagues (Parmesan et al. 1999, Parmesan and Yohe 2003, Parmesan 2006)), care is needed in interpretation because such coincidences have often misled ecological understanding.

The argument is mounting that the primary forcing mechanism of the recorded distributional shifts and abundance fluctuations is climate change (Hengeveld 1985a, Huntley et al. 1989, Huntley and Webb 1989, Huntley 1991, Preece 1997, Parmesan et al. 1999, Parmesan and Yohe 2003).

Whatever the outcome of such discussions, the image one gets from the above analyses of the local ecology of even familiar species, is one of ongoing change in abundance, and even in geographical occurrence. To consider all of this change as representing a return to stability, or equilibrium, raises the question of what stability actually is (Grimm 1996, Grimm and Wissel 1997), and how it could possibly be defined to accommodate the perspective of ecology we are advocating, on the basis of the examples we have presented.

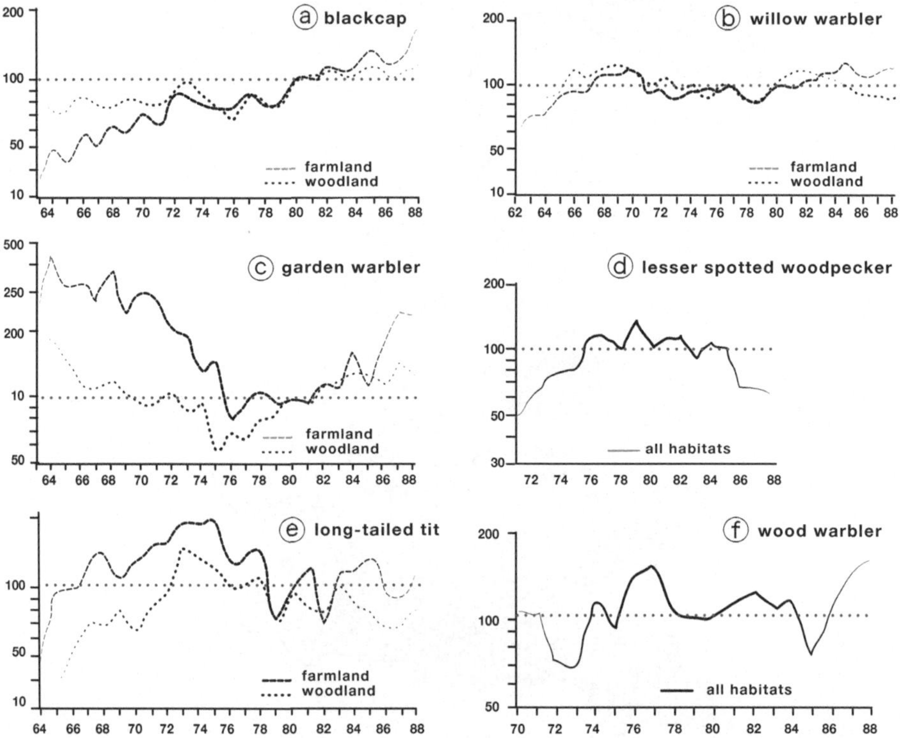

Figure 4.22 Comparison of abundance changes across a selection of species. All reproduced, with permission, from Marchant et al. (1990), and comments on more recent trends come from BTO data (British Trust for Ornithology 2012). **(a)** General increase in the blackcap (*Sylvia atricapilla*), which has increased since 1990. **(b)** General stability in the willow warbler (*Phylloscopus trochilus*), which has declined since 1990. **(c)** Recent increase in the garden warbler (*Sylvia borin*), which has remained steady but with variation. **(d)** Recent decrease in the lesser-spotted woodpecker (*Dendrocopos minor*), which has since been in steep decline. **(e)** General fluctuation in the long-tailed tit (*Aegithalos caudatus*), which has increased steadily since 1990. **(f)** The hazards in "classifying" or "typing" the population trends is illustrated with the data pertaining to the wood warbler (*Phylloscopus sibilatrix*), which has been said to have exhibited "no recent change", but has since been in steep decline.

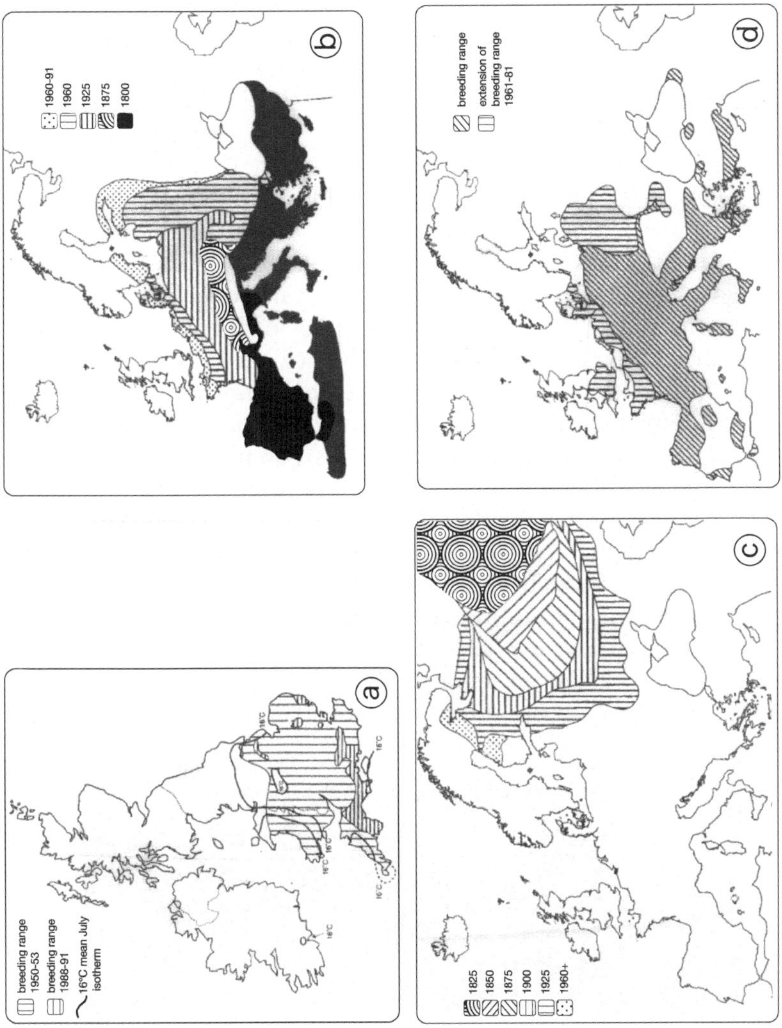

Figure 4.23 Recent major distributional changes recorded in birds. **(a)** The southward retreat of the woodlark (*Lullula arborea*) in England and Wales. **(b)** Northward expansion of the serin (*Serinus serinus*). **(c)** Westward expansion of the yellow-breasted bunting (*Emberiza aureola*). **(d)** Extension, mainly eastwards, of the firecrest (*Regulus ignicapillus*). Reproduced, with permission, from Burton (1995).

3.5 Tropical Systems

Much of the data presented so far in this chapter comes from temperate areas. Many may feel, though, that the tropics are more stable ecologically, and that ecological generalizations based upon the concept of stability are therefore applicable there. However, the temptation to categorize geographical regions for purposes of generalized ecological interpretation should be avoided. On the one hand, this subverts the search for general underlying principles. On the other, the ecological data that are available indicate that the tropics also experience considerable variation and that it is ecologically significant (e.g., Young 1982, Wolda 1983, Walsh 1996, Walsh and Newbery 1999, Potts 2003).

Tropical Marine Systems

Tropical oceans are dynamic and variable. The Pacific and Indian Oceans, for example, are subject to regularities in atmospheric pressure fluctuations, termed the Southern Oscillation. Although regularities in climatic and oceanic conditions follow, we also see the intrusion of what are seen as extreme anomalies, the most famous of which is called El Niño. The longer persistence of hotter than usual sea surface temperatures off the South American coast prevents the usual nutrient-rich cold upwellings from reaching the surface. The fish that rely on this usually cyclical provision of food are thus negatively affected, and therefore piscivorous bird numbers and fisheries also decline (Caviedes and Fik 1992). Extreme El Niño events even extend to the Pacific coast of North America, where they have been implicated in negative influences on salmon fisheries through effects on physiology, survival and migration patterns (Miller and Fluharty 1992), and even on the local diversity and abundance of cetaceans (Neumann 2001, Benson et al. 2002). El Niño anomalies affect the western Pacific as well, and even the strength of ocean currents is affected.

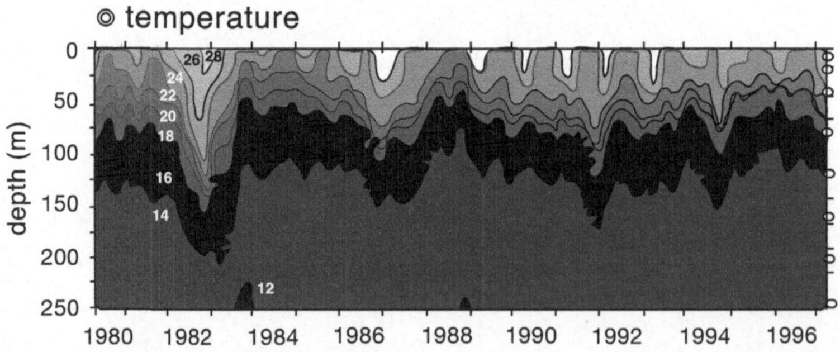

Figure 4.24 Pacific Ocean sea surface temperatures over time and at different depths at a single Pacific Ocean locality (0°, 110°W), at intervals of 2°C. Redrawn from Yu and McPhaden (1999a).

The implications for the ecology of marine organisms are clear. Pacific sea surface temperatures vary across seasons and years and temperature depth contours also vary (Fig. 4.24). Again, we see the stochastic imposition of environmental conditions on organisms, which are sensitive to temperature, salinity and pressure. The stochasticity relates to El Niño events, which themselves are subject to stochastic influence in terms of both onset time and region, as well as in intensity (McPhaden and Yu 1999, Yu and McPhaden 1999a, b).

These oceanic systems, despite any regularities that may be detectable, experience severe regional and temporal disruptions, because of external influences on various forcing variables. Cloud cover, for example, affects radiative heating and thus the amount of energy stored in a particular oceanic system. In turn, geographical weather patterns are altered. Associated with El Niño events, the extreme variability and anomalous warming of the sea in the western Pacific is so comprehensive that the question of the extent, regularity and predictability of climatic cycles, in general, has not yet been fully resolved because of the "awe-inspiring complexity of the climate" (Burroughs 2003).

Another source of variation is the interaction between weather systems. The Indian Ocean, for example, besides being influenced by the Southern Oscillation is also seasonally monsoonal, with these systems being driven by the differential between land and sea temperatures. Undoubtedly, the monsoons are affected by variation in El Niño-Southern Oscillation events, thus introducing another source of stochastic change. Understandably, monsoons vary in intensity, geographical extent, and duration. For example, the monsoonal system is bounded in the south by the Inter-Tropical Convergence Zone, and this typically varies across five degrees of latitude (10 – 15°S), with all its consequences for the African biota, including terrestrial species. Even the two inter-monsoonal periods that take place each year, and which are relatively calm, experience conditions that are variable.

Seasonal wind changes influence ocean currents, as well as oceanographic conditions. Consider tuna fish to get a feeling for how this affects organisms. Each tuna species has a specific temperature optimum, so the shape of the interface between the thin upper layer of warm water and the cold deep water, called the thermocline, affects them differentially as a consequence. And so does the depth of the thermocline. This, again, varies considerably across space and through time, which, in turn, defines the southern distribution of the tuna species, a limit that is, consequently, dynamic (Sharp 1992). Further, the deeper the thermocline, the less the fish are constrained in the water column, so spatio-temporal variation in these physical variables accordingly affects the behavior and hunting of the fish.

Tropical Terrestrial Systems

Terrestrial systems in the tropics are, in turn, affected by marine influences, perhaps mainly through altering patterns and intensity of rainfall. Even the most equable of tropical climates, those where lowland rainforest grows, are thus influenced. Dry-season rainfall events affect the abundance of some organisms (e.g., Wolda 1978a, b, Young 1982), and not only those reliant on vegetation for food (Fig. 4.25). For example, diapause-stage Lepidoptera of some species may be harmed by such rainfall events (van der Laan 1959, Ito 1961). Variation in the distribution and intensity of rainfall events (Fig. 4.26), as well as the onset and duration of wet and dry seasons all affect organisms, and this differentially for different species (e.g., Walsh 1996, Walsh and Newbery 1999, Potts 2003), as shown by the different seasonalities and population trends across several herbivorous beetles in "unseasonal" tropical lowland rainforests (Fig. 4.27).

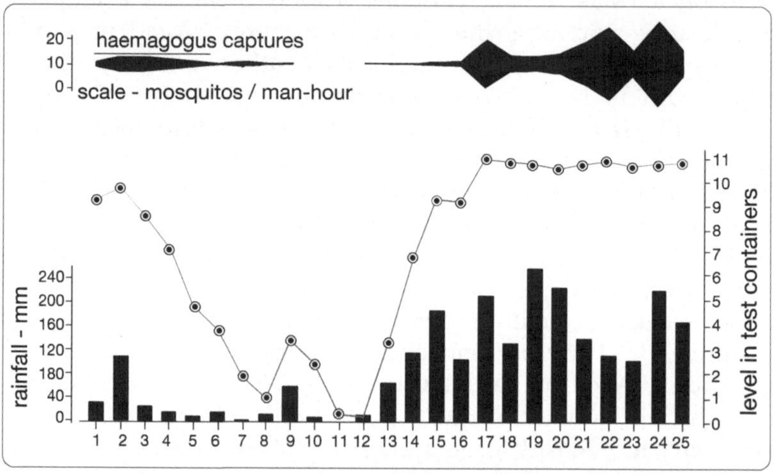

Figure 4.25 Weekly rainfall (bars), water level in experimental containers (line) and catches of the mosquito *Haemagogus capricornii* (kite diagram) in Colombia. Reproduced, with permission, from Young (1982).

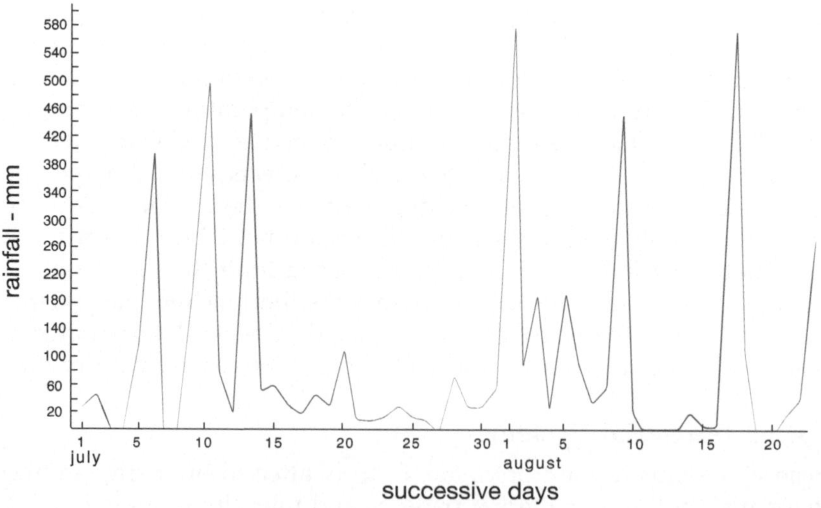

Figure 4.26 Daily variation in rainfall during the wet season, in a lowland rainforest in Costa Rica. Such variation in rainfall affects the activity of many insects, including butterflies and cicadas. Reproduced from Young (1982), with permission.

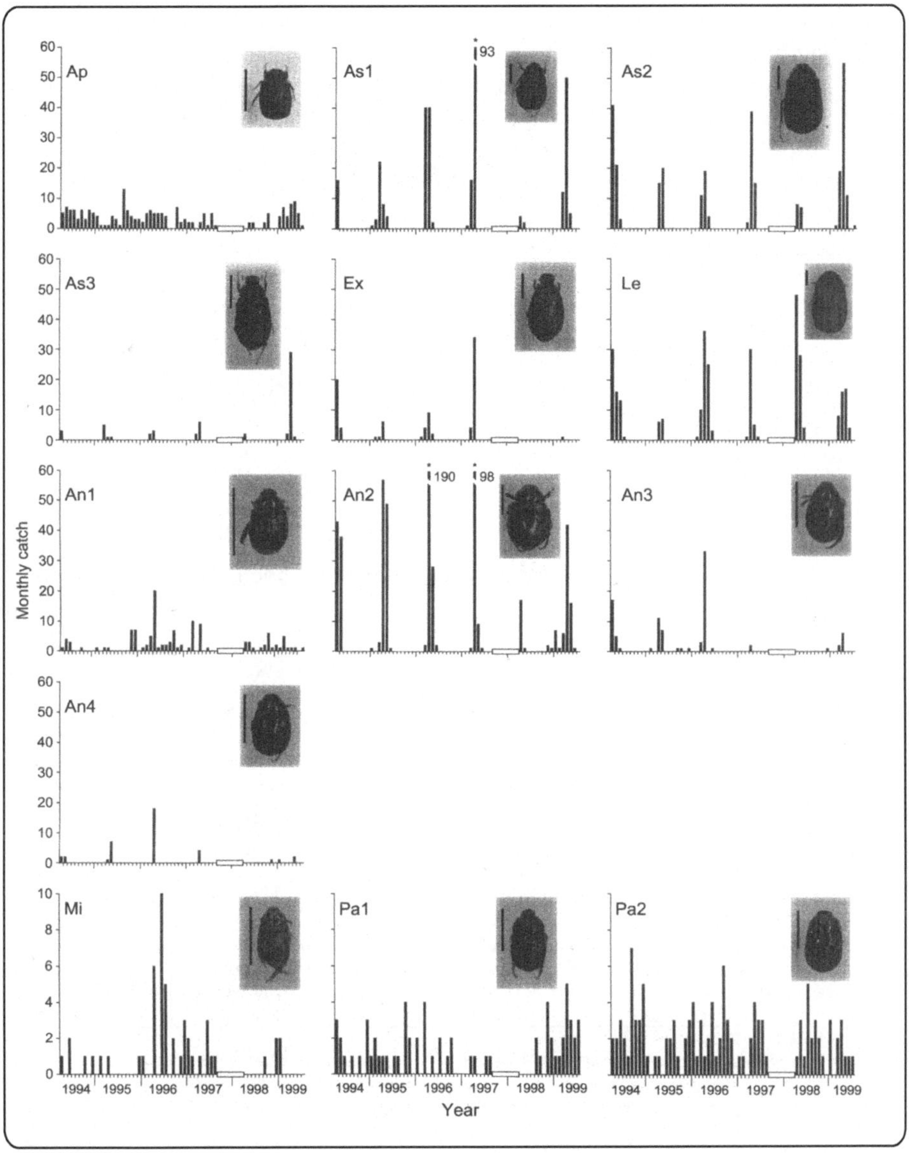

Figure 4.27 Population trends and seasonality of 13 species of herbivorous beetles in two subfamilies (Melolonthinae and Rutelinae) in a lowland Bornean rainforest that is commonly considered aseasonal. Reproduced, with permission, from Kishimoto-Yamada and Itioka (2013).

Climatic influences in lowland forests also vary spatially. Periods of peak rainfall, for example, differ regionally in response to local wind patterns and topography (Schwerdtfeger 1976). And even if regions are similar in their total annual rainfall, they are still likely to differ in their monthly distribution of rainfall, only in part because of differences in elevation

(Fig. 4.28). Features of the dry season also vary in the predictability of their beginning and termination dates (Young 1982). Cyclonic influences are stochastic and vary in frequency, intensity, and in the area they affect. Although these are often seen as disturbances (e.g., Pickett and White 1985), they are a normal part of virtually all environments. They alter reef structures and the ecology of reef organisms in the sea, as well as rainforest structure and composition on land (Connell 1979).

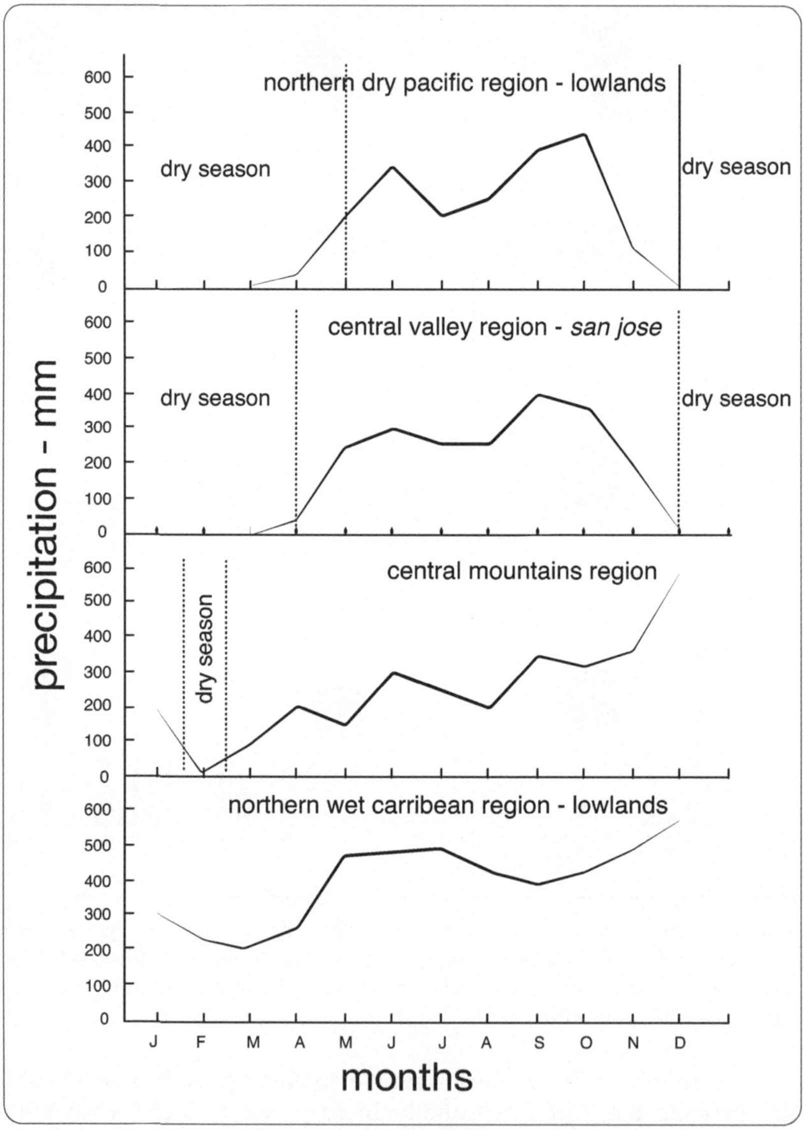

Figure 4.28 Variation in the duration and intensity of the annual dry season at some representative elevations in Costa Rica. Ten to thirteen years of data are included. Reproduced from Young (1982), with permission.

Historical information about the composition of vegetation in terrestrial tropical systems indicates that during the Quaternary there were periods of considerable change in tropical regions (e.g., Jolly et al. 1997), even in the wet tropics (Flenley 1979, Walsh 1996), which are often said to be particularly stable. Both altitudinal changes (Fig. 4.29) and distributional shifts have been documented, and enough evidence has accumulated now to demonstrate that such change is not restricted to montane areas, but takes place geographically. A close focus on tropical systems thus reveals that tropical studies should not be based on assumptions of stability, or on the expectation of ecological systems returning to stability. "Considering the rate at which fluctuations of state variables occur, tropical rain forest and coral reefs may be said to be the two most internally unstable ecosystems on earth" (Kikkawa 1977).

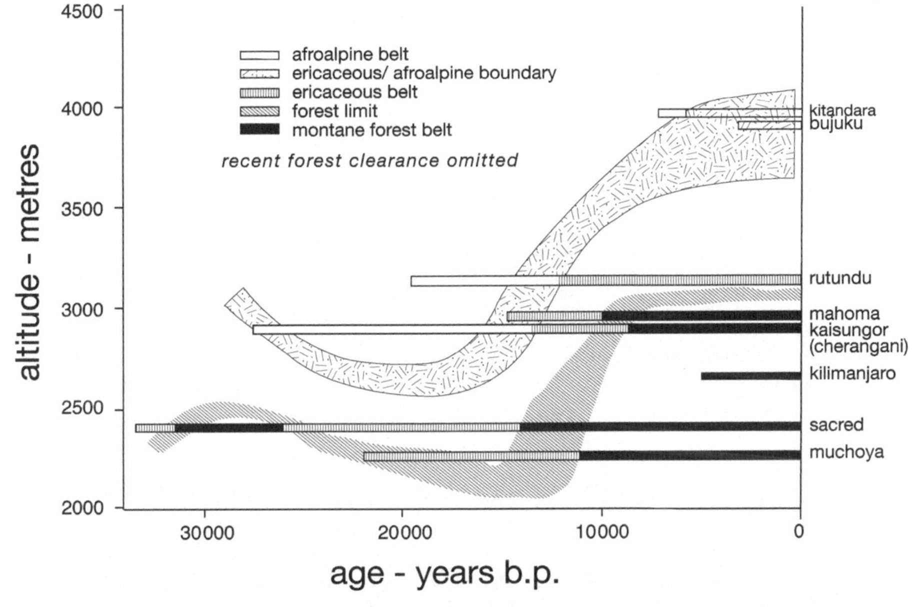

Figure 4.29 Summary diagram of Late Quaternary vegetation changes in the east African mountains. The light grey curve represents the change in altitude of the Ericaceous/Afroalpine boundary and the hatched curve the forest limit. Reproduced, with permission, from Flenley (1979).

3.6 Rates of Change

Historical evidence indicates that organisms respond within a short time to altered environmental conditions, whether these circumstances be caused by climate change directly, or indirectly through habitat change. Even substantial environmental changes, from one vegetation type to another, for example, may take place over a short time. Comprehensive changes between forest, wooded steppe, and steppe took place in Mediterranean localities within periods of less than 200 years (Allen et al. 1999).

The northern distribution of European butterfly species, normally considered to be "non-migratory" or "sedentary", have shifted at the same rate as isotherms have moved north during the twentieth century (Parmesan et al. 1999). Such rates of spatial change are consistent with the view that movement, stochastic in its direction and distance, is the norm among all organisms (see Chapter 9 for details), and also with the view that no such thing as a truly sedentary species exists (e.g., Taylor 1986). Contemplate, for example, the shifts in distribution achieved by land snails (Fig. 4.30), which seem to have taken place quite rapidly (Preece 1997).

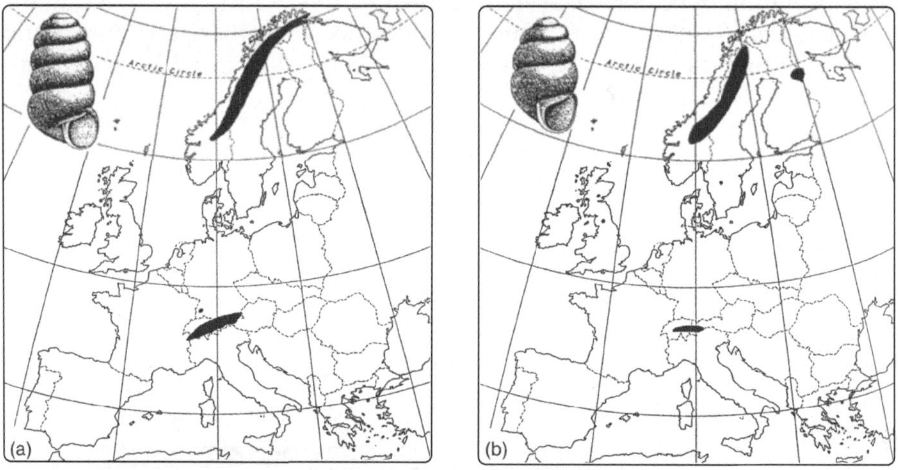

Figure 4.30 Modern distribution of two land snails that were common in Britain during the Late-glacial and early Holocene. **(a)** *Columella columella*. **(b)** *Vertigo genesii*. Reproduced, with permission, from Preece (1997).

4 ECOLOGICAL DYNAMICS AND AUTECOLOGY

The ecological situations examined so far in this chapter illustrate that stability could not be considered the usual state in ecological systems. Each of the situations examined above represents large numbers of species and this indicates that the recorded responses are not atypical. And they all show that the ecological responses of organisms to varying conditions, imposed primarily by climatic or biogeochemical variation or change, take place even over short time scales. Such short-term responses indicate that in ecology we are generally dealing with situations that are atypical according to demographic theory. All of this implies, in turn, that the application of concepts of stability or equilibrium to ecological systems does not reflect reality well, much to the detriment of ecological theory. These concepts fail even when used heuristically, for they misdirect and/or limit investigation substantially, as demonstrated below. More significantly, they show that theory needs to be developed from its very basis to cope specifically with the dynamic nature of ecological systems revealed by the examples covered above.

In this section, we therefore draw the most significant general consequences of these observations for autecological theory. As we deal with each point we also explain what the observations mean for the relevant aspects of demographic theory, because progress in ecology requires not only new theory but also the conscious rejection of those aspects of theory that do not work well. In the remaining chapters of the book, we develop these points into a theoretical structure for autecology.

4.1 Nature of Autecological Theory

General autecological theory must cover all species despite the fact that for each species studied ecologically we should also erect a corresponding theory of its ecology (Walter 1995). In other words, we need an overarching heuristic that guides the investigation and interpretation of the ecology of each species in its own right. This duality in theoretical requirements is necessary because the adaptations of organisms to the environment are species-specific to the point that each species is adapted to a particular subset of the vast array of environmental features (or axes of differentiation) that potentially impact on organisms (Chapter 5, Section 2). In providing a heuristic generalization for autecological theory we do not canvas a single process as a basis that generates expectations of what issues and questions are ecologically relevant, as seen with density dependence vs. density independence in population ecology and interspecific competition in community ecology. Instead, autecological theory focuses on the organism-environment interaction, with the environment defined broadly to incorporate all environmental influences on organisms, biotic and abiotic. The variables of potential relevance to particular species and situations can then be dealt with as required. Although the focus of autecological theory is on individual organisms and their requirements, tolerances and responses, species are of significance because the adaptations carried by individual organisms mediate their interactions with various aspects of the environment and are species-specific.

4.2 Spatio-temporal Dynamics of Species—The Primary Pattern for Explanation

The primary pattern to be explained in ecology is the spatio-temporal dynamic of the species under investigation. That requires tests of which feature(s) of the environment (be they temperature or rainfall patterns, or particular stages in vegetation dynamics) influence(s) the ecological intensity of a species across its geographical distribution. The examples detailed in Chapter 3 demonstrate that the more complete the understanding of the mechanisms of the species involved for interacting with their environment, the more likely the hypothesis proposed to explain the documented spatio-temporal dynamic of the species concerned will reflect reality accurately.

Investigation of that dynamic may also entail a consideration of what produces the environmental circumstances (e.g., vegetation type or edaphic features) that are conducive to the species in question and may well require historical investigation and interpretation. The focus of the investigation may well be local, but that cannot ignore the geographic context of the species in question and any movement that may be relevant, as demonstrated by the continual movement of the ground beetles in Holland (Section 3.2). Scales of investigation are considered in detail in the following chapter.

In contrast to the autecological emphasis on species-specific spatio-temporal dynamics, demographic ecology theory and interpretation has been strongly and consistently influenced by the view that stability in ecological systems is the norm. With any perceived disruption of that pattern, for example as a consequence of extreme weather events, ecological interest is then usually focused on the return to the original state (e.g., Pimm 1991), which is often assumed to represent an equilibrium level. An increase from low population densities is thus seen to be an automatic response to a release from density dependent pressure, although such increases are not inevitable. In autecological theory, by contrast, any increase in abundance (even from extremely low densities) and also "failures" to increase, is understood entirely in terms of the organism-environment interaction and what environmental conditions are favorable (Walter 2008).

Stability was built into the principles of demographic ecology theory, even if that was only tacit (Worster 1977), which helps explain the persistence of such views. Consequently, ecological stability is often not defined explicitly - it is accepted as being usual and generally understood. When stability is defined, though, the vast range of definitions gives away its arbitrary nature (Grimm 1996, Grimm and Wissel 1997). That we cannot generalize effectively about the stability postulated for ecological systems indicates further that it cannot be taken as given, and ecological stability therefore cannot be used as a basis for ecological theory.

4.3 Understanding Species and Ecological Interpretation

Autecological theory treats all species conceptually in the same way as each other because, for example, the interaction between their species-specific properties and the environment may result in one type of outcome under some circumstances, and another outcome under another set of circumstances. Outbreaks, for example, are more a consequence of context and the nature of the organism-environment interaction than reflecting a characteristic or property of the organism itself. The ecology of species therefore cannot be categorized typologically. These points are justified further and dealt with in more detail in Chapter 8, Subsection 3.2. In other words, categorical phrases like "outbreak species" and "invasive species" (e.g., Wallner 1987, Price 1996) are not explanatory; they do not (and cannot) reflect characteristics of the organisms concerned. Autecological theory therefore does not acknowledge "kinds" of population dynamics, an aspect that is frequently canvassed in population ecology.

Of more concern to autecological understanding is the accurate determination of species limits. The underlying reason for this relates to gene flow with respect to the species gene pool (or field for gene recombination) (Carson 1957, Paterson 1985). Adaptations are "contained" within the species gene pool and since those adaptations dictate the ecology of the species concerned, the limits to the species represent crucial understanding for accurate ecological interpretation. Too little attention is paid to this point in ecology and this is not a question of taxonomic "splitting" or "lumping" but one of correct interpretation (Paterson 1991, Walter 2003).

Autecological theory explicitly excludes community level considerations, in part for the reasons just given but also because individuals of each species (even if only as spores or seeds, for example) move and establish independently of those of other species, depending on their specific ecological requirements. Structured ecological communities or assemblages therefore do not exist (Walter and Paterson 1994, 1995). In many ways, ecological "communities" are similar to stellar configurations, which are not persistent entities, but are images we have formed, from our own scale of perception, of a brief part of a dynamic feature. They are both epiphenomena, although we can make more practical use of stellar configurations because of their enduring persistence relative to the history and lifespan of humans (see Peterson 1993). Further, the very concepts of community and guild are misleading (Walter 1988a, Walter and Paterson 1994). Those concepts focus attention on a subset of species that have in common only the reputation for "sharing a common resource" and thus for "having the potential to be competitors" (Abele et al. 1984) or, in broader terms, simply having the "potential to interact" (Lawton and MacGarvin 1986). The community approach thus effectively eliminates the ecologically significant, species-specific characteristics of the organisms because it concentrates on possible common demands, competition and perceived means of coexistence, and thus yields information that is at best rather trivial with respect to the central issues of the organisms' ecology (Sale 1988, Walter 1988a, 1995). This does not provide compelling reason to consider organisms in terms of communities or assemblages.

4.4 Climatic Influences

The climatic influences that act on the individuals of the species of interest need to be understood in terms of how they affect the organisms and thus their ecology. Average measures are likely to be useful for particular purposes, but specific diversions from average are also likely to be significant in explaining local situations. The challenge is to develop a generalization for climatic influences that holds across species. Undoubtedly, such a generalization will be unlike the demographic interpretation of climate, for the latter treats climatic variation as averaging out over the time frame typical of ecological studies, which, if true, would justify an emphasis on deterministic demographic factors such as density dependence. This is the underlying argument of Pimm (1991), and is widely accepted, perhaps usually tacitly.

4.5 Movement of Organisms—Spatial Adjustment to Environmental Change

The widespread occurrence and ecological significance of movement as a mechanism by which individuals adjust to spatio-temporal heterogeneity needs to be incorporated into the core of ecological theory (Taylor and Taylor 1983). The underlying basis for the spatio-temporal dynamics observed and measured in any species is essentially physiological and behavioral (Hengeveld 1985a, Walter 1995). This implies that ecological theory needs a stronger connection with the internal processes that are central to the functioning of individual organisms than has hitherto been available. It also needs a stronger connection with the environmental processes that influence the physiology and behavior of individual organisms. The consideration of these processes must be at dimensions that are more extensive than local, because the provision of conditions suitable for the species of interest is a significant feature in the spatio-temporal dynamics of the organisms concerned, for the following reasons.

Changes to the environment through time, within a locality, are strictly linear, simply because that is the nature of temporal change. Such change may be recurrent in some respects, as with seasonality, so is predictable to a certain degree. Spatial variation is unpredictable in this regard because it has no *a priori* direction, being imposed on the three dimensions of the earth's topography. That is, spatial dynamics have no inevitable directionality, and so are unlike the linearity of temporal change. These vectors of environmental change place severe limitations on the types of biological properties that can help organisms tolerate unfavorable conditions through time, and to locate favorable localities, but it does have the important advantage for individual organisms in that they can escape when conditions change. The adaptations for movement are generally related to the life cycle and way-of-life of the organisms concerned. They may be restricted to a particular stage in the life cycle and different stages may be differentially adapted for movement. In short, the movement of organisms overcomes heterogeneous conditions in space, and maintains the ecological constancy of organisms, and thus the continuity of their lineage. Organisms also have adaptations that help them span unfavorable conditions within a locality, through time, by hibernation for example. Temporal variation can thus be met locally, either by eco-physiological flexibility, or by different responses in discrete life stages.

Because the dispersion of particular subsets of environmental conditions is ever changing, the individuals that comprise a species must adjust to the spatial dynamic of the environment. They must, in effect, track suitable conditions continuously, from one locality to the next. In sessile organisms, it is the individuals of successive generations that move on, to the point that organisms are effectively always on the move. They do, however, tend to stay longer in favorable places than in unfavorable ones. The frequent "voluntary" movement of organisms away from suitable conditions (Chapter 9) surely indicates the significance of the challenge imposed by environmental heterogeneity in the ecology of organisms, even when their densities are not at debilitating levels. Even though breeding colonies or territories may be returned to year after year, these have inevitably shifted with time. Adélie penguin colonies have tracked suitable breeding beaches as climate has shifted since the last glacial maximum (Ritchie et al. 2004), and continue to do so now (Shepherd et al. 2005). Environmental conditions change and numbers of even common species decline at some stage, as have house sparrows in Europe (e.g., Robinson et al. 2005) (see also Fig. 4.22).

Despite the continual movement of individuals, more or less discrete aggregations do form, but these break down when conditions deteriorate. These aggregations are commonly referred to as "populations" in demographic ecology, and are treated as ecological entities. These do not necessarily equate to the Mendelian (or sexual) populations of genetics, mainly because they are the statistical by-products, or effects, of the behavior of individuals in space and through time, and so are defined arbitrarily (Walter and Zalucki 1999, Walter 2008).

The original focus, in demographic ecology, on temporal processes, particularly those of short duration, took the focus off movement as a crucial aspect to local ecology. Although ecologists were aware of the significance of movement, with renewed interest from the mid-1960s (e.g., Johnson 1969), species tended to be typed as migratory or not, so this body of knowledge did not impact on ecological theory (Taylor and Taylor 1983, Walter 2003). When aspects of ecological theory did consider the movement of organisms, as did island biogeography (MacArthur and Wilson 1967), movement was interpreted only with respect to the equilibrium that followed the colonization and extinction of species on islands, so the fundamentals of ecological theory were not affected at all. Although metapopulation theory was developed to include the obvious movement of organisms, it was built primarily around the concept of logistic growth. The movement of organisms was therefore inferred to stabilize systems at the metapopulation level, but with populations as the units, rather than individual organisms (Hanski and Gilpin 1991). The metapopulation approach effectively compounds the problems associated with local logistic theory, rather than solves the problem posed by the movement of organisms. In "spreading of risk" theory, movement is seen in the context of the statistical stabilization that is seen to come from the averaging of independently rising and declining local populations of a species, and which contributes in an unspecified way to population persistence (Andrewartha and Birch 1954, 1984, den Boer 1968). Even the explicit treatment of spatial diffusion, as a basis for spatial processes, is often assumed to be underpinned by logistic growth (Tilman et al. 1997). All of these ideas, therefore, impose a pre-existing set of expectations on the movement of organisms. They do not investigate the causes of such movement and the consequences that this individual behavior has for the way in which ecological systems actually operate. Furthermore, the properties of the organisms themselves, and their ecological requirements, are virtually entirely omitted from these theoretical developments.

5 CONCLUSION—GENERAL AIMS FOR AUTECOLOGICAL THEORY

An important aim we have for this book is to develop a logical connection between the physiological and behavioral processes that are inherent in the survival and reproduction of individual organisms, and the spatio-temporal dynamics of organisms that make up the subject matter of autecology. These goals are pursued in the chapters that follow. A significant practical consideration in ecological investigation is the scale of investigation and the scale at which the subsequent interpretation must be developed. Autecological considerations suggest that two scales need to be considered, that relevant to the species of interest, which is geographical, and that relevant to the individual organisms of that species, which is much smaller. This is the topic of chapter five, on environmental matching.

Chapter 5

Environmental Matching
Individuals, Species and Scales in Ecology

CHAPTER OUTLINE

1 INTRODUCTION

In autecological theory the interpretation of the ecological intensity of organisms through time and across space is developed from the perspective of an understanding of the interactions between organisms and their environment at large. These are the interactions that support the continuity of the lifeline and reproduction of the individuals of the species in question, as well as the interactions responsible for movement, as explained in Chapters 3 and 4. For organisms to operate effectively and reproduce, their species-specific requirements and tolerances must match the spatio-temporal dynamic structure of the environment. This chapter expands on how this is achieved and how an appreciation of this process affects our understanding of ecological systems.

Chapter 4 demonstrated that organisms are influenced directly by a range of environmental variables, by the temporal structure of those variables, and by changes in their intensities and durations. The primary drivers of the variables concerned are external, mainly climatic and biogeochemical, although these influences can be transmitted indirectly. For example, climate may force a change on vegetation structure, but it may be the resultant decrease in shading that affects the spatio-temporal dynamics of small terrestrial organisms within a particular locality. Such local influences from the environment are increasingly recorded in the ecological literature and involve organisms as diverse as snails, butterflies, rats, plants and even humans (Jenkins 1974, Dobkin et al. 1987, Baur and Baur 1993, Weiss et al. 1993, Madsen and Shine 1999, Petru et al. 2006).

Environmental changes induced by the activities of organisms, such as soil enrichment by colonizing legumes are also significant as they influence vegetation structure (Vitousek et al. 1987). Mortality factors like predation and parasitism are also obvious, for they result in immediate death or readily measured debilitation. Competition, strictly defined to exclude processes such as facultative predation and changes to the environment (Walter and Paterson 1995), may also influence local ecology, although its impact is undoubtedly more sporadic than other ecological variables, in part because it is subject to subtle contextual intrusions as well (e.g., Andrewartha and Birch 1954, Buss 1990). Competition cannot, therefore, provide a sound basis for developing generalizations in evolution and ecology (Walter 1988a, 1995, 2008).

The autecological approach to understanding the spatio-temporal dynamics of organisms, and thus their changing local abundance, provides a broader context than that circumscribed by an emphasis on local population equilibria and resource-based communities. It achieves this through its primary emphasis on the broader, geographic context across which environmental processes act and it defines ecological interactions much more broadly than just in relation to common resource use. For each species investigated ecologically, two levels of spatial abstraction become relevant, one related to the species as a whole, the other to the individuals of that species. The distribution of the species is indicative of the spatial limits within which the individuals of the species will generally live, move and reproduce, and is also likely to indicate which environmental circumstances are crucial to the existence and reproduction of individuals of that species. The edges of the distribution in different areas are likely to be imposed by different environmental variables (see below), and thus provide additional clues as to the different requirements of the species in question. The individuals of that species will dictate the scale of investigation at local levels, for their behavior will indicate the scale at which they range and interact with the environment.

The issues outlined above are tackled as follows. First, we consider the nature of environmental heterogeneity and climatic influences (Section 2). We do so with the aim of specifying a means by which to define the "structure" of the environment of particular organisms. Attention is then switched to the organisms themselves, to justify the view that the adaptations that underpin environmental requirements and tolerances of organisms through the various stages and phases of their lifeline are not only species-specific but also complex (Section 3). The ultimate function served by such complex adaptations almost invariably requires input from several sub-components, each of which serves a subsidiary function. A generalized concept of how these environmentally relevant properties of organisms function within the environment is then developed (Section 4). In brief, organisms must effectively "match" their species-specific life cycle and adaptations to the temporal structure of the environment if they are to survive and reproduce. This concept of environmental (and habitat) matching allows ecological scales of investigation and interpretation to be specified much more explicitly than previously (Section 5), and the reasons for doing so in relation to both individuals and species can be justified further. The autecological background presented in this chapter provides a basis for returning to the process of adaptation. It considers what an autecological appreciation of organisms in nature implies about interpreting the adaptive process and provides insight about the particular conditions under which evolutionary change, and ultimately speciation, is likely to take place (Section 6).

2 THE ENVIRONMENT IN AUTECOLOGY—VARIABLES, STRUCTURE AND HETEROGENEITY

Not all influences within any local environment impinge on the wellbeing (or otherwise) of the organisms of any (and each) particular species living there. Those aspects of the environment that are of direct relevance to the existence and reproduction of the organisms of the focal species make up the "operational environment" of that species (Spomer 1973). This subset of the surrounding environment represents the living conditions required by the organisms (of the species of interest) and is specifiable in terms of the requirements and tolerances of those organisms. These influences may be thought of as the primary influences from the environment. Predators, herbivores and parasites impact negatively on organisms, sometimes substantially so, but these are still probably best considered secondary influences within local environments. (Note that this abstraction differs substantially from that of Andrewartha and Birch (1984), who work from the proposal that "the environment of an animal consists of everything that might influence its chance to survive and reproduce", and thus gives equivalent status to resources, malentities, predators and mates.) The organisms' physiological and behavioral requirements, and their physiological tolerances, as well as any secondary influences that might be detectable, seem to be what influences organisms when they assess a locality (at least for those that have the capacity to do so, and with "assess" used broadly to include all senses and behaviors that might influence settlement of organisms). This process entails habitat recognition.

The components of the operational environment are those involved in the exchange of mass or energy with the organisms of interest, such as heat and light (which are energy variables) and various inorganic factors (e.g., gases and liquids) and organic ones (e.g., nutrients, hormones, and infochemicals) (which are mass factors). Spomer (1973) provides a much fuller list and justifies this stance in the conceptual development of the environment in ecology. In practical terms, ecologists represent the operational environment of organisms in terms of proxy measures (Spomer 1973). The relevant axes of the immediate environment include physico-chemical factors as well as biotic ones, with the former including temperature, rainfall, relative humidity, insolation, a large variety of chemical gradients, edaphic conditions, substrate structure, physical currents, and so on. These may be ecologically specific in terms of life stage, time of year and the dimension of the variable. The axes that represent the biotic factors among the primary influences of the environment include such features as vegetation structure, host organisms, and vector organisms. Thus, several axes of ecological differentiation are discernable (Appendix). The operational environment should be thought of primarily in terms of the requirements of the species, and then in terms of its tolerances. These two aspects are not mutually exclusive of one another, and the species' requirements for one life stage (e.g., vegetative growth in lime trees) might be available in an area in which the tolerances of another life stage (e.g., growth of the pollen tube in lime trees) are exceeded (Pigott 1974). Autecological information of this nature is not merely observational or correlational. Obtaining it requires insight about the natural world and appropriately designed tests, as indicated by the research of Pigott (1975).

Before elaborating on environmental structure, we note that in demographic ecology the abiotic is treated as if it were entirely unstructured, so the ecological impact of abiotic influences is usually factored into explanations and models as a single average value, if such factors are included at all. Effectively, the abiotic is treated as single dimensional, so ecologically relevant information on the statistical attributes of these variables is missing (e.g., amplitude, frequency, period and sequence of variation). The biotic, in contrast, is seen as structured, but virtually exclusively through the feedback loop anticipated from density-related demographic influences (Turchin 2001, Coulson et al. 2004), for this is in principle the only way in which populations can be regulated in the way required by the logistic equation, its underlying premises and its elaborations.

In autecology, by contrast, the abiotic is typically seen as structured, with its composition quantifiable in terms of frequency of occurrence of particular measures of the different variables (requirements and tolerances) that are significant to the species of interest. The incidence of particular conditions across time periods or across space can thus be quantified and expressed in frequency diagrams. The environment is thus structured in specific ways despite the conditions along each axis of differentiation (Appendix) changing stochastically, and despite them even changing independently of one another. The autecological perspective is thus irreconcilably different from the demographic approach with respect to the conceptualization of the environment, and these points are developed in the rest of this chapter.

Density-related (or demographic) variables cannot be quantified in the way described for autecological variables in the paragraph above. In the field, they frequently have only a small impact and often are not even measurable, but this debilitating weakness is largely ignored despite the extent and intensity of the criticisms (e.g., Andrewartha and Birch 1954, Dempster 1983, Dempster and Pollard 1986, den Boer 1986a, 1987a, b, 1988, Taylor 1986, Stiling 1987, 1988, Murdoch and Walde 1989, Cronin and Strong 1994, Murdoch 1994, Wolda et al. 1994, Chitty 1996). Further, these demographic variables are themselves open to so many influences that even when they are operative they cannot be meaningfully quantified, but are still expected to exert a more or less constant pressure. Perhaps most telling is that these demographic parameters were formalized within the context of the logistic equation, which has no spatial context at all (Andrewartha and Birch 1954).

Ultimately, it is the organism-environment interaction that dictates the ecological outcomes that are usually expressed in terms of the distribution and abundance of organisms. These outcomes are more accurately summarized and expressed with respect to the spatio-temporal intensity of occurrence of the species of interest. To specify the nature and intensity of the interaction, each environmental axis of differentiation that is relevant to the organism in question has to be identified and quantified in a particular way. Ecologists have typically found it convenient to deal with temperature in terms of mean values, and consequently have tended to downplay the ecological significance of temperature because these averages tend to correlate quite poorly with the spatio-temporal dynamics of organisms. However, each axis that represents a particular temperature profile presents a range and pattern of variation that is continuous and extensive, and organisms within any locality thus experience temperature in terms of various patterns of maxima, optima, and minima over daily and seasonal cycles (Billings 1952). Autecologists measure levels of intensity, mean maxima and mean minima, optima, and levels of variation. As time proceeds (whether daily or seasonally), mean values and their associated measures of variance shift relative to one another.

Stresses can come in various ways, perhaps as winter minima that are too low, or even as winter maxima that are not high enough to compensate for the negative effects of winter minima, and so on. A stress in this context is not necessarily (or only) a biochemically or physiologically stressful extreme, but may be a deficiency in thermal energy over a given period. In other words, deficiency in an ecological requirement also generates stress. Measures of physical and chemical variables also vary across space, at local or geographical scales. Sometimes they vary gradually, and in other instances abruptly. Furthermore, the periodicity or wavelength of the exposure periods is subject to variation as well, so that periods of warmth may be longer or shorter across space or seasons. The other environmental axes, such as moisture levels, will show similar variation, but variation along one axis does not necessarily correlate with variation along others. Variation is also typical of the biotic axes. Vegetation structure varies with species composition and local conditions, and such variation may be seasonal or spatial. Host organisms and vectors also vary, among conspecific individual organisms, but mainly across species. Furthermore, interactions between variables may ameliorate environmental circumstances or render them even more intense, and such interactions also warrant consideration.

Organisms in nature have to face the environmental conditions just described on a daily basis. A significant aspect of this organism-environment interaction is the differentiation across species, even closely related ones, in terms of the ecological tolerances and requirements of their constituent individuals (e.g., Figs. 3.1, 3.2 and 4.6). Autecology is designed to cope with this specific variation, as follows.

3 THE REQUIREMENTS AND TOLERANCES OF ORGANISMS ARE SPECIES-SPECIFIC AND SPECIES-WIDE

In dealing with the organismal side of ecological interactions, we have written so far mainly in relation to single organisms of a particular species and life stage. But even closely related species have different requirements and tolerances in relation to the physico-chemical and biotic environmental variables they may encounter, and the successive life stages within a species show similar variation, with each stage responding to the environment in its own way. The multidimensionality of interactions thus becomes apparent, for as individuals pass through the successive life stages that typify their species, they interact with a number of variables, each of a particular stage-specific maximum and minimum intensity, duration, frequency and wavelength. If individuals of a particular species are to persist, they must effectively match their requirements against the various environmental axes, in terms of the combinations of variables that are ecologically significant to the species of interest, and they must do so at each life stage. Furthermore, the temporal sequence in which appropriate conditions for the successive stages of the life cycle occur, within the localities occupied by the organisms, is critical to the survival of the individuals through their entire life (Chapter 3).

The lifetime requirements and tolerances of plants of a particular species, for example, can be structured and depicted in temporal graphs (as in Figs. 3.17, 3.19 and 3.20) because of the nature of species. Across the geographical distribution of a species we observe stability in the adaptive properties that we believe to be environmentally relevant to them.

That is, conspecific individuals have similar adaptations to one another, and measurements of those properties and of their ecological effects generally follow the normal distribution. This explains why, in different parts of a species' range, alternative environmental variables are likely to be more significant ecologically, for example rainfall at the southern limits of a species and temperature at the northern ones (e.g., Castro et al. 2004). This is so because conspecific organisms in different localities of the species' distribution have the same adaptive mechanisms or properties, but environmental variables change in intensity and duration across that geographical distribution, so ecological outcomes involving that species will vary across space (and also through time).

Such stability in species is demonstrated temporally and spatially by the lack of detectable evolutionary change at the margins of species' distributions (Hoffmann and Blows 1994, Bridle and Vines 2007), by the species-wide distribution of the major adaptations of species (Walter and Paterson 1994, Rajapakse and Walter 2007, Popple et al. 2008), the tracking of suitable environments by organisms (Coope 1994) even at the scale of decades (Hengeveld 1985a, Parmesan and Yohe 2003), and the pattern of species stability in the fossil record, the latter of which is now widely accepted (Eldredge et al. 2005). With respect to the last point, the punctuated equilibrium of the fossil record thus has a contemporary ecological equivalent that is spatial (e.g., Fig. 5.1) rather than temporal (e.g., Rajapakse and Walter 2007, Popple and Walter 2010).

Species stability is attributable to several features, mainly to do with the environmental relationships of the organisms, the specifics of their reproduction, and the nature of their genetic and developmental systems. More particularly, the mechanisms that support the environmental interactions of an organism support the persistence of that organism through its lifeline. This means that the subset of interactive mechanisms that is crucial to each given life stage (or set of environmental circumstances) must work effectively (Chapter 3). These requirements attract stabilizing selection, as does the nature of the adaptive mechanisms that underpin virtually all interactions, for most of them comprise properties that are integrated into functional complexes (called complex adaptations or character complexes) (Frazzetta 1975, Hengeveld 1981, Paterson 1985). The nature of complex adaptations is returned to below and expanded more fully in Chapter 6, Subsection 4.3 in relation to understanding how they operate and how that understanding informs ecological interpretation.

The requirements of sex also impose rather strict limits on change (Paterson 1981, 1985). Individuals of each sex must have identical habitat relationships during the reproduction phase, if mate location, mate recognition and fertilization are to take place successfully. Moreover, the sexes must be attuned physiologically to each other in time, which is a complex procedure in the endocrinology of both vertebrates (Young 1975, especially pp. 466-468) and invertebrates (Highnam and Hill 1977, Klowden 2007), and involves several interacting genetic pathways

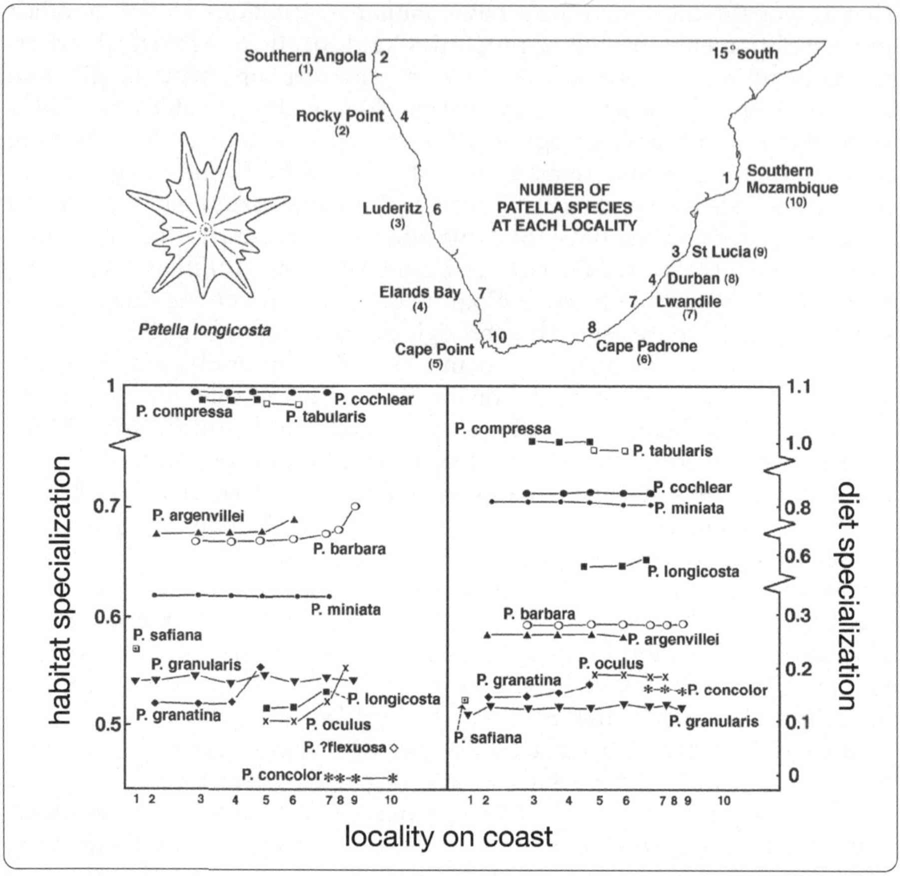

Figure 5.1 In southern Africa, limpet diversity increases towards the south. The map indicates the localities where limpets were sampled and the number of species recovered at each. Below are plots that specify which species were present at each locality (locality number on the x-axis matches that, in brackets, on the map) and the habitat **(left plot)** and diet **(right plot)** specialization of each. No matter how widespread a species, where it occurred or which other species were present locally, its habitat and diet remained the same across localities. Such ecological stability represents a general pattern in ecology, but remains little investigated. Redrawn from Branch (1985).

in plants (Hepworth et al. 2002, Sung and Amasino 2004, Valverde et al. 2004). Reproductive synchrony across the sexes in all of these organisms involves sensory interactions with the environment. Given success in synchrony, individuals of opposite sex must then recognize one another as potential mating partners (Paterson 1978, 1980, 1981, 1985, 1988) and be capable of localizing (Huber 1985, 1990, Moore et al. 1993) and/or reaching one another within a heterogeneous environment that is spatially extensive. Cellular recognition must also take place between the sex cells, which can involve intricate mechanisms (e.g., Brewis and Wong 1999,

Boavida et al. 2005, Vieira and Miller 2006). This level of precision places demands on the underlying biochemical, physiological and behavioral processes and this, in turn, confers further stabilizing selection across the individuals of a particular gene pool (or species). The mechanisms of interaction carried by individuals during the reproduction phase are species-wide spatially (e.g., Popple and Walter 2010) and constant through time (e.g., Vrba 1984), because all of the features mentioned above act in concert towards a single end, fertilization (Paterson 1981, 1985). That is, we are dealing with complex adaptations.

Further stability is evident in the genome. Aspects of the genetic system have remained stable over extensive periods, to the extent that the genes involved in the regulation of crucial processes have remained essentially unchanged (but for minor differences in DNA sequences) across countless generations and through numerous speciation events (e.g., Arthur 2004, Carroll et al. 2005a, Carroll 2008a, Gilbert and Epel 2009, Brakefield 2011). They have done so because they have served the same specific function throughout, so change at that level of operation would be deleterious. Co-adaptation within the genome leads to further stabilization. The mutual adjustments of the different components of the organism and its genome ensure that the development of each feature of the organism follows its appropriate temporal trajectory relative to the others. Failure in this integration translates into failure to function (Arthur 2004), and the considerable variability detected in genetic systems (e.g., Gilbert and Epel 2009) may exist to maintain developmental function. This would allow corrective adjustment to inappropriate shifts in the developmental process, as well as different responses (plasticity) to varying environmental influences. A specific adaptation acquired early on in a lineage may persist through future diversification within that lineage and thus eventually persist as a general property across species at higher taxonomic levels.

The requirements, features and processes detailed above explain why the primary adaptations of species are fixed (i.e., are species-wide in their distribution) and why domesticated organisms return to their ancestral state (or wild type) as soon as breeding programs are stopped or when they establish wild populations (e.g., Sol 2008). Given these strictures, the question of how adaptive change takes place, and under what circumstances, takes on particular prominence.

The genetic basis for variation and change is now much better known through recent advances in understanding the genome (e.g., Gilbert and Epel 2009). Major changes may be coded mainly through the duplication of genes, with the duplication "freeing" a copy of the gene to the extent that a new function coded by adjustment to that copy is able to evolve (Carroll 2005), and by alterations to the timing of genes being switched on and off. For example, even morphological features as simple, in relative terms, as butterfly wing spots are coded by a homeodomain containing a transcription factor (called *Distal-less* (*Dll*)) whose expression is associated with that of genes in the Hh pathway, a regulatory circuit also coopted into wing development (Brakefield et al. 1996, Carroll et al. 2005a, pp. 167-170). The diversification of *cis*-Regulatory elements may be involved, but how this takes place is not yet clear although transposable elements may be involved (Carroll 2008b). Although allelic substitutions also play a role in adaptation, that role is far more limited than previously envisaged, and is not responsible for the major adjustments we see, across species for example (Carroll 2005, Werner et al. 2010, Brakefield 2011). The key issue, though, is what ecological circumstances disrupt stabilizing selection and enable adaptive changes of such magnitude to take place. We justify, through the rest of this chapter and further into this book, why an understanding of the autecology of organisms is central to interpreting evolutionary change.

Although claims for having demonstrated local adaptive change and incipient speciation are made routinely, we feel that the examples offered warrant deeper scrutiny. A persistent preoccupation of evolutionary biology has been the exposure of sources of variation and the documentation of recent or contemporary genetically based change, because evolutionary ecologists still expect ecologically meaningful short-term change at a local level (Mopper and Strauss 1998, Thompson 2005, Bridle and Vines 2007). Despite intensive search, examples of such local change are few (Merila et al. 2001, Ehrlich 2004). Nevertheless, attempts are made to reconcile such anticipated change with patterns of species stasis in the fossil record (Eldredge et al. 2005). The empirical evidence that is presented in favor of local change of this nature has, however, to be treated with caution, as it has been derived almost universally through attempts at verification that significant evolutionary change does take place locally within the distribution of a species. Few genuine tests among real alternatives are available. In any case, claimed examples usually entail such minor change (e.g., Carroll et al. 2005b) that this cannot be taken as indicative of the environmental circumstances under which speciation and the evolution of new lineages takes place.

The autecological perspective therefore suggests that (i) claimed cases of adaptive change need closer scrutiny (e.g., to ensure cryptic species are not being conflated and that change in prokaryotes is not being used as a model for eukaryotic evolution) (Walter 2006) and (ii) the degree of the change demonstrated is meaningful in terms of the extent and the range of differences between even closely related species (see Carroll (2005) and Jeong et al. (2008) for examples). Many demonstrated changes are minor, usually just incremental, but the assumption is that even such minor divergences indicate that the populations concerned are on the path to significant future change. This intersection between ecology and evolution surely warrants deep introspection, because the autecology of organisms, the requirements of reproduction and the nature and workings of the genome itself indicate that the envisioned adaptive divergence in lineages is unlikely. This stability also explains why adaptation to the relatively inferior environmental conditions often encountered at the edge of a species' distribution, as measured in terms of survival and reproductive output and usually referred to as fitness, has proved so elusive to demonstration (e.g., Hoffmann and Blows 1994, Bridle and Vines 2007).

In short, a strong argument can be made for stability being the general pattern in the life of a species, with change restricted to particular times, localities and circumstances. The observable stability in species lasts at least until the next major environmental disruption that initiates adaptive change in a relatively small population and, perhaps, results incidentally in the evolution of a new species (Paterson 1985, 1986). We still need to understand how negative pleiotropic effects are overcome (see Carroll 2005), and how the range of features that typify the new species is fixed in the species' gene pool. Interpreting adaptive change is an intricate and controversial issue that is pursued in Section 6 below.

Interpreting the genetical limits to species accurately, in terms of which individuals are potential mates, is therefore necessary for understanding the distribution of the ecological properties among individuals, whether within a locality or geographically. In other words, the adaptive mechanisms of sexual forms are related to the species gene pool from which they derive. This is the primary relevance of the species concept to ecology. Because positive assortative mating, in nature, dictates the distribution of adaptive mechanisms, combining more than one species unintentionally when interpreting ecological phenomena distorts understanding immediately. This mistake is, however, frequently made, even with organisms that are important economically or are of environmental significance. Examples of where the accurate resolution of species limits has solved such problems are now more widely available (e.g., Paterson 1991, 1993a, Fernando and Walter 1997, Bickford et al. 2007, Malausa et al. 2007, Rafter et al. 2013). Although this point about the imperative to interpreting species' limits correctly may be widely accepted, it is not widely practiced (Paterson 1991, Walter 2003). Different populations with different ecologies are frequently treated uncritically as locally adapted intraspecific variants, as pointed out by Clarke and Walter (1995), and presumably occurs in this way because evolutionary change in populations is seen as ineluctable and part of a continuum.

Such local adaptation and divergence is expected under the demographic ecology framework, because competitive selection is expected to drive differentials that affect "efficiency". Tests on species status might therefore not be considered particularly relevant. Increasingly, though, cryptic species or sibling species are being found under a single name. For ecology, the important consequence is that the different species virtually all have unique ecological aspects. Thus, "species" once considered to be "generalist" have been found to comprise a group of species each with much more specific requirements (Paterson 1991, Walter 2003, Bickford et al. 2007, Rafter et al. 2013). That this has even involved such prominent organisms as killer whales (Pitman and Ensor 2003, Morin et al. 2010, Pitman et al. 2011) signals with incredible strength the importance of this issue. Numerous "host races", "local races" or "ecotypes" (among other sub-specific categorizations) will probably prove to be separate species and this prognosis may hold for at least some (and perhaps many) of the claims of populations diverging sympatrically (see Walter (2003)). The tendency in evolutionary biology to dismiss species, even deride their objective existence (e.g., Ehrlich et al. 2004, Mallet 2008), appears to follow logically from the common belief that local populations adapt continuously and so readily that the idea and appreciation of species limits are irrelevant. This interpretation is logically consistent with Darwin's obligation to demonstrate that species merge imperceptibly one into another and with his view that natural selection is fundamentally competitive (Paterson 2005). Also, a widespread lack of appreciation is still evident that species defined taxonomically (i.e., species taxa) are often not co-extensive with species defined in evolutionary (i.e., population genetics) terms (Paterson 1981, 1986).

The temporal scale of evolutionary change is not often considered directly in ecology. It is rather taken for granted through the assumption that evolutionary change is unremitting and locality specific, to the extent that it is seen as an optimization process (e.g., Ghiselin 1974, Mayhew 2006). The measurement of evolutionary fitness against ecological situations is therefore common. Should data indicate evolutionary stasis, situation-specific constraints are invoked or sought. But such explanations only state why optimizing selection is not having the effect it is expected to generate. No surprise, then, that when the evolutionary pattern of punctuated equilibrium was proposed (Eldredge 1971, Eldredge and Gould 1972), the relatively brief phase of change was portrayed as being readily explicable, whereas explaining the extended period of stasis was seen to pose a problem (Maynard Smith 1981).

The autecological consideration of ecological scales and the responses to them of individual organisms suggest that the evolutionary fitness of organisms is an end-product of many ecological influences, is informative of the performance of organisms under a particular subset of environmental conditions, but does not explain or predict the evolutionary trajectory of the population concerned. A further problem in using such fitness measures to explain or project evolutionary trajectories is that they usually assume abiotic environmental constancy and the fidelity of organisms to a particular locality. They consequently take no account of the monumental loss of fitness that undoubtedly occurs as a result, for example, of organisms migrating and dying before reproduction. This happens to large proportions of individuals at times, even large vertebrates (e.g., Duffy and Rudolf 1986).

Organisms are mobile so as to face the dynamics of the environment through habitat and environmental tracking, and the stability observed within species means that local optimization could only ever be selected within extremely narrow margins, if at all. Indeed, only limited variation is recorded within the bounds of species-wide properties. The implications of these strictures for other areas of ecology are fairly obvious. Local adaptive adjustment to co-occurring species within a community context is highly unlikely (Walter et al. 1984, Walter 1991), for example. The concept of the ecological niche, which is the demographic or competitive template against which populations adapt to ensure their coexistence, is also not useful for predictive or practical purposes (Hengeveld 1988b). Because species respond individualistically to environmental influences at scales specific to the individuals that make up these species, the various scales of variation for the different hierarchical levels postulated to make up ecological systems do not relate well to the scales of activities of individual organisms. Furthermore, the proposed ecological hierarchy routinely omits species (Eldredge 1985, p. 165, Price and Waldbauer 1994, Schowalter 2000, Mayhew 2006, Krebs 2008), presumably because the local population or metapopulation is considered to substitute for species. Even when individuals are included it is primarily done so through their numbers (because density is given such prominence in demographic ecology) rather than through species. Consequently, ecological processes specific to populations, communities and ecosystems, or to local, landscape, regional and geographical scales are unlikely to be found.

For example, the differentiation of local from regional scales for interpreting community structure (Ricklefs 1987, Lawton 1999, Mora et al. 2003) has not resolved the issue of ecological scales of investigation and interpretation, principally because such scales are designated arbitrarily and numerous species are covered irrespective of their independent evolutionary origins and geographical distributions, different habitat associations, and their idiosyncratic ecological responses and behavior. Studies of the source of local species have proved useful, however, for tracking historical movements of species into the localities in question, such as in the mown meadows of the Baltic coast (Holzner and Immonen 1982, Zobel 1992) and the organisms at the edge of their distribution on the British Isles (Mabey 1995). These observations pre-dated the conceptual distinction between local and regional scales in community ecology, and the latter has added little more to an understanding of the origins or local ecology of the species concerned (Walter 1993a). The "local-regional" distinction is applied irrespective of the species-specific traits of organisms, a criticism that applies just as strongly to the hierarchy of spatial scales proposed for ecological investigation. The same problem commonly besets the application of modern remote sensing techniques, for their capacities and limitations tend to be dictating scales of measurement and interpretation in ways that are not directly derived from the ecological responses and behaviors of organisms (Mitchell and Powell 2003) and may therefore be irrelevant to them.

Because the members of any one species differ from those of all other species, and because all have specific ecological requirements and tolerances along one or another environmental axis of differentiation (Appendix), the variety of combinations and permutations in environmental factors to which organisms may have to respond (and to which they may have adapted) is vast (Billings 1952). This virtually boundless variation in environmental conditions translates into an infinite number of possible species' ecologies.

In other words, the nature of the genetical, developmental, organismal, and sexual systems sets limits to the variation in the ecological abilities of organisms. Surprisingly, ecologists and evolutionary biologists have tended to focus on the variation they can detect, or feel they have detected, and have omitted the essentials of the system (Haukioja 1982). Undoubtedly this is a legacy of Darwin's focus on intraspecific variation as a means of crushing the then current views of species stability. But Darwin was working against biblical concepts of stability, not modern evolutionary concepts of stability (Paterson 1986, Turner and Paterson 1991). However, some flexibility in the environmental responses of organisms is evident, in reaction norms and other forms of phenotypic variation (e.g., Gilbert and Epel 2009), but this flexibility is small in comparison with the range of environmental influences, and the range of variation in intensity in each of them.

Together, the concepts of species specificity, species stability, environmental structure and dynamics, and the spatio-temporal adjustment by organisms to this dynamic structure lead to the concept of "environmental matching" (Walter and Hengeveld 2000). From there, the issue of ecological scales can be resolved with more precision.

4 STRUCTURED PROCESSES AND ENVIRONMENTAL MATCHING

The only way in which organisms can cope with environmental heterogeneity and climate change is through matching their attributes with the environment across space and through time, a process that is not only dynamic, but is influenced by stochastic events. We conceptualize this relationship between organism and environment as the "environmental match" of the organism with the environment. Any action by the organism to alter that match, whether developmental, physiological, or behavioral, we refer to as the process of environmental matching. How organisms achieve a match is the central issue in autecology, for this is the primary determinant of observed distributions of organisms and the intensities of occurrence of species (the latter of which are inappropriately treated as local abundance in demographic ecology). Environmental matching is specific to autecology and central to it and should not be confused with the "ecological fitting" of species into communities, which is entirely demographic in its primary focus on empty niches and competitive relationships (Walter 2013). Chapter 6 explores the diversity of mechanisms used to achieve environmental matching. The ecological measurement of environmental matching and its consequences are detailed in Chapter 7.

The autecological view conceptualizes the environmental template as the entire dynamic surface of the globe, across which organisms shift about in their various ways and in relation to that subset of the environment to which they are adapted (Walter and Hengeveld 2000). Organisms respond rapidly, but with spatio-temporal dynamics that vary across species. Individuals, unaware of where conditions that are favorable to them may be, generally move more-or-less randomly across the landscape until they encounter an area that matches their requirements, a process called environmental tracking or habitat tracking. The latter is a subset of the former and involves the use of sensory mechanisms to detect and localize suitable habitat. Many organisms fail in this regard; reports of vagrant birds in unusual localities are legion. Some individuals may well move on to a more successful match, but it is not inevitable, and vast numbers of insects die when their flight takes them across the coastline, for example (e.g., Elias 2010, p. 46). These requirements together reflect a species-specific complex of physico-chemical and biotic factors. These factors must also be available in a particular temporal sequence that matches the changing requirements of organisms as they age and reproduce. Once an area has been accepted, in part through the recognition of environmental cues, the individuals involved may remain in the location or area for some time. This retention time may encompass the entire breeding season or life of the individual, or it may be of relatively short duration if the organism is in an appetitive phase of its life.

When numbers of individuals find a particular area and remain in it for relatively long periods, the density of that species in that area will consequently rise, and successful recruitment by those individuals may take place and supplement the local aggregation further (Fig. 5.2). These "Markovian populations" thus assemble in a partly stochastic way, in that individuals arrive and depart fortuitously (Hengeveld and Hemerik 2002). In other words, their retention times are also influenced stochastically (Chapter 9). As a result, such aggregations constantly undergo a turnover of individuals, at a rate that varies with time and according to the suitability of conditions. It follows from this that turnover even occurs in localities where conditions match the requirements of the individuals, because of the continual movement of organisms (Chapter 9). Depending on the suitability of local conditions, the ecological intensity (or local abundance) of the species will vary from place to place, so the more suitable the locality (per unit of area), the greater the ecological intensity (or larger the population) (Fig. 5.2). A much more distant perspective of the ecological landscape might give the impression of an ongoing local build-up in the intensity of individuals of a particular species in various localities (local "populations") with their eventual subsidence (local extinction) (Fig. 5.2). This rise and decline can also reflect the coming together of reproductive individuals in localities favorable for reproduction and for their offspring to hatch, germinate or be born, and then develop through their initial stages of growth. Furthermore, the changes may sometimes appear synchronous and sometimes they may be out of step.

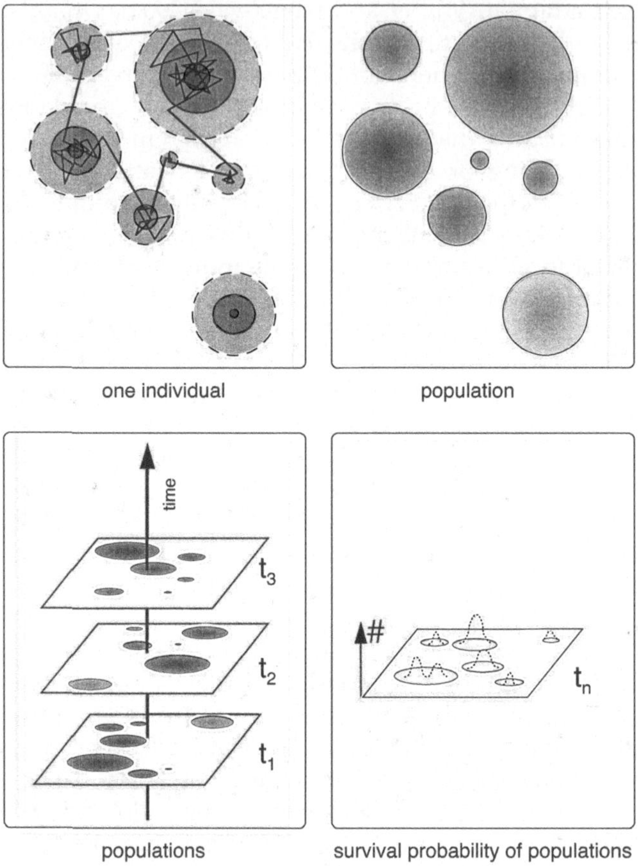

Figure 5.2 Diagrammatic representation of the spatio-temporal dynamics of a species. **Top left.** A single individual moves across a landscape. When it encounters an area of suitable habitat it spends more time there (in this case because it moves more slowly and turns more frequently, but it could represent a pause to breed). **Top right.** Consequently, the number of individuals (a measure of ecological intensity) is greater in areas of suitable habitat, because the retention time of individuals is greater there. **Bottom left.** Suitable environmental conditions shift geographically through time, under the influence of environmental change. Consequently, local aggregations ("populations") shift with time as the individuals find localities in different areas that are suitable. **Bottom right.** Representation of the probability of survival of different local aggregations ("populations") at one time; the probability is an estimate of how long the environment remains favorable for individuals of the species concerned. Reproduced, with permission, from Walter and Hengeveld (2000).

5 ECOLOGICAL ADJUSTMENT—SCALES AND LOCALITY OF ECOLOGICAL INVESTIGATION

For individual organisms to persist they must respond to environmental change when it is inappropriate to their requirements, or they will fail.

They may either move elsewhere or respond through physiological adjustment, or they may respond in some combination of these processes. The physiological approach usually entails slowing or shutting down metabolic systems. A limit exists, however, to the duration for which they can maintain themselves (see Anekonda et al. 1999, for a comparative study of plant respiratory mechanisms and the differential ecological consequences), so ultimately the organism or its offspring must be mobile. Either the parent organisms carry their offspring to another area, as gametes or eggs, or the early life stage must be capable of movement. Despite their small size and early appearance in the life cycle, eggs, hatchlings, seeds and spores are individual organisms that have their own specific environmental requirements. Indeed, these early stage organisms are the agents of movement in many organisms, particularly sessile ones. They have mechanisms to aid persistence and, in many (e.g., planktonic larvae), mechanisms for directed locomotion (e.g., Steinberg et al. 2002).

Motility of individuals is characteristic of organisms, occurring in at least one of the stages of the life cycle (Chapter 4). The sessile stages of animals and plants cannot move in response to environmental changes, and are restricted to such responses as expansion by growth into areas adjacent to their environment, shutting down, or activating physiological mechanisms (for barnacles to survive desiccation and for motile organisms to enter hibernation, for example). Ultimately, though, the organisms adjust spatially in a random fashion, because they have no knowledge of conditions elsewhere or at any other time (Rapport 1991), another premise of autecology. Thus, the interaction between individual organisms and their heterogeneous and dynamic environment provides the framework for establishing the spatial and temporal scales of environmental variation that are relevant to ecological interpretation.

The following subsections consider these scales in each dimension, but only after dealing with the way in which individual organisms, in general, respond to environmental circumstances. For the convenience of explanation, we consider spatial responses independently of temporal ones, despite space and time merging (into locality) in the consideration of any individual organism's ecology or, more specifically, its environmental matching. The ecological significance of species as biological entities is then reconsidered, for species are of direct relevance in specifying the scale at which ecological processes are to be understood.

5.1 Spatial Scales of Variation

The motility of an organism and its propensity to move define the spatial dimension across which the organism may interact with the environmental variation imposed upon it through climatic variation and the spatial heterogeneity of conditions. These are species-specific features, for even closely related species differ in these regards, as shown by studies on the two major pest bollworm species in Australia (as summarized by

Walter (2003)) and investigations of closely related bird species (and even subspecies) (e.g., Perez-Tris et al. 2003, Arizaga et al. 2006, Baldwin et al. 2010). The distance of movement of samples of conspecific organisms, taken together, defines the frequency distribution of distances they have covered, or their contact distribution (see Fig. 7.4 for an example). The latter term relates to the frequency (and therefore the probability or chance) of an organism making contact with a suitable locality (with regard to it providing favorable conditions, the localization of a mate, or a breeding site). The statistics characterizing the shape of this distribution vary, depending on the mobility of the organisms, the spatial variation and dynamics of environmental conditions, and perhaps on the interaction between these various aspects. Also, in some species, the contact distribution may be influenced by the way in which density interacts with the social system or territoriality of the species, through the frequency of contacts between individuals.

Over generations, two outcomes become evident. One, under conditions of climate variation or change, the spatially dynamic environmental matching by individuals in successive generations ultimately results in a shift in the geographic distribution of the species (Figs. 3.1 and 4.18). This is also true of species that have been, on the evidence available, considered to be non-motile or sedentary, with a few exceptions (Chapter 9). Two, the movement of individuals in the way described will work against any effects of local adaptation and is likely to contribute further to the genetic homogenization of species throughout their distributions. Spatial expansion is likely to continue under these circumstances, unless physical features or the spacing of favorable localities prevent further movement or settlement. The physical features may be geographical barriers, such as mountain ranges, seas and rivers, or the spacing of unfavorable environmental conditions within a region. The higher the proportion of unfavorable environmental circumstances, the more impenetrable the area, and, hence, the greater the environmental barrier (Hengeveld and van den Bosch 1997). In some cases two subspecies, defined as such for their being observably discernible from one another and because they mate at random when in sympatry, are distributed differentially but do have a zone of contact that may be stable (as in the willow warbler's migratory divide in Scandinavia (Bensch et al. 1999)). Explaining these differentials, and their shifting pattern as climate changes, has yet to be achieved in ecology but a useful start has been made with Appalachian salamanders (Walls 2009).

The usual environmental conditions of particular species tend to be specific, because the physiological responses that determine survival rates, reproductive output and mobility of the individuals that make up the species tend to be specific (with some variance, depending on species). The three parameters listed define the rate of spatial expansion of that species; they are partly inherent to the species and partly influenced by the responses of the organisms to their environment (see collared dove example (Chapter 8)). Thus, the rate at which individual organisms avoid unfavorable conditions and their chances of encountering favorable ones is determined by their rate and pattern of movement, and thus by the spatial dimension of their movement, that is, by their contact distribution.

Through the specificity of the environmental responses of the individuals of a species and their spatial dynamics, species colonize new areas of suitable habitat that appear within their range (Eber and Brandl 1996) and they may also extend their distribution in this way (Hengeveld 1989a), with rates that are species-specific. Also, the directions in which distributions shift are particular to each species, and represent the advection term of spatial diffusion. This directionality results not only from the spatial distribution and dynamics of the prevailing climatic conditions but also the localities of the organisms initially. For instance, when species start expanding geographically from local refugia that sheltered them from extreme conditions, the particular refugial localities tend to be species-specific as do the geographical tracks followed by each species (e.g., Davis 1981b) because these relate to their species-specific adaptations, although they may be influenced by historical factors (so there is also a stochastic element to be considered). Different subsets of species tend to survive in different refugial areas, or even in sets of refugia (where they are spread geographically but provide the same environmental circumstances (e.g., Bennett 1997, Sommer and Benecke 2005, Schonswetter et al. 2006, Sommer and Nadachowski 2006)). All of this implies that the individuals of one species will be collectively out of step, in their spatial adjustment, with other species, and temporally out of phase with them, as seen in many Quaternary ecology data sets.

5.2 Temporal Scales of Variation

Time as a physical dimension is different from space in being one-dimensional and directional. In many respects, it is also recurrent because of the structure and dynamics of the solar system, which generates diurnal or seasonal variation, various temperature or rainfall cycles, glacial periods, and so on. Such cyclicity provides the basis for a certain degree of recurrence of environmental conditions, to which organisms may be specifically adapted, particularly to those of shorter duration such as diurnal and seasonal cycles. Although many temporal aspects relevant to the ecology of organisms are "cyclical" in this way, not all of them are so. In any case, these influences can be disrupted stochastically, as a result of changes in air currents and winds, cloud cover, and so on, and such influences may alter the amplitude of the oscillations or their period.

Specific physiological mechanisms allow the exploitation of seasonal variations, even if a specific set of conditions becomes manifest only irregularly. Irregularity demands that organisms must be capable of maintaining themselves across the maximum interval during which suitable conditions are not available. For example, established plants temporarily shut down their physiology and flightless aquatic invertebrates, such as notostracan and other crustaceans, encyst themselves as resistant stages (e.g., Thiery 1997). Frogs of some types burrow during periods of drought (e.g., Kayes et al. 2009) and predatory birds that rely on outbreaks of their prey for rearing their young skip breeding in lean years, as recorded for such species as the rough-legged buzzard (*Buteo lagopus*) and other raptors in Finland (Sundell et al. 2004).

Organisms have thus adapted to seasonal and irregular variation. They survive disruptions of periodicity through mechanisms that allow persistence despite the variance in amplitude of conditions (extremes), and across the variance in the duration of adverse conditions to which they are exposed. This implies that species are not adapted to the prevailing average climatic conditions envisioned in demographic models. They are adapted not only to a particular variance, skewness and kurtosis in the variables, but also to their sequence, recurrence and cyclicity. That is, the adaptations of organisms work to minimize the effects of environmental extremes that are experienced in the usual habitats occupied by the organisms in question.

An understanding of the functioning of these adaptations allows us to determine the ecological risks that the individuals of a particular species run in a stochastically influenced environment. Thus, the variance should be stable within certain limits set by the adaptations of the organisms. Such recurrence, even with its disruptions, allows living systems to remain within a locality for protracted periods, to the extent that their presence in one spot spans several environmental cycles. Sessile organisms, in particular, rely on such "predictability" for their survival. When, however, the temporal variance in living conditions does exceed the variance tolerated by the organisms they have to track the conditions to which they are adapted, across space, if they are motile. This reduces their mortality risk, unless the risks of movement are relatively much higher. If they are not motile the equation that describes the movement of propagules may be somewhat different, but such movement is built into the life cycle of most sessile organisms and takes place regardless, and usually seasonally.

The degree of recurrence of particular conditions and the typical pattern of variance associated with them will eventually be disrupted, through climatic change, and perhaps expressed initially with a more frequent occurrence of conditions more extreme than usual. Although individuals may not themselves be capable of shifting in response to such changes, their propagules usually establish at some distance from their parent. Over generations, the distribution of the organisms thus shifts with the shifting conditions. In some cases, the sessile adults may survive for several generations within a locality, perhaps persisting through vegetative reproduction, despite the unsuitability of the local conditions for reproduction or for the juvenile stages of the species (Pigott 1992). When the propagules cannot follow shifting conditions, extinction comes with the demise of the last reproductive individuals.

5.3 The Relevance of Species in Ecology—The Specification of Scale

The properties of individuals that are relevant to them in their interactions with the environment are acquired through the hereditary mechanism carried by their parents. For sexual organisms, the concept of the species gene pool is therefore of paramount importance in our understanding of such properties and their distribution among organisms and in relation to the spatio-temporal distribution of species (Walter 1995). Organisms derived asexually present different problems, but are likely to be less complex because the interpretation of their properties is not clouded by the subtleties introduced by sexuality.

In either case, the unique relationship between each of the adaptive mechanisms of a species and the structure and spatio-temporal dynamics of the environment dictates that both an adaptation-specific and a species-specific approach be taken in defining spatial and temporal scales of environmental variation. They must be specified with respect to the requirements and tolerances of each mechanism that is relevant to the environmental interactions of the species in question, and in relation to the environment itself. Even closely related species will differ from one another in these interactions with the environment and with respect to the relative weights of such interactions in their effect on the survival and reproduction of the organisms concerned. Species thus differ inevitably from one another ecologically. They do so not to promote coexistence (Sale 1988, Walter 1988a, van Klinken and Walter 1996), but simply because complete coincidence between them in the relationship between their properties and the structure and spatio-temporal dynamics of the environment is improbable given the diversity of axes of environmental differentiation (Appendix), and thus the vast array of combinations possible.

The various properties of a species operate at different scales of variation. Those that relate to escape from a predator operate at a finer, more immediate scale than those relating to the recognition of an appropriate mating partner, because this latter process has to be synchronized hormonally, for example. Also, temperature and humidity often vary at broader temporal and spatial scales than precipitation and thus require response rates in individuals that relate to this scale of environmental variation. After a species' response mechanisms have been decomposed, mentally, according to their response rates and the nature of their relationships with the relevant environmental variables has been established, the component parts can be reintegrated mathematically into a mechanistic model that accounts for the interrelationships between the various components and their relative weights during the response process as a whole. The ecological individuality of species shows up in the differences in the parameter values that represent the respective properties of the individual organisms of different species. Consequently, differences across species in the structure of the models that define their respective responses will be evident.

The global context within which environmental matching is conducted, by all individuals that make up a particular species gene pool, results in the observed geographical distribution of species and also the "lifespan" of species. An appreciation of the geographical distribution of any particular species can be used to develop hypotheses as to how the interactions of individuals influence the observed locality-specific responses of organisms to the environment. This means, conversely, that with the information derived from organisms about the way in which they respond to particular environmental variables, we can deduce 1) the range of conditions that are appropriate for the species, and 2) the spatial variation in environmental matching, since environmental limitations may differ from one locality to another. The species' distribution may also provide insight as to the environmental limits that can be tolerated by members of the species. Similarly, the temporary existence of a species in an area provides information on the duration for which particular conditions persisted, such as the occurrence of a fauna of mega-herbivores at higher northern latitudes in the now extinct grass steppe vegetation (Guthrie 1990).

5.4 Autecology and Scales of Variation

The underlying issue in autecology is the way in which individual organisms maintain their energy budget, and thus survive and reproduce (Kooijman et al. 1989, Kooijman 2010). This maintenance takes place by means of the deployment of the adaptations or properties of organisms that are relevant to the environment they usually inhabit, and which are deployed for different durations at various stages of the individual's lifeline (Rose 1997), and at various points in the lifetime movement of the individual, its lifetime track (Baker 1978). These functional interactions thus indicate the temporal and spatial scales of observation required to understand the spatio-temporal dynamics of the species concerned, and these are indicated by the responses of the individuals concerned.

From local study sites, we can move to questions of distribution. Where is a species distributed geographically, and how do the internal dynamics (as defined by the species-specific adaptations of the organisms) and the external dynamics (imposed climatically and biogeochemically) influence that distribution? The spatial dimension is thus expanded to the species' distribution, and the temporal dimension now becomes historical. The latter has to be reconstructed with what can be gleaned from molecular analyses and what can be resolved through paleontology. The information derived from investigations targeted at the two sets of dimensions, and relative to individuals and the species to which they belong, can now be employed to explain such phenomena as the local existence and intensity of occurrence of a particular species. From this, it is evident why the explanation of the local **coexistence** of species, relative to differential resource utilization functions (e.g., Mouquet et al. 2002, Zhang et al. 2004), targets the wrong process. The alternative approach focuses on explaining the local **existence** of each species in the area (van Klinken and Walter 1996, Walter 2008). This is done relative to the "provision" of suitable environmental circumstances for the species of interest. What dynamic provides suitable vegetation for a particular bird species, for example, or the open ground required by rainforest plants for seedling growth (e.g., Connell 1978)? Even rare species and those that appear only sporadically can thus be included, which cannot be done in community ecology (Walter 1988a). There are no shortcuts to these levels of explanation, as sought in the generalizations developed under the demographic paradigm. That is why conservation priorities, applied insect ecology and fisheries management all demand species-level treatments.

6 AUTECOLOGY, ADAPTATION AND SPECIATION

In the preceding chapters and sections we have explained why and how autecology focuses on the spatio-temporal dynamics of organisms in nature, as determined by the way in which they interact with their environment and match with its dynamic structure. In other words, organisms adjust spatially to that subset of the environment to which they are primarily adapted. They do so by movement across space and physiological adjustment through time, as required and as is possible for the organisms, for they depend on their species-specific adaptive mechanisms for these purposes.

The perspective on organisms in their environment, as outlined above, also has implications for evolutionary understanding, for it reveals which features of organisms are acted upon by natural selection in nature and how the environmental context influences its actions (Paterson 1986, Walter and Donaldson 1994, Finlay-Doney and Walter 2012a). This, in turn, reveals how and why evolutionary divergence takes place, and what leads to speciation. In brief, autecological considerations indicate that stabilizing selection usually predominates and that directional selection, significant evolutionary change, and speciation require special circumstances of geography and environment, as detailed and justified below.

In different localities or on different resources, individuals of the same species frequently experience differential survival rates or reproductive outputs (or both of these). Measures like these are usually portrayed, in evolutionary ecology for example, in terms of differential fitness outcomes that are expected to lead inevitably to directional selection and thus population divergence. In other words, adaptive change is considered to be ongoing in extended populations of species in their usual environment. Undoubtedly, selection for particular alleles does take place in extended populations and changes in allelic frequencies are commonly evident, but these changes are relatively minor in that they do not affect the mechanisms that mediate the primary interactions of the organisms concerned. Rather than seeing such relative declines (in survival or reproduction) as indicative of strong selection pressure within a geographical mosaic of different evolutionary trajectories within a species (Thompson's (2005) "interactive norms"), autecology sees these influences as simply reflecting a poor match between the organism and its environment in the place and at the time of evaluation.

Significantly, the adaptive mechanisms of organisms are species-specific. They are restricted to the species' gene pool and, conversely, have a species-wide distribution. How are complex mechanisms, in intricate combination, spread across extensive geographic areas? The traditional view is that a slow accumulation of small changes builds with time. However, the spreading of a substantive adaptive change across the entire distribution of a species is too unlikely a process to account for the observable fixed differences across closely-related species or the species-wide distribution of adaptive mechanisms. This last point raises a further logistical problem. Given that adaptive changes are considered to arise in different populations of a species (i.e., in different parts of its geographical distribution), how are such changes coordinated to produce a complex mechanism that is species-wide in its distribution? This should be a central question for evolutionary ecology, but no explanation has ever been offered. Such co-ordination across geographical space is simply not possible, and any interpretation offered would inevitably be unacceptably teleological. Also, reliance on the "on average" argument brushes too much aside and an explicit mechanistic model is required. The Recognition Concept of species developed by Hugh Paterson provides a model that addresses all of these concerns (Paterson 1981, 1985, 1986, Turner and Paterson 1991). All of this supports the view that adaptive change requires special circumstances.

Adaptive change to complex mechanisms requires an appropriate spatio-temporal context. The autecological perspective implies that the generation of selection pressures that force substantive evolutionary changes (such as those evident in the differences across closely-related species) requires, instead, environmental circumstances that are quite different from those evident in extended populations. Evolutionary change must be forced under particular circumstances of (i) reduced population size and geographical restriction, so that the genetic basis of the change to the complex adaptations can spread to all individuals in the population and become fixed in the population, and (ii) environmental conditions that challenge the adaptations, lifeline and reproduction of the organisms concerned, so that strong selection pressures are generated to drive the change through the entire population. (Darwin's analogy of artificial selection illustrates the process, for a small pool of isolated individuals, the desirable genotype, is subject to strong selection by all undesirable variants being removed entirely from the protected gene pool.) Should a population that has been through such a process evolve into a new species (as a by-product of that adaptive process, for the production of a new species cannot be selected for directly (Paterson 1982a, 1985), as envisioned in "ecological speciation" for example), the new mechanisms are species-wide in their distribution. They will be carried to all parts of the species' distribution when it expands geographically.

The nature of the selection is also significant. It must be directional and strong. Fixation is much less likely to take place if selection is based on differential reproductive output (fitness in competitive terms) than if differential survival is imposed by environmental circumstance. Such "hard selection" (Wallace 1968) results from the population under consideration being confined under environmental conditions to which the organisms are not primarily adapted. Known cases include small numbers of individuals being transported adventitiously to an isolated habitat, on an island for example, or being "trapped" in a small pocket of their usual habitat as the main body of that habitat retreats through the influence of climate change. Small relict populations on mountains or in valleys reflect this process. If ongoing climate change affects the environmental structure or dynamics of the isolate, it could set up a selective regime that is directional. Isolation, on its own, is insufficient, as indicated by the return of individuals from different refugial populations unchanged (but for slight genealogical divergence that can be measured phylogenetically) to a panmictic distribution over a broader area (e.g., Sommer and Benecke 2005).

Most isolated populations are likely to suffer extinction, but some survive. Pre-adaptation is likely to play a role in these latter cases. Individuals that carry adaptive mechanisms that are extreme relative to the main body of the species' gene pool may survive better, on a new host plant for example, and thus contribute to intergenerational persistence (Walter 1993b). Once the adaptive mechanisms have been adjusted to the new environment, probably usually with the added intrusion of pleiotropic effects, the spatio-temporal dynamics of the environment then dictate the ecological patterns we see around us. Environmental structure and dynamics impose on how organisms, with their adaptive mechanisms, can interact with their local circumstances, and this then dictates to a large extent where they will reach geographically and where they will survive and reproduce.

The above outline explains evolutionary stability and also diversification. It does so without the logical contortion of saying that populations are adapting rapidly, virtually always and everywhere within the context of a geographical mosaic of environmental structure (Thompson 2005, 2009a, b), but without overall change taking place in the species as a whole (as accepted by Eldredge et al. (2005)). These latter treatments find themselves in the position of having to differentiate "species characters" from "variable characters", but without providing an *a priori* method for identifying these differentials within the body of an organism.

The speciation model associated with the Recognition Concept of species also helps to explain why adaptation cannot be detected at the edge of a species' distribution, despite serious efforts to measure the expected change (e.g., Hoffmann and Blows 1994, Bridle and Vines 2007), for it does not take place there. The organisms at the edge of the distribution are in a relatively inferior environment and suffer reduced reproductive output and survival as a consequence (Angert and Schemske 2005). The Recognition Concept sees such differentials as a consequence of the organism-environment interaction, and not in terms of the fitness seen in evolutionary ecology to drive selection inevitably. "Fitness" thus reflects the conditions under which adaptive mechanisms evolved; it does not (and cannot) predict the future course of evolution.

The adaptive change that has been recorded in contemporary time in nature has been relatively trivial. The most convincing cases involve relatively simple genetic changes, and include the allelic substitutions and duplications that underpin insecticide resistance. Many other putative cases of local adaptation that are more substantial than insecticide resistance have been described in the literature. Many of these have sunk from view because they were simply cases in which more than one species had been mistakenly included in the interpretation. Perhaps the classic case is that involving the leopard frog *Rana pipiens*, but dealing with this requires an appreciation of certain aspects in the development of species theory.

The polytypic species concept (Mayr 1963, pp. 334ff.) was developed to cope with species as complexes of populations rather than as invariant typological entities. In particular, it dealt with the variation within species that could be expected from Darwin's competitive model of natural selection. This addition to species theory effectively added a geographical dimension to dealing with species and so complemented the isolation concept of species (which was then called the biological species concept (see Glossary)). Although a generalized "leopard frog theme" could be recognized (Moore 1975), regional differences and clinal variations were evident. Darwin's de-emphasis of species thus ultimately led to the elevation of local populations as the foci for understanding evolutionary change, which was anticipated to be contemporary rather than historical.

Differences were therefore expected across species, even significant adaptive differences, and allopatric populations could be related to one another in terms of their being conspecific populations in the process of divergence. Consequently, the number of acknowledged species dropped dramatically (Mayr 1963, pp. 341-342). The decisions leading to this reduction were largely taxonomic, although any amount of gene flow through cross mating was taken as a signal of conspecificity and therefore that the populations were indeed diverging in contemporary time. The different variants were seen as incipient species, although this perspective remained mainly tacit. The prevailing view of the leopard frog thus developed: "... *R. pipiens* extended from Canada to Panama, with each population adapted to local conditions. These populations were thought to represent a reproductive continuum, although specimens from distant localities or distinct habitats might be reproductively incompatible as a result of local adaptations" (Hillis 1988).

In subsequent investigation of various aspects of the leopard frog's biology, much more attention was paid to the precise geographical context and details relevant to each study. More precise appreciations of the morphology, vocalization, reproductive behavior in contact zones, and karyotype of the various populations were obtained, and gene flow in local areas of overlap could be assessed. Over two dozen independent gene pools, or species, are now acknowledged and these are genetically, physiologically and behaviorally quite distinct from one another (Moore 1975, Hillis 1988). An appreciation of hybridization in this context is important. Hybridization is frequently enough seen to "erode" species' integrity. But this view sees species as having "integrity" or "cohesion" (Petit and Excoffier 2009), as if this were a feature that could be protected by natural selection. Such thoughts represent a return to the typology inherent in the isolation concept of species (Hengeveld 1988b, Walter 1995). The important point with regard to hybrid zones is the stability of adaptive features on either side of them (e.g., Packert et al. 2005), and it is the relative independence of the allopatric zones that is of primary significance. This is true even if hybridization takes place at random in the overlap zone. In such situations we are dealing with subspecies defined in population genetics terms (Ford 1974, Walter 2003, p. 162-163). In other words, the adaptive process in small populations ends (or stabilizes) when the complex adaptations and life cycle of the organisms in that population function effectively in the new environment. The change has no endpoint other than that, so we can expect to see all degrees of differences between populations of organisms in nature, from full species to almost complete similarity. The status of these populations, relative to one another, can be assessed by human observers in terms of their potential to share in fertilization with their respective parental populations and, where there is geographical overlap, the relative frequency of hybridization.

Numerous such "polytypic species" have been re-investigated since. With regard to the leopard frog, this revelation came a long time ago, but the same mistake in interpreting species' limits is still being made. The problem is that such erroneous cases are more than likely to provide the examples of contemporary cases of incipient speciation, reinforcement and ecological speciation. Such mistakes can be anticipated (or foreshadowed) from the understanding of species generated by the Recognition Concept of species and an appreciation of autecology.

7 OVERVIEW

Ecologists must question, investigate, and interpret ecological situations at the appropriate spatial scale individually for each species that is involved. The ecological picture is considerably simplified if species are not combined for such investigative purposes, unless one of them contributes to the environment of the other. If co-occurring species are combined to answer questions of spatio-temporal dynamics and local diversity, we have to deal simultaneously with organisms that operate at widely different scales of variation. This is impractical to the point of being impossible, and could be considered ecology's "two-species problem", a legitimate equivalent to astronomy's three-body problem (Peterson 1993). Such an approach creates the impression that nature is more complex than it actually is.

In this way the relevant environmental variables and scales of environmental variation to which the individuals of the species of interest are adapted can be defined appropriately and accurately. This approach makes sense of the almost limitless variation across species by which they match the environment and, hence, of the boundless variation with which they interact with the environment. Despite the variation in ecology across species and the way in which individual organisms cope continuously with ever-different conditions, such situations can be generalized in theoretical terms as well as statistically, and these aspects are expanded in the last few chapters.

In the meantime, the two types of "structure" identified in this chapter, structured adaptive mechanisms and spatio-temporal structure of the environment can be brought together and discussed in terms of "structured processes in ecology" (Chapter 7), which allows expansion of the concept of spatio-temporal environmental matching by organisms. First, however, we look much more closely at the mechanisms by which organisms interact with the environment, for they are integrated across their morphological, physiological and behavioral components to achieve a particular end. We conceptualize such integrated wholes as "environmental response systems" in Chapter 6.

Chapter 6

Environmental Response Systems of Organisms
Co-determinants of Spatio-temporal Dynamics

CHAPTER OUTLINE

- Introduction
- Autecology in relation to ecophysiology, ecomorphology and ethology
- The questions that define autecology
- Environmental response systems
- Species-specificity and ecological context
- Conclusion

1 INTRODUCTION

Earlier chapters covered the adaptive mechanisms of individual organisms. These are the adaptations or properties of organisms that are involved in matching the structure of the organism's life cycle to the dynamic structure of the environment. This chapter deals more precisely with the intricate nature of these mechanisms and how they mediate the environmental responses of organisms and thus contribute to the spatio-temporal dynamics of species in nature. Such mechanisms are intricately structured. They simultaneously involve physiological, behavioral and biochemical processes, as well as anatomical features, and we tend to see their actions expressed in terms of organisms responding behaviorally and physiologically to environmental signals and influences. Since other organisms form part of any organism's environment, these signals may be of a social nature.

The mechanisms with which organisms respond to their environment represent combinations of various facets of their internal workings, mainly biochemical, anatomical, physiological and behavioral. Despite the integrated actions of all these contributing processes, they are traditionally assigned to different sub-disciplines in biology. The sub-disciplines, in turn, tend to remain somewhat discrete from one another in practice, to the extent that the specific aspects of the mechanisms investigated by each are often not dealt with as truly integrated systems. Fertilization mechanisms provide a good example. The intersexual signaling of reproductively mature adults is usually dealt with independently of the coordinated maturation of the gonads or the structural compatibility of the sex organs, and so on, despite each playing a significant role in ultimately achieving fertilization and syngamy. We thus lose important aspects of the organisms' synchronization with and adaptation to particular aspects of the environment.

The reductionist-based understanding described above is essential for interpreting how and why particular components of complex adaptations operate. Nevertheless, a synthesis across the components is required relative to the ultimate function(s) of the complex mechanisms, for it is these functional features that dictate the life and life cycle structure of organisms and thus their ecology. Although such synthesis should fall within ecology and evolutionary biology (or population biology), very little integration of this nature is available yet, probably because ecological theory currently pays relatively little attention to individual organisms and their place in ecology, and evolutionary ecology focuses mostly on perceived fitness consequences of improved efficiency in the working of adaptive mechanisms. Autecology provides a way around this problem because it focuses on individuals and the integrated responses within them that direct the way in which they interact with their external environment.

Each particular type of response of organisms to their environment, however simple it may appear, relies on this concerted operation of a number of very different mechanisms. For example, escape responses are outwardly simple, but typically include sensory, behavioral and locomotory components as in cockroaches (see Manning and Dawkins 2012, pp. 7-11), and also hormonal systems in mammals (see Moyes and Schulte 2006, pp. 337-338), with the specific outcome related perhaps to the state of the organism's energy budget, water balance or thermal circumstances. We can generalize such concerted sets of mechanisms in terms of them being "Environmental Response Systems", a concept introduced earlier as "Ecological Response Systems (Walter and Hengeveld 2000) but we now emphasize that the interaction is with the environment as a whole. This chapter details the mechanisms involved in environmental response systems, with examples, to demonstrate how the effects of the operation of such mechanisms dictate the spatio-temporal dynamics of organisms.

In preparation for this, we first examine the relationships between the various biological sub-disciplines that are relevant to developing meaningful questions in ecology and thus to ecological investigation and interpretation. The main point of discussion is the articulation between autecology and the sub-disciplines that underpin it (Sections 2 and 3). We then extend this perspective to consider how ecological generalizations can be developed with regard to the species-specificity that is evident in the way organisms interact with their environment. To do this we deal with the role of individuals in ecology, and the nature of their species-specific ecological properties (Sections 4 and 5). These considerations lead to an expansion, in the following chapter (Chapter 7), of how information about environmental response systems is extended and quantified to provide specific information about survival probabilities and reproductive outputs.

2 AUTECOLOGY IN RELATION TO ECOPHYSIOLOGY, ECOMORPHOLOGY AND ETHOLOGY

Physiological, morphological and ethological mechanisms have long been seen to have important ecological impacts, hence the disciplines of ecophysiology (e.g., Lambers et al. 1998, McNab 2002, Karasov and Martinez del Rio 2007), ethology (e.g., Gould 1982a, Burkhardt 2005), and ecomorphology (e.g., Wainwright and Reilly 1994). Because physiological processes do not operate independently of organismal structures, and behavior is inevitably tied up with physiology, and *vice versa*, we use the single phrase "environmental response systems" for the composite mechanisms, of mostly singular function, that characterize the interactions of organisms with their environment.

Such environmental response systems cover the complex, functional processes that relate the organism to its surrounding environment and support it within that environment. Information on the modes and circumstances of operation of such mechanisms is critical to ecological understanding, for these processes determine how organisms match with environmental conditions and thus influence directly their survival chances. For example, the presence of a particular insect herbivore on any particular species of plant, even if it is considered a "generalist", is likely to be a consequence of the structure and operation of its host-finding mechanism (Rajapakse et al. 2006) relative to its environmental circumstances. This statement is not intended, however, to downplay the other adaptations of these organisms to their host plants. The ecology of vertebrates is also influenced in this way. Ground woodpeckers (*Geocolaptes olivaceus*) feed on only a subset of available ant species, for they search out nests that lie just beneath stones and feed almost entirely from them (Oatley et al. 1989). Their sensory physiology, foraging behavior and other adaptations combine to direct their feeding in a particular way, even when other ants are common on the surface of the ground.

The sub-discipline of ecophysiology deals with the complex mechanisms that underpin such processes as thermoregulation and osmoregulation, usually in association with biochemistry, physiology, morphology, and behavior (or ethology). Such mechanisms and their consequences for survival have been studied mostly in relation to extreme local conditions by ecophysiologists (e.g., Schmidt-Nielsen 1990, Alexander 1999). "Extreme" is usually interpreted in relation to human tolerances and perceptions, so emphasis is placed on desert conditions, hyper-salinity, extremely low temperatures, and so on. Appreciating this bias foreshadows the need to define concepts like extreme and stress in relation to the adaptive mechanisms of the particular species concerned, or even to particular developmental stages of individual organisms and their degree of acclimatization. The species-specificity of "extreme" in terms of environmental stress means that the specifics are not readily transferable to other species or life stages.

Ecomorphology is the sub-discipline that relates the internal and external morphological structure of organisms to their environment and way of life, as seen in Frazzetta's (1975) deep appreciation of the relationship between the tree dwelling habit in some python species and the adjustments to their body and prey grabbing structures to cope with arboreal life. The functional requirements of the mouth for grasping, holding and swallowing prey are provided by subtle mutual adjustments to the skeletal, cartilaginous and muscular structures of the head. Ecophysiology and ecomorphology thus deal with the internal organization of organisms and the direct interface of the organism with the environment, including that provided by a range of sensory aspects.

Ethology covers the spatial and temporal adjustments of the organisms to external conditions, as informed by endocrine and neurosensory input and feedback. Although each of these sub-disciplines also deals explicitly with the survival of organisms, the scale of investigation is relatively short-term in looking at immediate effects or responses under particular circumstances. The questions addressed thus often relate primarily to the way in which particular organisms cope with "extreme" environmental conditions (e.g., when crocodiles enter the marine environment (Taplin and Grigg 1981)).

Such studies do not, therefore, aim to relate their findings explicitly to the chance of survival of the organisms throughout their entire lifetime, across generations or across different biotopes. They do not consider the long-term survival of sets of organisms nor the persistence of organisms of a particular species over many generations; they concern themselves only with the properties and processes within the lifetime of an individual organism, and often only part of its lifetime. Although behavioral ecology (*sensu* Krebs and Davies 1997) often projects the long-term consequences of certain types of behavior, such as the finding of food, it does not do so by considering the lifeline and lifetime track of individuals. Instead, the questions of behavioral ecology generally focus on relative efficiency among individuals and, from such considerations, the projected costs and benefits to the individuals concerned and this is expressed in terms of fitness, and even in terms of future fitness. The shortcomings of this approach are considered later (Chapter 10), as are the distinct differences that exist between ethology and behavioral ecology.

To relate the operation of environmental response systems, as a whole, to the lifetime survival and persistence of organisms across generations requires different information from that gathered under the auspices of the sub-disciplines dealt with above, and it is derived from both a different approach and a broader scale of technical investigation. This point is expanded next, to illustrate the relationship between the various sub-disciplines of biology, to show how they relate differentially to the same subject matter, and to clarify and emphasize that none has priority in that all are equally relevant to a full understanding of any biological system and to the quantification necessary for ecological understanding.

3 THE QUESTIONS THAT DEFINE AUTECOLOGY

3.1 Relationships between Sub-disciplines in Biology

In science, the various disciplines focus on somewhat different phenomena and on different questions. They nevertheless link with one another logically through their dealing with common subject material, to the extent that distinct boundaries do not exist between the disciplines. Thus chemistry overlaps substantially with physics, both overlap with biology, and so on. Within biology, as a discipline, the various sub-disciplines are also defined in terms of their subject material, which again overlaps.

To understand how the sub-disciplines of biology are related to one another, at least two further points need consideration, and they entail considerable subtlety. Both relate to the questions that are given primary importance in the various biological sub-disciplines. The first aspect has to do with how one relates the questions of one sub-discipline to those of another. The second relates to paradigmatic differences within sub-disciplines (as exemplified by the differences between demographic ecology and autecology). In this section we examine the principal questions of ecology from these perspectives. This enables these particular ecological questions to be related to questions investigated within other sub-disciplines of biology. Our main aim in this respect is to link the subject material of the sub-disciplines in mechanistic terms. The impact of paradigmatic difference is subtle, but deep, and is also elucidated in this chapter. Later, in Chapter 10, we use the framework developed here to examine the place of autecological theory in relation to that of the other sub-disciplines in biology, and to compare it with the equivalent scheme for demographic ecology.

3.1.1 Links across sub-disciplines

Following Rose's (1997) view of the structure of biological science, we first draw the distinctions that characterize the different sub-disciplines within biology. We do this by examining the differences in how biologists from different sub-disciplines typically investigate the escape response of a frog, say, from a viewpoint characteristic of each sub-discipline. Thus, a physiologist might be expected to ask questions relating to the gathering of sensory (environmental) information, its transfer to the brain, its interpretation within the brain, the subsequent transfer of a response to the muscles, and the integrated contraction of those muscles. Biochemists analyze and interpret the biochemical properties of the retina, axons and muscle, to determine the molecular and ionic train of events that leads to the transduction of environmental signals, generation of nerve action potentials, synaptic transmission and muscle contraction. Molecular geneticists question the relationship between particular enzymes and the genes that code for them. At the whole organism level, behaviorists investigate aspects relating to why the frog jumps, perhaps with regard to the other options available to the organism, such as crypsis. Finally, the historical element is introduced by evolutionary biology. Why does the frog jump? How did it acquire that ability? At what stage in the evolutionary process did frogs acquire that ability? And so on. Despite the above outline being something of a caricature, it conveys at least two revealing points.

One, each sub-discipline focuses on somewhat different aspects of the same event (or type of event). In the example above, it is jumping in frogs. Biological systems and the study of biology are thus not hierarchical (Rose 1997), as commonly claimed. Biologists in different sub-disciplines have different foci and techniques, but they all contribute to the synthetic interpretation of how and why frogs jump. As Rose (1997) states, the sub-disciplines are largely epistemologically separated but contribute to a single ontology or interpretation. There is no hierarchy. This argument holds for ecology, as well, but the hierarchical view of the organization of organisms, as well as ecology as a discipline, persists (e.g., Price and Waldbauer 1994, Huffaker et al. 1999, Schowalter 2000, Mayhew 2006, Pickett et al. 2007, Jagers op Akkerhuis 2008, Krebs 2008).

Two, the lack of a hierarchy is confirmed by the "blending" that is becoming more apparent between the sub-disciplines than was evident in their earlier, formative days, with the broader application of molecular techniques for example. It is even evident in the widely appreciated "four questions of ethology", which specify the need to understand the causation of behavior (perhaps as influenced by environmental and physiological processes), the function of that behavior (or its goal relative to survival and reproduction), its evolutionary origins (phylogeny) and its development as the organism grows (ontogeny) (see Manning and Dawkins 2012). Each question thus has a specific task within its particular sub-discipline and each such question clearly has no more relevance than any of the others (Rose 1997). For a complete understanding of any ethological phenomenon, we require information from all of the sub-disciplines. The articulation of the questions in this way effectively provides a perspective on the differential focus of the different sub-disciplines (see Manning and Dawkins 2012).

3.1.2 Paradigmatic differences within sub-disciplines

The specific questions that are typically addressed within each sub-discipline shift according to the interpretation of adaptation that is invoked. Two different approaches are thus evident in studies of behavior, for example. On the one hand, ethologists analyze complexes of behavior in relation to their ultimate function(s), with an emphasis on the subsidiary functions and their transitions (see Figs. 3.11 and 3.12). The whole (or at least a substantial part) is thus analyzed as an adaptive structure, a mechanism by which an end is achieved by organisms of the species of interest within a particular environmental setting. That mechanism thus contributes in a specific way to the survival or reproduction of the organism in question. The herbivorous insects and ground woodpeckers dealt with above provide typical examples. Although the behavioral structures under consideration comprise regular sequences that can be represented in ethograms, alternative routes within the ethogram are evident, with such variation in behavior and its stochastic influences being quantifiable (e.g., Morris 1970, Haccou and Meelis 1994, Rafter and Walter 2013). Stochastic influences on these systems can thus be realistically included.

In ethology, as just outlined, the behavioral structures are assumed to have evolved in the past, in relation to selection pressures from the environment. An understanding of the structure thus allows one to determine how and why that mechanism evolved, and from what it evolved. The mechanism is treated as representative of the species, with alternative behavioral options being built into particular parts of the behavioral sequence. The tacit assumption would thus be that the mechanism is under stabilizing selection, for it evolved to fulfill a particular function under particular environmental conditions (Paterson 1986, Walter and Donaldson 1994). Variants that are too extreme would not fulfill that purpose and the individuals involved would likely suffer reduced survival or reproduction rates, which results in stabilizing selection.

This strictly ethological approach does not invoke or rely upon optimizing selection and so provides a strong contrast with the approach taken by behavioral ecologists (Walter 2013). The behaviors and their transitions given significance by ethologists are treated as relatively unimportant by behavioral ecologists, mainly because they are seen as proximate mechanisms whose study does not reveal much in an evolutionary sense (e.g., Waage and Hassell 1982, Kokko and Jennions 2010). Behavioral ecologists elevate the "ultimate" side, the evolutionary explanation, but it is an evolutionary explanation of a particular slant. Since behavioral ecology was developed on the axiom that natural selection is optimizing selection (see Ghiselin 1974, Collins 1986, Walter 2008), understanding of the actual mechanisms has been sacrificed in favor of some form of quantification of the postulated efficiency with which organisms perform certain tasks (such as in the host-searching behavior of parasitic wasps (e.g., Murdoch and Briggs 1996)) and the inclusive fitness benefits achieved through organisms taking certain options (such as in becoming social and maintaining the social habit (e.g., Abbot et al. 2011)). In both cases, the fitness benefits are seen to be accrued immediately, because the system is considered to run deterministically on competition (hence the emphasis on maintenance of adaptations rather than on origins (Walter 1993b)). That is why kin selectionists interpret a particular sex ratio "strategy" of a parasitic wasp or social hymenopteran in terms of the number of grandchildren it will produce (Hamilton 1967, Wilson 1975, Hölldobler and Wilson 1990, Cronin 1991). Such analyses are not necessarily even related to survival rates or actual reproductive output, but are related rather to the reproductive output of other individuals, which are seen as competitors. An allele for greater efficiency is thus assumed to displace a "less efficient" allele, as in the workings of the Hardy-Weinberg equation.

3.2 Ecology as a Sub-discipline in Biology—Patterns and Questions

Ecology asks questions as to why organisms of particular species occur where they do and about the intensity of occurrence of species in particular areas. These are both questions of dynamics. Demographic ecology and autecology tackle these issues in different ways from one another, because each identifies and emphasizes different ecological patterns for explanation, and each has its own set of assumptions. The assumptions do not overlap across the two approaches and cannot be reconciled with one another, as illustrated in Chapter 2. These differences are so fundamental that they demonstrate there are two paradigms in ecological interpretation (Hengeveld 1997, Hengeveld and Walter 1999, Walter and Hengeveld 2000, Walter 2008).

By specifying the assumptions of each paradigmatic approach to ecology we can also start to draw relationships between autecology and the other sub-disciplines in biology, as well as those for demographic ecology (with its two main arms, population and community ecology). The deep, consistent differences between autecology and demographic ecology can thus be specified further.

3.2.1 Demographic ecology

Demographic ecology recognizes patterns at two levels, the population and the community. Although the occurrence and structure of such patterns are persistently questioned, they still drive investigation in ecology. Essentially, the approach is deterministic, as reflected in its associated assumptions. A feature of current ecology is to publish "consensus" papers, ones that claim general agreement that past debates of these issues have been somewhat diversionary and that we can now move on by accepting some combination of the original determinism and stochasticity (e.g., Turchin 2001, Coulson et al. 2004). Consensus of this nature tends to maintain the original demographic emphasis, for it simply adds a generalized stochastic "correction factor" to bring model output in line with observed patterns. This approach thus fails to extend understanding of how the adaptations of organisms mediate their interactions with the environment and how these, in turn, influence the spatio-temporal dynamics of the species, as explained through the remaining chapters.

Situations in which demographic theory apparently works well relate mainly to the application of demographic principles to the study and interpretation of the dynamics of common species. This theory was developed mainly with respect to species that are numerous and widespread (Walter and Zalucki 1999). The wide distribution and high densities of such species may well have convinced ecologists that physico-chemical conditions are not of primary importance in ecology. Also, density-related (demographic) influences may well be evident in the local ecology of the species, such as through territoriality. But field data reveal little consistency in such demographic influences and demographic trends, and interpretation came to depend more on model output (Hassell and Godfray 1992), which clearly indicates a deep problem. And when organisms in the field are considered in relation to their physiological requirements and tolerances, and how their sensory and behavioral interactions with environmental variables may influence their survival and reproductive output (see Section 4.1, below), it becomes clear that many factors influence the spatio-temporal dynamics of organisms. Further, these factors may act interactively (Section 4.3, below), and different ones may predominate in different areas and at different times. Such an approach applies to both common and rare species. That demographic theory may seem to work well in some cases is perhaps not surprising. Its failures, though, need to be considered in any assessment of its general utility, as does the direction provided by its questions.

Instances in which the demographic approach applies are, therefore, those in which the local abundances are high enough and consistent enough that the densities of the organisms themselves seem capable of representing ecological factors in their own right. These are situations in which all else does, indeed, seem equal. Agricultural pests and biological invasions sometimes seem to fall into this category, if only for relatively short periods. In such cases, though, the environmental conditions in areas that host these species have been simplified to a large extent and rendered spatially uniform, at least in relative terms. Such conditions tend to be conducive for only the handful of species that match with them. The extent to which they match with the prevailing conditions, and the spatial extension of these conditions will dictate how abundant and widespread they become. Pests and weeds, therefore, do not represent particular biological categories, as some have inferred, but their ecological "performance" reflects the spatial extension and uniformity of conditions to which their adaptations and requirements match well.

Invasions are often human-induced but not inevitably so (Botkin 2001). They become obvious, even menacing, when the net rate of reproduction of the invading species remains consistently higher than the replacement value that would keep density stable. This happens when environmental conditions after the incursion or introduction are favorable for them or somehow become more favorable. In all these cases, the numbers of individuals may be so high that they appear to become an ecological influence in their own right, although the primary ecological influences that allowed them to become so high in the first place should not be forgotten. Further, density effects at such population levels tend to be rather unsubtle (Fig. 7.5) and they still need to be interpreted in relation to the backdrop of seasonal and other changes in the environment (e.g., Chapter 2 on psyllid outbreaks) for as soon as the environmental conditions deteriorate, numbers lose any forcing quality they might have had.

The dynamics of some other processes may also be influenced by extremes of densities of organisms. These are primarily biotic interactions between individuals, such as predator and prey, mating partners, and so on. These are functional interactions (Walter 1993a) and are mediated by behavioral mechanisms so they are unlike competition that takes place through resource consumption. Also, high densities of organisms sometimes alter the usual behavior of some individuals. When the availability of foliage of a particular host plant species, for example, falls to particularly low levels, whilst insect herbivore numbers are still high, the insects sometimes move to nearby plant species upon which they do not typically feed (e.g., Marohasy 1996). But again the numbers are secondary driving forces in the ecological process, being themselves determined by primary factors that influence the survival of individuals in the first place. The effects of density can, in such cases, be studied independently, but such densities can be understood meaningfully only when they themselves are interpreted in relation to the factors that elevated them originally (Walter 2008).

We emphasize again that in most species numbers are so low that they do not exert a demographic force by themselves; most species are rare, even in their distributional center where conditions are, on average, at their best for them. In short, the demographically based interpretations leave out too much information on the individuals' environmental response systems that is important in understanding the ecology of any particular species. Although demographic influences may seem to be operative, close examination will inevitably reveal a wealth of ecologically significant information that is necessary to interpret the distribution of local abundances realistically and as accurately as possible, as the psyllid example of Chapter 2 illustrates. Even when numbers, through density effects, appear to operate as driving forces entirely of their own accord, they (and their effects) can be explained only in relation to the primary influences around which autecological interpretation is developed.

3.2.2 Autecology

Autecology differs fundamentally from demographic ecology in emphasizing the primary pattern in ecology as being the species-specific spatio-temporal dynamics of organisms. The species-specificity inherent in such spatio-temporal dynamics has achieved some notoriety in ecology through Gleason's (1917, 1926, 1939) ideas on the "individualism" of species. Such individualism or ecological idiosyncrasy is generated by the interactions between the species-specific adaptations of organisms and a multifaceted and structured, but dynamic and stochastic, environment. The questions of autecology relate to these interactions. How are the interactions mediated and structured? How are they influenced by environments (or biotopes) that differ from one another? How do the interactions impinge on the lifetime track and lifeline of the subject organisms? What are the consequences for the geographical representation of the species and for its intensity of representation across space and through time?

In addressing these questions autecological investigation clearly draws on information from physiology, behavior and other sub-disciplines, as outlined above. But the way in which the information and insight is used must be adapted to the requirements of ecological understanding. So the consequences of behaviors, for example, are quantified to reflect an individual's survival chances and reproductive output, which are specified probabilistically. These chances reflect the likely fate of an individual organism within its environmental setting. Thus, the specific questions asked elicit the particular types of information required for understanding the ecology of a species. When behavioral aspects are under consideration, it should be expected that the information seen as significant to autecological understanding may be seen by ethologists as somewhat superficial. The autecological information is given a different context from that of ethology by the other ecological information supporting it, and so is enhanced through the combination into a broader, ecological, interpretation.

3.2.3 Ecology, other biological sub-disciplines and ecological hierarchies

The assumptions of autecology are consistent with those outlined for ethology in the previous subsection. This conjunction of basic premises helps define the relationships, within biology, among the different sub-disciplines. Clearly, autecology is strongly related to the ethology of Lorenz, Tinbergen and others, but has relatively little in common with behavioral ecology and optimizing selection. Logically, the latter way of interpreting behavior and selection is consistent with the premises and practices of demographic ecology.

Demographic ecology is almost universally portrayed as hierarchical, which is consistent with the hierarchy envisaged within organisms, from the molecular level to cells through organs to the individual. Extending the hierarchy into ecology, above the level of the individual, introduces serious disjunctions into ecological interpretation. It separates the actions of individuals from interpretations of populations, for example, and populations from local diversity (referred to as communities). Each such level is portrayed as a legitimate area of ecological endeavor, but without the levels being well integrated through mechanistic connections between them. Logical connections do exist between populations, metapopulations and communities, in terms of their underlying principles of equilibrium and density related regulation (see Walter 1995), but integration among all levels remains weak, for the following reasons.

1. The biological identity of the individual organism as the basic unit of response to ecological circumstances has been lost entirely, for adaptive mechanisms are excluded from interpretations of abundance (Walter 2008).
2. Metapopulations are portrayed as populations of populations, with each population still treated in logistic terms (as pointed out by Caughley (1994)) and in relation to the premises of population regulation theory. The acceptance of metapopulation theory has been enhanced, despite its flaws, through many studies of individuals and their properties being given a metapopulation context, often enough retrospectively, but without explanation or justification (e.g., Ehrlich and Hanski 2004).
3. In community ecology, species become the units within the perceived structure, but such treatment is unacceptably typological (Hengeveld 1988b, Walter 1991, 1995), and even were populations to be substituted for species, the group selectionist overtones are unacceptable (Walter 1991).
4. As units of analysis, group statistics are used, such as N_1 and N_2, and these combine the effects of more than one process on the survival and reproduction of organisms. For example, temperature and predation can affect the dynamics of a particular species in a single locality, yet N combines these influences into a single parameter. Further the multiple processes involved are usually not identified in the general models.
5. Species are commonly neglected, even denied, as ecologically relevant categories. Yet, many ecological processes are most effectively and realistically analyzed at this particular level (Chapter 5).

Ultimately, demographic ecology becomes separated as a sub-discipline in biology from the related biological sub-disciplines that should underpin it, for there is no mechanistic connection between physiology, for example, and the processes emphasized in demographic theory. This mechanistic basis has been omitted in favor of such epiphenomena as population equilibrium, density-related processes, community structure and macroecological patterns (Walter 2008).

Autecology, in contrast, proposes a different view from the hierarchical one outlined above. Ecology is, instead, seen as a single sub-disciplinary entity that brings a specific epistemology to the development of a unified ontology of organisms. Its mechanistic continuity with developmental biology, molecular genetics, biochemistry, physiology, morphology and ethology is strong, for it focuses on integrating the insights and information derived from these sub-disciplines into ecology, by providing a framework on which to arrange their ecological relevance. In other words, the operation of the mechanisms investigated in these various sub-disciplines is related to the chances of survival of individual organisms within the environmental contexts that are relevant to them, and ultimately to the fate of local "populations" (Walter 2008). Such autecological information is, in turn, valuable within the context of evolutionary and phylogenetic theory, as it allows species' environmental requirements, interactions and relationships to be interpreted relative to one another. Significantly, autecology does not result in ecology being broken into a hierarchy as explained more fully in Chapter 10.

The questions of autecology are thus aimed at the lifetime survival of organisms in the short term, and at their persistence across generations in the long term. Such temporal scales, covering at least the lifetime of an individual organism from fertilization and birth to its ultimate demise, generally fall outside the scope of ecophysiology, ethology and ecomorphology. Moreover, these sub-disciplines are usually not concerned with processes as they relate to spatial scales (although movement is dealt with as a behavioral phenomenon). Spatial processes are essential to the survival probabilities of individuals and, hence, to autecological understanding. They relate primarily to an individual's mobility with respect to the spatial kaleidoscopic dynamics of environmental conditions, as well as to the accessibility of any localities that become suitable environmentally. This spatial adjustment is the primary means by which organisms deal with the vagaries of the environment, for genetic adaptation takes place only under specific circumstances (Chapter 5, Section 6).

4 ENVIRONMENTAL RESPONSE SYSTEMS

4.1 Understanding the Environmental Response System

The environmental response system of an individual organism comprises a suite of interdependent and interactive mechanisms. Each of these mechanisms serves particular, sometimes divergent functions, including osmoregulation, thermoregulation, digestion, vision, movement, sexual signaling and defense, among others. Together, they operate in relation to the particular environment inhabited by the species of interest, and to the dynamics of that environment. Making up the environment are other individuals, both conspecifics and members of other species.

Autecologists measure the survival and reproductive probabilities of organisms under the range of environmental conditions to which they may be exposed, whether through experiments, observations or field sampling. They need to understand enough about the mechanistic basis of the adaptations involved, regardless of whether they are related to thermoregulation and water balance, or to the sensory basis of behavioral responses, to interpret the results of their observations. Usually, and often necessarily, these mechanisms are studied in isolation of one another, to establish their mode of action and function.

Autecological research is designed to understand these mechanisms sufficiently well to reveal their significance to the life and reproductive capacity of individual organisms across localities and over time, with the purpose of synthesis into ecological interpretation. Ecologists confront a range of difficulties in fulfilling such an aim.

1. Various such mechanisms operate simultaneously within an individual, in various combinations and sequences.
2. They may well operate only within a particular stage of the life cycle, again in combination with other mechanisms.
3. In most animal species, the relevant mechanisms have to be evaluated relative to the motility of the organisms within an ecological context, and this is often of extensive spatio-temporal dimensions.
4. Ecological influences operate interactively. For example, the influence of changes in salinity (within an estuary, for example) needs to be investigated in relation to temperature profiles if survival probabilities are to be evaluated accurately (e.g., Haefner 1970, Bjorkman 2000, Sardella et al. 2004, 2008, Thomas et al. 2007). Some interactions are even compensatory, so a high water table, abundance of fog and even low temperatures may compensate for low rainfall (Daubenmire 1974, pp. 346-348, Fischer et al. 2009).
5. The dynamics of the environment add another layer of complexity. Climate change, vegetation dynamics, biogeochemical shifts and anthropogenic influences all impact on ecological systems and need consideration. These are the primary influences on the structure and spatio-temporal dynamics of environmental conditions, and provide the subject matter of ecosystem ecology (see Chapter 10).
6. Climate is a primary motor in ecology. It dictates the geographical distribution of potential living conditions for each species, where different vegetation types can occur, as well as the temporal and spatial scale of various ecological influences. The unpredictability of weather conditions and climatic changes, both temporally and spatially (e.g., Mathieu et al. 2004, Rodwell et al. 2004, Sutton and Hodson 2005, Sutton et al. 2005, van Oldenborgh 2005), are obstacles to anticipating environmental conditions elsewhere or in the future.

All of the above factors and influences operate directly through the environmental response systems of individual organisms and need consideration when survival probabilities are measured and estimates of reproductive output are made with respect to locality, environment and environmental dynamics.

Many types of environmentally relevant mechanisms of organisms are fairly well understood in ecophysiological, ecomorphological, and ethological terms (see Gould 1982a, Schmidt-Nielsen 1990, Alexander 1999, Burkhardt 2005, Moyes and Schulte 2006). However, many remain poorly understood. Perhaps the most significant of these to autecological understanding are ones that involve sensory processing and the movement of organisms (Dusenbery 1992). As an example, the polyphagous thrips *Frankliniella schultzei* uses multiple host species, but its abundance varies across plants of different species, with most individuals and the highest rates of reproduction being associated with the primary host species in the local study area (Milne and Walter 2000). But why individuals of this species are associated with host species other than the primary host has yet to be determined. Does it relate to individual thrips being forced to move among host plants, through senescence of flowers, or because adults emerge into localities that have become unsuitable since they pupated? And what information (or combinations or sequences of information (Hebets and Papaj 2005, Yaku et al. 2007)) leads to individuals landing on flowers of a particular plant species? Is it just sensory information from the plant or from the flower itself, or does the thrips' physiological condition also play a role in that stressed individuals, which have been unable to locate a suitable (primary) host for some time, are more inclined to visit flowers of other species because of falling behavioral thresholds? Further, what are the retention times of individuals on flowers of particular species? The ecological consequences of these occupancy rates are not well known, although the species is predicted to be abundant only in localities in which primary hosts are growing (Milne and Walter 2000). And how much of the occupancy of inferior hosts can be attributed to chance? Also, we still need to determine how these occupancy rates affect the rates of development, survival and reproduction, and to what extent the movement among flowers affects the survival probabilities of *F. schultzei* individuals. Clearly, an understanding of what attracts individuals to a flower of a particular species, and the context of its attraction to flower species that usually have relatively low occupancy rates, would provide the appropriate background for investigating and interpreting the autecology of *F. schultzei*. Further, the species limits of these thrips across host plant species needs to be investigated thoroughly, since only a start has been made (Milne et al. 2007). Thereafter, the host associations across geographical space and the extent of movement of these insects also need to be resolved.

The cotton bollworm (*Helicoverpa armigera*) is notorious for the broad range of host plant species on which the larvae have been recorded feeding. Pigeon pea (*Cajanus cajan*) is an important primary host species for *H. armigera*, and this has been demonstrated across a substantial proportion (i.e. Australia and India) of its broad geographical distribution (Rajapakse and Walter 2007). Electro-antennography coupled with gas chromatography has demonstrated that pigeon pea emits a range of volatiles to which adult female *H. armigera* individuals respond, and that they are present in this species at high concentrations relative to levels in the lesser host species tested. Also, each of those other hosts has only a subset of the range of physiologically active volatiles that is present in pigeon pea (Rajapakse et al. 2006). Thus, relative rates of host-use correlate with the nature of the volatile signal associated with each.

That the sensory physiology of a species influences not only its abundance on a particular host species, but also probably its abundance overall, has been demonstrated with the membracid bug *Aconophora compressa*. Introduced into Australia for the biological control of the weed lantana (*Lantana camara*), *A. compressa* became abundant only on fiddlewood (*Citharexylum spinosum*), another verbenaceous species. In glasshouse choice tests, female bugs settle at almost equal frequency on lantana and fiddlewood, and they develop and reproduce at relatively high rates on these two host plant species (at least in comparison with other hosts used in Australia). In the field, however, the frequency of *A. compressa* on lantana is only 15% of its frequency on fiddlewood (Manners and Walter 2009, Manners et al. 2010).

Field tests in Australia with sentinel plants placed at equal distances from *Aconophora*-infested fiddlewood plants showed colonization rates on lantana at only 15% the rate at which the fiddlewood sentinels were colonized. Furthermore, female bugs placed on fiddlewood and lantana plants deployed in the field remained where they were placed and produced egg masses at equal frequencies across the plant species (Manners 2008). *Aconophora* has vast additional resources available to it in Australia in the form of lantana, but its sensory system directs it away from lantana. These bugs use lantana mainly when they "spill-over" from fiddlewood when densities become extreme there, a phenomenon well known in weed biological control (Marohasy 1996). Typically, the numbers on lantana subsequently dwindle. Ultimately the local distribution and abundance of the species is affected to a considerable extent by the way in which the insects interact with their various host plant species through the capabilities of their sensory organs (Manners 2008).

Comprehensive autecological information represented by the range of research questions detailed above is not at all common (see Walter 2003, Chapter 7), perhaps mainly because the ecological perspectives developed for organisms that use multiple hosts ("generalists") are mostly demographically derived, and the organisms are assumed to optimize in general terms under variable conditions (see Finlay-Doney and Walter 2012a). Clearly, the autecological approach provides different insights as to the changing abundance of organisms compared with demographic ecology, as foreshadowed in Chapter 2.

4.2 Autonomic and Active Response Systems

In considering the various physiological, morphological, and behavioral mechanisms of organisms from the ecological perspective, it is informative to separate out passive and active components of the environmental response system. The passive we label the autonomic response system. The autonomic components are essentially physiological and act whilst the organism is in any locality, and they operate regardless of conditions—virtually continuously, even if at a low rate or intensity. The operation of autonomic mechanisms may feed back to the neural and endocrine system of the organism and may thus invoke behavioral responses. Their ecological impact can be measured in terms of tolerances and reproductive output.

The active response system operates in relation to incoming environmental sense data (Dusenbery 1992). In animals, sensory input is processed as pattern (or template) recognition (Huber 1985, Cunningham 2012), with the organism adjusting its behavior accordingly. Individuals may even migrate in response to sensory information. Once an organism has established residence, feedback from the autonomic component will return information that impacts further on residence duration. A response could well be continued residence. In plants, active responses are more subdued. Their stems may grow towards light if they are shaded, or their roots may grow differentially towards a source of water or nutrients. Even the movements of their propagules are not under direct control, but effectively constitute an ongoing, active response to the environment.

Since the influences of both passive and active systems are integrated, they are distinguished from one another only for convenience. Measurement of the ecological consequences of each system therefore tends to be made in different ways, but in modeling ecological systems their outcomes have to be integrated, as explained more fully in Chapter 8.

4.3 Individuals and Adaptive Mechanisms

The concept of environmental response system represents a different perspective of ecophysiology, ecomorphology and ethology, for it explicitly relates the mechanisms of potential ecological relevance, which are internal to the organism, to the questions that are given prominence in autecological theory. The central autecological questions relate to the survival and reproductive rates of organisms and thus to the persistence of species within a spatio-temporally dynamic environmental context, expressed as a function of the properties of that particular species as these relate to its requirements and tolerances. This explains why the precise operation of these internal, short-term mechanisms are not so central to ecologists as they are to ecophysiologists, for example, who seek to understand how these specific mechanisms operate under particular ecological circumstances. For ecological understanding, it is the implications of the workings of these mechanisms for the survival and reproductive probabilities of individuals, and the persistence of species within localities that are important. Clearly, knowledge of how these internal mechanisms of organisms work improves ecological interpretation and understanding, but knowledge of this nature cannot be equated with ecological understanding, in much the same way as biochemical knowledge aids in physiological understanding, but remains a different field of investigation (Rose 1997).

The spatio-temporal scales covered by ecophysiologists, ecomorphologists and ethologists are typically small, and relate to the specific physiological or ethological question of interest. For example, how do the individuals of a particular species survive the extremely high daily and seasonal temperature fluctuations and associated low humidities in their desert habitat? Anatomists and physiologists have unraveled the structure and function of the mechanisms for water retention (e.g., Schmidt-Nielsen 1990, Willmer et al. 2000, Moyes and Schulte 2006), and behavioral and physiological studies have revealed the significance of diurnal patterns of crevice use, torpor and basking for winter survival in minute desert marsupials (Geiser and Pavey 2007), for example. The temporal and spatial scales necessary for investigating the immediate function of a particular mechanism, such as the movement between sun and shade by thermoregulating ectotherms, can be dealt with over the course of a day, and within the relatively small area that allows it to move between the alternative thermal environments (e.g., Schultz 1998). In short, the scale of interest of physiologists or ethologists covers just a brief part of the organism's lifespan, and only a small part of its lifetime track in space.

Ecological interpretation requires information about the systems outlined above, but it is information of a somewhat different nature. Essentially, the probability of failure (and its converse, survival) is of special interest. What is the probability that the subject organisms gain protection from severe temperatures in a particular biotope? What is their chance of surviving to reproduce across the different biotopes inhabited by the species? How does the structure of the thermal environment impact on the physiological status of the organisms, through their energy budget, and how does that thermal structure interact with the environmental response system of the organisms? Such ecological interactions affect profoundly the probability of reproduction of the organisms concerned, as well as their reproductive output. For ecological understanding, it is crucial to determine how these probabilities change through time and across space.

Ecological understanding cannot focus on single environmental variables, like temperature, for organisms interact with other variables, particularly moisture in terrestrial environments (e.g., Edney 1982, Davies et al. 2004) and salinity in water (e.g., Sardella et al. 2004, 2008), usually with profound results. The interaction of the relevant response system of the organisms with the structure of the thermal and moisture environments is bound to be influenced by yet other variables, perhaps on a local scale. Topography and agronomy, for example, can modify moisture gradients. Yet other ecological influences are also likely to be relevant. Periodic flooding, for instance, and the subsidence rate of floodwater could increase mortality probabilities directly, and also indirectly through rates of disease infection.

Even in environments that are suitably structured physically to provide adequate shelter from thermal and moisture extremes, the actual structure of the thermal and moisture environment may be such that the time available for growth or activity of the organisms is inadequate for them to fulfil the seasonal requirements of their energy budget. So, for example, the slightly higher specific temperature of chalk relative to granite evidently allows some southern ground beetle species to complete their life cycle on the island of Gotland, but not on the nearby granite mainland of Sweden (Lindroth (1949), and see Hengeveld (1990, p. 179) for further examples).

The way in which ecological variables are interpreted above illustrates how autecology relates ecologically relevant mechanisms and physico-chemical tolerances of individual organisms to much broader scales of time and space than are covered in other biological sub-disciplines. The entire lifespan and lifetime track of the individual organism is related to the relevant variables and their dynamics and, further, is extended across generations. Although paleoecology deals with evolutionary time frames, covering many generations, and biogeography covers spatial scales across the entire geographical range of species, autecology relates information on the environmental response system and its interaction with the environment to these more extended scales. For example, it is only autecological principles that can be applied to the investigation of the beetle species of Fig. 3.1 to understand their present and past distributions in mechanistic terms. To interpret it in terms of disturbance, non-equilibrium or communities adds unnecessary steps to the interpretation, and these interpolations have more potential to mislead than to enlighten.

The edge of a species' distribution is particularly revealing in terms of its environmental response system. That the environmental response system has to operate within the context of these marginal conditions suggests that one is likely to detect failure of the environmental response system in such situations and this will provide insight as to the ecological conditions required to sustain the organism's energy budget. This is a fundamental point to physiologists (McNab 2002, p. xxv), but remains a puzzle in demographic ecology, for the optimization argument expects adaptive compensation, even though efforts to demonstrate this have proved frustrating (Hoffmann and Blows 1994, Angert and Schemske 2005, Bridle and Vines 2007). Working from the perspective of individual organisms and their species-specific adaptations allows the intensity of occurrence of the species in a particular locality within its geographical distribution to be determined more realistically. And once that has been established, any secondary influences (such as predation or competition) can be more meaningfully investigated and interpreted.

Should environmental conditions change, through altered climate or encroaching vegetation for example, questions related to the retention of individuals in the area are raised. Do all individuals remain in the area, or do they move away, or do adults remain, with mainly their offspring leaving the area? Density may play a role, but that has to be interpreted relative to the environmental context (Andrewartha and Birch 1954, pp. 23-24). For example, bateleur eagles (*Terathopius ecaudatus*) are evenly spaced in protected areas, with large territories (3.1 nests/100 km^2) (Watson 1990b). Outside of such areas, where the birds once bred, they now fall readily to jackal baits and it is usually young birds that are now seen in these areas (Watson 1990a), as they leave the relatively small areas of high bateleur density and intense territoriality that remain (Watson 1989). Their mortality rates are evidently high.

In general, therefore, we need to determine the consequences of a particular environmental structure or environmental change for the survival and reproduction of the subject species, and perhaps even for the probability of the sexes meeting and achieving syngamy. Also, what are the consequences of climate change for the rates of survival and reproduction of individuals living in different parts of the species' distribution? How can such consequences be expressed in an ecologically meaningful way, and how can one generalize about ecological responses? In this respect, the ways in which survival and reproduction are influenced stochastically by such external influences as weather, movement across unsuitable areas, or predation have to be incorporated. These responses all need to be made amenable to calculation in some way and so to generalization.

Autecological theory thus expands the scale of investigation of physiologists and ethologists, for example. Questions must therefore be phrased relative to physiological and behavioral mechanisms in a somewhat different way and within a broader context. In short, the internal mechanisms of organisms and how they work are not the direct focus of autecological research. Rather, the overall outcome of the operation of those mechanisms, in terms of survival rates and reproductive consequences, provide the focus of autecological research, and this is investigated and quantified in relation to the variously structured environments in which the individuals of a species habitually live. Work of a physiological and behavioral nature thus enables the extension of these insights to the life of the organism across the spatial domain of the species and through its temporal trajectory. The overall structure and performance of any organism's environmental response system encompasses a range of operations including, for example, thermoregulation, osmotic regulation, various sensory and behavioral processes and so on. These systems and their responses may be interactive, and they may operate with a certain degree of flexibility, all of which ultimately determines the probability of survival of the organism under the conditions concerned.

In autecology, the individual is thus an open system involving several environmentally relevant mechanisms, with the interactions, interconnections and coordination between them defining their joint operation as a system. Consequently, the operation of one element must not detract from the appropriate functioning of any other in the system. Rather, their joint operation must be coherent, with the effects of one mechanism enhancing that of the others. This is yet another reason to believe that stabilizing selection is the norm in natural systems. Any deviations in the expression of these mechanisms are likely to be maladaptive, especially since negative pleiotropic effects are so common.

As an example of the coherent action of the different properties of the environmental response systems of organisms we take the example of pythons (Frazzetta 1975). Since terrestrial pythons are physically supported by a stable substrate they can grow to large size, and extreme mass is not a problem. They strike prey rapidly and heavily and can immediately throw their body coils over the prey to envelop and constrict it. By contrast, habitually arboreal pythons, such as the emerald python *Chondropython viridis*, are necessarily much more slender and weigh much less than terrestrial python species. Living in trees has further consequences. Stalking, capture and subjugation of prey in the restrictive environment of tree branches imposes different requirements from those of terrestrial ambush. The strike, for example, must be more precise and certain, for a lost grip can mean loss of that prey item. Arboreal pythons therefore have a more flexible jaw apparatus to push the longer and more slender teeth further forward, and thus penetrate the prey more securely on contact. These structures, although weaker than those of the other pythons, could be modified in this way because the collision forces, when the strike makes contact with the prey, are lower than in other pythons, because of their lighter build. The extremely complex skeletal structures associated with the mouth are thus adapted to hold struggling prey and bring it back to the coils of the body for constriction (Frazzetta 1975).

Such complexes of co-acting properties have consequences for the ontogenetic pathways required for the development of the organism. Advances in developmental genetics cast doubt on the traditional emphasis assigned to the relatively minor structural mutations and allelic substitutions that have been the focus of evolutionary ecology (Hoekstra and Coyne 2007). Instead, adjustments to regulatory genes provide a more complete and convincing interpretation of how such change takes place (Chapter 5, Section 3).

4.4 Species and Adaptive Change

We have emphasized how individual organisms embody environmental response systems and how the interactions of individuals of a particular species with their usual environment determine the ecological patterns seen in natural systems. In demographic ecology, however, higher levels of integration than the individual are usually recognized as being more significant in dictating ecological pattern, principally in the form of populations and communities, and the processes that are assumed to operate at those levels. However, the species-specificity with which individual organisms respond to the environment provides little ground for developing robust ecological generalization at these levels (Gleason 1939, Walter 1995, 2008, Walter and Paterson 1995), as expanded below.

The conceptualizations embodied in "population" and "community" are essentially local in spatial terms. They were developed in relation to the ecological processes that were early given prominence in natural selection theory and ecology. Although the promulgators of these ideas, starting with Darwin, have consistently acknowledged the diversity of ecological influences (and thus influences on natural selection), from climate to predation, the generalizations they developed were uniformly focused on demographic processes through the agency of density and its consequence, competition. Such processes would almost invariably operate only at a local level. That this is a problem is evidenced by the inevitable logical contortions in attempts to extend the influence of density dependence to more extensive spatial scales (e.g., Hassell et al. 1987, Latto and Hassell 1988).

Natural selection was originally conceptualized by Darwin as an inexorable process driven primarily by intraspecific competition (Brady 1982, Paterson 2005), presumably with different selection pressures being exerted in different local populations of a species (although Darwin seemed not to have considered geography when writing about selection or explicitly differentiated local from species-wide). Darwin's (1859) focus on resources and competition (Paterson 2005) was carried over to ecology, and the tacit assumption is that any advantageous character, one that confers greater efficiency in competition with other organisms, will be selected locally and ultimately spread to other parts of the extended population. Complex adaptations are considered to be built up by this process, but in an unspecified way. Considerations of co-ordination of changes to the mechanism both spatially and temporally have never, to our knowledge, been addressed. The competitive efficiency presumed to be involved is deemed to be sufficient to explain all these aspects of the adaptive process. In terms of the ecological consequences, the logistic equation embodies this bias towards density, competition and efficiency, and thus retains generalization at this local level, which helps to explain why species are played down or neglected in demographic ecology theory.

The neglect of species is nevertheless surprising given that they exist as natural entities that are objectively defined through their mechanisms of reproduction. Sexual species have a complex mechanism that ensures fertilization under the usual conditions of environment and habitat in which individuals of that species are found (Paterson 1985, 1986). Sexual species are thus self-defining, in the memorable words of Lambert et al. (1987), and they therefore exist independently of human interpretation, which is not true of ecological populations and communities, or genera and other higher taxonomic groupings. Despite this special status of species they tend to be disregarded in the ecological literature. We therefore detail the ecological relevance of understanding species and the advantages of including species explicitly in ecological theory in the following paragraphs.

In a sense, asexual organisms can more accurately be said to reproduce than do sexually propagating organisms in that their propagation is clonal (except where automixis allows some degree of gene "shuffling").

The adaptive mechanisms carried by the offspring of asexual females are essentially those of the parent. Stability through time is expected, since their mode of reproduction will not disrupt the complex adaptations of these organisms through recombination and mutational changes would tend to be selected out through negative pleiotropic effects. Selection would work in this way because each generation gets exposed to the same suite of environmental circumstances through females depositing their eggs, larvae or seeds in the usual habitat of the species. Across geographical space, as a consequence, asexual species are essentially consistent in the primary features of their morphology, physiology, behavior and ecological associations (Walter 2003, Najar-Rodriguez et al. 2009), and they have not contributed much to eukaryotic diversity (Foottit 1997).

In sexual species, the mechanisms and ecological processes that bring the sexes together in space and time are somewhat distinct from those that operate with respect to the survival of individuals, although some elements may relate to maintenance and growth of individuals as well as to aspects of mating. Brown planthoppers of rice (*Nilaparvata lugens*), for example, are species specific in their host plant use, communicate sexually by means of vibrational duets through the plant stem, and so do not interact with congeneric planthoppers on nearby *Leersia* plants (Claridge et al. 1985). Indeed, at the time of investigation the planthoppers were all included within a single species. Their host specificity appears to play a role in their sexual communication since viable hybrids are produced, apparently quite readily, in the laboratory, and they are fertile. Field sampling yielded no individuals with hybrid songs, which indicates that their habitat associations and specific host plants form part of the fertilization mechanism.

The fertilization mechanisms of sexual organisms are co-adapted across the sexes, and each, in turn, is adapted to operate effectively within a particular environmental context. Vocal calls in rainforests thus have different physical properties from those used by organisms adapted to existence in more open environments (Morton 1975, Padgham 2004). The consequences of the specificity of such mechanisms for ecology are fundamental. First, these mechanisms delimit the field for gene recombination (as defined by Carson (1957)), and thus set the limits to the species' gene pool (when defined in population genetics terms (Paterson 1985)). They thus confine the other environmentally-related adaptations of organisms and consequently dictate how the adaptations of organisms will be distributed. This helps explain the species-specificity of adaptive mechanisms and the species-specificity in the ecological responses of organisms.

Organisms must have a mechanism with which to localize and recognize one another in a heterogeneous environment if they are to achieve effective fertilization. Inevitably this involves intersexual communication. The planthoppers mentioned above first locate a host plant; their host searching mechanism takes them to plants of a single species, and the different cryptic species of planthoppers associate with different host species. On the host, the sexually mature insects perform a duet that enables the male to locate and then home in on the female. The interaction continues through courtship, intromission and sperm transfer, each of which involves a suite of co-adapted signal and response mechanisms. Pollination in plants can be interpreted similarly through the adaptive use of pollen vectors and the mechanisms of gamete recognition (e.g., Clarke and Knox 1978, Dumas et al. 1984, Brewis and Wong 1999, Vieira and Miller 2006).

The mechanisms involved in sexual reproduction are thus finely tuned across the sexes and relative to habitat and environmental circumstances, and the system is, as a consequence, under stabilizing selection. Individuals that send inappropriate signals, signal at the wrong time, or have receptors that are not tuned to the signals of the opposite sex, for example, are liable to fail in securing a mating partner. These adaptations are thus maintained in stable form within the species gene pool or "field for gene recombination", through the requirement that they function effectively (Paterson 1985). This explains why, despite intense search spanning several decades, recorded change in wild populations is extremely limited (and includes such things as insecticidal resistance, slight changes in flower color or alteration to beak length in bugs) and is most convincingly demonstrated in situations in which spatial adjustment is restricted (Galapagos finches and guppies) by space being limited and movement restricted (see Thompson 2005, Walter 2006).

As detailed earlier (Chapter 5), significant evolutionary change is restricted in time and space to small populations exposed to intense selection pressures, which explains why good examples of contemporary adaptive change are so hard to come by (and are, consequently, so eagerly sought in evolutionary ecology), and why adaptive mechanisms are species-wide in their distribution. That adaptive change is restricted in this way, and takes place in response to contingent circumstances, means that autecological generalizations will take a very different form from those anticipated in demographic ecology, which is underpinned by the almost universally held expectation of ongoing adaptive change across all populations. These aspects are expanded next.

4.5 Ecological Generalization and Prediction

The concept of environmental response systems developed in this chapter provides an individual-based mechanistic explanation for the spatio-temporal dynamics that dominate ecological systems. This brings us back to the ideas of ecological scale in general, and to those of environmental matching in particular. We thus derive a basis for ecological generalization and prediction that differs fundamentally from the way in which predictive ability in ecology is pursued in the literature of demographic ecology.

Demographic ecologists do acknowledge stochastic influences (e.g., Hassell and May 1985, Coulson et al. 2001), and generally interpret them as disrupting the underlying deterministic processes that are driven primarily by density, for their foundation statements and general models are embodied in the logistic equation. Generalization and prediction are sought both typologically and in terms of process, with both approaches manifest together in the relevant claims and models that have been published (e.g., Southwood and Comins 1976, Southwood 1977). For example, typological generalizations are evident in statements that encapsulate how species of type *X* will respond to particular environmental circumstances in *Y* fashion (e.g., Orians 1962, MacArthur 1972a, Colwell 1986, Lawton 1999). The deterministic approach embodied above reflects the anticipation of a systematized general chart of ecological relationships, something of the nature of the periodic table of chemical elements (Southwood 1977) or the Hertzsprung-Russell star chart (Keddy 1994). The problem is that species cannot be meaningfully grouped for such purposes. Species-specificity stands in the way. In any case, the primary dynamic is provided by the production of new individuals and the rate at which they are lost, both processes that are influenced by interactions, through the entire life cycle, with numerous environmental axes of differentiation. Whereas the recurring structures and the predictability in the outcomes of interactions involving chemical elements, for example, provide the regularity we see in the periodic table, these simply have no equivalents in ecology. Demographic ecologists nevertheless seek generalizations and understanding to extend across all circumstances within each level of the ecological hierarchy. The advantage of such generalizations, should they have proved effective, would have been their transcending the idiosyncrasy of species.

Autecological generalizations are built at a level that accommodates the idiosyncrasy of species, and are based upon the nature and universal occurrence of environmental response systems. The generalization encapsulates the nature of species and the way in which individuals interact with their environment through their life cycle. As such, it applies to all organisms, but it does not transfer solutions from one species and set of environmental circumstances to other species, in the way sought in demographic ecology. Rather, the general insight provides direction for instigating research into the solution of ecological problems (Walter 2013). Because the autecological responses of organisms to environmental conditions can be formulated and quantified in terms of survival probabilities, generalizations can be developed about the various interactions and movements of organisms of a given species within any particular ecological context. As risk represents the inverse of survival chance, this approach essentially follows that of the risk analysis applied in human society (as detailed in Chapter 9). Projections can be made, and predictions generated, but these are of a different kind to those that are sought in demographic ecology, being species and context specific.

All aspects of the environmental responses of individuals of any particular species can thus be quantified in relation to the survival probabilities or risks within a particular ecological context. This provides a general ecological method for all species and all ecological circumstances. Although this methodology is general and holds for all species, it is not of the form anticipated in demographic ecology. Within limits, it is species-specific and can predict future states. The limitation of prediction is recognized, especially since we cannot predict weather conditions with any certainty, and especially not in relation to the range of axes of ecological differentiation that pertain to organisms in nature.

5 SPECIES-SPECIFICITY AND ECOLOGICAL CONTEXT

Individuals of each species require particular environmental conditions that must be satisfied for their survival and reproduction. They thus occupy particular biotopes that may differ perceptibly from one another, although each fulfills the habitat (or environmental) requirements of that species. The species-specific environmental requirements are met through the complementary processes of environmental and habitat tracking. Those organisms that have sensory capabilities use sensory information from the environment (about vegetation structure, substrate chemistry, and so on) to detect and localize suitable habitat conditions. This defines the sensory process of habitat recognition. Individuals of the species concerned thus achieve environmental (and habitat) matching.

Across the geographical range of the species, different biotopes may well be occupied, all providing (to some degree at least) for the environmental requirements of the organisms concerned. At least, they match the organisms' sensory perceptions of their habitat requirements. For example, in the northern hemisphere, ground beetle species can be found in open biotopes towards the north, whereas they find the humid and cool conditions to which they are adapted only in the woods of the southern parts of their range. They thus match with the same habitat conditions across their range by occupying the different biotopes that provide those conditions (Hengeveld and Haeck 1981). This phenomenon is known under various terms, but has been most directly expressed by Walter and Walter (1953) as "Relative Habitat Constancy". Species with a patchy distribution tend to be exposed to similar environmental conditions across localities; their individuals thus experience "constancy in factor space" (see Hengeveld and Haeck (1981) and Walter (2003, Fig. 9.11)).

When environmental conditions change, organisms track suitable habitat across geographic space according to their species-specific adaptations. Such environmental tracking may either take place actively, or by differential reproduction across space (Hengeveld 1985a). Not all biotopes occupied by a species are equally good, in terms of survival and reproduction. Many individuals of the ground beetle *Amara plebeja*, for example, settle in unfavorable heathland when they leave their winter diapause sites. Although they cannot reproduce in this biotope, they end up remaining there, apparently because their flight muscles are autolysed on arrival (van Huizen 1977).

Different stages of the life cycle of a large proportion of all organisms also have different ecological requirements from the other stages through which they develop. This means that ecological influences need to be considered for all stages in the life cycle separately to understand the distribution of the local intensity of occurrence of a species across its geographical distribution. In all of this, species specificity is the norm. This interpretation offers an entirely different perspective on local diversity from that encountered in community ecology. Because each species matches with different subsets or aspects of the local environment, we cannot consider their match with specific habitat conditions from a community perspective. Since any locality provides habitat conditions that satisfy the requirements of several to many species, local diversity can be explained only with respect to the requirements and axes of ecological differentiation of each of the inhabitant species. They match specifically the various dynamic and spatial aspects of the environment (including their interactions with co-occurring species).

The consequences of the operation of the environmental response system of any species are readily conceptualized in relation to organisms matching the environment where conditions are appropriate, principally through ongoing spatial adjustment. The consequences of mortality and movement are that we can visualize ecological dynamics in terms of individuals confronting continuously varying spatio-temporal environmental conditions. Individuals remain longer (relative to the time spent in unsuitable conditions) in localities that match their sensory perceptions of a tolerable biotope, even if it does not prove suitable in all respects or in the longer term. The number of individuals of a particular species within a locality thus provides a spatial indicator of the extent of the dynamic match between their environmental requirements and the external "provision" of suitable environmental conditions. Yet, such an indicator can never be considered absolute for environmental quality in a locality because of the species specificity of ecological requirements and responses, and also because secondary modification of the primary pattern may take place, through predation for example.

Ecological scale is also species specific, so the species' distribution defines the scale of ecological interpretation, but needs to be considered from two perspectives, that of the individual and that of the environment. Each individual organism inevitably follows a spatial course during its life, called its lifetime track. Although there will be great variation in lifetime track among conspecific individuals, the behavior associated with movement tends to be species-specific. This dictates the local scale of ecological investigation. By contrast, the environment provides the global context for the ecology of the species, and dictates the limits to its geographical distribution, as well as the level and distribution of environmental heterogeneity within the overall distribution.

From the perspectives on scale just developed, we turn back to the organism, for now we can add the entire lifespan to the framework of the species-specific interaction of organisms with the environment. Each stage of the life cycle also has specific requirements and tolerances, and these may differ significantly from one another. For a species to be able to complete its life cycle at a locality, the requirements and tolerances of each stage must be locally satisfied. The various phases in the life cycle, with their individual environmental responses and requirements, have to match the seasonal timing with which the environment changes as well as the particular facets of environmental variation.

Similar requirements hold for each season or year that the organism is reproductive. Each reproductive episode must fit into the seasonal pattern of temperature conditions as well as any other environmental variable that is relevant to their requirements, including such aspects as the growing conditions of the vegetation on which the animals feed, the reproductive phases of their prey, and so on. The environmental match is one of degree, and is flexible to some extent as influenced by the norms of reactions typical for the species of interest. The degree of matching that is a consequence of this flexibility is indicated by the local intensity of representation of the species and the age distribution of the organisms in the locality, as well as their reproductive output, survival probability, rates of immigration, and the retention times of individuals before emigrating.

6 CONCLUSION

Overall, this chapter provides the basis of an entirely mechanistic approach to understanding the major questions of ecology, namely the dynamics of the local distribution of species and their intensity of occurrence (usually referred to as local abundance) relative to the heterogeneity and dynamics of their external environment. This approach also provides the basic information for interpreting the local diversity of species, but it does so somewhat obliquely, or indirectly, because the question of local diversity subsumes specific questions about the local existence of each species present (van Klinken and Walter 1996). By contrast, the coexistence question is fundamentally one of common resource use and competition among species for that resource. The focus of the coexistence question has the effect of blurring the species-specificity of ecological requirements and tolerances, and therefore generates information that is relatively superficial (Sale 1988, Walter 1988a). This is true even if the focus is shifted from common resource use to common habitat requirements, for example.

So far, we have written about ecological interpretation only in qualitative terms. We did so to expose the assumptions and principles, as well as consider them in relation to organisms in their natural environment. This approach should not be seen to imply that autecology is not quantitative, merely that the aspects requiring measurement are different from those given prominence in the demographic models. The quantitative aspects of autecology are considered in detail in chapter seven.

Chapter 7

Quantifying Autecology
Survival, Reproduction and Movement

CHAPTER OUTLINE

1 INTRODUCTION

Population equations are mathematical representations of bodies of ecological theory. They reflect the assumptions that are made about the primary influences of the environment on organisms and they frame ecological interpretation to make it more explicit. Nevertheless, they also have the potential to constrain interpretation, and this point is also dealt with below, in a general way.

The equations that are core to demographic ecology theory are represented by the logistic equation, which we introduce below and which reflects the central value given to balance or equilibrium in demographic ecology (as argued by Cooper 2001). Populations are perceived to be in a state of balance or equilibrium because they are seen to be regulated against excessive growth by density dependent processes—competition, predation, herbivory and the like. However, density dependence is written directly into the logistic equation virtually entirely in the form of intraspecific competition, as described more fully in Section 3. The logistic equation has, however, also been expanded in various forms to incorporate other density dependent influences, interspecific competition and predation, each in their own general equation. The concepts embodied in the *S*-shaped logistic curve, and the equation itself, are evidently linked deeply and inextricably within the concepts and practice of demographic ecology, although the realism of the patterns and concepts, the relationship between them, and their extension to specific interpretations in ecology have been contested vigorously (e.g., Kingsland 1995). This point holds true even though different models are now frequently used to model population dynamics, including simulation and matrix models. An understanding of demographic ecology therefore needs an appreciation of the structure and workings of the logistic equation, as well as its deficiencies with regard to its central role in ecological theory and research.

The logistic equation and its associated models, mainly the Lotka-Volterra competition equations, were developed a long time ago, and their history has been elegantly detailed together with the contention associated with their assumptions and applications (Hutchinson 1978, Kingsland 1995). Significantly, they are represented in virtually all contemporary ecology texts, often at or near the beginning (e.g., Roughgarden 1998, Turchin 2003, Begon et al. 2006, Mayhew 2006, Rockwell 2006, Rose and Mueller 2006, Ricklefs 2008, Townsend et al. 2008, Cain et al. 2011, Price et al. 2011, Smith and Smith 2012). Whereas most ecology texts do not identify core theory explicitly, nor discuss it, the principal implication of the material they cover and the way in which it is structured reflects the belief that equilibrium and density effects are central influences on organisms in the field, must therefore lie at the heart of ecological theory, and must be understood from the outset. More mechanistic models than the logistic have been developed, including structured matrix models (Caswell 2001), and these provide extremely useful tools for ecologists. They tend, however, to be used almost exclusively to model situations with respect to equilibrium dynamics, despite their recent use in modeling "transient dynamics" (Ezard et al. 2010). The latter are considered far less typical of natural systems than equilibrium dynamics, which confirms in many ways that the balance of nature remains a strong but tacit guiding metaphor in contemporary (demographic) ecology (Cooper 2001, Walter 2008). Most pertinent in this regard, therefore, is the current sanction of the logistic equation as reflecting demographic ecology's core premises, as they apply to populations, metapopulations, and communities (Huffaker et al. 1999, Turchin 1999, 2001, 2003, Mangel 2008).

Many authors tag a proviso to their mention of the logistic and Lotka-Volterra models, claiming that although they are not necessarily accurate representations of population processes they remain useful starting points (Turchin 2003, Rockwell 2006, p. 7), that they "help us to picture the way in which competition may act" (Varley et al. 1973, p. 40) and that they "get at some extremely basic feature of trophic interactions" (oscillations) (Turchin 2003, p. 34). These equations must, therefore, be seen to encapsulate the organizing principles in demographic ecology, which are equilibrium or regulation and density dependence (usually coupled with the self-regulation of populations (Turchin 2001, 2003)).

These equations also comprise the central element of the more recently constructed metapopulation theory (e.g., Caughley 1994, Moilanen 2000), which was developed explicitly for those organisms considered to have a patchy distribution. Essentially, a metapopulation is treated as a spatially balanced population of populations, but in which real migration rates and the consequences of migration are not quantified or measured. Essentially, patch occupancy through local movement is seen as core to understanding the persistence of populations (Rockwell 2006, p. 115). Movement is seen simply as a linkage that balances local populations across their patchy distribution, but is not considered a central aspect of ecology.

With this background we generalize the approaches to studying population dynamics (Section 2) because not all such research is entirely constrained by the logistic equation. Nonetheless, a more detailed examination of the logistic equation is required. Section 3 focuses specifically on how r fails in ecological theory, mainly because it is not mechanistically derived, and deals with the constraining influence that density dependence is seen to have on exponential population increase. We are then in a position to review criticisms of the logistic equation, with an emphasis on the most debilitating ones (Section 4). Section 5 starts the development of an alternative mathematical approach to dealing with the dynamics of organisms in nature. To achieve this, we demonstrate how and why the proposed relationship between r and R_0 (the net rate of reproduction) is flawed biologically and thus unhelpful to the development of general theory as well as the investigation of specific ecological situations. As a practical parameter estimate R_0 should be considered independently of the logistic equation. Used in this way it is directly relevant to autecological measurement and understanding. This approach anticipates the development of general autecological models for quantifying and understanding the spatio-temporal dynamics of organisms, in Chapters 8 and 9, but before that some preliminary remarks are made about ecological modeling (Section 6).

2 STUDYING POPULATION DYNAMICS

Not all practicing ecologists trace their empirical work directly to the equations at the core of demographic ecology. Indeed, population data are collected in various ways. This section therefore outlines three general empirical approaches to data collection in relation to understanding population dynamics in nature, and to its interpretation.

Those who have sought to demonstrate directly the basis and mechanics of the logistic equation in the field represent the first approach. To this end organisms assumed not to undergo migration have generally been selected, so that immigration or emigration would not intrude unduly in attempts to verify the density dependence that is assumed to regulate populations (e.g., winter moth (Varley et al. 1973) and viburnum whitefly (Hassell et al. 1987, Southwood et al. 1989)). On the available evidence these assumptions are unlikely to be valid (Holliday 1977, Byrne and Blackmer 1996, Isaacs and Byrne 1998, Loxdale and Lushai 1999, Walter 2003, p. 244) (see Chapters 4 and 9). In any case, and despite strong density dependent processes not being evident in any of these systems, the ecological significance of density dependence was not challenged. The same is true of other such studies, including that on rodent cycles (see Chitty 1996). Such *ad hoc* adjustment has helped maintain the foundations of demographic ecology.

The second approach to data collection and understanding seems more removed from the demographic equations. In such studies, data have been collected without apparent theoretical influence, or at least this influence is not claimed. Nevertheless, the investigation of density effects is often at the heart of the investigations, for the concept of population regulation remains central to the interpretations offered. It persists, perhaps, because regulation is seen by many to be required if we are to have general ecological knowledge or understanding (as explained by Cooper (2001)). At the very least, the empirical work of most ecologists attests to their acceptance of density dependence playing a central role. In some cases, the properties of individual organisms and environmental influences on them have been investigated in relation to population change. Investigations of this type have often been very revealing as to the way in which the attributes of individual organisms mediate the organism-environment interaction (e.g., Shapiro 1986, Thomas and Singer 1998). Some of them have even been geographical in extent and overtly focused on large-scale movement (Gatehouse 1987, Rainey 1989, Wilson and Gatehouse 1993, Wellington et al. 1999), but despite their apparent success in explaining the ecology of the species concerned they have not been able to move theory away from its focus on population regulation and the logistic equation.

The problem seems, again, to come from the demographic theory that implicitly underpins the interpretations offered. In particular, the features of individual organisms that are given prominence in the theory related to density dependent population regulation do not include the species-specific adaptive mechanisms that support the life cycle within its spatio-temporal environmental context (as detailed in Chapters 3 to 6). Rather, they concern variables given relevance to population change in demographic models, as explained by Schoener (1986a), Lomnicki (1988), Kingsolver (1989), and Sutherland (1996). Similarly, weather is portrayed in demographic terms as "... a limiting factor that determines the relative importance of various biotic factors for population dynamics and is a source of environmental variation that affects physiological rate processes and mediates interspecific interactions" (Kingsolver 1989, p. 330). Attention is thus subtly removed from individual organisms, their adaptive mechanisms and the interaction between these mechanisms and the environment.

The fascinating feature of the empirical studies that exemplify the second approach is that they have been developed and conducted without the use of demographic theory, and yet they stand as landmarks for understanding the ecology of particular situations. In some interpretations, however, and especially more recently (e.g., Ehrlich and Hanski 2004), metapopulation theory has been invoked in explanation, apparently for its spatial element. This development overlooks, however, the logistic backbone to the mathematical basis of metapopulation theory, its deterministic structure, reliance on population regulation theory, inability to include climate as a primary driver influencing demographic parameter values and environmental tracking, and restriction to that subset of organisms with a patchy distribution.

The third and final approach to data collection and ecological understanding includes studies in which direct tests of demographic theory have been conducted or in which alternative approaches have been developed to that defined by the logistic equation (e.g., Davidson and Andrewartha 1948a, Wellington et al. 1950, den Boer 1970, 1979a, b, White 1970a, b, 2008, Dempster 1982, Wellington et al. 1999). Few studies fall into this category and, despite their attracting considerable attention, they have not shifted the prevailing dominance of the logistic approach in understanding ecological systems. Indeed, their major general achievement seems to have been the eventual inclusion of stochastic parameters into deterministic population models (e.g., Godfray and Waage 1991, Coulson et al. 2001, Stenseth et al. 2004). Reasons for believing that this adjustment to theory is inadequate are presented below and in Chapter 8.

Certain aspects of the demographic equations have even been elevated to the status of principle and general law (Turchin 2001), and they have persisted in this way despite the long list of debilitating problems identified in them (e.g., Andrewartha and Birch 1954, Pielou 1977, Hall 1988). Because these equations still hold center stage in ecological theory, we dismantle their structure in the following section to demonstrate why they are inadequate for expressing ecological situations accurately, even in modified form. We therefore begin with a consideration of the intrinsic rate of population increase (r), the value of which represents the primary response of organisms to their environment according to the logistic equation. Indeed, r is even equated with Darwinian fitness (e.g., Kingsolver 1989, Pough 1989), which gives the false impression that fitness is rather easily measured. Also, the assumption of *cetera paribus* (other things being equal) in the interpretations of fitness, measured in this way, is bound not to be valid (Rapport 1991).

3 THE INTRINSIC RATE OF POPULATION INCREASE, r, AND CONSTRAINING EXPONENTIAL GROWTH

The intrinsic rate of population increase, r, reflects the rate at which the local population increases in size, and is one of the main parameters of the logistic equation (as in the equations below). As such, r in the logistic equation has no mechanistic inputs that would explain its value in relation to the ecological influences acting on the organisms involved. The value of

r is defined, instead, as the difference between the birth and death rates of a population, $r = b - d$ (Andrewartha and Birch 1954, p. 35). The value thus derived for r does not even deal with individual organisms. Even though the equation $r = b - d$ may seem mechanistic, it is only descriptive in that b and d simply constitute r. Because r is not measured, being derived indirectly from measured values, it cannot explain the values of b and d. All that this representation does, therefore, is suggest what to count so that r can be determined from quantitative data. That is, $b - d$ defines r so cannot explain it.

By contrast, mechanistic models analyze the various components that contribute to b and d, and thus help confirm which variables and processes are involved and how their observed values are derived. Such an approach is consistent with autecological principles and so can be extended to lead directly to a more detailed formulation in which the species-specific properties and lifeline of the individuals of the species are represented. Further extension would include consideration of the degree to which these properties match the individuals' local environmental conditions. The mechanistic connections just outlined define the aims of all autecological research, which is involved with the matching of organisms to environmental circumstances, and is developed further below (Section 5).

Immigration and emigration of individuals also influence the local ecological intensity of organisms (or growth or decline of populations), and these key ecological processes are therefore sometimes incorporated into an expansion of r, with $r = b - d + i - e$ (where i reflects the immigration rate and e the rate of emigration). The rates at which individuals move into and from the population are, like b and d, influenced by prevailing environmental factors, such as temperature, rainfall, wind speed and direction, humidity and so on, depending on species. Further, migration intensity varies between species, as well as between individuals of the same species, depending on local conditions, or on physiological differences between life stages or stages of reproduction, and migration frequently takes place independently of density (Andrewartha and Birch 1954, pp. 86-101, and see Chapter 9).

Later in this chapter we examine more closely how a mechanistic conception of the rate of reproduction of a set of individuals can be developed (Section 5). Before that we specify how this net rate of reproduction (R_0) cannot be directly related to r, the intrinsic rate of population increase, and thus demonstrate further why r, as used in the logistic equation, fails as a meaningful ecological measure. In this, we follow the reasoning developed by Birch (1948) and Poole (1974) (see also Andrewartha and Birch (1954)). This reasoning clearly exhibits some of the debilitating assumptions of demographic ecology, which must be avoided in the development of a more mechanistic, autecological approach.

The reproduction of organisms, in demographic ecology, is represented simply by the exponential equation

$$N_t = N_0 e^{rt} \tag{1}$$

in which the number of individuals at any given time t, is N_t, and this follows directly from the number at the initial time, N_0, multiplied by the natural logarithm e to the power of rt. Here, r represents the intrinsic or innate rate of increase, which is constituted by $b - d$, and time t, which is always greater than zero. Equation (1) can be reformulated:

$$\frac{N_{t+1}}{N_t} = e^r = \lambda \tag{2}$$

where λ is the finite rate of increase, and is here defined only for the time interval $t = 1$, although any time interval could be specified in the formulation. Thus, e^r is the same as the ratio of the number of individuals at time $t + 1$ relative to that (N_t) at the previous time interval, which is not necessarily N_0. The parameter λ therefore represents the rate of increase of the population from one interval of time to the next. In this equation, t falls away in the exponent because it is equal to 1. It therefore does not change the value of the exponent, which is then dependent only on r.

In ecological texts the exponential equation is commonly given in the differential form

$$\frac{dN}{dt} = rN \tag{3}$$

which denotes the rate of change of N over time t. When the intervals of time t are again taken as equal to 1, the differential equation becomes a difference equation. It is obvious that the rate of growth rN is variable because N can have any value for any single value of r, so that the potential increase in population growth is unlimited. The next mathematical development was therefore to impose a limit on such unconstrained population increase to represent more accurately what was believed to take place in the field, so the parameter α was introduced, and defined:

$$\alpha = \frac{\lambda - 1}{K} \tag{4}$$

The limit to population increase is then specified by the constant K, introduced in the 19th century as the "upper limiting population" of humans that could be supported on a particular area of agricultural land (Kingsland 1995, pp. 65 and 74). It was subsequently, and independently, defined more generally as the maximum number of individuals a particular environment could sustain. This mathematical construct was then interpreted ecologically as the carrying capacity of the environment, a concept introduced to field ecology through range management and which even there was seen as difficult to define (Young 2002, pp. 40 ff). In nature, which is not analogous to the situation on farms, the carrying capacity is impossible to measure and is certainly not equivalent to the observed maximum population density (Chitty 1996, p. 39). This stricture is quite clearly debilitating for the logistic equation.

In any case, the equation for regulated population growth rate then becomes

$$N_{t+1} = \frac{\lambda N_t}{1 + \alpha N_t} \tag{5}$$

This is the differential form (or the difference equation for $t = 1$) of the logistic equation and is the one commonly used in theoretical demographic ecology to describe an *S*-shaped growth curve with upper level *K*. It differs only from exponential growth in that it contains the parameter α, which slows population growth to the point at which no further growth can take place. The influence of this parameter is referred to somewhat misleadingly as the "environmental resistance" and is represented by $K - N/K$ in the logistic equation. The parameters indicate clearly that it is not the environment, as a whole, that is imposing on the population, but the density of the population itself relative to the amount of available resources. At best, the array of remaining influences in the environment (which is extensive) is included as an indirect and implicit influence. The equation is not mechanistic because *K* is derived from *r*, which is in turn derived from the observed values of *b* and *d*.

In biological terms, the most notable aspect of the five equations outlined above is that individual organisms are treated as identical to one another. Further, the adaptations of organisms are not taken into account; they are effectively assumed to be irrelevant ecologically and therefore irrelevant to the development of fundamental ecological generalizations. The number of organisms, taken together and relative to the carrying capacity, is all that is ecologically relevant in the logistic equation.

Extensions of the model lead to the Lotka-Volterra and Leslie equations in which two or more species that maintain a stable logistic equilibrium relative to one another are characterized by values that reflect their population densities, namely N_1, N_2, and so on, and their competitive influences on one another (see Poole 1974). Different species thus remain divested of their specific adaptive mechanisms, and the influences of these properties on population numbers and change are also ignored. The species are effectively treated as equivalent to one another, and indeed the Lotka-Volterra competition coefficients actually convert numbers of one species into equivalents of the other competing species, so that population density can be adjusted accordingly. Algebraically, this deliberate conversion of parameter values is not permissible and, in any case, such variables should not be treated as constants.

In short, the core demographic population equations, the ones generally considered to justify the direction and emphasis of current ecological investigation and interpretation, are based on the following flawed assumptions, among others.

1. Carrying capacity, *K*, is not constant over time and not uniform across space.

2. Carrying capacity cannot be predicted quantitatively in any realistic way, which means it cannot be quantified independently of the numbers of organisms under consideration, thus introducing circularity.
3. Carrying capacity can be estimated only from existing data, and then only iteratively by alternatively adjusting the values of r and K. It cannot be estimated independently, which excludes scientific test of the model.
4. Most data sets are incomplete relative to the logistic curve, as they tend to represent only the initial part of the curve, the intermediate part, or the end. Statistical testing is impossible under such circumstances. In the few cases in which the logistic and other relevant models have been tested against one another, the logistic model consistently performed worst, and even showed no resemblance to the data in some tests (e.g., Feller 1940, Pielou 1974, Cohen 1995).
5. All conspecifics are treated as equivalents, despite their obvious sex, age and physiological differences in their reproductive output, mortality rates, behavior, motility, and so on. More recent developments (including matrix and individual-based models (Caswell 2001, Grimm and Railsback 2005)) do, however, incorporate some of these aspects but have not influenced theory to any deep extent.
6. The adaptations of organisms, the mechanisms or properties with which they confront and respond to the environment, are excluded from consideration despite their obvious ecological significance.
7. Species are treated as ecological equivalents, which is patently untrue, for even closely related species have different spatial distributions because of their different adaptations and tolerances (Figs. 3.1, 3.2 and 4.6).
8. Numbers of one species cannot be converted by means of a constant into numbers of another species.
9. Density dependent effects, generated mainly by intraspecific competition, are assumed to play the central role in influencing population dynamics.

These are not the only problems with the logistic equation and its derivatives, as is explained elsewhere in this chapter, but they do illustrate the extent to which demographic ecology and even its more recent extension is built upon faulty premises.

4 THE UNDERLYING ASSUMPTIONS—THE PARADOX OF RETAINING THE LOGISTIC

In Eq. (1) only the number of individuals N_t and time t vary. All other variables are treated as constants (Andrewartha and Birch 1954), including even the intrinsic rate of increase, r. Therefore, b and d, which define r, are thus automatically also treated as constant over time, or they compensate each other perfectly such that *r* remains constant. This is an unrealistic reflection of what takes place in ecological systems, because births usually predominate in nature during favorable seasons and deaths predominate when conditions are less favorable. Further, conditions in different seasons are unrelated so it is unimaginable that the rates of birth and death across different seasons balance one another in all cases and species, and will do so across different locations. Because ecologists do want to know exactly what is affecting their subject group of individuals through time, and which ecological factors are most prominent, the assumption of constant and species-wide birth and death rates is paradoxical, as is their being treated as strictly compensatory with respect to one another.

The assumption of constant or compensatory birth and death rates has further undesirable consequences. In particular, the different age classes that inevitably exist within any population of organisms, and each of which has its own rate of fecundity and mortality, are treated as stable across successive generations. In technical terms, one assumes a stable age distribution within the study population during the time interval chosen, however long that interval might be. Ecologically, this implies that the external conditions that typically affect such a distribution remain constant over that time interval, that the individual organisms do not develop during that interval, and that they have the same requirements and threats during that interval, whatever its duration. If this were the case in natural systems, unstable age distributions should inevitably become stable with time. Such assumptions and their logical consequences do not stand scrutiny with respect to real ecological systems. Even more important, such an approach ignores the variable environmental conditions that typify ecological systems (see Chapters 3 to 5). Ecological explanation developed in the terms outlined above thus suffers a lack of realism. This influence is worryingly pervasive in that the views embedded in the original equation are imposed on ecological systems in modern extensions of demographic theory and practice.

In all treatments of populations outlined so far in this chapter, whether population numbers remain unconstrained or are constrained according to the constants in the logistic equation, the group of individuals (the "population") is closed and spatially uniform. This assumes that the numerical dynamics of the population are entirely dependent on the birth and death rates. Immigration and emigration are thus excluded, even under crowded or spatially changing conditions, which introduces further complications in terms of ecological reality. The structure of the equations ensures not only that all life stages and individuals are affected equally, but also that the effects are instantaneous and spatially uniform, a situation unknown in real systems (Andrewartha and Birch 1954, p. 377). This is the "fallacy of instantism" (Ginzburg and Colyvan 2004, p. 70), because changes in density would take time to impact on lifespan, fecundity, development rate, and so on. These assumptions are also implicit in equations of unrestricted growth, but they are more apparent when growth is constrained by density. Such assumptions can be avoided only if the variables that are directly relevant to the ecology of individual organisms are used to structure ecological models with respect to age, seasonality, mortality, and so on, and if movement and the stochastic influences on each of these processes are included.

Ultimately, the rich array of species-specific responses of individuals to continually changing environmental conditions has been reduced in the logistic based equations to a few biological constants and an unvarying and spatially uniform environment. The ecology of individual organisms is thus encapsulated entirely in their density and their projected numbers at any specified future time. Raymond Pearl saw no problem in such an approach, for he considered such statistics to represent realistically even environmental variables in the "... integrated end effect upon population size of the aggregated forces tending towards increase on the one hand, and decrease in numbers on the other hand ..." (cited by Kingsland 1995, p. 80). Clearly this is inappropriate because the different variables relevant to the system under investigation need to be treated independently of one another for a full understanding.

Because the logistic and associated equations treat immigration and emigration as irrelevant to local population ecology, the number of individuals essentially becomes confined to a closed unit, a population, characterized by an equilibrium number, a particular stable age distribution, and certain fixed rates of birth and mortality. In contemporary terms, the definition of convenience for a population is "... a group of individuals of the same species that live together in an area of sufficient size to permit normal dispersal and migration behavior, and in which population changes are largely determined by birth and death processes" (Turchin 2003, Rockwell 2006). That is, the population is defined to exclude any influence of movement, and tends to be investigator specific and therefore somewhat arbitrary (Rockwell 2006, p. 6, Krebs 2008, p. 104). The exclusion of movement in this way is still sanctioned, even though "without movement the individual has no behavior and the population has no cohesion so that distribution in space is isolated from distribution in time ..." (Taylor and Taylor 1983).

Not surprisingly, the terminology associated with demographic ecology and its overall approach has parallels in the history of the physical sciences. The intrinsic or innate rate of population increase, r, is so named because it is considered typical for and inherent to the population, and is derived by combining the birth and death rates. Aristotle assigned a "natural" place to falling objects, or "an innate pulse to change" to moving bodies. It was not until Galileo and Kepler started describing these phenomena quantitatively, and until Newton formulated the movement of falling objects and orbiting bodies mathematically by using several variables independent of the outcome of the process that such movement was understood. That is, after a period of inductive quantification, the models were structured mechanistically by Newton, thereby allowing deductive studies to follow

(Dijksterhuis 1969). Ecological theory thus seems, in relation to physics, quite poorly developed and out of phase with the general methodology of the natural sciences.

5 STRUCTURING THE ECOLOGICAL RESPONSE PROCESS

Each species has a unique set of adaptive mechanisms and, as a consequence, each species is unique (or idiosyncratic) ecologically (Chapters 3 and 4). These properties are not accounted for directly in demographic population models. Instead, the numbers of individuals are given special prominence. This implies that numbers, devoid of any real biological information, have been used to develop general statements that are believed to represent fundamental ecological generalizations. For example, the establishment of a particular level of abundance reputedly follows an optimization process towards the logistic level, K, with the system assumed to be closed. This virtually ensures that populations remain more or less the same over time, independent of the vagaries of the environment. Populations are thus seen to follow fixed external rules or laws and numbers in ecology have been given an existence independent of the organisms whose dynamics they are supposed to model. We argue that for fundamental ecological generalizations to reflect reality better, they must relate directly to the properties of organisms and the way in which these adaptations mediate interactions with the environment. We deal with this stipulation through the remaining chapters in this book.

The properties of species have presumably not been included in the principles of demographic ecology because their inclusion would be inconsistent with the law-like expectations that are held for ecological systems and would thus work against that avenue to generality (Walter 2013). This approach to understanding ecological systems prevailed despite species being known to have evolved specific mechanisms for enhancing the survival and reproduction of individuals within the context of the heterogeneous and dynamic environments that each inhabits. Autecological theory focuses on these adaptations of organisms and provides a place for them in ecological modeling. Specifically, the models have to be structured to incorporate the details of the mechanistic aspects of the interactions of organisms with the environment. The solution involves an emphasis on the net rate of reproduction, R_0, and elimination of the intrinsic rate of increase, r.

Before considering the way in which ecological models should be structured, we consider the relationship between r and R_0. The most direct way in which to estimate the innate rate of population increase is through the estimation of the net rate of reproduction, R_0, which is the production rate of females from one generation to the next. To estimate R_0, data on the life history of the individuals are required, primarily their age-dependent survival and fecundity (see Keyfitz (1968) and Keyfitz and Flieger (1968) for the finer categorization of parameters used in human demography). We thus see that R_0, although portrayed as a singular estimate, is derived from sequences of estimates of survival probabilities and fecundity projections, and that these are averages across representative samples of individuals of different ages. The intrinsic rate of increase, r, is then usually calculated from the natural logarithm of the singular value derived for R_0, by dividing this latter value by the mean duration of the generation time t:

$$r = \frac{\ln R_0}{t} \tag{6}$$

This manipulation produces immediate problems. As soon as R_0 is divided by the mean duration of a generation (as required in Eq. (6)) it eliminates biological information that is of immediate relevance to understanding the processes that underpin the environmental interactions, survival rates, movement and reproductive rates of the particular organisms of interest. For instance, a vole may be born at either extreme of the age of its mother, ranging from about the fourth week of her life to her seventy-second, as Andrewartha and Birch (1954) point out. The mean, by contrast, effectively concentrates the births at one moment in her life. The generations thus become neatly (but artificially) separated by units of time t. The same happens when R_0 is calculated and this one attribute, which really amalgamates into a single number a sequence of age-dependent events in the lifeline of numerous individuals, is then in danger of being seen as accurately reflecting the intrinsic growth rate, r, and thus being treated as a constant.

5.1 R_0 and Structuring the Ecological Response Process

The procedure outlined above for determining r has the significant implication of actually discarding the ecologically relevant information from which both R_0 and t have been calculated. The consequence is that the analytical and mechanistic power of the equation is sacrificed and with it goes any hope for developing a mechanistic interpretation of the ecological process. In terms of mechanistic understanding in ecology, it is more realistic to use the estimates of the age-dependent survival and fecundity rates that define R_0 directly. The ecological factors that affect the dynamic variation in R_0, plus the factors that determine it in each age group, can thus be identified (e.g., Hemerik et al. 2004). The important consequence for ecological generalization is that the equations that are needed to formalize ecological theory can, accordingly, be structured mechanistically. Outcomes derived through this approach, moreover, remain independent of the constraining assumptions of the logistic equations dealt with previously. An alternative general mathematical formalization for ecology is therefore available and predictions can be deduced for setting up rigorous tests of theoretical interpretation against independent data.

To calculate R_0, a life table is drawn up for a cohort of individuals, all of which are initially in the egg or seed stage in the same time interval (and this can be done for different time periods and localities). The first column in the table contains the age groups from birth to the maximum age achieved by individuals of that species (Table 7.1). The second column contains the numbers of individuals that constitute each of these groups. These numbers, commonly expressed as percentages of the initial numbers of individuals, diminish continually towards the older age groups as a result of mortality. Mortality rates are usually highest for individuals in the youngest age classes, and fall off with age. The resulting distribution of relative (percentage) frequencies, or survivorship distribution, can also be expressed graphically in relation to time. The shape of the resultant plot depends on the biological properties of the organisms concerned, so is related to species, and is also influenced by the degree to which those properties match with the prevailing environmental variation. Under constant environmental conditions the resultant survivorship curve, for insects, is often *J*-shaped (Fig. 7.1). The plot may also be in the form of a survivorship surface for two or more factors that vary and exert their influence interactively and differently across the different life stages as shown in Fig. 7.2. This diagram includes data for the egg stage alone, whereas a survivorship surface response would include the other life stages.

Table 7.1 Life table of the weevil *Sitophilus oryzae** in an environment of 29°C and 14% grain moisture content. Alternative environmental conditions will support different rates of survival and fecundity. See text for further explanation. Adapted from Birch (1948).

AGE (WEEKS) x	*SURVIVAL (%)* l_x	*NO. OF OFFSPRING* m_x	*PRODUCT* $l_x m_x$
4.5	0.87	20.0	17.400
5.5	0.83	23.0	19.090
6.5	0.81	15.0	12.150
7.5	0.80	12.5	10.000
8.5	0.79	12.5	9.875
9.5	0.77	14.0	10.780
10.5	0.74	12.5	9.250
11.5	0.66	14.5	9.570
12.5	0.59	11.0	6.490
13.5	0.52	9.5	4.940
14.5	0.45	2.5	1.125
15.5	0.36	2.5	0.900
16.5	0.29	2.5	0.800
17.5	0.25	4.0	1.000
18.5	0.19	1.0	0.190
			$R_0 = \Sigma = 113.560$

* Referred to as *Calandra oryzae* (small strain) in the original. These insects have since been placed in the genus *Sitophilus* and the large and small "strains" recognized as different species. The one involved in this study was probably *S. oryzae*.

Both methods for representing the data can express the age-dependent mortality or survival of the cohort of individuals under inspection, from birth to death. In cases in which a cohort cannot be followed in this way, the percentage of females within the respective age groups can be used, but here the data derive from different cohorts of different ages. Consequently, the data lack information on age-dependent mortality relative to the properties of the organisms so that a survivorship curve cannot be drawn.

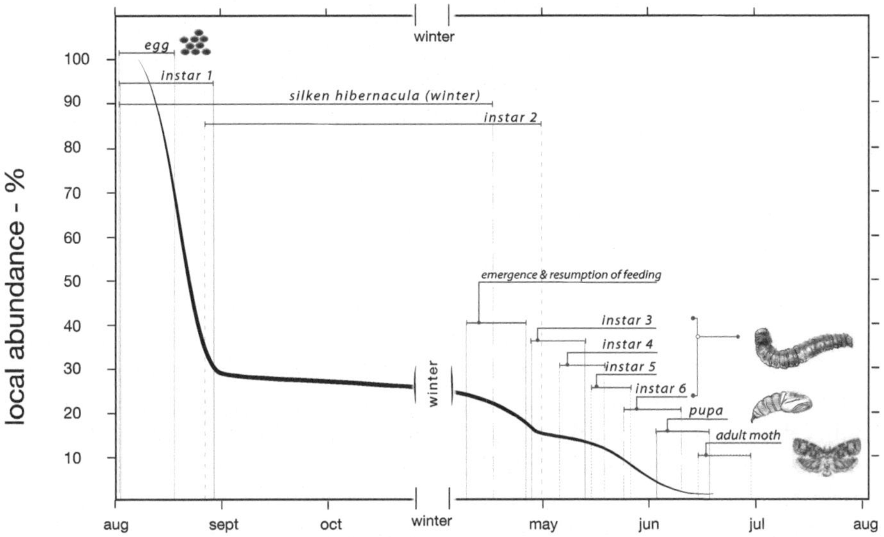

Figure 7.1 Survivorship curve for various life stages of the spruce budworm (as redrawn from Morris (1963)).

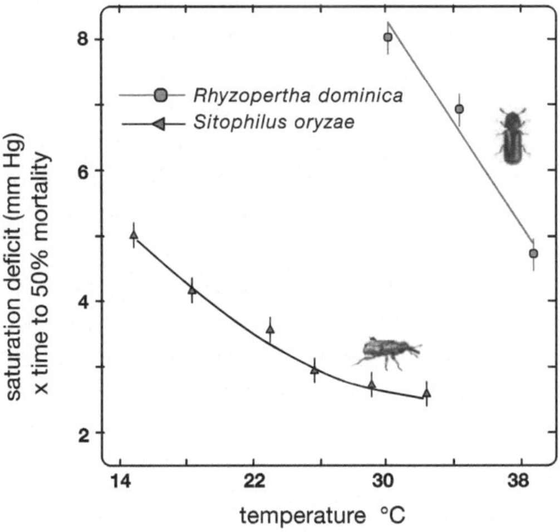

Figure 7.2 Fifty percent death rate curves for the eggs of two beetle species, *Sitophilus oryzae* (see footnote to Table 7.1) and *Rhyzopertha dominica*, for various combinations of two environmental factors, saturation deficit and temperature (redrawn from Birch (1944) and Andrewartha and Birch (1954)).

The third column in the life table contains the absolute number of offspring produced by the females in each of the various age groups, or their age-dependent fecundity, which decreases for physiological reasons with regard to each individual. When the percentage of individuals in each age group is multiplied by their respective fecundities, the actual reproductive output for all individuals in each cohort is obtained, for entry in the fourth column of the life table (Table 7.1). The sum of all figures in this fourth column yields the value of R_0, the net rate of reproduction across all females in the cohort concerned, under the environmental conditions that prevailed through the period of observation. The data are thus locality, time and cohort specific, as described above. Under the experimental conditions of Table 7.1, weevil numbers would multiply 113.6 times in each generation (Birch 1948).

The discussion, above, has been confined to the minimum data needed for calculating R_0. Life tables are, however, invariably more complex than this, and include other age-dependent information about the subject individuals (see, for example, Keyfitz and Flieger (1968) for global human data and Keyfitz (1968) for a mathematical treatment). The value of R_0 represents the number of females in the generation that follows the study cohort. Thus, when $R_0 = 1$, the number of females in the new generation is the same as in the previous generation, implying that the population size remains the same, given that no immigration or emigration takes place. For values of $R_0 > 1$, population size increases, and for those where $R_0 < 1$, it decreases.

A comparison of values of R_0 over a range of successive years, or between species, yields valuable information on the major environmental influences on the abundances of one or more species in the field. That is true also of comparisons of the related survivorship curves and fecundity distributions. Environmental factors that may have influenced these distributions through their effect on the net rate of reproduction over the years or between species can be identified in this way. The effects of such factors on the fecundity of a particular species could even be experimentally determined in the laboratory, and the predictions thus obtained be tested in the field, and so on. Independent tests of ecological interpretations thus become possible.

The advantage of the approach just outlined is that it does not discard biological information on age-specific survival (as in the calculation of r). A quantitative basis is thus available for formulating and addressing ecological questions in a direct way that relates to individual organisms of selected species within their particular environmental setting. For example, what imposes such high mortality on younger individuals? Is this mortality the same under all conditions and how does it vary across the distribution of the species? We know from recent human history, for example, how much child mortality is decreasing because of improved medical treatment, but considerable variation still exists in the level of improvement across countries, so it still varies considerably on a global scale. The decrease in child mortality had a strong impact on the shape of the survivorship distribution and, hence, on the value of R_0. The consequent increase in R_0 has led to problems of overpopulation. This example (see McNeill 1990) illustrates the particular ways in which local densities may be influenced, and the extent to which they can fluctuate.

The perspective developed above illustrates further why ecological systems are not balanced (or in equilibrium). Abundance levels vary considerably across the geographical distribution of species and they vary through time (Hengeveld and Haeck 1981, 1982). Understanding the processes that contribute to the observable variation is crucial to interpreting the dynamic structure of ecological systems. To this end, the factors that drive this spatio-temporal variation need to be separated out so that their influence can be quantified in relation to age-dependent survival rates and age-dependent fecundity rates, or both. Also, the variation in numbers across space and through time should not be considered "noise", to be dismissed as awkward simply because the mean value has been given special significance in demographic ecology. Indeed, variation of this nature is inherent in ecological systems, which explains why the subject matter of ecology is so different from that of astronomy, physics and chemistry, for example, and why the derivation of equations and setting goals for ecology from experience in these sciences has not been as helpful as anticipated by some.

The variation outlined above can be appreciated more readily when ecological systems are considered from the perspective of species-specific differences. Species' survivorship surfaces tend to be specific, as Fig. 7.2 shows for two species of insects, where the rates of survival (or mortality) vary not only with temperature at one humidity level, but also with different temperatures at various levels of humidity. Clearly, R_0 varies in rather complex ways, and cannot be averaged in any meaningful way. That is, its variation has an undeniable significance in the understanding of ecological systems. With ecological theory, the aim is to explain and thus predict the dynamic patterns in the geographical distribution and local abundance of organisms, so the sources of variation just outlined must be accommodated. The only way in which this can be achieved is through the focus provided by the most reliable generalization about the system, that the species-specific adaptations of organisms interact with environmental conditions and that the outcome of this interaction sets the local presence and intensity of occurrence of organisms. That is, organisms have to take advantage or make do with their properties and cope with the variability in the environment through them. Autecology thus focuses on the specificity of the responses inherent in the species-related adaptations of organisms to environmental structure, dynamics and variation.

When considering the value of R_0, species are of interest because their constituent individuals have adaptations that positively influence their survival within the usual environment of that species. Since each stage of the life cycle is exposed to the environment, the entire life cycle needs to be included in such investigations and calculations. Some species have physiological properties that delay their development so as to reallocate their energy within a limited budget, or they may be able to hibernate or enter diapause to tide them through times when environmental conditions challenge their energy budget. Conversely, survival probabilities are reduced in those environments where the organisms and their life cycle do not match the structure and dynamics of the environment so well, as at the edge of the distribution of most species.

A significant but under-rated aspect of the environmental response system of organisms involves the ability of individuals to move around within their environmental setting, and thus adjust their ecological circumstances behaviorally. Such spatial adjustments constitute an empirical challenge because they do influence R_0. That is, ecological studies become more subtle, and the quantification of ecological processes and outcomes more difficult. Movement is likely to be specific to life stage and season. In turn, this will influence the situations and conditions under which the individuals of the next generation will emerge. To be effective, the seasonal timing of the life cycle must allow sufficient time for building up reserves for bridging the unfavorable period or season that can be expected periodically.

Autecological quantification therefore begins with the way in which the life cycle is structured. What properties do the individuals of each life stage possess? How are these properties distributed across the individuals in question, across space and over time? And how do these properties match the prevailing environmental variation? The degree of that match is expressed in terms of the probability of survival of the individuals through their life, their fecundity, and ultimately in the females' replacement rate R_0, for it is this that sustains the persistence of any species in a locality (in combination with movement). For all of these reasons, the information contained in well-worked life tables must not be sacrificed for r. That is, R_0 must remain the focus of autecological attention. The problem with the focus on the intrinsic rate of natural increase r ($= \ln R_0/t$), for theoretical purposes, is that r eliminates all the information that goes into making up an accurate life table, so the significant ecological information (including that on the stochasticity of the matching process (see Chapters 8 and 9)) is lost. The consequence, for theory, is that population density is seen to be readily justified as the sole driving force in population dynamics, for all other factors are excluded by the design of the parameters and equations, and stochastic influences on population numbers are simply seen as disruptive of the balance that is expected in populations (and to which the population is expected to return as it recovers from the disruption). The practical consequence is that investigation of specific ecological situations focuses primarily on density for insight, interpretation and extension to theory.

The problem for ecologists is, instead, to develop an explanation of how individuals are coping within the environment in which we usually find them, rather than how a postulated population equilibrium is maintained. It is the degree of matching of the organisms' properties with the structured variables that make up the environment that is significant, and this matching takes place in different ways, as shown in the next section.

5.2 Consequences for Ecological Understanding and Applied Ecology

The outcomes of several different processes in the lives of organisms determine the final estimate of R_0, some more than others, including stage-specific survival rates, post-dispersal survival rates, and offspring production rates. An understanding of the influence of each subsidiary variable on the ultimate value of R_0 is essential if ecological processes are to be understood, and so too the way in which they change spatially and temporally.

To illustrate, R_0 for a particular species might be the same across localities, but the contributing rates of survival and reproduction could differ across those localities to the extent that they incidentally "compensate" for one another. Figure 7.3 shows that sparrowhawks in the Netherlands, Scotland and England (East Anglia) have comparable values for their net rate of reproduction R_0, as indicated by each of these groups of birds falling near one of the four dark lines that cross the graph. In each case, however, it is

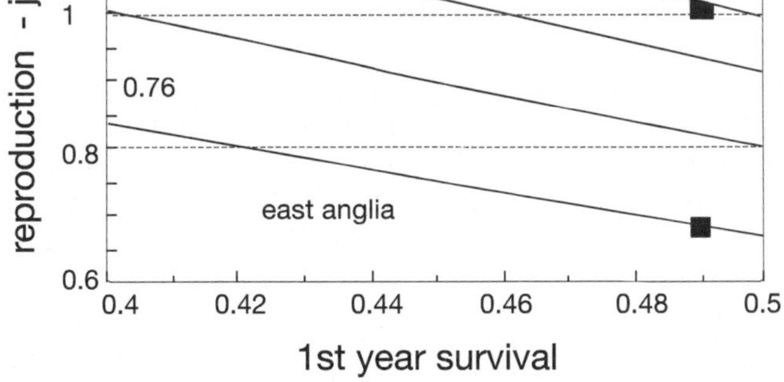

Figure 7.3 Relationship between first-year survival (x-axis), the number of females fledging per breeding pair (y-axis), and adult survival (the four curves with adult survival rate given above them) for R_0 = 1. The position of each of the three geographical areas in the western European part of the range of the sparrowhawk is indicated. Reproduced, with permission, from Lensink (1997).

different combinations of first-year survival and number of fledged females produced by each breeding pair that contributes to $R_0 = 1$. The birds in the Netherlands clearly have to produce more fledglings than the birds in East Anglia to achieve $R_0 > 1.0$ (Lensink 1997). The particular survival values represented in Fig. 7.3 were obtained in different geographical regions, but the same relationships among them hold for survival and reproduction values across different seasons, vegetation types, and so on. In all localities with $R_0 = 1$ in the figure (see lines in the graph) there is a reduction of 1 per cent in first year survival "compensated" for by an increase of 0.5 per cent in the number of female fledglings per breeding pair. This quantified representation of the components of R_0, for sparrowhawks, shows that different values of juvenile survival and numbers of female fledglings per breeding pair, as influenced by different environmental factors, may well result in the same outcomes across localities for the net rate of reproduction of the birds.

Individuals of a particular species, at various localities and over time, interact with the various environmental variables as they impact them. The outcome of these interactions indicates the degree to which the individuals match the varying conditions within particular localities. The impacts of the important environmental variables on the survival and reproduction rates of the organisms, and thus on R_0, can be quantified, expressed and interpreted independently of the observed pattern of population change. This pattern is, after all, the end result of all the processes involved. This means that the impacts of the ecologically significant environmental variables are independent of the changing population densities because they are virtually always factors other than population density itself, and because they are largely physico-chemical in nature (albeit with biotic influences such as vegetation structure). These factors, as a consequence, are usually not influenced by immediate population size, or even by ultimate population size. This is so despite the significance conferred on density by demographic ecologists and by evolutionists who invoke optimization (for population density is perceived to drive the evolution of increased efficiency through intraspecific competition).

The independence between ecological processes and population growth has the consequence that in a certain area a species may be abundant for some time before "crashing", and thereafter remaining rare (as in small mammals in arid environments (Greenville et al. 2012)). These very different abundances may be a simple consequence of the factors influencing mortality or fecundity in the various life stages remaining the same before and after the crash, with the result that R_0 fluctuates around 1 during both periods. The practical consequences of such understanding extend deep into applied ecology, perhaps especially for biological control and fisheries (Walter 2003, p. 265). That the density consequences of ecological processes do not inevitably feed back on those processes also explains the changing abundance of species across their geographical distribution, for example, where the highest abundances (estimated statistically with long time series) typically occur in the central areas and the lowest ones at the margins (Hengeveld and Haeck 1981, 1982). Figure 7.3 provides an illustration of the mechanistic influences responsible for such patterns.

The variation in population density of a species across space has been attributed to the chance coincidence of a high number of independent factors (e.g., Brown 1995), a process encapsulated in the phrase "risk spreading" (den Boer 1968). This approach fails to provide an obvious reason for the highest abundances typically being in the center of the species distribution. In the demographic ecology literature geographical variation in abundance is generally not given much attention besides noting that population densities towards the edge of a species' distribution are generally lower because of abiotic stresses predominating away from the center of the distribution, and elaborate models have to be constructed in explanation (e.g., Guo et al. 2005, and see Chapter 9).

The analysis above illustrates the alternative consequences if ecological theory focuses on r or on R_0. An intermediate position could hold that both r and R_0 are, indeed, part of ecological theory. They do both appear in many ecological texts and could be portrayed as part of the same development. However, practice in ecology suggests otherwise. The logistic equation, with its total reliance on r, and its emphasis on density relationships and population persistence, represents the central point of focus in demographic ecology to the extent that even when the quantification of R_0 has yielded insights that reduce the significance of the logistic, these insights have not been exploited. The treatment of the Whytham Wood winter moth study provides a revealing example, for Varley et al. (1973) persisted in offering an explanation based strongly on density dependence despite the contrary evidence.

In conclusion, general ecological models must include detailed life table data that spans the species' entire life history. Such information is not only species specific but also context specific, as the dynamics of ecological systems vary spatio-temporally. This ecological "idiosyncrasy" may be seen as unfortunate, but this is the nature of the subject matter. The adaptive mechanisms or properties of species and their consequent ecological expressions are different across species because of the way in which adaptation takes place and the way in which new species arise (see Chapters 5 (Section 6), 6 (Section 4) and 10 (Section 5)). Occasionally, the structure of ecological processes is the same across species, as in some invasions across relatively uniform agricultural landscapes (see Chapter 8, Section 6), although the parameter values themselves tend to differ across the species and environmental circumstances, as the study by Hemerik et al. (2004) shows.

The advantage of the autecological approach we advocate is that it preserves the meaningful biological information and helps prevent ecologically relevant information, subtleties and complexities from being ignored or missed. The use of R_0 in ecological investigation, along with the autecological analysis of the variables that influence its values, offers a way forward, not only in ecological understanding, but also in the development of more realistic generalizations. The autecological generalization does not yield the sort of predictive answers, or general projections that drive the structure of the general demographic models, but it does raise significant questions that are essential for the acquisition of data that are critical to ecological investigation and understanding.

5.3 Extending R_0 through the Diffusion Parameter *D*

The biological detail within a life table involves the rates of survival and fecundity according to the age of the individuals. Such a structuring reflects the operation of the adaptive mechanisms of the different life stages of the species of interest relative to the variation in their environment. These adaptive mechanisms can thus be interpreted in relation to the specific structure of the life cycle and the way in which it matches the temporal or spatial variation of the environment, as described in Chapters 3 and 4. Individuals can, however, adjust their environmental circumstances by moving physically. They can thus avoid unfavorable conditions and track more favorable ones, whether these pertain to food, temperature, moisture, breeding sites or any other environmental variable that is relevant to them. Even plants and other sessile organisms can adjust in space, but this entails a generational delay for it involves the movement of propagules, which have adaptive mechanisms to promote their dispersal. Growing plants can adjust, to a limited extent, through the differential growth of roots, rhizomes or branches, for example. Mostly, though, sessile organisms adjust spatially through the movement of their seeds, eggs or larvae.

The movement of individual conspecific organisms, taken together, results in an overall rate of collective dispersal. A term can thus be added to the temporal replacement rate, R_0, to represent this dispersal rate, and this effectively adds a spatial component to our general view of population dynamics. We therefore refer, in autecology, to the spatio-temporal dynamics of the species in question. In the simplest case migration or dispersal is included in the form of the diffusion parameter *D*. The value of this parameter can, again, be estimated from a frequency distribution of dispersal distances that individuals have covered during some relevant time period, often that between their birth place and the site at which they first reproduce, as quantified in the geographical expansion of the collared dove (van den Bosch et al. 1992). Generally, most collared dove individuals move only short distances, whereas a few cover much longer distances, thus resulting in a *J*-shaped distribution (Fig. 7.4) (Hengeveld and Hemerik 2002). Again, specific ecological questions arise. What factors determine the shape of this particular distribution? What distances are covered under different environmental circumstances? And what properties do individuals have to enable such movement and to guide it?

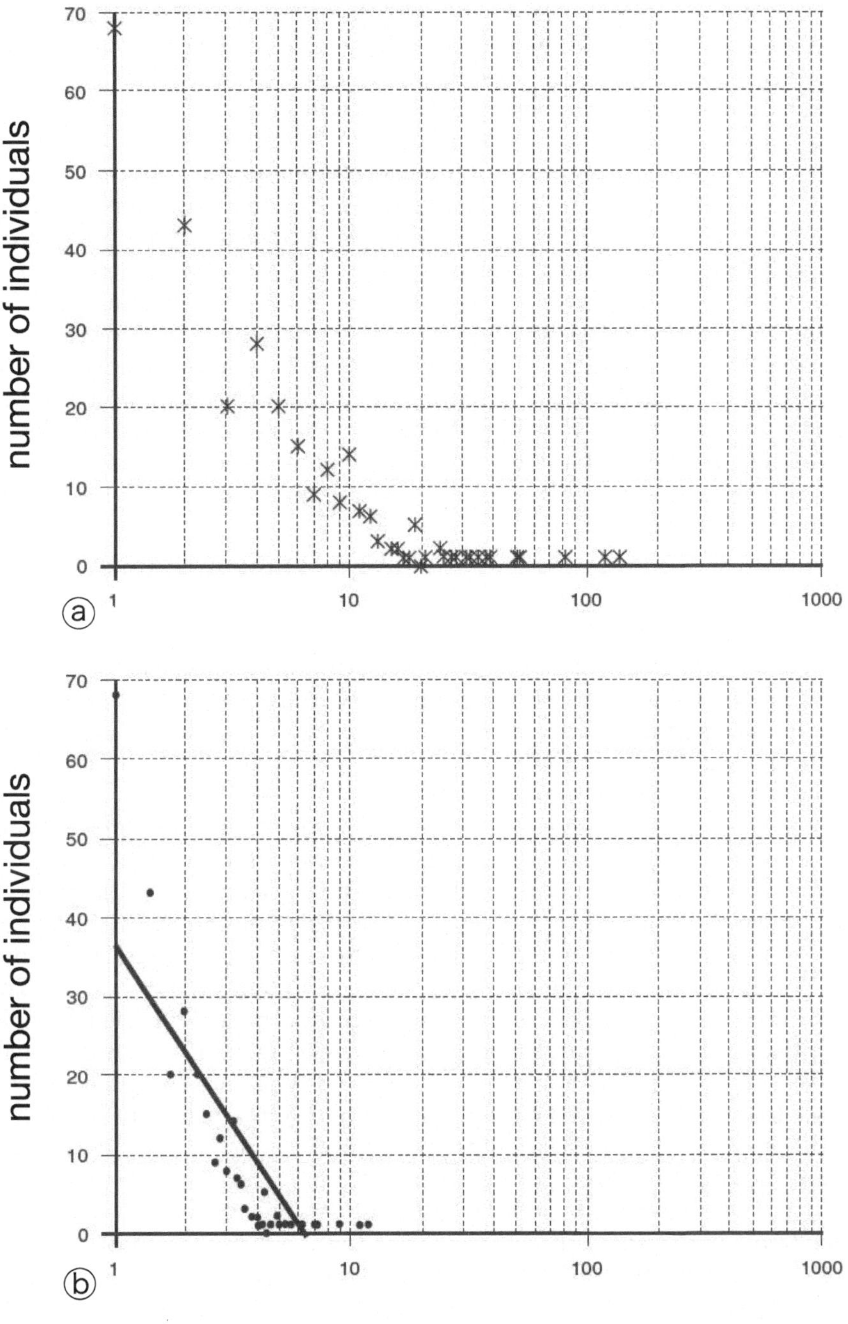

Figure 7.4 *J*-shaped dispersal curve for the collared dove, as represented by the number of breeding birds as a function of the distance from their parents' nests. **Top** is the logged distances, and **bottom** the square roots of the logged distances. Reproduced, with permission, from Hengeveld (1989a).

Integrating R_0 and D yields an estimate of the spatial adjustment rate across all individuals and, hence, an estimate of their overall survival rate as they track their species-specific environmental requirements, which are distributed heterogeneously and shift with time. Understanding the spatio-temporally dynamic structure of the environment relative to the requirements and adaptations of the species of interest is required for realistic quantification of the movement of individuals. Unfavorable areas have to be crossed and temporal climatic dynamics change favorable areas into unfavorable ones, and these aspects have to be accommodated in the equations. The equations thus become more complex and detailed, and the fieldwork that is required to satisfy the data requirements more elaborate. These demands for more quantification than is required by the more simple approach of the logistic equation (and its developments) are essential because these variables contain the biological detail needed to explain the overall probability of survival of the individuals concerned and, ultimately, the persistence of the "populations" and, thus, the species that they constitute.

5.4 Ecological Insights

The ongoing movement of organisms is superimposed upon a heterogeneous landscape of local areas of varying suitability to the species of interest, and areas that are varyingly unfavorable in this regard (Taylor and Taylor 1983, Hengeveld and van den Bosch 1997). This conception of the dynamics of environmental variables yields novel ecological insights.

Of particular significance for ecological understanding is the detection and location by individual organisms of environmental conditions that are suited to their requirements and tolerances, and the resultant spatial tracking by organisms of suitable conditions. This draws attention to the spatial extent of the areas that are unfavorable to the survival of organisms, which is highly significant in understanding their spatio-temporal dynamics. That is, the exclusion of spatial variation and migration from the basic principles of population dynamics and from the associated general equations means that we lose sight of an important set of mortality factors and physiological stresses that are themselves variable in intensity. Organisms are unlikely to spend time feeding or nesting under unfavorable conditions, and are likely to keep moving until they locate sites better suited to their requirements. Movement within unfavorable conditions implies increased chances of exhaustion, desiccation, drowning, overheating, infection, predation, and so on. That is, the greater the extent of unfavorable areas relative to favorable ones, the greater the risk of mortality. Mortality rates are therefore predicted to be proportional to the extent of the unfavorable areas. Under extreme environmental conditions virtually no areas may be entirely suitable for a species; under such circumstances individuals in even the most suitable pockets or refuges will be subject to severe stress and mortality rates.

If mortality rates are, indeed, related to the area of unfavorable environment, they will not increase linearly, but exponentially (Hengeveld and Hemerik 2002). Because reproduction is a multiplicative process that results in an exponential increase in numbers at each reproduction event, some factor must presumably reduce the value of the exponent of population increase. Demographic ecologists corrected for the exponent in terms solely of local density, with specific reference to the local carrying capacity, *K*. That is, they focused strictly on the area in which the population was seen to be restricted. The alternative that we suggest above is not mechanistically related to density in the way proposed for population regulation, although it may well be density related. We suggest that the mortality related to crossing the unfavorable areas that separate areas of favorable habitat could well be the feature that reduces such "overproduction", thus countering exponential reproduction. Conversely, the massive reproductive output of organisms (that is, the value of the exponent) is likely to have been selected in relation to the probability of survival against such a demanding ecological backdrop. A further consideration in this regard is the provision of the appropriate sequence of environmental conditions for the species in question.

The spatial distribution of the unfavorable areas, as well as their extent and precise nature, have also to be taken into account, as these aspects will seldom be uniform. The favorable areas are likely to have a random or Poisson distribution, so that the distances between the entire set of favorable areas will be exponentially distributed. Consequently, the extent of the unfavorable area that makes up the matrix between them increases accordingly, and with it the rate of mortality associated with the movement of organisms. And this, in turn, increases mortality exponentially such that exponential reproduction is again countered. Although movement may, at times, be related to density dependent population expansion, it will mostly be unrelated to density as most species are relatively rare (Williams 1964, Pielou 1975, Walter and Zalucki 1999). In any case, some organisms even move from localities that are apparently suitable to them and are not overpopulated (Andrewartha and Birch 1954, Walter 2003, Chapter 9), so the process of redistribution of organisms tends to be ongoing.

A final complication is that localities do not remain favorable (or even unfavorable) indefinitely, because climatic conditions are variable, whether seasonally or sporadically, and climate does change continuously. Sooner or later, conditions will change and, from an individual organism's viewpoint, improve or deteriorate. Again, the Poisson is the simplest distribution to describe such random fluctuation through time, so the times between favorable conditions will once more be exponentially distributed. For individual organisms, this means that the chances of mortality related to unfavorable conditions will also be exponentially distributed through time. In turn, this implies that the chances of finding favorable conditions, with its negative consequences for the organism's probability of survival, will also be distributed in this way. Yet again, the overall mortality rate for a number of individuals will increase exponentially, and this takes place on top of the other exponentially distributed causes of mortality discussed above, in this section. This contributes further to the reduction in the exponential effects of reproduction. Again, this aspect is left out of consideration when using r instead of R_0. Indeed, the rate of settlement in a heterogeneous area is, of itself, already exponentially distributed, even before the Poisson distributed distances and times are taken into account (Hengeveld and van den Bosch 1997, p. 219, Eq. (3)).

The various situations, detailed above, that lead to exponentially distributed mortality rates are those that are most straightforward to appreciate and to deal with practically. Their effects will vary, in less straightforward ways, with the location within the range being sampled, the centers being more uniform and constant than the margins, for example. Moreover, climatic variation causes their distribution to expand, contract, and shift continually. Consequently, the degree of spatio-temporal uniformity associated with any particular locality will vary over time. As a rule, the various chance processes are superimposed, each one having a different temporal or spatial dimension. This means that the overall chances will have exponents of some fractal dimension instead of having integer values (Mandelbrot 1977).

General expectations about the probability of survival of the individuals, or about those of a species' persistence therefore require an intimate knowledge of the species-specific requirements as they relate to the spatio-temporal dynamics of their environment. Models that involve data loss, as in the logistic equation, will seldom predict these probabilities or explain abundance satisfactorily.

5.5 Risk Analysis

The survival rate of conspecific individuals under the environmental circumstances outlined in the previous subsection is best dealt with conceptually and quantitatively in terms of risk analysis (Parry 1978). To do this one needs an understanding of the specific requirements and tolerances of the species of interest, in terms of their environmental response systems and the relevant parts of their environment. A risk analysis for such a group of individuals can be developed by quantifying how often, how well, and for how long these requirements are met. A comparison of the frequency plots representing the organisms' requirements and the dynamic structure of the environment can then be used to estimate the proportion of time that the two overlap and for which conditions are favorable for the organisms. The reverse of the probabilities that these requirements are met defines the extent to which the survival of the individuals is at risk. A detailed picture of the chances of survival of the individuals and of the factors influencing these chances can thus be developed. Life table data reveal the risk that individuals run through their lifetime, but such risks can also be estimated from such things as the frequency distribution of dispersal distances (Hengeveld and Hemerik 2002). Are the maximum distances that individuals can cover within a certain time period, for example, large enough relative to the minimum distances across the unfavorable areas between suitable sites? It is possible that only a small fraction of the individuals can cover these distances, which implies that the short-distance dispersers would be heavily selected against.

Autecological analysis, as outlined above, provides a direct, mechanistic connection to natural selection, and thus to evolutionary theory. Demographic ecology cannot do this, simply because it has no direct connection to the biological features on which selection might be acting. The use of r instead of R_0, means that all the biological traits on which selection could act have been eliminated. When risk analysis pertains to a feature of the life cycle as a whole, as in the requirement each year of a minimum number of degree days to produce seed (e.g., Parry 1978, and see Chapter 3), this connection with evolutionary theory cannot be made unless information is available on the number of individuals that needed less than that minimum. With risk analysis the survival probabilities of individuals within a stochastically varying environment can be predicted, as detailed further in Chapter 9. And adaptive mechanisms reduce mortality risk in that they counteract the effect of environmental stochasticity. Therefore, risk analysis is most meaningfully conducted on the ecological requirements that are associated with particular properties of the species in relation to the structure and variance of the relevant environmental features. As such, risk analysis is pivotal for autecological analysis and understanding, and for interpreting adaptation.

6 SOME METHODOLOGICAL REMARKS

6.1 Modeling Aims

One of the aims of the demographic models discussed at the beginning of this chapter is to predict N_t, the number of individuals, N, at a future time t, from the population density at an earlier time (N_{t-1}). To this end, the intrinsic rate of increase, r, must be known, and it is assumed to be constant over the given time period. Furthermore, a particular cut-off level for population growth, the carrying capacity K, is invoked, for this provides the regulatory capacity to slow population growth to a stable level. This implies that the ecological factors most relevant to dictating abundance are r, N and K, and that other factors are relatively less important. The control of population numbers is therefore implicit and is written into the equation *a priori*. Thus, the foundational assumptions of demographic theory ensure that other ecological factors or processes have, at best, only passing significance, and are mainly of observational, local or applied relevance. Only those factors defined in the equations are given general relevance. The processes and parameters included in these equations are not related to the specific properties (or adaptive mechanisms) of organisms, but are assumed to hold for all species, with only the values of the parameters having to be estimated for substitution into the equation. This implies that any differential ecological influences (across species, localities and time periods, for example) are obliterated. Once the overall average has been determined, the development of the process determining N_t for any future value of t can be given with certainty.

In the field, however, the processes that are ecologically influential may vary from one situation to the next, from one locality to the next, or between localities. Furthermore, they differ across species. These differences between species, even closely related ones, are inevitably dictated by their different adaptive mechanisms, and are evolved stochastically and independently of other such species (Chapters 5 (Section 6), 6 (Subsections 4.3 and 4.4) and 10 (Subsection 5.2)). We know they evolved independently of one another because their traits are species wide (e.g., Fig. 5.1), whereas the overlap of the species ranges is only partial (Figs. 3.2 and 4.6) and varies over time. Often the species of interest do not even overlap spatially because species ranges expand, contract and shift independently of each other (Fig. 3.1) according to their species-specific responses to physical factors, which operate at a broad, geographical scale.

Consequently, the assumptions made by the demographic equations, whether explicit or implicit, simply do not hold for organisms in nature, so alternative general mathematical formulations are necessary. In short, the aim of ecological research demands redefinition, from the extrapolation of numbers of individuals from previous values, to constructing survival probabilities from the suite of the environmental response systems of organisms in relation to the environmental variables that influence them. This is the field of autecology. Since species all differ from each other in their environmental responses, through their specific requirements, sensory responses and tolerances, they express individualistic ecological behavior through their specific environmental response mechanisms. Autecology aims at understanding this individualistic behavior, and attempts to reconstruct survival probabilities from the interaction between these response mechanisms and the environment. Rather than extrapolating the future development and persistence of populations as the target of interest, it is the survival chance of individual organisms under particular (and changing) environmental circumstances that is of primary concern.

6.2 Retrospective *versus* Prospective Explanation

The constraints imposed on the interpretation of ecological systems by demographic ecology theory inevitably leads to the definition of populations and communities, seen in terms of hierarchical units in ecology, as closed systems. Such an approach constrains not only research aims, techniques and interpretation, but even disarticulates ecology from a direct relationship with other biological sub-disciplines. This particular disjunction results from the adaptive mechanisms of individuals being excluded from ecological theory, except, perhaps, in anecdotes. This perspective partly explains the emphasis, in evolutionary ecology, on the process of adaptation as a competitively driven optimizing process and the simultaneous de-emphasis of adaptive mechanisms.

Calculating parameters like N_t is only possible when we know past values of r, K, or N_{t-1}, etc. We can extrapolate our knowledge of their values to future situations, under the assumption that they remain the same throughout, which is a form of retrospective explanation. It is often based on the statistical analysis of existing data, and so excludes the exact structure of the relevant ecological process from deduction and quantification. Furthermore, the approach is simply descriptive because the same parameters and parameter values as used in the initial models provide the basis for extrapolation. Although variability in parameters may be included, they are seldom (if ever) derived from biological data. Because such extrapolation simply projects past events into the future in exactly the same terms as before, we have no basis for understanding the underlying process(es). A major advance over this methodology is achieved with the application of the methods of prospective explanation.

The autecological methodology proposed in this chapter has the advantage, over demographic methodology, of providing a more detailed and complete understanding of the spatio-temporal dynamics of species, both in general and in specific cases. With an emphasis on the species-specific morphological, developmental, physiological, and behavioral mechanisms of the environmental response systems of individual organisms, it provides a more realistic approach to ecological understanding. And, overall, ecological systems are appropriately treated as open and dynamic. The requirements and tolerances of the organisms, relative to their environmental response systems, are then matched against the relevant features of a spatio-temporally structured environment that imposes its dynamic on the organisms. Expectations can thus be generated of what to expect under particular regimes of environmental variation. An unsatisfactory match indicates that a particular proportion of organisms is likely to encounter problems in maintaining themselves and so will die or fail to propagate. The autecological approach is prospective because it predicts, in this way, future states of the ecological system under consideration. It does so from independent information about the level of variation of environmental matching and explanations are, on this basis, testable.

Mathematical equations are designed explicitly to reflect the structure of the process under consideration. Data appropriate to each parameter is then used to run the model, and independent predictions about particular environmental circumstances can be made. That is, the impact of particular changes to the system can be anticipated. For example, the influence of low rainfall or high temperature (or a combination of these) can be predicted. These outcomes can then be compared with observations from the field, to determine the accuracy of the model. Such an approach is deductive and prospective.

Not only can we predict what we may expect under future conditions, but we can also make predictions about different conditions elsewhere in a species' distribution. Range expansions, contractions, or range shifts can therefore be predicted. That is, the environmental matching of organisms, if understood in principle, will allow predictions of the organisms' environmental tracking if the structure and spatio-temporal variation in its environment is known (Hemerik et al. 2004). Theoretical understanding thus connects directly with ecological application. This is only possible if we know the properties of the species concerned, how they match with the characteristics of the environment concerned and the degree of that match. See Vors and Boyce (2009) for the application of such an approach to understanding how climate change is impacting reindeer and caribou. This approach provides a clear alternative to demographic ecology for it does not seek general solutions to postulated ecological characteristics like "invasiveness". The generalization in autecology relates simply to the good fit between the environmental response systems of the species in question and the structure and spatio-temporal dynamics of the environment.

Although demographic models are usually deductive, they are not similarly prospective, because the ecological process itself is not constructed and the parameter values that are included contain insufficient information to allow independent predictions to be made. By contrast, the life table analyses and analyses of dispersal that are central to autecology provide the information necessary for making independent predictions as to the cause and outcome of the processes concerned, that is, to formulate deductive, prospective models. This approach to modeling is extended substantially in Chapter 9.

7 CONCLUSION

Demographic ecology is strongly associated with the foundational premise that ecological systems persist because they are balanced (e.g., Cooper 2001). This theoretical perspective maintains the focus of ecologists almost exclusively on the broad issues of population regulation and population persistence, through the inevitable action of density dependence.

Although density dependence is detectable in nature, it is sporadic and localized, and tends to have a stepped effect (Fig. 7.5) rather than it influencing ecological processes in linear fashion as expected for population regulation (Dempster 1983). Demonstrating density dependence in a field study cannot, in any case, be assumed to explain any trend detected at the same time in population densities because such studies have generally not been designed to distinguish cause from effect and thus have limited explanatory value (Chitty 1996, p. 174). In any case, equilibrium density and carrying capacity are impossible to define in any real sense. Carrying capacity is not equivalent to the observed maximum population density and population equilibrium is not defined by the average of a sequence of samples, as is often implied (see Chitty 1996, p. 39). These difficulties, among others (e.g., Andrewartha and Birch 1954, Dempster 1983, Wolda 1989, Wolda et al. 1994, Chitty 1996, den Boer and Reddingius 1996,

Murray 1999), imply that the expectations of ecologists derived from an acceptance of population regulation as a regular process in nature is misplaced. Nevertheless, the idea of ecological balance is still directing attention at aspects of the system that are ephemeral, impossible to measure accurately, and are clearly virtually irrelevant to the external features and biological variables that represent the basic workings of the system. These problems are still evident in the basic demographic equations (Section 3, this chapter) as well as in their more modern equivalents and perspectives for theory.

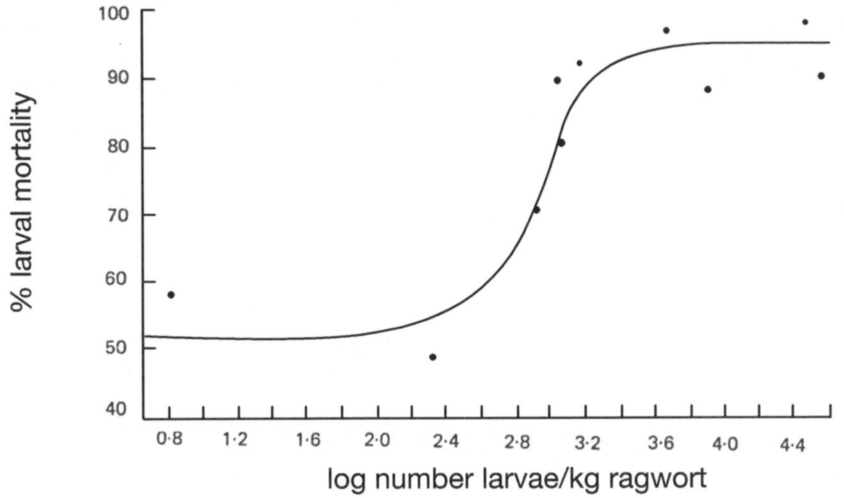

Figure 7.5 Relationship between larval mortality and density in *Tyria jacobaeae*. Reproduced, with permission, from Dempster (1983).

The ultimate consequence is that organisms and their adaptive mechanisms are largely excluded from ecological generalization in the demographic approach to ecological modeling, and generally tend to be overlooked in particular studies and syntheses. Efforts to be more inclusive of organisms have been made, but success can be limited, at best, when the properties of organisms and their primary environmental influences are tagged onto unrealistic foundational principles (Section 1, this chapter). Since most ecologists learn what foundational concepts to focus on from their early exposure to ecology text books, which tend to be steeped in population regulation theory (Section 1, this chapter), it is not surprising that demographic ecology dominates ecological perceptions, despite the long history of criticism it has attracted. Although debate has been intense at times, resolution has never been achieved, the debates have ultimately been seen as diversionary, and the traditional view has persisted even if moderating qualifications have been added.

Ecologists have noticed that the systems they study may not be in equilibrium. Indeed, the idea of non-equilibrium systems is a step forward

(Rohde 2005, 2013), but still leaves ecology with two difficulties. First, non-equilibrium can be defined only with reference to equilibrium, but this is the equilibrium simultaneously being denied (Walter 2008). This logical impasse leads directly to the second problem. Non-equilibrium still requires a positive direction as to how to characterize ecological systems, and this is provided by autecology, for it avoids focusing on local populations as the basic unit of interest, one in which the movement of organisms is ignored (e.g., Turchin 2003). The assumptions and methods of autecology also do not reduce all the information in a life table to a single constant, by calculating $r = \ln R_0/t$. This emphasis on r means it is impossible to avoid models of exponential population growth, which need the exponent to be moderated at densities that are too high or low, to avoid outbreaks or extinctions. And by excluding habitat tracking in a spatially heterogeneous and dynamic environment, it is virtually impossible to see any alternatives to density-based counter-effects to exponential population growth. The extreme data reduction imposed by the assumptions of demographic ecology therefore leaves little choice other than mathematical deduction and analyses from retrospective data. The real risk is that the models become untestable and theory becomes speculative.

The autecological approach focuses on the organism, the structure of its life cycle and its species-specific properties as they change throughout its life. These features of the organism are interpreted, quantitatively, against the spatio-temporal dynamics of the heterogeneous environment. In this context the environment is specified in terms of those axes of differentiation (Chapter 5) that are relevant to the subject species. The general mathematical representation takes the basic form of $N = R_0$ with some relationship with D, and with the appropriate mathematical expansions to accommodate the relevant environmental heterogeneity, life cycle structure, and so on. The advantage of the autecological approach is that it places ecology in direct relationship with the biological sub-disciplines that are closely associated with it, including physiology, ethology, morphology, and so on. Furthermore, autecology shares direct links with evolutionary theory through its focus on the survival value of individuals as conferred by their biological properties. In other words, individuals operate only in their own immediate ecological interest, not with respect to future fitness.

This analysis is extended in Chapter 8 to develop a deeper consideration of ecological patterns and the different approaches to modeling, with an emphasis on the mathematical reconstruction of ecological processes to test ecological interpretation. This perspective is further expanded (Chapter 9) into an explanation of how stochastic influences can be incorporated into ecological modeling through a consideration of environmental gradients and risk analysis.

Chapter 8

Research in Ecology

Patterns and the Scientific Exploration and Reconstruction of Ecological Processes

CHAPTER OUTLINE

1 INTRODUCTION

Fundamentally different conceptualizations of ecology are possible, as illustrated in Chapter 2. Ecological theory has, however, largely been committed to a single direction, as if no alternative approach were possible. This direction has been maintained despite intense criticism of the underlying assumptions and a genuine alternative viewpoint from which to develop theory being long available (Andrewartha and Birch 1954). This commitment to a single set of interrelated ideas and aversion to deep discussion in ecology may stem, at least in part, from methodological flaws not having been noticed in the initial development of the theory. The discipline is not only complex, and the subject matter subtle, but the scientific basis of ecology was set early, when it was readily open to influence from natural theology (Worster 1977, Mabey 1995). Reliance on inductive reasoning has also played a considerable role, as explained below.

Modern ecology had its origin with Darwin (1859), whose environmental perspective on evolutionary change and the mutability of species was developed within a broad geological and geographical context. Darwin knew that species were associated with particular climatic regimes and that they moved, unaltered, into and out of localities as climate changed. Charles Lyell had ensured Darwin knew this through a letter he wrote to him, but it was an issue that Darwin took no further (Bennett 1997, p. 9). Instead, his attention was drawn more strongly to organisms in their "web of complex relations" (Worster 1977, p. 153). Especially influential was the competitive world revealed to him by de Candolle and Malthus (Eiseley 1958, p. 106, Himmelfarb 1959, p. 161, Worster 1977, pp. 149 ff, Paterson 2005). This view held that fecundity had to be brought under control (Worster 1977, p. 153), primarily through competitive feedback, at all localities and at all times. Competitive natural selection was therefore seen to be operating everywhere and all the time, so evolutionary change had to be taking place everywhere and continuously, leading to diversification. That, according to Darwin, ameliorated competition (Worster 1977).

Darwin thus developed the equivalent of geological uniformitarianism for population biology. The agents of change were everyday observable processes, continuously in action and open to our view (Kingsland 1995, p. 9). The focus of natural selection, and therefore ecological processes and the principles of ecology, was seen as locality specific and contemporary. Darwin's uniformitarian view naturally extended to species. They were seen to be made up of local varieties (which today are referred to as biotypes, strains, genotypes and the like) that were in a continuous state of divergence and inevitably on the path to full species status. The latter was achieved when "divergent natural selection" (see Glossary) led to gaps in the distribution of morphological features (Mallet 2008). The prospect that change to adaptive mechanisms might have required special circumstances, such as a small population being geographically confined to conditions inimical to its requirements and tolerances, was thus excluded. Darwin tacitly saw the world as relatively static in a spatio-temporal sense, with the primary dynamic being a localized, competitive one. Competition was not his only option in developing natural selection theory, but it was the one he settled on and the one he promoted consistently in his writing (Brady 1982, Sylvan 1994, Paterson 2005).

Darwin did not confine himself to intraspecific competition. He also set the scene for development of the competitive exclusion principle. His statement about the mistle thrush causing a decrease in the song thrush in parts of Scotland (Darwin 1859, p. 76) focused entirely on competition through these species being seen to have similar ecological requirements to one another. Methodologically, Darwin's view on this was flawed. On the basis of virtually no data, simply the description of a pattern of one species becoming scarce and another similar species increasing, he launched a conclusion in which he considered no alternative possible explanation. We now know, for example, just how sensitive the song thrush is to change in vegetation structure (Peach et al. 2004). Others, with no more evidence than Darwin, followed his lead, a clear case of precipitate explanation (Walter 1988a) against which Chamberlin (1897) warned students strongly (and is an issue seen as so crucial as to have warranted a reprinting of Chamberlin's paper in *Science* (1965, vol. 148: 754-759) and a rewrite for modern students (Railsback 1990, 2004)). Also, population biology was developed initially under the intellectual sway of inductive reasoning. Although induction is critical to scientific development, particular forms and uses of the logic can be extremely stilting.

Baconian induction, also referred to as empiricism, holds that scientists gather observations or data objectively (Lovtrup 1984). The accumulation of such data is seen to reveal patterns from which hypotheses emerge. Supplementary data sets build and help to verify the hypotheses until we have robust theories and ultimately laws. The practical consequences of such a view is that the fundamentals of a body of science are seen to have been derived rigorously and objectively. The derived understanding could therefore be trusted as robust (Walter 2003). This was the general view of science, although it was never the only view, when Darwin developed his interpretations, and it remained strong in many quarters into the twentieth century when population biology was being developed. Karl Popper's writings contributed substantially to such views being scrutinized in the biological sciences (Popper 1963, 1983, Magee 1973, Chalmers 1999), and we see this in the greater emphasis on the word "test" when methods and approaches are being described in ecological and evolutionary papers. Nevertheless, not all claimed tests are "strong tests", by which Popper meant they do not have the ability to falsify an interpretation if, indeed, it is false. When realistic alternatives are not even considered in the tests, theory development is weakened even further. Popper's falsification criterion, although not as clear-cut as once thought (Lakatos 1970, Chalmers 2003), still provides us with a strong edge when designing scientific tests as meaningfully as possible.

In short, population ecology had its fundamentals laid down during the period when the empiricist views outlined above held sway. These fundamentals still provide the basics in ecology texts (Chapter 7, Section 1). Since most population biologists learn ecological theory first and encounter organisms later (Drury 1998, Kohn 2004, p. 313), the ground is set for these fundamentals not only to persist, but for them to be defended vigorously, even against criticisms that would be seen to be debilitating in other disciplines. None of these points demonstrates that the fundamentals of ecology are faulty, just that they do warrant serious reconsideration. In this chapter we not only specify what a more rigorous inspection should entail, but also detail the methodological and statistical procedures that should help to overcome the problems outlined above.

This chapter details how ecological understanding must be rigorously developed through three phases of research if theory development is to be robust. Section 2 introduces these phases by presenting a general outline of the variables and processes that influence the ecology of organisms and what ecologists aim at when they target process in terms of ecological interpretation. The question of which ecological variable(s) and process(es) to study, whether in specific cases or for more general theoretical purposes, is usually not obvious. Initially, ecological patterns attract attention, which represents **phase one of ecological research** (Section 3). The problem is that numerous different patterns have been identified in ecological systems, some of which are undoubtedly epiphenomena or by-products and do not warrant direct explanation. Which is to be given prominence? Which is the most fundamental to general ecological understanding? This section exemplifies the problem faced in the justification of the most basic of ecological patterns. In other words, phase one involves description through the assessment of the realism of the problem. Once a pattern is accepted as worthy of explanation, the variable(s) deemed significant and the nature of the ecological process(es) that generate the pattern are investigated. This is **phase two of ecological research** (Section 4) and is exploratory in that it works through the inductive estimation of the identity and relevance of the variables that may determine the process under investigation. It has its basis in statistical analysis. This is followed by reconstruction (Section 5) in **phase three of ecological research**, through the development of analytical mathematical models that are entirely mechanistic. Phase 3 is deductive in that the model incorporates the variables identified in Phase 2. The variables are given a relative weight to postulate how they may operate in relation to one another in nature. Output from the model is generated by running it with independently collected field data on the survival, reproduction and movement of individuals of the species concerned. Those outcomes are then compared statistically against field data, again collected independently, on the process that has been modeled (e.g., a range expansion) to test the accuracy of interpretation. Such models inevitably need refinement, in part because environmental factors with impacts at smaller or larger scales than those considered originally are likely to need further consideration and perhaps incorporation. Those at smaller scales are likely to operate randomly with respect to the scale modeled and those at larger scales would operate more consistently, but would require attention to reflect the process more accurately by integrating these effects.

The first two phases are retrospective for they focus on historical data, whereas Phase 3 is prospective in dealing with future expectations of the survival of organisms and thus the spatio-temporal dynamics of the species. In principle, models may be deterministic, chaotic or stochastic. We argue that stochastic models will best serve ecological theory and exemplify the advantages of this approach with a worked example (Section 6). Chapter 9 extends this perspective in detailing the structure of such models and their data requirements.

2 PROCESS STRUCTURE AND APPROACHES TO ITS ANALYSIS IN ECOLOGY

In principle, scientists work towards understanding an ever-greater number of phenomena in the world by the smallest number of basic processes. These processes explain, in mechanistic terms, how the deep regularities (or patterns) of the world are generated (Plotkin 1994, pp. 78-79). The population biology literature tends to use the term process and mechanism interchangeably. In evolutionary biology, however, "mechanism" has a particular meaning and is used virtually synonymously with "adaptation" (Williams 1966), as in the properties of organisms rather than in the process of adaptation, so generally we should speak about process in ecology and reserve mechanism for reference to adaptations. Although many ecological processes are mediated by adaptations, they almost invariably have significant external input, from the climate for example, and some ecological outcomes are "unintended" consequences or by-products of the actions of organisms and so, again, should not be seen as the direct action of adaptation or mechanism. Process is therefore, by far, the most neutral term for what are often called ecological mechanisms.

Ecological processes have an underlying biological basis, which we have identified as the species-specific complex adaptations or properties of organisms, and the workings of these mechanisms are influenced by a potentially large number of variables. The environment imposes both a temporal and spatial structure on ecological processes, especially with regard to physico-chemical and biogeochemical variables. The different variables and different measures of these variables influence different species and different life stages of these species differentially. A range of other potentially important variables also exerts various influences. These latter include organisms of other types (herbivores, predators, parasites, pollinators, and so on), organisms that affect the environment (vegetation structure, nitrogen enhancement, decomposition, and so on), and density effects on resource availability. Although the list just given is general rather than exhaustive, it does give a good idea of how multifarious and subtle the influences may be in any ecological process, and also hints at how difficult it may be to abstract meaningful generalizations in ecology. We therefore need to consider which variables are most important in any given process and how accurately we can generalize about these variables and about ecological processes as a whole. Clearly this is not a trivial exercise because one can emphasize different variables and reconstruct processes in different ways. This chapter therefore justifies in some detail the methodological approach that should be taken in such a quest.

The investigation of ecological processes has at least three phases, as outlined above. First comes pattern confirmation. The pattern that is to be explained has, itself, to be investigated to ensure it is real. The criterion of reproducibility is a universally accepted cornerstone of science (Neufeld 1987, Jasny et al. 2011), but is seldom applied rigorously in ecology.

If spurious patterns are given prominence, progress in understanding ecological processes is undermined. Next is the exploration phase, the investigation of the variables that might contribute to the process that generates the pattern of interest. The aim is to reduce the number of explanatory variables or processes to a minimum, for science seeks simplicity in this form of generalization. Finally comes the reconstruction phase, in which a general model of the process is developed. This final step provides the test of the validity of the overall interpretation, for now the calculated result can be compared against independent observation to judge its accuracy.

The basic methodological distinctions just outlined, although fundamental, are not often raised in ecological discourse. Interpretation suffers because the ongoing conflation of these procedures continues to work against progress in population biology (Hengeveld 1992), especially when verification of an expectation is sought without serious consideration of alternatives. A sharp distinction between these sequential procedures is needed if population biology theory is to be developed effectively. Surprisingly, perhaps, pattern confirmation has never been rigorously pursued to the point of resolution in ecology. The reason, we suggest, is partly historical and partly the result of the general approach in ecology today giving prominence to particular demographic processes, as outlined earlier. The frustrated search for ecological patterns that are consistent with the processes that are emphasized in demographic ecology is therefore ongoing (e.g., Lawton 1999).

The exploration phase (Phase 2) is the search for variables that potentially contribute to the documented pattern. This entails inductive statistical techniques in the quest for significant correlations or associations. Positive outcomes are taken to indicate which ecological variables are potentially important. Although Popper (1959) excluded analyses of this nature from science, such a rigid view seems unhelpful because this is a necessary step in developing theory. To then ascertain which of the variables contribute to the pattern of interest, experiments or observations are constructed into risky tests and this approach does embrace Popper's falsification criterion, although it has been implemented rather weakly in ecology. If the results of the treatments fail to meet prediction, the interpretation has failed and is rejected. Acceptance of an interpretation, however, comes through supporting evidence as much as robustness of that interpretation in the face of strong test (Anderson 1987).

Exploratory research in ecology, whether experimental or observational, has almost inevitably been confined to a small subset of the variables that could influence the pattern of interest. The ecological relevance of each such variable is initially investigated only as an isolated parameter. Such tests are important, but on their own have the real potential of strengthening any original belief in the ecological significance of the variable being tested. They do not test for the ecological significance of the variable relative to other possible influences on the pattern of interest, as explained by Chitty (1996).

For example, tests of the importance of interspecific competition seem not to have resolved the issue that inspired its intense investigation, assertions to the contrary notwithstanding. That is, ecological theory is still resource and competition based rather than based on the interactions between adaptations and the environment. This situation persists even though serious doubts about competition providing a sound basis for developing ecological theory have often been justified (Andrewartha and Birch 1954, Heck 1976, 1980, Rohde 1979, 2005, Connell 1980, Strong and Simberloff 1981, Simberloff 1982, 1983, 1984, Andrewartha 1984, Walter et al. 1984, den Boer 1986b, Sale 1988, Walter 1988a, 1995, Walter and Paterson 1994), but see Simberloff (2004)), and even though non-scientific influences may be important in this persistent emphasis on competition (Simberloff 1984, Walter 1995, Walter and Paterson 1995, Paterson 2005). In any case, the tests of competition as an isolated variable cannot provide a clear-cut answer. The problem in investigating competition in this way is compounded, for example, by study sites in the field being selected for the likelihood that competition is indeed important there. A comparison across randomly selected sites within the overlapping distribution of the species that are allegedly competing would be far more acceptable methodologically. The history of research conducted to verify the significance attributed to density dependence provides another good example.

Reconstruction procedures (Phase 3) build upwards from the basic information provided by the research conducted in Phase 2, usually through deductive mathematical modeling. To test the significance of any variable in ecology, estimates of the different parameters have to be combined into a model that represents the ecological effects of the variables as well as the effects of their interactions. The whole is thus designed to reflect the structure of the process of interest. Such tests involve a complete reconstruction of the process, and this means that the results from the model can be compared statistically with independently collected data from the field to provide an independent test of the importance of the variables. An appropriate fit between calculated and observed infers that the variables identified as ecologically significant in the exploration phase of research do indeed operate as reflected in the structure of the model. A general model that predicts outcomes accurately, especially when environmental circumstances are varied, provides an independent test of the original interpretation, the one derived through exploration.

The reconstruction phase of theory development is not well advanced in ecology. Although general models have been developed, they have not been of the type most appropriate for general theory. In Section 4 we consider various curve fitting and simulation procedures, and explain why they are inadequate for this particular purpose (although they do have utility in areas more specific than theory). Section 5 considers the general process-based models that are available (deterministic, chaotic and stochastic) to point out that the most appropriate approach in this regard has languished in the development of ecological theory.

3 PATTERN IDENTIFICATION AND JUSTIFICATION (PHASE 1)

Essentially, patterns suggested or assumed to be representative ecologically must be tested statistically to ensure they really do represent real phenomena. The methodology entails sampling to test the veracity of the pattern(s). For simple claims (e.g., involving two distributions of beak size in birds), the situation is straightforward. Are they statistically different from one another, or do the samples represent two realizations of the same distribution? If any difference is not significant statistically, the samples are considered the same and no explanation of any anticipated differences is justifiable. Patterns inferred in ecology also need to be tested in this way, and only if they differ significantly is a biological interpretation warranted.

Claims of patterns relating to more complex phenomena, particularly local adaptation within species, population equilibria and community structure, which are central to theory and expectation in current population biology theory, have generally suffered a different fate from that just outlined. Sometimes observations are made and samples taken, but generally pattern justification has not been given much methodological prominence and certainly the data available are seldom scrutinized as deeply or seriously as the situation warrants. The consensus that is evident in population biology does not seem, therefore, to be based on strong evidence. We illustrate this below with two examples, the existence of species as discrete and real entities in nature and the question of population stability or balance (Subsections 3.1 and 3.2). Clearly, persistent research dedicated to understanding the underlying mechanistic basis of a "pattern" that is not representative would not be the most economical or rational approach to understanding in population biology, and testing the pattern must be free of assumptions about process (Chapleau et al. 1988), a point that is still ignored routinely in this pursuit.

We then turn to the one approach (Subsection 3.3) in which a statistical procedure, the use of null models, was specifically developed to achieve the goal of pattern justification, this time in community ecology, and consider why that approach failed. We also point out how null models have also appeared under a different guise in ecology, as mechanistic models (which are detailed in Section 5), a role to which they are not at all suited. Mathematical models have, nevertheless, been used quite extensively in the development of ecological theory, but their application has mostly had the unintended consequence of misleading rather than informing. Again it is the logistic equation that draws attention. We therefore finish this section with a consideration of the roles of curve fitting and simulation in ecology (Subsection 3.4).

3.1 The Pattern of Discrete Species

Speciation has always been seen as a crucial process in the history of life on earth. Without it, diversity could not have continued to increase in the face of all the extinctions that have taken place (Fig. 8.1). With the sexual process, governed by a precise mechanism to achieve fertilization under the conditions usually occupied by the species in nature, came the potential for the indefinite initiation of daughter species (Paterson 1989). The evolution of a new sexual species has the consequence of acting like a ratchet, "locking" the adaptive changes that took place at speciation. Given the appropriate ecological setting and the environmental conditions that cause the fertilization mechanism "lock" to be reconfigured (Chapter 6, Subsection 4.4) yet other species can evolve from that one. Speciation can thus be seen, retrospectively, as incidentally having provided additional raw material for further diversification.

Darwin (1859) set out to explain the origin of species in terms of everyday mechanistic processes. In doing so he was confronted with the biblical stability of species, and he had to break the hold, in peoples' minds, of that perspective. He did so by arguing that species are fluid, and not immutable (Mayr 1982, p. 269). Look how they vary across the different areas they inhabit, and look how difficult it is to distinguish mere races from species. These imperatives again seem to reflect his uniformitarianist outlook for evolution (see Mallet 2008, and Section 1, above). He was arguing for the imperceptible change of one species into another in different geographical settings but seemingly within the overall distribution of the single "parental" species. This is perhaps the only way in which Darwin could have argued effectively against the prevailing view of a deity-ordained permanence of species' structure and station (see Ellegård (1958), for how strong that influence was in society at that time). Furthermore, his concept of natural selection, developed entirely around the concept of competition (Paterson 2005), provided a mechanism that would drive change in that way, slowly and imperceptibly across a broad geographical front.

We still live with Darwin's original concept of natural selection, the one in which competitive superiority (now often interpreted in terms of optimization) drives change ineluctably (Paterson 2005). Evolutionary ecologists therefore expect local adaptation within different geographical areas of a species' distribution. Cases of such local adaptation are therefore ardently sought (see Blanquart et al. 2013), evidently to verify a particular concept of selection. Thus, a world of gradual, ineluctable change is perceived, one in which species boundaries have little meaning, and the original pattern of distinct species and their significance in diversification have been pushed aside. Belief in the strength and importance of a process now has prominence over observable pattern. For example, hybridization is often seen as evidence that species are not significant to understanding ecology and evolution, as when sympatric speciation is advocated or defended (Bush 1994). And species are relegated to the vague state that Darwin imposed on them to suit his aims at the time. To Darwin, species were those entities that were distinguished as such by a good naturalist, and this approach has been recommended for current use (Mallet 1995), undoubtedly to fit in with current views of selection, which align with the competitive model advocated originally by Darwin.

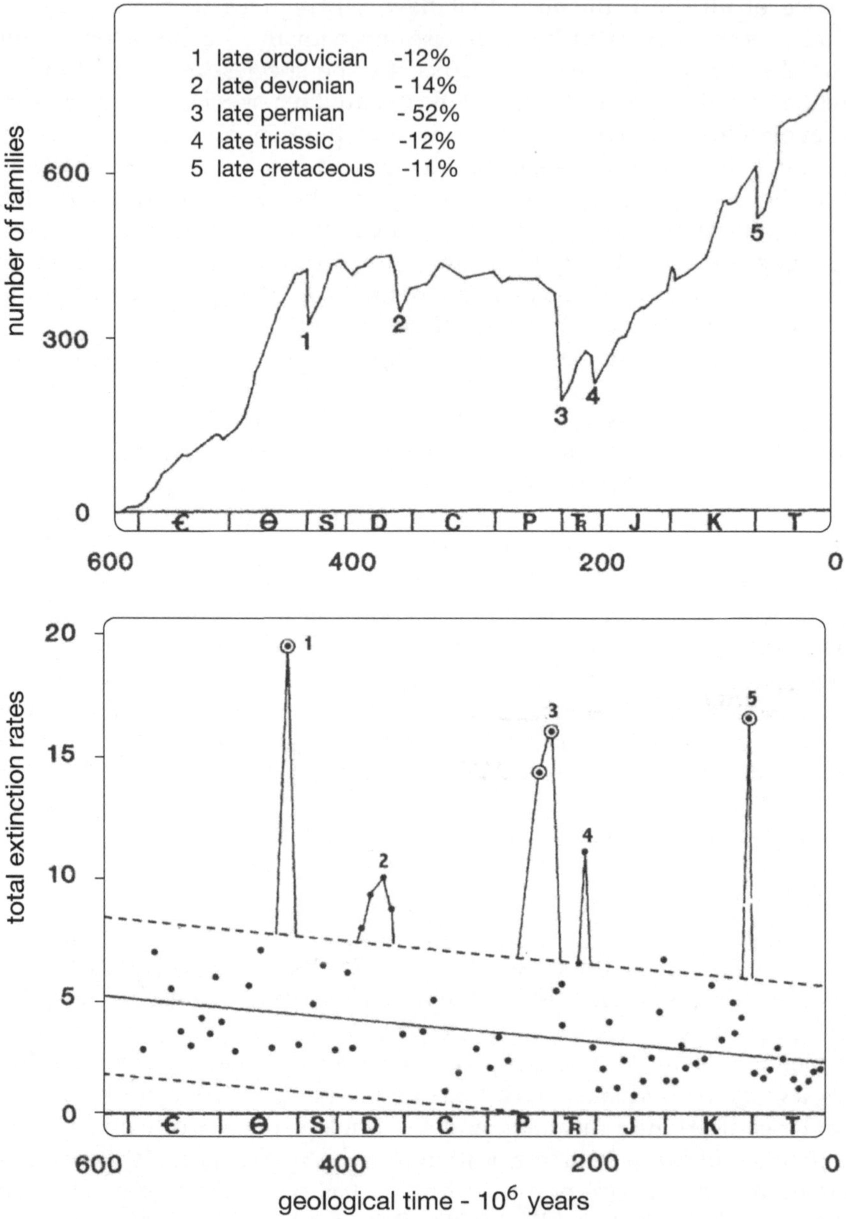

Figure 8.1 Eukaryotic diversity has increased through time, despite ongoing extinctions. **Top** graph shows the diversity through time for families of marine invertebrates and vertebrates. **Bottom** graph shows total extinction rate (extinctions per million years) through time for these same organisms. The major mass extinctions are numbered (1 = Ordovician, 2 = Late Devonian, 3 = Permian, 4 = Triassic and 5 = Cretaceous). The solid line is the regression line taken to indicate the background extinction rate, from which the mass extinctions stand out, and the dashed lines the 95% confidence intervals. Redrawn from Raup and Sepkoski (1982).

In other words, an anticipated process (competitive selection) is placed ahead of the pattern that demands explanation. The real pattern is that of discrete species, perhaps the most prominent pattern in nature. A rich literature supports the existence of species as objective entities, for they have complex functional mechanisms that ensure fertilization and thus set the limits as to which individuals can participate in the gene pool (Paterson 1993c). Furthermore, species are not arbitrarily defined, as are local populations, and they are stable through time and across space (Eldredge and Gould 1972, Coope 1978, 1979, 2004, Vrba 1980, 1984, Henderson and Lambert 1982, Paterson 1985, 1986, Lambert and Henderson 1986, Walter and Paterson 1994, Vrba and DeGusta 2004, Popple et al. 2008), a feature now acknowledged even by evolutionary ecologists, who nevertheless still explain the pattern of stability in terms of competitively driven ongoing rapid change in local populations, all of which is somehow expected to cancel out over time (Eldredge et al. 2005). The molecular basis of sexual recognition processes supports the view that stability is the common pattern (e.g., Ferris et al. 1997, Liss et al. 1997). In effect, species are essentially self-defining through their sexual behavior, which is underpinned by the complex species-specific adaptations that ensure fertilization (Lambert et al. 1987). They are thus real entities that can be defined objectively, which is significant given that species diversification is largely underpinned by sexual reproduction. Asexual organisms have not generated the same diversity, and they cannot be interpreted in the same way as sexual organisms. They have to be defined morphologically and in terms of their ecology (DeBach 1969, Paterson 1981, Walter 2003, Najar-Rodriguez et al. 2009).

A strong push is evident in the current literature to show continuous variation (e.g., Mallet 2008), and visual representations are clearly evident in illustrations of bird species, for example. A relatively small subset of bird species appear to show fine grades of differentiation across their geographical distribution, as in the red-tailed black cockatoo (*Calyptoryhnchus banksii*), yellow wagtail (*Motacilla flava*) and great tit (*Parus major*) (see, for example, illustrations in Handbook of the Birds of the World, Lynx Ediciones, volumes 4, 9 and 12). The visual appeal is clear, but may well be misleading, as revealed in recent analyses that take a geographical perspective on the pattern and the processes involved, and as detailed in the paragraph that follows the next. Nevertheless, most studies of species diversification target situations in which the process of speciation is assumed to be in progress, as in the investigation of putative cases of sympatric speciation.

Such cases have recently been incorporated under the indiscriminate generalization "ecological speciation" (e.g., Rundle and Nosil 2005). The phrase is unhelpfully expansive; it incorporates all aspects related to speciation as well as the adaptive process, but without the guidance of a species concept. Again, the accepted competitive process conferred on natural selection is allowed to drive all related interpretation and data collection. Although pluralism in use of species concepts is often advocated (e.g., Hausdorf 2011) there is but one function related to sexual species that can legitimately focus interpretation, that of individuals achieving fertilization in their usual habitat (Paterson 1981, 1985, 1993c). Reproductive isolation is a group feature and selection can therefore not target it as an adaptation. Adaptation comes as a by-product of selection acting on survival and fertilization as it relates to individuals, and it is these other processes that the concept of species should focus on (as in the Recognition Concept of species (Paterson 1993c)). We have a chance then to understand species formation. "Ecological speciation" goes even further than detailed above, however, in having each new species evolving to fill an empty niche through "diversifying selection". This not only implies a teleological role for natural selection in selecting to produce a new species, but also gives a prominent role to the competitive concept of resource partitioning (Walter 1991) through an empty niche having to be available. The determinism and hope inherent in this whole approach to generalization is clear from a recent title: "Possible incipient sympatric ecological speciation in blind mole rats" (Hadid et al. 2013). In brief, ecological speciation is a superficial generalization that hides far more than it reveals.

Fortunately, some investigations of apparently variable species have taken a broader, geographical perspective. Here the initial focus is on the geographical pattern of the distribution of the organisms' adaptive mechanisms and the potential for gene flow within areas of sympatry. Revealingly, a pattern of spatial stability of characters and ecological relationships (such as habitat associations) is usually evident, as in the species complexes associated with *Parus major* (Packert et al. 2005), *Pauropsalta annulata* cicadas (Popple et al. 2008), alpheid shrimps (Duffy 1996), bitterling fish (Zaki et al. 2008) and even in aphids that are predominantly asexual (Najar-Rodriguez et al. 2009). In areas of sympatry, some of the variants (whether typified originally as subspecies or "types") show no evidence of gene flow or extremely low levels of gene flow (e.g., Packert et al. 2005). These "variants" are undoubtedly species, and the stability of characters across geographical space indicates strongly that each arose in a relatively small population within a restricted geographical area (Paterson 1981, 1985, Popple et al. 2008). Their differential adaptations to the environment and largely independent geographical distributions indicate, too, that they arose in allopatry which, in principle, is the way in which diversification takes place (Paterson 1981, 1985). Separating the putative process of sympatric speciation from local hybridization after secondary contact by means of population genetics analysis is more difficult than often claimed or implied, and this is why the geographic pattern of stability across the allopatric zones is the key. In short, claims of sympatric speciation in such cases must invoke too many unlikely intermediate steps for it to provide a realistic interpretation of the situation.

In some situations, two such "variants" mate at random in the area of overlap. Despite the observable differences between them, the two are essentially conspecific and, by definition, are subspecies (Paterson 1956,

1964, Ford 1974, Popple et al. 2008). The origin of subspecies is in no way different from that of species, having taken place in a small geographically isolated population (see Walter 2003, pp. 162-163). To say that two species hybridizing in such situations provides evidence that species are not objective entities in nature (e.g., Mallet 2005) clearly places subjective taxonomic criteria ahead of population genetics criteria. The former are arbitrary, the latter deals with the limits to gene flow within gene pools, and besides hybridization being expected between subspecies, occasional hybridization elsewhere is neither unexpected nor a challenge to species defined in terms of positive assortative mating through the mediation of the species-specific Fertilization Mechanism (Paterson 1985, 1991).

The independent ecologies, adaptations, and origins of subspecies are explained in the same way as the origin of species. With subspecies, adaptation in allopatry resulted in organisms that differ ecologically to some degree, but which have undergone no change (or little change) to their fertilization mechanism during the period of adaptation. When studies seek to verify sympatric speciation or ongoing divergence of populations treated as conspecific, they tend to focus on areas in which the two types overlap and where differential resource use is evident. Such circumstances, and any hybridization between the types (if it occurs), are given as evidence for competitive selection driving adaptation ineluctably to full species status. The determinism and teleology in these views is dealt with above. Those studies that have documented the geographical distribution of characters show up dramatically the limitations of such local studies (Popple and Walter 2010, Hereward et al. 2013, Walter 2013).

In summary, this section illustrates that the agenda for scientific interpretation must be set by the observable patterns and these must be firmly established through adequate research that is methodologically sound.

3.2 Population Stability

A consideration of what pattern typifies ecological populations is deceptively complex and subtle. The debate has therefore been contorted – a good selection of the contrary perspectives on offer has been compiled by Rohde (2005) and this reflects accurately the contemporary lack of resolution in population ecology. These difficulties reflect fundamental problems in approach to the science (e.g., Murray 1999, Cooper 2001, 2003, Walter 2008). Furthermore, local populations are arbitrarily defined and the relationship between the individuals that comprise the local population and the conspecific individuals in other parts of the species' distribution is not dealt with mechanistically or, more frequently, is ignored. Consequently, a general pattern is sought from a fragment of the system within which the individual conspecific organisms are born, develop, live, move about, reproduce and die. Excision of part of a system in this way is inimical to developing robust generalizations, and provides fertile ground for disagreement. We therefore see why the explanation of modern population ecology still relies on an implicit and unstated set of premises that derive from the vague notion of "balance of nature" (Cooper 2001, Walter 2008), and why there remains indecision about the primary pattern to be explained.

Field data have long been available on outbreaks of species, local extinction of species and the movement of individuals in and out of "local populations" (and here the winter moth provides a salutary lesson (Chapter 2, Subsection 3.2)). These discrepancies against expectation should have been enough to warn ecologists that local population stability is not a representative pattern in ecology. The problem has been side-stepped in several ways. Species, for example, have been categorized as "outbreak species", "low density species", "keystone species", "supertramp species", "engineer species", and so on, for the purposes of ecological interpretation. Typological treatment of this nature is unacceptable (Mayr 1982, Hengeveld 1988b) and such classification is in any case not explanatory (Walter 2003, pp. 115-128).

The debate about how much variance in population density reflects general stability may satisfy some, but again indicates that we have deeper problems. Too much prominence is yet again given to belief in a process, density dependence, over pattern, as shown for species in the previous subsection, where the prominence given to competitive natural selection is allowed to push aside even the reality of species. In ecology, an implicit belief in the power of density dependence sustains belief in local population regulation, despite observation to the contrary (Andrewartha and Birch 1954, Murray 1999). Criticisms have cut deep, but have not lead to resolution. Instead, the power of density dependence is supported by even more extravagant claims, from the one extreme of not needing to test its alleged power at all because of its self-evident ecological importance (as summarized by Murray (1999)) to notions that leave mechanistic connections between ecological processes and consequences unspecified, as in claims that density dependence must be important somewhere sometime. "Since logic dictates that some density dependence must occur, at least occasionally, in the dynamics of every species, though not necessarily in the dynamics of every population ... at least some populations of all species are affected at least occasionally by density dependence, which is required for long-term persistence at the regional scale ..." (Hanski 1998).

The significance of interpreting the ecology and adaptations of "local populations" in relation to the species as a whole (and across its entire distribution) is crucial (Moore 1957, Paterson 1982b, Walter et al. 1984, Popple et al. 2008). It is not encompassed, however, by the "local – regional" dichotomy (e.g., Ricklefs 1987, Mora et al. 2003) or the landscape hierarchy of spatial scale (e.g., Perry 2002, Wiens et al. 2007b) proposed to take account of spatial scale in such circumstances. Problems naturally ensue, and these tend to relate to the failure to work from an explicit understanding of the mechanistic relationships between organisms and their environment, and are compounded by the persistent faith that local populations need to be interpreted in relation to temporal stability or regulation, which again reflects the deep-seated philosophical reliance on ideas associated with a "balance of nature" (Cooper 2001). The hold of such ideas on modern ecology should not be underestimated, their lack of empirical support and arbitrary nature notwithstanding.

The other way around the obvious variations in population density has been to accept population stability (in terms of the logistic equation) as an idealization that has been built on the basis of sound principles. Should this approach be accepted, then exceptional variations, such as those outlined above, alert us to what is interesting in population ecology (Turchin 2001). Even so, the movement of individuals into and out of "local populations" (subjectively designated) is a debilitating problem. Metapopulation theory has been developed to incorporate such interchange between local populations, with stability being rescued by supply of individuals from productive local populations to declining local populations, but the stability is now spread across the so-called metapopulation. This solution, though, only compounds the problem as detailed in the following subsection on null models.

Ultimately, the decision about patterns in population biology is not straightforward and seldom unanimous. Clearly, however, the issue warrants introspection, but we have argued earlier that the only consistent ecological pattern across all species is the dynamic spatial adjustment of the distribution of species in response to altered ecological conditions, and the fact that this adjustment is species-specific (Chapters 4 and 5). This interpretation sees species as real entities with unique sets of complex adaptations, and with the primary ecological interactions being those between these adaptations and the structured, but dynamic environment.

3.3 Null Models in Tests of Pattern

Null models represent an approach in which the reality of suggested ecological patterns is tested statistically. They were topical late in the 1970s and for some time after (Connor and Simberloff 1979, Strong et al. 1979, Harvey et al. 1983, Gotelli and Graves 1996). Essentially, null models ask the question of whether claimed ecological patterns result from biological processes and so require a biological explanation, or if they represent the product of random processes and are epiphenomena (or by-products) and do not warrant direct explanation. The main focus of null models in ecology was on the question of the relevance of "communities" of organisms in ecological explanation. These models seem to have been deployed in exasperation at the lack of progress in ecology, mainly because of the devotion of some ecologists to particular deterministic patterns. The "competition school" saw structure across the local groups of organisms that were deemed to share a common resource (and thus had the potential, in principle, to compete (e.g., Abele et al. 1984, Lawton and MacGarvin 1986)), and this was under scrutiny. The advocates of null models said that local diversity was made up of a random collection of species, with any resultant pattern that was detectable not having been mediated by competition.

The problem with the null model approach in ecology is fundamental. The patterns that are tested are complex in that they include aspects that are functional (or adaptive) and these, by their very nature, are not random, having been shaped by evolutionary processes in relation to environmental circumstances. Different species, in multiple species tests, even if not directly related to one another phylogenetically, may respond to the same environmental signal, vegetation type for example, and co-occur because of that. They are therefore not independent of one another in a statistical sense. Moreover, phylogenetic influences will also be evident across the data points (i.e., the species), a feature that again violates the assumption of independence of the data points. Congeneric species, for example, share ecologically relevant features through their common ancestry. Systems comprised of multiple biological components, such as the diversity of species on a group of islands or the number of conspecific individuals in a locality through time, cannot be represented reasonably by null models. For, example, artificially combining species into a "community" on the basis of common resource use generates a pattern that is an epiphenomenon, for it is the explanation of each species' independent evolutionary origin and its presence within the locality of interest that is required (Walter et al. 1984, Walter and Paterson 1994, Walter 1995, van Klinken and Walter 1996).

The serious attempt to subject ecological patterns to statistical scrutiny through null models, although concerted, seems to have failed. Null models of community structure never resolved the issue at all, probably because such models are inappropriate for the purpose, and their use in this regard seems to have faded for this reason. The hypothesis that a statistical (or random), non-biological explanation underpins the claim is effectively the null hypothesis, H_0, which means that the use of null models in this way is essentially statistical in approach. The biological explanation is the alternative hypothesis, H_1. These two hypotheses are thus mutually dependent on one another methodologically. The one is simply formulated as the statistical alternative of the other. Although they are not independent theories null models have frequently been given misplaced prominence as mechanistic models (Section 3.4 below), usually as an equivalent among several alternative models, as illustrated next.

Some critics of population regulation theory have asked whether the extent of population fluctuation can be explained by chance, or whether some biological process is indeed responsible. The explanation that is based entirely on chance is thus regarded as an alternative, testable theory, for example the theory of population stabilization by risk spreading (den Boer 1968, Reddingius and Den Boer 1970, Reddingius 1971, den Boer and Reddingius 1996). Many independent and essentially unknown ecological variables are thus seen to influence the numbers of individuals in a population to the extent that the population is stabilized. To test this proposition, the question becomes whether the "bounds between which the population fluctuates are, or are not, significantly narrower than one would expect given the circumstances *X*" (Wolda 1989). Although this is simply a return to the original question of whether populations are regulated, "stabilization by risk spreading" was treated as an independent explanatory theory, with a status equivalent to any other biologically based alternative (e.g., Andrewartha and Birch 1984) and therefore independent of them methodologically. However, risk spreading by its very nature is merely the chance process constituting the null model to population regulation and so is not methodologically independent.

Risk spreading has also been given spatial form. The densities of subpopulations are seen to fluctuate independently of one another in space to the extent that their differences in abundance cancel each other statistically (Hanski 1982). The population as a whole thus shows statistically "stabilized" fluctuations across space, but without movement of individuals among populations. This, though, is the null model for various metapopulation models, including Levins' (1970) original such model and those metapopulation models in which large productive populations "rescue" declining ones from extinction through migration (Brown and Kodric-Brown 1977, Hanski 1982). Risk spreading thus effectively becomes the test for whether migration should be adopted as a biological explanation for there being restricted population fluctuations across space.

If migration was seen to be the regulating process at the level of metapopulations, yet another biological process would need to be invoked, that of density-dependent migration. This is necessary to provide the migrants that move in the appropriate direction between populations, from high to low densities. And this effectively invokes the physical mechanism of logistic growth underlying metapopulation dynamics (Gilpin and Hanski 1991). The chance process that operates across space, Markovian in character and outlined above, would thus serve as the null-model for directional density-dependent migration.

A similar misuse of null models is evident in biogeography. Here, many independent and unidentified variables have been invoked to explain the spatially non-uniform distribution of local abundances within species ranges (Brown 1984). In a further development, this null model was united with risk spreading into a combined theory in which several variables were weighed subjectively relative to each other (Brown 1995), and biological mechanisms were sought to explain this statistical risk spreading (den Boer 1999), despite its group selectionist overtones. This approach was taken despite the spatial homogeneity that is considered to result from risk spreading, and the observation that the distribution of local abundances is almost inevitably spatially heterogeneous.

An evolutionary example entails the search for an explanation of perceived variation in speciation rates. Vrba (1987) examined five hypotheses with these emphasizing: (1) birth rates, (2) gene flow, (3) resource use, (4) extrinsic factors, and (5) random factors, respectively. The random factors hypothesis considers that there are "so many different proximal causes that the resultant patterns of species diversity show no consistent correlation across clades with any characters of organism or species" (Raup et al. 1973). As such, it represents the null model of each of the other four biological hypotheses, although it is presented as an alternative hypothesis in its own right. This implies it represents a testable mechanistic model, and must therefore be methodologically independent of the other four. This would be acceptable only if the biological mechanism that was to be tested in the first place actually served as an alternative to the statistical explanation. The random-factor model would be an effective null-hypothesis only if it had been compared with each of the others separately rather than the five having been treated as equivalent alternatives, but this was not done.

A significant negative consequence of null models being given equivalent status to mechanistic models is that the lack of biological realism in null models tends to work in favor of the paradigmatic model, which effectively remains untested and thus tends to be taken for granted.

3.4 Curve Fitting and Simulation Procedures

Curve fitting is a procedure often used by ecologists. Its sole purpose is to describe, as closely as possible, the pattern of the data points derived from observation, sample or experiment. The consequent diagrammatic representation should be achieved in the simplest terms and with the lowest number of parameters possible. Significantly, the parameters need not reflect any underlying process, so this lack of restriction allows wide choice. Because the parameters need have no ecological or biological relevance there is no requirement they be interpretable mechanistically. They simply represent arbitrary constants that best fit a curve to the data points of interest. They also have no constraints of dimension, which poses no problem because they do not have to represent any inherent qualities of organisms or their environment, and there is no need for mechanistic connections between the parameters. Ultimately, the equation itself has no interpretation in terms of ecological processes. The sole criterion for judging the suitability of such models is the closeness of fit to the data points and the simplicity of its application. Sometimes, however, the "observations" are not even derived from raw data, but simply represent the pattern that is presumed to reflect the situation in question, which is true of the logistic equation and its derivatives (see below).

A consequence of this freedom from parameter choice and model structure is that several very different models may be used to describe exactly the same pattern of data points. For example, to describe an S-shaped growth pattern, the logistic equation is just one of several descriptive equations available (e.g., Pielou 1974). The logistic does not actually fit the available data better than the other equations that have been applied to *S*-shaped curves (Feller 1940, Cohen 1995), but is portrayed in ecology texts as the deterministic equation that alone accounts for *S*-shaped population trajectories. Despite this lack of theoretical or observational justification, the logistic is seen to encapsulate the foundational principles of population ecology (Turchin 2001). The problem for ecology at large is that the parameters given prominence in this way have not been identified from a consideration of the biology of the system. Rather, the parameters have been imposed on the biology of the system. The special significance accorded an inappropriate equation in ecological theory suggests its other limitations warrant scrutiny (Andrewartha and Birch 1954, Pielou 1977).

The aim of simulation equations is to obtain a process outcome that best fits a set of observations (Grant and Swannack 2008). As such, they lack generality for they mimic specific situations (e.g., Godfray and Waage 1991). Should the fit seem insufficient, the choice of parameters or the structure of the model can be altered until the outcome looks satisfactory. In this lack of constraint, they resemble the simple curve-fitting procedures mentioned above and thus often include circular reasoning. Circularity in such a situation is justifiable since the sole purpose is to obtain the best fit between calculated results and observations. When the fit is deemed sufficient, the model can be used for extrapolation of future states of the system, but such prediction is not based on understanding and is thus limited in terms of general theory.

Mechanistic models, in contrast, comprise interpretable parameters that are testable in that they must relate to the biology of the system as well as relate algebraically to one another. This is the major difference between mechanistic models and curve fitting procedures. To reflect the algebraic connections demanded of mechanistic models, the dimensions of the parameters must be the same. Weight, for example, has a different dimension from time, length, or surface area, and cannot be subtracted from variables of other dimensions or multiplied with them.

To make accurate ecological predictions, one needs to know which variables are most important ecologically and how they are integrated into the overall ecological process under investigation. This process is dealt with next.

4 EXPLORATION OF CANDIDATE VARIABLES (PHASE 2)

Statistical analysis forms the core of the inductive exploration of which variables in nature may be ecologically significant in the process responsible for the observed pattern. Its approach is detailed in the first subsection, below, and the subsection after that argues that demographic theory in ecology is still in this exploratory phase. Section 5, which follows, deals with the reconstruction of the relevant variables into a mechanistic model that is as unencumbered as possible with needless variables whilst retaining accuracy.

In principle, the exploration phase of research should be kept to a minimum, simply because it is not fully explanatory on its own. The tests involved are statistical and simply ask whether a significant correlation is evident between variables x and y, how variables a and b (two plant species, perhaps) affect outcome y (survival of a herbivore species, for example), whether climatic variables *g* and *h* correlate with reproductive output or offspring survival, whether population density affects survival rate or reproductive output, and so on.

The exploration phase in the investigation of complex biological systems, particularly ecological systems, will almost inevitably extend across time (for decades perhaps). The variables responsible for the pattern under investigation, the ones that can justifiably be included in the equation(s) intended to reconstruct the process considered to have led to the pattern under consideration, are seldom clear from the outset. Furthermore, the number of potentially important variables may be large, often too large for practical purposes in developing analytical models. To reduce this number as much as possible and as realistically as possible, the relative contribution of the various candidate variables must be assessed in the reconstruction phase of research (Section 5).

4.1 Statistical Analysis

Statistical analysis in ecology is initially inferential and begins with classifying or ordinating the data that represent the ecological subjects of interest. The aim is to uncover relationships that may reveal which variables are relevant to the ecological pattern under investigation. Specifically designed tests are then required to establish if that factor or variable is, indeed, causative. The statistical analysis involved in such tests (e.g., regression or ANOVA) differs from the exploratory statistics used initially, in having an independent variable. A statistically significant outcome supports the view that the factor originally inferred to be ecologically important is indeed statistically and ecologically significant. This factor can be considered causal, although this interpretation remains hypothetical because further experimental and observational investigation is required to test that view. Although statistical models are used to estimate future outcomes probabilistically, as in the insurance industry, they do not operate from a basis of understanding (Pilkey and Pilkey-Jarvis 2007). That is, the outcomes of such analyses are only suggestive.

The above also holds for the variable represented by the X-axis. The relevance of that variable is inferred from the data, but that does not mean that its significance is understood mechanistically. A statistically significant outcome only supports the view that the factor originally inferred to be ecologically important is indeed statistically and ecologically significant. Although this factor may be considered causal, this does remain hypothetical. Popper was therefore mistaken in not including inductive techniques in scientific research, because it inevitably plays such a significant role in the initial stages of understanding causal phenomena (see Walter 2003, p. 33). Statistics quantifies inductive reasoning by suggesting the possible relevance of variables postulated to play a mechanistically causal role, whereas deductive reasoning quantifies the algebraic and interactive relationships between those variables. Statistics thus also differs from model construction in its assuming the independence of the parameters that were initially inferred to play a role.

Statistical analysis of the nature outlined above represents the exploratory phase of research into the pattern of interest and the process(es) that may be responsible, because any causal relationship that is inferred between variables, even if supported statistically, is not informative of any interactive effects between the parameters or of any relative weighting in their ecological importance. The relative contributions of the variables to the process(es) responsible for the pattern under consideration requires the relevance postulated for them to be tested further before they can be incorporated into a mechanistic model that specifies explicitly the relationships between them (as expanded in the following subsections).

Statistical analysis involves curve fitting, but it has specific aims that are quite different from those of curve fitting for modeling purposes. Both approaches do, however, aim at achieving the closest fit between the output from the simplest and most general model and the observational or experimental data. Statistical analysis is, however, a more quantified and less interactive sort of curve fitting, with a linear or polynomial model representing the "curve". The subject data in statistical models have no explicit spatial or temporal context, so are highly context or site dependent although this is often ignored, when such results are generalized broadly for example. Also, when experimental results are involved the different dependent variables are implicitly treated as equivalent ecological influences. These are assumptions that are almost invariably violated, for ecological processes in the field are almost inevitably different in this regard. Statistical models also treat the effects of the various ecological factors independently of one another. They are treated additively in the linear equations that are typically used and are thus assumed not to interact with one another. Moreover, the dependent variable(s) (y) are also assumed not to interact with the independent variable(s) (x). The structure of the statistical equations is not intended to reflect the structure of the process under investigation, and indeed cannot do so because the ecological variables are likely to be non-additive and comprise non-linear components.

The early stages in the statistical analysis of an ecological process ask whether a data set reflects a real pattern or process. In t-tests and analyses of variance, for example, two or more frequency distributions are analyzed to establish whether they are statistically the same, and thus likely to have been produced by the same process, or whether they represent outcomes of different processes. Regardless of the statistical outcome, the process(es) that gave rise to the distributions, and any interactive effects between them, remain(s) to be investigated independently, for at this stage they can only be inferred indirectly. In other words, the process structure remains unclear, even though the statistical procedure may be quite sophisticated. The statistical model used to test the data will not necessarily bear any relation to the mechanistic model that is ultimately developed to represent the structure of the process that generated the ecological pattern. Statistical models, used in this way, remain static and cannot describe the dynamics of ecological processes.

The variables thus identified as important in statistical descriptions represent inferred deterministic components of the stochastic ecological process and, should mechanistic models be developed subsequently, may become the parameters that will structure the model. Any variation in their measurement is almost always expressed through an error term that is assumed to result only from inadequate observation. Variation of this nature cannot be included realistically in terms of stochasticity inherent in the ecological process. The distribution of this error term is usually taken to be normal or Gaussian, although it may also be Poisson distributed (Kooijman 1979) or follow any other probability distribution.

Statistical analysis also gives insight into the extent to which the variables that influence a process considered to be responsible for an ecologically significant pattern co-vary, and the way in which they do so. Similar ecological situations are then assumed to be explicable with reference to the same statistical equation and output, by extrapolation. That is, statistical methodology is inductive (Fisher 1948) and retrospective in principle, because it can only describe past processes, not explain or predict future developments of a process. Although extrapolations may well hold if they are made on the basis of statistical analysis, this does not imply that the nature of the processes that underlie the phenomenon under study are explained, nor how they may operate under different conditions.

Any understanding derived from statistical analysis remains insufficient for ecological prediction, principally because ecological systems are influenced by such a diversity of factors that they cannot be treated as if all other things were equal. When ecological situations vary in this way, the only common feature across such situations may well be the entire structure of the process that generates the pattern, which, for real understanding, must be reconstructed in terms of its primary variables or parameters. The invasion of the collared dove into western Eurasia has been modeled in this way, for example, and is detailed in Section 6. The values of the parameters are likely to vary across ecological situations and localities (Chapter 9, Subsection 3.2 and Fig. 7.3) and will therefore require independent quantification across those situations if predictions are to be made on the basis of understanding.

The ability to predict outside the bounds set by the initial conditions reflects scientific understanding, because even the best fitting curve that summarizes a set of observations in the simplest terms cannot be used to make predictions. Statistical curve fitting is methodologically inadequate for this purpose. Statistical analysis can only help identify the variables that are relevant to the ecological process that is under scrutiny. To make accurate predictions, one needs to know which variables are concerned and how they are integrated into the overall ecological process under investigation.

4.2 Population Dynamics and Community Ecology are Still in Exploration Phase

Research on population dynamics has not yet advanced to the reconstruction phase. Theory and practice remain in the exploration phase. Although the existence of general models may give the appearance of reconstruction, their reliance on concepts of carrying capacity, population equilibrium and density dependence are emphasized to the extent that individual organisms, their adaptive mechanisms, and environmental dynamics are not considered explicitly. Even approaches claimed to be "organism centric" in perspective (Wiens 1976, Wiens et al. 2007a), although helping to address this shortcoming, do not relate the organisms explicitly to the environment through their sensory and other physiological mechanisms, but do so indirectly in relation to more generalized abstractions of the environment like "patch", "grain" and "contrast".

In population ecology the mean density of individuals that make up the local population is calculated from a time series of samples. The levels of fluctuation of the local population are then interpreted in logistic theory relative to this mean value, almost invariably in terms of density dependent predation (or parasitism), or relative to the non-mechanistically defined carrying capacity that is assumed to feed back on population density through competitive interactions. The existence (or reality) of an equilibrium state is thus assumed, not tested. The abstraction of "population equilibrium" in this way, deemed acceptable by many (e.g., Paine 2002), inevitably shapes investigation and interpretation. Although we may discover more information about specific processes from research of this nature, such discovery is inevitably limited and any synthesis that ensues is likely to be stilted. The claimed advantages of pluralism in this context (e.g., Schoener 1986b, Aarssen 1997, Paine 2002) thus promote the status quo (Walter 1995, 2008). Many see no need, in ecological theory or practice, of deep change, the latter of which is often referred to somewhat pejoratively as "revolution" (Colwell 1984, Ricklefs 1987, 1989, Paine 2002). The calls for pluralism in the discipline therefore seem calm, rational and sufficient, even though scrutiny of the foundational statements does seem to be needed. Pluralism is therefore likely to ensure there will be only limited adjustment to theory rather than the deeper change that is required.

The densities that make up local populations tend to be interpreted independently of conspecific individuals moving into the specified locality or leaving it, because the logistic equation specifies such a stance. Within the local area of interest, all other ecological variables are considered to be constant and uniform. That is, all other things are assumed to be equal. The classic studies of the winter moth (Varley et al. 1973), viburnum whitefly (Hassell et al. 1987, Southwood et al. 1989) and vole population cycles (Chitty 1996) illustrate that such arbitrary designations cut across ecologically significant processes and invalidate the assumptions that underlie the design of the studies and their interpretation.

The mathematics associated with the logistic equation and its derivatives remain central to demographic theorizing (Chapter 7) and help ensure that local population densities constitute the most significant element of the process of demographic interaction, through their local regulatory feedback. This situation persists, in all likelihood, because a true reconstruction phase did not take place in relation to these equations, for this would have led to the development and testing of a mechanistic model to represent the key processes considered responsible for the local equilibrium value (not just the mean value). These processes would have required formulation in terms of the properties of the individuals concerned, the environment, and how individuals compete relative to one another. To our knowledge, such a model has never been developed and applied to field data. The logistic equation, by contrast, deals with the density limitation of a local population of a single species through restrictions imposed on them by the carrying capacity (K) of the area that makes up the local arena. The real difficulty that this presents is as practical as it is conceptual, for K is impossible to specify directly in a mechanistic way. The structure of the equation constrains K to indirect estimation by iteration for a certain range of r (Poole 1974). These two parameters, central to the equation, are thus effectively mutually dependent, not only because of their empirical circularity in the iteration, but also because of their underlying methodological circularity. In principle, though, if K were estimated mechanistically in terms of a set of environmental variables that were independent of the resulting demography, it could indeed be explanatory in terms of its limiting the numbers of individuals to the maximum that the local area could sustain. However, K is effectively dependent on r, for it is specified in terms of r. Thus, any reconstruction potential is aborted and the analysis remains in the exploration phase.

The logistic equation abstracts the processes considered to be ecologically significant in the determination of organism abundance. Technically, the biological idiosyncrasies of individuals (through their suite of species-specific adaptations, life stage, physiological status, and so on, the nature of their environmental requirements (Chapters 3 and 4) and the condition of the local environment (Chapters 4 and 5) are bypassed because the equation focuses entirely on the numbers of organisms relative to the number capable of being supported, the carrying capacity, despite this having no characteristics, not even a spatial dimension. The practical consequence of these difficulties is that carrying capacity simply cannot be defined realistically. Consider also the temporal dimension. The definition of an instantaneous carrying capacity would be "the population size that would obtain if the conditions of that moment would persist long enough for population size to stabilize", which has little practical meaning in terms of research and interpretation. One proposed solution suggests it would be "more profitable to define the equilibrium value in relation to the conditions over a longer period of one year" so that the definition would read: "the value a population size would obtain if the conditions of the past year were identically repeated year after year, long enough for the population to stabilize" (as reviewed by Wolda (1989)).

Methodologically, the demographic variables used to characterize the temporal pattern of abundance in the exploration phase of research are also used to reconstruct the process that is considered to generate this pattern. Although the theory has not therefore been tested, the model formulated in this way forms the foundation on which population dynamics theories, through the logistic equation, and the theory of community ecology, through the Lotka-Volterra equations, have been built. Although both population and community ecology have seen recent developments, their logical foundations remain unchanged. A sounder approach would have seen the development of a mechanistic model that reconstructed the processes seen to be significant in the changing abundance and distribution of a species, through the inclusion of independent, non-demographic variables. These would represent the properties of the individuals of the species of interest and would be complemented by the dependence of their values on environmental conditions. Without the inclusion of such independent variables, reconstruction is not possible and this explains why population ecology still remains in the exploration phase of research, despite its apparent recent advances (e.g., Turchin 2003, Coulson et al. 2004, Coulson and Godfray 2007). The lack of true testing perhaps explains why empirical ecological research shows that populations are influenced by a diversity of environmental factors, yet the theory that predominates is still monopolized by the competition-based logistic equation.

5 RECONSTRUCTION (MODELING OF ECOLOGICAL PROCESSES TO TEST INTERPRETATION) (PHASE 3)

Ecological processes can be reconstructed mathematically by combining the candidate variables in a mechanistic model. Of particular significance to ecology is that undue emphasis of one variable can be circumvented through this approach. Certain ecological variables or processes, particularly competition, contribute mostly noise to the overall process, especially when the analysis focuses on broad geographical scales. Many studies, by contrast, claim to indicate that competition is ecologically very important, but such studies almost invariably focus on small-scale plots, mostly selected because they are expected to yield positive results (Walter 1988a, 1995). Competition should not be emphasized in this way. Also, variables that are subsidiary to ecologically more significant variables need to be treated accordingly in general models. The nesting behavior of birds, for example, is subordinate to the net rate of reproduction, although it does contribute to that rate. It is not necessary to include this subsidiary behavior in an equation that has the net reproductive rate as a major variable. That behavior and its variants should, however, be part of any model designed to explain the net rate of reproduction in its own right.

In any equation intended to model the process(es) leading to a particular ecological phenomenon or pattern, the mutual relationships between the variables and the relative weights among them must also be accounted for. These relationships and weights are expressed mathematically by the signs that link the variables, the estimates of the parameter values, and by the shape of the function, which may be polynomial or exponential, for example. Such process reconstruction procedures entail mechanistic modeling. The variables involved, and their interactions, are thus combined and weighed relative to each other in the form of a mathematical equation. The functional relationships of the variables must be justified mathematically and biologically so that the equation, in its condensed format, mirrors the process of interest.

The procedure of mechanistic modeling proceeds from basic principles. In ecology, for example the structure of the process of interest can be deduced from the properties of the individuals involved, the structure and dynamics of the environment and from the interactions between them. The process that is modeled is therefore the interaction between organisms and their environment. Autecological theory therefore does not see systems in contrasting terms of "bottom-up" and "top-down" processes. The models are predictive because they are constructed independently of observations of the process that is under analysis (a spatial shift in distribution, for example), because they focus on the properties of the individual organisms involved and the environmental variables that affect them. This abstraction of principles thus has the strength of the model being applicable in entirely new situations. The validity of the abstraction can thus be subject to independent test (e.g., van den Bosch et al. 1992, Kooijman 2010), as illustrated in Section 6.

5.1 Mechanistic Modeling

The scientific understanding of ecological processes requires an explicit understanding of the way in which the significant variables operate jointly to contribute to the process that underpins the pattern of interest. An understanding of the operation of the constituent variables in isolation is only ever a first step (Section 4). The various ecological variables that contribute to the structure of the process under consideration must be reconstituted by means of mechanistic modeling, which needs to be structured explicitly to show how that particular process operates in nature. The variables must be combined mathematically and be weighted accurately relative to one another, all of this justified empirically and mathematically, to express their mechanistic relationships as realistically as possible. The dimensions of the various parameters in the equation must also match one another to be acceptable algebraically, and they specify the way in which the various parameters and their connections operate together in the process under study.

The demand for the precise representation of the ecological process of interest in mechanistic modeling is that only one particular model can reflect a particular process realistically, as determined by the closeness of fit to observable outcomes in nature. Models with different fundamental structures simply cannot reflect reality equally; all but one will inevitably misrepresent the underlying process in one way or another. In this, mechanistic models contrast sharply with computer simulations of ecological processes, for the latter are not constrained by the demands of realistic representation, and the choice of parameters or the way in which the process is structured in the model do not have to be justified. Different simulation models can therefore be equally adequate for the purpose.

Mechanistic models reconstruct the structure of processes and weigh the contributions of the various parameters and their values relative to each other. They are therefore the only models that can be used outside the initial bounds of observation to provide general theory. In other words, they are applicable to ecological situations other than the one for which they were originally developed, even for those in which the parameter values may differ from the original situation, because the model structure remains valid across equivalent ecological situations. If, however, any assumption that provides a basis for the model is found to be invalid, the model must be reconstructed. Such mechanistic models provide a basis for generalization in ecology. They can, for example, be used for comparing the process structures that influence other species and such comparisons can be made across vegetation types and climatic conditions. The strength of such models in terms of theory development in ecology is that they are true generalizations; they lead to scientific understanding for they may be extended to other general circumstances or they may be adjusted to deal with specific circumstances.

The technical and methodological issues that relate to curve fitting, statistical analysis, and mechanistic modeling are not the only features that differentiate them from one another. They differ also in their relationship to the very aims of scientific analysis. Scientific research *sensu stricto* is not concerned primarily with forecasting future states of a system, although the ability to forecast or predict future states through mechanistic understanding is desirable. Success in mechanistic understanding demands the rigorous approach of mechanistic modeling described above. Such prediction in scientific understanding relates to the technical issue of comparing calculated and observed values. Use of the term prediction can be ambiguous as it has two somewhat different applications in science, although they do overlap, sometimes considerably. Prediction is used in the sense of scientific testing, where a prediction is made on the basis of a model (which may be verbal or mathematical, and which represents understanding of the system), and in terms of forecasting the future state of a system. The latter, of course, may also constitute a scientific test.

Scientific testing involves not only comparison of the similarity of the calculated outcome with observed data, but also the testing of the various assumptions and components of the mechanistic process. In terms of ecological theorizing, the initial mathematical models, such as those expressed by the logistic and Lotka-Volterra equations, are inadequate, despite their various elaborations (through the simple addition of a stochastic term to represent variable climatic influences) and their widespread use (see Chapter 7). New mechanistic models are needed to provide the system of scientific understanding represented by autecology. Methodologically, mechanistic models are predictive in the sense that from first principles, and from data independent of the process outcome, forecasts can be made as to the future state of the system in question, as exemplified in Section 6 below. Models of this nature are intended to replace the single model that underpins demographic ecology theory, the logistic equation.

5.2 Types of Models for Analyzing Ecological Processes

Chapter 3 described the differences between deterministic and stochastic processes in ecology, and justified why stochastic models represent ecological processes more accurately than deterministic models do, and are therefore more relevant to developing ecological understanding and ecological theory. The introduction of chaos dynamic models into ecological theorizing does not alter this perspective. These points are given context before each type of model is discussed in the subsections that follow.

Probably more ecologists than ever accept that processes in the field are indeterminate or unpredictable (e.g., Drury 1998, Grant and Grant 2002). Essentially, we can never be certain of the outcome of any particular process, even should the starting conditions be the same. Such acceptance is reflected in the development of ideas about ecological systems being non-equilibrium systems (e.g., Rohde 2005, 2013). Unexpected intrusions are frequent, and these inevitably interfere to at least some extent with any deterministic ecological process under investigation, and thus impose another trajectory upon it. A variety of outcomes is thus possible, for each such process, although each may fall within a certain range of expected values. Each outcome thus represents a single realization from a great number of possibilities. Together, the suite of realized outcomes forms a frequency distribution that represents the relative probabilities of each potential outcome. Such frequency distributions can be characterized by several statistical attributes, including the mean, variance, skewness, and kurtosis. The mean value, or expected value, can be seen to represent the deterministic component of the process under consideration, whereas the other measures mentioned represent variation in the way in which the process has run.

The models that have predominated in ecological theorizing are entirely deterministic. The general models that stem from the logistic equation incorporate only the deterministic component of the process under consideration, density dependent regulation of local population density. They thus disregard not only variation in the observations but also the range of processes that influence population dynamics. Stochastic models, in contrast, do include measures of variation around the mean, this latter of which represents the deterministic aspect of the process. They do so to represent the indeterminate aspects of the overall process, and they include them throughout the model, wherever appropriate and even in the intermediate steps that make up the final process.

Chaos-dynamic models were introduced into ecology relatively recently, ostensibly to accommodate and explain indeterminate outcomes of deterministic processes. However, after the indeterminacy of the initial step, the intermediate steps are all deterministic, which provides a strong contrast with the structure of stochastic models. The indeterminacy in the overall outcome of chaos-dynamic models is entirely a result of an inflation of slight differences in the starting conditions of the model as the model runs, even though these differences may be extremely slight (and are widely known as the "butterfly effect"). These models are also labeled models of deterministic chaos, because they represent the underlying process entirely by means of a deterministic structure, and this is overlaid by the chaotic indeterminism at their initiation.

After discussing each type of model, below, we expand on the implications of stochasticity in ecological processes, not only for ecological theorizing and application, but also in terms of making ecological measurements and developing general mechanistic models.

5.3 Deterministic Models and Their Limitations in Ecology

Deterministic models include only the deterministic component(s) of the process under scrutiny, and represent no variation. This is valid when there is no variation discernible in the data, or when this variation is negligible. For example, some physical and chemical systems can be accurately and comprehensively treated as deterministic, as in Newtonian mechanics and optics, and in the chemical mass equations that deal with the interactions of many atoms or molecules, at least over time spans that are long relative to the individual interactions. Their outcomes are thus predictable with practically 100% accuracy under a broad range of conditions. This situation contrasts strongly with that in nature as captured by ecological data.

Deterministic models in ecology are underpinned by five assumptions—(1) the individual organisms are identical and discrete, (2) numerous individuals are involved, (3) the environment is constant and uniform, (4) effects of distance do not need to be included, for all individuals are assumed capable of interacting with one another simultaneously, and (5) no variation from outside intrudes between the different steps of the process. The rate of population change in demographic ecology is described along these lines by differential equations, when population growth is considered to be continuous, but by difference equations when the process is periodically disrupted, as in the case of discrete generations, seasonal effects, and so on. These equations include no variation in the data, for they lack an error term. Indeed, their very structure excludes the representation of any such measure of variation in interpretation of the data. If variation is added as an error term, the rest of the process continues as it would have before the intrusion.

Once a particular process structure has been represented mathematically in a deterministic model, the inclusion of the initial conditions in the system is sufficient to calculate the outcome of that process. In practical terms, a single measurement, or just a few, will suffice to run the model, as any variation in measurement is assumed to be insignificant, and outcomes are not expected to be amplified during the process. The outcome then only needs to be confirmed by a final observation. The tests that are conducted typically consist of a simple comparison between the calculated and the observed values at the termination of the process. True statistical testing becomes necessary only when errors in measuring the outcomes of the process yield too great an error term to allow discrimination between the calculated outcomes from those derived from alternative models.

Modeling of deterministic processes is relatively simple because it effectively removes variation from any consideration. Such models therefore tend to use historical data, as the process is expected to operate similarly in the future. Deterministic processes are therefore by far the least demanding of all processes from the viewpoint of acquiring data, but unfortunately they have little relevance in ecology because stochastic processes play such a prominent role in ecological systems. This requires a shift from models that require only relatively few input data, to the autecological requirement of substantial data sets across one or more matrices, to allow variation to be estimated in the output. Autecological models thus incorporate life history or other biological data, derived from individual organisms, and these models therefore have the potential to predict outcomes of the process concerned, whether at a later time or in a different locality, and these predictions can be tested with the values of independently obtained observations on the system (whether at a later time or in a different locality).

Non-mechanistic models still underpin demographic ecology theory even though they have been subject to virtually no true testing, despite their relative simplicity. The consequences have deep implications for theory and practice in ecology. In particular, ecological interpretation and practice is still governed by the view that density dependence is central to ecological systems and should have a central place in ecological generalizations. The place of the deterministic logistic equation in such generalizations as well as in ecology textbooks has been critical to maintaining this perception. Soundly criticized, and with no empirical protection, it persists as the centerpiece in ecology even though stochastic influences are now widely accepted to alter the course of ecological processes significantly (Coulson and Godfray 2007). And it retains its central focus in the face of there being no clarity on "how stochasticity should be introduced into the model, onto which demographic parameters, and with what correlation structure"

(Coulson and Godfray 2007). We return to stochastic influences in ecology after dealing with chaos dynamics.

5.4 Chaos Dynamics and Determinism

The consideration of chaotically influenced ecological dynamics in ecological theory has provided diversionary interest, but has contributed little to ecological understanding. Essentially, models that include a chaos component do not model those chance influences that are measurable in a way that helps understanding through the development of robust generalizations in ecological theory.

Chaos dynamic processes have been formulated in terms of the deterministic differential and difference equations introduced in the preceding subsection. These equations result in non-linear dynamics (or chaotic influences) yet still have no terms that incorporate variation that is biologically relevant to the course of the ecological process they model. The outcomes of the process are not predictable in a deterministic way because even a minute difference in the parameter values, entered as input into these models at the start of a run, is going to be amplified substantially during the course of the process. Huge variation in the output data generally ensues, because of the sensitivity of the process to initial conditions (Ruelle 1991)(see Fig. 8.2).

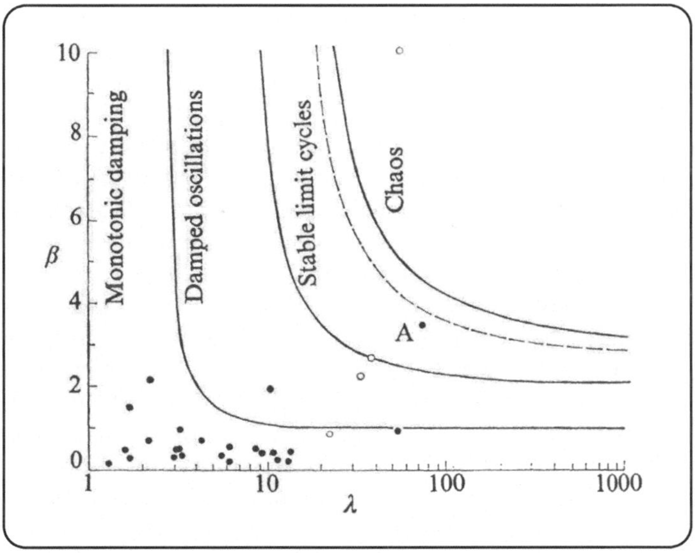

Figure 8.2 The solid line demarcates the stability domains for the density dependence parameter, β, and the population growth rate, λ, in the equation $X_{t+1} = \lambda X_t[1+X_t]^{-\beta}$, which has been used to fit a considerable amount of data on insect populations. The dashed line shows where 2-point cycles give way to higher cycles of period 2^n. The solid circles come from analyses of life table data on field populations, and the open circles from laboratory populations (from May (1976)).

Chaos dynamics is widely known through the so-called "butterfly effect". A butterfly flying in Brazil, for example, is considered capable of generating a storm elsewhere, perhaps in Texas. The minute turbulence of the wings could be the slight, unexpected, non-linear input into the weather system, which, all other things being equal, gets amplified during the subsequent deterministic course in the development of that system. That is, the determinism of the underlying process is believed to underpin (and explain) this process entirely. However the random, or chaotic, input renders prediction impossible. Within the context of ecological theory the typical example is the process of population growth as modeled by deterministic equations, such as the logistic equation, but in which small variations in the intrinsic rate of reproduction, r, constitute the variable element in the input (May 1976).

Qualitatively, the source of chaotically influenced variation is most easily visualized by considering the rather simple, physical origin of a binomial distribution, with reference to the Galton board (Fig. 3.34). Beads, contained initially in a funnel at the top of the board, are simultaneously dropped down the Galton board. The number of beads that rolls into each of several adjacent containers at the bottom of the board is counted at the end of the process. The resulting discrete, binomial distribution of beads across the containers is bell shaped, and is similar in appearance to the continuous normal or Gaussian distribution from mathematical probability theory.

The beads are identical to one another in size, shape, weight, and elasticity. The pins against which they bounce are identical to one another, are fixed in position, and are not affected by the relatively light beads when these hit them. These basic elements of the apparatus allow no variation, so this structure represents part of the deterministic structure of the process by which beads fall. This is complemented by each contact of bead on pin, or bead on bead, and the subsequent path of each bead following quite strictly the deterministic principles of Newtonian mechanics.

The chaotic input in the example above is illustrated in greater detail with reference to Fig. 3.35, where two billiard balls start moving in a straight line from precisely the same spot, but at slightly different angles to one another (Ruelle 1991). They collide with fixed, convex obstacles, but not with one another (since they leave at different times). After only a few collisions, the two trajectories have nothing in common, for the location and angle of each impact is unique. The initial minute difference in direction of the moving balls is amplified, to the extent that the outcome is entirely unpredictable. Indeed, the outcome can be represented only in statistical terms even though the parameters of the course of the balls and their impacts are deterministic and they follow Newtonian mechanical principles that can be described by a set of differential equations.

Identical processes are at play in the Galton board example, but with the additional stochastic input of the beads striking other beads or pins. That is, the path that the beads take after leaving the funnel at the top of the board is determined by (1) the slightly variable position of the beads relative to each other in the funnel, (2) the variable rates at which they roll, and (3) the different ways in which the beads bump against each other in the funnel and during their trajectory down the board. These three influences do not affect the process structure or its deterministic nature. In terms of modeling the system, they represent the variable part of the process, and this relates to the starting circumstances of each bead. The variation in the path taken by beads is enhanced at each step during the deterministic part of the process, in the way depicted in Fig. 3.35. This renders the process, as a whole, very different from that in a strictly deterministic system; its outcome is indeterminate, although each step in the trajectory of each bead is deterministic. The indeterminacy results from none of the beads participating in any of these steps exactly as they would have if the system had been entirely deterministic. Nevertheless, all of these interactions, taken together, result in a rather smooth frequency distribution of beads in the container at the bottom of the board, because the system is closed and despite the fall of each bead being discrete. An entirely open system would be unlikely to generate such a distribution (Ruelle 1991).

Despite the unpredictability just outlined, the final path of each bead can, in principle, be described by Newtonian mechanics in terms of a set of differential equations for each of its bounces against pins and other beads. But such description is not predictive and the eventual bell-shaped distribution of beads is typical for chaotic processes; the slightly variable input of the beads leaving the funnel is amplified in one way or another during the subsequent deterministic processes in which they participate as they roll down the board, hitting various pins and other beads. The miniscule differences in initial conditions with respect to position, velocity at the time they separate from the other beads in the funnel, and the direction in which they leave the funnel mouth remain unknown. Despite the fact that these differences in initial conditions are usually infinitely small, none of the pathways is likely to be repeated exactly; each pathway is unique. Yet, it is precisely these small initial differences that start dominating the deterministic component of the process; soon, the chaos dynamic component completely obscures the effects of the deterministic component of the process.

How do models of deterministic chaos fare in ecology? Tracing the probable cause of a phenomenon from observation of that phenomenon alone is not possible in systems influenced by chaos or chance. Huge variation in the data simply cannot be related to a particular process or to minute levels of variation in one or more initial conditions. A storm in Africa could not be traced back to the flight of a particular butterfly, whatever its locality and whatever its influence. Moreover, we cannot discern the conditions of a continuous process in which each stage influences the next so significantly. That means we cannot include such influences in our explanations of phenomena or our predictions of outcomes. Indeed, such postulated influences are not considered in climatological studies or projections, for the emphasis in climatology is firmly fixed on discernable patterns and general processes. Should external influences intrude, they are also considered, but chaotic influences are never invoked to explain why we do not see what we expect to see, as recommended for ecology when a chaotic element, such as variation in r is added as input to the logistic equation (see May 1976).

Large bodies of ecological work still consider that populations are regulated by density dependence, even though evidence of such a process is not readily or consistently detectable. That is, data on population trends and density effects from the field generally do not meet expectation. Chaotic interference has been said to explain such variation, to the extent that it is admitted that density dependence, as the causative process, though operative, may not be detectable (Subsection 3.2). Such a view obviates the need to investigate the phenomenon because the outcome becomes irrelevant. Such a destructive approach to further research must be avoided (Murray 1999), for it removes the explanatory power of ecology as well as the testability of ecological hypotheses.

The application of chaos dynamics in ecology, to safeguard the assumption that particular deterministic ecological processes, such as logistic growth, do operate under variable conditions, raises many problems without resolving any. Knowledge of the final statistical outcome of the beads' descent in the Galton board is, on its own, inadequate for reconstructing exactly the process that generated the pattern, and which therefore dictated the path of any particular bead, at least if that process is assumed to be influenced chaotically. More than one process could, in principle, have given rise to any particular frequency distribution, which means that determining which was causal without further information is impossible (Pielou 1974). In short, chaos dynamics hides the scrutiny of the structure and dynamics of ecological processes and, as hinted at before (Chapter 3, Section 4) and expanded in the following section, it does not model stochastic influences in ecology in a realistic way.

5.5 Stochastic Processes

5.5.1. Contrasting deterministic chaos with stochastic influences

Stochastic models include deterministic components as well as representation of the stochastic influences that impinge on the operation of each aspect of the deterministic "skeleton". The way in which the stochastic influence is included depends on the way in which the deterministic aspect of the system is seen to operate. As described above, demographic ecology has simply added a single stochastic descriptor of extreme weather variance to their Markovian simulations of age structured populations, which are still considered to be influenced primarily by density dependence (e.g., Coulson et al. 2001, 2004, Stenseth et al. 2004, Lande et al. 2006). The resultant fit of modeled data to time series field samples is then taken as justification that the deterministic density dependent skeleton is an accurate representation of the workings of ecological influences on organisms.

Reconsider the physical model that generates a bell-shaped, binomial distribution of numbers of beads across the various containers of the Galton board (Fig. 3.34). We explain initially why the process that leads to the distribution of the beads is not truly stochastic, before considering an example that illustrates one that is so. The frequency distribution of the beads across the catching containers at the bottom represents the result of the amplification, within an otherwise deterministic process, of slight initial variation in starting conditions (or input data). This initial variation is a relatively minor part of the entire process, and the deterministic element dominates the rest of it. This domination is reflected in the formal process structure being represented by the equations that represent the trajectory of beads as they roll, bump into one another or against other obstacles, and bounce off them as they make their way down the arena, in a similar way to Ruelle's (1991) billiard balls (Fig. 3.35). We therefore speak of the process as being deterministic, and, hence, of deterministic chaos; the generation of the negative binomial distribution in this way is deterministic with respect to the process structure. It represents chaos dynamics because of the mere amplification of the initial variation in the starting conditions or input data. As such, it is not an example of a stochastic process, despite the widespread use of the binomial as representative of stochastic outcomes in statistical texts for students. We therefore contrast, below, the generation of the binomial in this way with truly stochastic processes, to demonstrate that the two types of processes logically merge with one another, their difference being defined by the relative "dominance" of the deterministic part of the process over the stochastic aspect.

Consider once more the Galton board that generates a bell-shaped frequency distribution of beads. To convert the deterministic process of falling beads to a stochastic one, both the beads and the board have to be modified in various ways, such that the original process structure is altered qualitatively. Ecologically, the beads represent individuals in their environmental setting, and the board itself represents their environment. For the beads to portray individual organisms more accurately, they now differ from one another in size, shape, weight, and elasticity, rather than being equal in all these respects, as before. These differences are continuous, not discrete, which changes the resulting frequency distribution, as in the difference between a discrete, binomial distribution and the continuous Gaussian distribution. Furthermore, the variation in each property of the beads now follows a particular frequency distribution, and each of these properties is not necessarily distributed according to the same frequency distribution as that describing each of its other properties. Individuality or specificity is thus introduced into the properties that dictate how each bead rolls, bounces and bumps.

The resulting variation in bounce among the beads thus mimics the variation that is observable in the ecology of individual organisms in the field, as they respond to environmental conditions and to each other. Individual organisms, moreover, do not even remain the same throughout their life, but also change with respect to physiological condition, sex, sexual condition and age, or according to their previous experience or location. Such transitions are often distinct in organisms with discrete larval stages, for example. The variation in ecological responses of organisms, both within and between species, and whether among individuals or through the lifetime of a single individual, is obviously much more elaborate than we can ever simulate by conferring different properties on the beads that bounce their way down a Galton board, however complicated that may be. We therefore need to structure our models according to each trait and introduce a chance distribution for each of them throughout this structured model. However, the principle will be clear from the metaphorical Galton board.

The Galton board metaphor also needs to be modified in relation to the ecological setting of organisms and their environment, as well as the heterogeneity and dynamics of that environment. Originally, the beads rolled down a funnel that remained unchanged through that process. The pins were identical to one another, and their properties and positions remained fixed. Also, the original board was smooth and flat, and held steady in one position. To make the Galton board analogy more representative of ecological systems, these features must also be altered. Bumps, hollows and sticky patches are added, even variations in time, and the whole is made to turn and shake whilst the beads roll down. Such variation in each of the properties of the apparatus can be represented by particular chance distributions. More significantly, in ecological terms, they mimic to some extent the spatial heterogeneity and dynamics of the environment to which organisms must respond continuously through their life. The variation in the natural environment can be mimicked further by also varying the quality (e.g., size and elasticity) of the pins and their configuration, as well as by varying the dimensions of parts of the apparatus. To complicate matters even further, all of these elements can be varied whilst the beads are rolling. Finally, if all the beads are allowed to roll simultaneously, they also bump against one another, a feature that adds one more stochastic element, this time mimicking biotic interactions between organisms.

Thus, very simply, a suite of environmental variations is added to the possible pathways of the simulated "organisms", each with its own consequences for the further course of the process. In the normal (deterministic) set-up of the Galton board all beads end up at the bottom. By contrast, the variations in the stochastic system result in some beads (or some types of beads) not arriving there at all, which represents death and possibly selection. This outline of the stochastic nature of ecological and environmental variation is repeated from one year to the next as well as differently in different locations among which, moreover, random dispersal and migration can occur.

The outline above illustrates why stochastic processes should never be confused with deterministic chaos. The latter does not allow variation to enter the structure of the process itself, which means that it is deterministic in essence. Strictly stochastic processes, by contrast, are defined by chance influencing each step of the entire process, as described above. Although the determinism inherent in the system seems to be obliterated and therefore irrelevant to understanding the system, this is not the case. The underlying deterministic structures and processes that underpin the system are crucial to understanding how the system is structured, how it works, how stochastic influences impinge on it, and what aspects warrant quantification. In ecological terms, these underlying features are the individual organisms, the species-specific properties of their environmental response systems and the environmental axes of differentiation that represent the elements of their environment with which they interact. We view and conceptualize such systems in a fundamentally different way from that dictated by a deterministic perspective or that of chaotic determinism.

In stochastically influenced processes, the operation of each component can be effectively represented only by a chance distribution of values, so here the stochastic component "dominates" the deterministic one. Although each bounce at each instance could, in principle, be described deterministically, the process as a whole is impacted by chance from one incident to the next. The relevance of this approach to ecological systems is clear. Such stochastic models represent ecological systems far more realistically than do the deterministic or deterministic chaos models dealt with earlier.

To illustrate, consider any process that concerns the survival or reproduction of an organism in nature, such as the post-fledging wandering of a bird or the host searching behavior of a parasitic wasp. Each step of such processes represents the instantaneous response of an individual organism to prevailing environmental conditions and the dynamics of those conditions. Each such step is characterized by a different chance distribution and can, in principle, follow one of numerous different possible pathways of stochastically influenced events. In the case of the newly fledged bird, the precise direction in which it leaves its natal area is essentially stochastic, as is the distance it flies and where it lands for the first night, and so on. Although the species-specific adaptive mechanisms of the individual, and the exercise of those mechanisms, can be described deterministically, the interaction with the natural world imposes stochasticity at each step.

5.5.2 Stochastic modeling

Because many influences perpetually redefine the course of ecological processes, as outlined above, such processes are best viewed as stochastic in principle. Consequently, ecological variation should not be included in ecological equations simply as an error term, as is done in regression equations. Variation is inherent to ecological processes and shapes them, and should be expressed relative to a deterministic framework or skeleton, and in terms of the variance, covariance, or kurtosis around the mean, for example.

Single runs of a chaos-deterministic process, with stochastically varying input, are depicted schematically in Fig. 3.36, and this illustrates how stochastic models can be used in causal analysis. At each time interval, a new position in a chance distribution is realized and this, in turn, influences the next position within the chance distribution of the elements along the contorted path that they ultimately follow. Mechanistic analysis requires that the sequence of events be followed from beginning to end, with each step in the course of events analyzed probabilistically.

In ecology the beginning of the analysis could be the egg stage of the organisms of interest and the reproductive output of the individuals that reach adulthood the end stage. What is relevant to ecological understanding is the accurate identification of all the relevant processes within the system in question, for this is where data collection must focus. Such models do not allow the reconstruction of previous stages in the overall process. Future such stages can be forecast, however, but only in stochastic terms, but such predictions are strictly restricted in terms of how far ahead one can predict, with limitations related to our weather forecasting abilities. To model the entire process so as to provide predictive abilities on the basis of understanding, the causal pathway must be reconstructed in terms of the changes in "direction" evident at each step in Fig. 3.36. This is the prospective reasoning that underpins the analysis of processes that are defined by chance, be they chaos dynamic or stochastic. In deterministic processes, by contrast, as soon as the process structure is known, we can re-construct past events and predict future developments from few observations.

Stochastic models demand a large volume of accurate data, so elaborate measurement schemes are required in the field. Such schemes must be carefully structured according to a good understanding of the process structure, as derived from the original exploration phase of research (Section 4). In ecological systems, data collection must be organized around the duality of environment and organism. Each of these is, in turn, structured and dynamic, the environment over time and across space and the organism through the life cycle and according to its species-specific adaptations (Chapters 3 – 5). Added to this complexity is the movement of organisms. All of this must, again, be taken into account in designing the data collection system.

The responses of organisms to environmental circumstances may differ between the sexes or across age groups, or between individuals inhabiting different localities, and so on, which again extends measurement schemes. To follow temporal shifts in this variation across individuals in all these categories requires knowledge of the species' physiology and behavior, as well as an understanding of how such traits develop in ageing individuals, along with causes of temporal and spatial variation in ecological conditions. Not surprisingly, such detail is not yet available, at least to any comprehensive degree, for any species. Presumably, the amount of effort required has not appealed to many ecologists, who have in any case focused on deterministic processes, mainly those related to density dependence and aspects of optimization. Examples of positive advances that have been made in accumulating the appropriate data and analyzing ecological processes are covered in Section 6.

5.5.3 Testing stochastic models for realism

In the structured models outlined above, the veracity of the assumptions made about each variable, and the way in which these variables interact during the course of the process as a whole must, in principle, be tested. The testing required is therefore much more elaborate than for deterministic models, because stochastic models reflect ecological processes in the form of continuous and anastomosing flows of many mechanisms, processes and influences. Further, each species has different ecologies and life expectancies, and they respond at different spatio-temporal scales of variation. In short, all the parameters in the equations, together with the effects of their interactions, must be physically measurable and mathematically independent of each other. The development of a model of a stochastic ecological process, in effect, requires the confrontation of its calculated outcome with outcomes from observations on the process in the field. That is, they need to be tested. Although structured stochastic models are much more complex than deterministic and chaotic models, they do provide a more accurate, mechanistic understanding of ecological processes, as well as an understanding of their most sensitive components. Such models represent a considerable conceptual advance over demographic models, as they are built explicitly around the life cycle, adaptations and environmental relations of organisms. For all ecological purposes, in which the emphasis falls on survival and reproductive processes, these models provide the only analytical method that is realistic. Of course, for practical reasons of mathematical simplicity, one uses only the deterministic component of the stochastic process. That is, the model structure was designed so as to retain the ability to use variable data as input and thus be able to calculate measures of variation in the output.

In effect, this approach entails the analysis of the ecological consequences for individual organisms of the adaptive mechanisms (or suites of adaptive mechanisms) they embody if they are to survive in an uncertain, stochastically varying environment (see also Elsasser (1987)). The approach thus leads readily to the causal analysis of such related issues as (i) the consequences of having to survive and reproduce in stochastically changing environments in terms of the behavioral mechanisms, physiological flexibilities and mobilities of the organisms, (ii) the consequences of complex adaptations being fixed at the species level across geographical space and through geological time (Chapter 5, Sections 5 and 6), (iii) the particular ecological contexts in which genetically based adaptation or spatial adjustment of species by habitat tracking takes place, and (iv) how adaptation and speciation takes place in relation to such external influences.

6 A WORKED EXAMPLE—THE RECONSTRUCTION OF BIOLOGICAL INVASIONS

The reconstruction phase of ecological research follows the exploration phase logically, as shown in the following example in which a model of the stochastic invasion of Western Europe by the collared dove (*Streptopelia decaocto*) is developed. Specific details and references are available elsewhere (Hengeveld 1988a, 1989a, Hengeveld and van den Bosch 1991). Notice that the model relates to a single species. The ecological generalization lies in the type of model, for it is this that is generally applicable but it does require an understanding of which variables are most significant as well as specific parameter values of the species concerned (Hengeveld and van den Bosch 1997).

The exploration phase in this case identified the relevance of the age-specific survival and reproduction, as well as dispersal, as ecologically significant parameters. The reconstruction aspect of the research related these parameters to one another mathematically to reflect the structure of the ecological process as accurately as possible. Finally, the validity of the reconstruction was tested by comparing estimates of such features as invasion velocity and local increases in bird density with independent counts made in the field. Before detailing the reconstruction we point out some of the difficulties inherent in the exploration phase and in the transition to the reconstruction.

Explorative research continues to reveal much about the ecology of invasions. The influence of external environmental conditions, including temperature and moisture, in allowing a species to invade a new geographical region and establish there often provides a starting point. These conditions must be understood and interpreted in relation to the environmental requirements and tolerances of the species in question (Chapters 3 – 6). Then there is the question of biotope conditions in the newly invaded area, often in relation to food and reproduction. Native species in the area may also be influential, through predation, disease, parasitism, modification of the environment or provision of suitable hosts. Finally, the role of life history parameters, such as the species' age distribution, ages of main reproductive output, and the dispersal kernel must be taken into account as influenced by the interaction of the individual organisms with their environment. The dispersal kernel is the mathematical representation of the rate of movement of the individuals of the subject species in the geographical area of interest, and thus its rate of spread through that area (Chapter 7). Yet other factors are typically investigated during the explorative research phase.

The problem for ecological investigation is therefore one of differentiating between these different processes in terms of their relevance to the invasion process. With regard to the analysis of the collared dove's north-westward spread across the Himalayas and into Europe from India, and from Europe eastward into Siberia (Hengeveld 1988a), the life history parameters mentioned in the paragraph above were chosen because the dispersal kernel represents the invasion potential, and is underpinned by recruitment. The survival distribution expressing age-dependent mortality rates (thus expressing the age distribution), together with the age-specific fertility rate, "compensate" numerically for local reductions in densities as individuals contribute to the spatial spread of the species over an exponentially increasing surface area as the invasion continues. The other variables and processes mentioned are all in one way or another subsidiary to these few parameters of the life cycle and reproduction of the species within the current context. Different species and different contexts, especially more heterogeneous ones than the farmlands of western Eurasia, are likely to require different considerations.

The most relevant parameters determining the velocity of range expansion are, therefore, the distances covered by the individuals per unit of time, as well as their net reproductive rate. In the model that was developed, the first was measured by the diffusion rate D and the second by the net reproduction rate R_0. The first of these was estimated from the frequency distribution of individuals having been found at different distances from their birthplace, whereas the second was considered to be age-dependent. The age-dependent variation in net reproductive output was estimated from both the mortality rate of the various age groups and the rate of fertility of those age groups. These rates were multiplied, and the products summed to get the overall value of R_0. This information was obtained from the literature and assembled in a Leslie matrix. Both parameters are therefore considered to vary, with the diffusion component as a typically stochastic process, although the model itself, for reasons of mathematical tractability, has been worked out with regard to its deterministic components only. The variation in the input, though, allows variation measures to be calculated in the output. As the final step, the calculated rate of expansion velocity, combining the values of both parameters, was compared with a series of observed geographical expansion rates over time. Values for these latter rates were obtained from independent information in the literature.

The same has been done, independently of one another, with the few other species for which adequate data were available in the literature, including the muskrat, starling and house sparrow. For each of these species, the calculated rates of expansion were specific to each species. These rates were calculated from life history data and then tested against the rates observed in geographical space to be certain the models are explanatory. In all cases, except for the house sparrow in North America, the points for each species fell on a straight line of 45° (van den Bosch et al. 1992). For the house sparrow, however, the skewness of variation in the calculated rates of diffusion and net reproduction had to be estimated. An additional parameter had also to be estimated to express the variation in reproductive output on top of the deterministic value, to improve the fit. This general model is, therefore, applicable to all the species tested and means that the expansion rates in these invasions are influenced primarily by their respective specific rates of dispersal and reproduction in the areas concerned. Note, though, they are all peri-urban or associated with agricultural areas.

In the reconstruction phase, direct observation and quantification of particular parameters thus served as input into mathematically derived, mechanistic equations based on a reaction-diffusion model. The parameters were estimated at the level of individual animals rather than at the level of the geographical spread itself, which, indeed, they have to explain. Furthermore, these observations were quantified in the native distribution of the species to avoid the circularity of reasoning that is often inherent in explorative research. These two aspects of the approach therefore meant that the mathematically calculated or predicted rate of invasion is derived independently of the observed rate of invasion. As the parameters used pertain to individual organisms as units of spread, on the one hand, and the extent of geographical expansion, on the other, there are parameters to serve as independent, explanatory, variables (the first category) and a dependent variable, or explained variable (in the second category).

This type of model is a mechanistic reconstruction because (1) the structure of the biological mechanism of the process is justified independently of the observed spatial dynamic (or invasion pattern), and (2) the data used for the calculations are independent of the data that reflect the observed expansion in distribution of the collared dove, which is to be explained ecologically. This prospective procedure is quite unlike that seen in statistical procedures, where an arbitrary, non-specific, additive model structure is imposed retrospectively on the data by the test selected. This is not to criticize statistical procedures, just to provide their appropriate context, for in statistics the test involves the fit between an observation (with a biological deviation) and the expectation of chance. These descriptive models of exploratory research use parameters extracted at the same level of the phenomenon they describe. By contrast, in the explanatory models of the reconstruction phase of research the parameters included in the model (reproduction or movement, for example) are subsidiary to the level of the phenomenon to be explained (invasion velocity, for example).

The distinction just made has serious implications for current demographic ecology modeling, for it shows the discipline is still mostly in the exploratory phase of research. The primary concern of these endeavors is still parameter estimation. The numerous investigations of competition and density dependence, for example, justify this point, as does the frequency with which these processes are invoked in general explanations as well as in specific studies. Nevertheless, the estimation of the relevance of these parameters and their weight relative to one another in a mechanistic equation designed to model a particular ecological process has yet to be attempted. A further problem is that the parameters considered to explain a phenomenon are usually from the very same level as one another. For example, birth and death rates are simply constituents of the overall numbers of organisms and so cannot be explanatory of these numbers. Whereas the equation $n = b - d + i - e$ is logically correct, it cannot explain any particular value of *n* because these variables constitute *n*. Similarly, a time series of numbers of individuals in an area, cannot be explanatory. When the numbers of individuals in a time series are replaced by parameters of the life history of the individuals, such as their age distribution and their age-dependent fertility rate, as in the analysis of invasion rates discussed above, the model containing these parameters becomes explanatory in the real sense and is therefore predictive and can be tested.

Mechanistic models can be extended logically should prediction differ from observation. This facility represents another way in which mechanistic models differ from the descriptive models of exploratory research. Although this procedure does involve a form of circularity, empirical circularity, this is acceptable in that it is quite different from the methodological circularity in which the demographic variables used to characterize the temporal pattern of abundance in the exploration phase of research are also used to reconstruct the process that is considered to generate this pattern (Subsection 4.2). If the predicted rate of geographical expansion differs from the observed rate, the other parameters of potential relevance in the invasion process, predation, parasitism, bad weather or soil conditions for example, may be affecting the survival chances of the invaders, reducing the number of offspring and, thus, the invasion velocity. These would need investigation and quantification, with an additional module added to the relevant parameter in the model. The same holds for the dispersal kernel. The initial collared dove model was designed for invasions through ecologically uniform areas, for the collared dove's habitat includes agricultural settings (Hengeveld and van den Bosch 1991, van den Bosch et al. 1992). The model was then extended to the more usual situation in nature, in which the environment is not uniformly suitable to the species of interest. This was achieved through the inclusion of parameters that accounted for mortality related to dispersal and for the proportion of habitable area in the region (Hengeveld and van den Bosch 1997).

This latter reasoning is the same as that underlying the distinction often made between effects of long- and short-distance migration in recent invasions. Initially, the analysis distinguished these two modes of migration during the collared dove invasion (Hengeveld 1988a). Later, however, the rare long-distance migrants, which have the noticeable ecological impact of forming bridgeheads ahead of the closed front defined by the movements of the short-distance migrants, were seen to represent the tail end of a continuous, J-shaped contact distribution (Hengeveld 1993). The entire distribution is in fact determined by two different processes, an exponential decay with distance and the spatially random movement of individual organisms. Together, they constitute a Bessel function (Pielou 1969) in which two chance processes together replace an inferred biological distinction (the two postulated migration modes) that ultimately proved to be somewhat misleading.

Dealing with ecological models as outlined above has precedents. Physics, for example, has progressed through generalizing, specifying, and extending models, and has thus unified numbers of seemingly disparate phenomena. The mechanistic model that describes the velocity of range expansion has followed this procedure. Although it was originally developed for application to spatially uniform ecological conditions (van den Bosch et al. 1992), it had to be extended to deal with the more representative circumstances of ecological heterogeneity (Hengeveld and van den Bosch 1997). From this basis the internal structure and dynamics of geographical distributions could be developed to include not only range extension processes but also range contractions and distributional shifts. Local ecology could thus be integrated with dynamic biogeography in a mathematically consistent way.

The rate and extent of geographical progression of an invasion can be predicted from data on the life history of the individuals concerned, as in the collared dove invasion model above (van den Bosch et al. 1992). Experimental data on the temperature sensitivity of the various larval stages of an insect species, the western corn rootworm for example (Hemerik et al. 2004), can be used to do the same thing, so this modeling approach and these examples demonstrate that autecology does provide ecological generalizations and general ecological models.

7 CONCLUSION

Ecological models need to incorporate stochastic influences, for such influences impact on the course of the ecological trajectory of all individual organisms. Current approaches tend to add generalized stochastic influences in the form of demographic or environmental stochasticity, with the former designed to account for the influences associated with small population size and the latter usually representing the influences of extreme climatic impacts. When ecological theory is considered, these influences tend to be added to population regulation models, because density dependence is believed to be the primary driver in population change because local persistence of populations is seen to depend ultimately on such biotic adjustment. In this perspective, survival and reproduction are treated as readily accomplished, so exponential population growth and the overproduction of individuals is expected to be a problem. Small population size is therefore expected to be alleviated simply through the lifting of the density-imposed challenges, but without any recourse to the environment itself or the environmental relationships, requirements and tolerances of the organisms concerned (Walter 2008).

Autecology, by contrast, sees survival and reproduction as an ecological challenge. The adaptations of organisms are species-specific and function best within the environmental setting to which the species became adapted at the time of speciation. These adaptations support persistence through bad ecological circumstances (as well as they are able) and account not only for local persistence but also for rapid population increase when environmental circumstances suit the species in question (Walter 2008). Density comes into play only under extreme circumstances, is not a primary driver in ecology, and is simply not adequate as a basis for generalization in the development of theory. The changing numbers of organisms in nature is accounted for by the interaction between organisms, through their species-specific adaptations, with their spatio-temporally dynamic environment. Mathematical models therefore have to include the impacts of stochastic influences at each major step in the life of organisms, through structured mechanistic models, despite these being demanding in terms of data.

The collared dove example illustrates how autecology includes stochastic influences in theory and in practice. Although mathematically advanced, ecologically it is a rather simple example, but it does introduce the autecological approach to understanding organisms and their environment. Naturally, there is far more subtlety in most ecological systems relative to that in the invasion of a highly motile bird across the relatively homogeneous farmland of Western Europe. The following chapter therefore deals far more explicitly with those aspects of organisms and the environment that are relevant to ecological understanding. Ultimately, Chapter 9 brings these two aspects together in terms of the dynamic match between organism and environment, and thus refocuses attention on the central questions of autecological theory. We thus develop the methodology for solving the autecological questions given prominence in this and previous chapters, and we do so by formulating them quantitatively in terms of risk analysis. Risk analysis is functional and prospective, using information on independent parameters that involve the survival value conferred on individuals by their biological properties within specific, and varying, ecological contexts.

Chapter 9

Organisms Across Space and Over Time

Deterministic Structures, Stochastic Influences, Environmental Gradients and Risk Analysis

1 INTRODUCTION

The mechanistic models championed in the previous chapter demand a perspective from ecologists that is somewhat different from the approaches suggested by demographic theory. They typically demand a lot of data, derived principally from measurements on individual organisms, and the incorporation of data on the various stochastic factors that influence critical aspects of their lifeline or ecological trajectory. Ecologists of all persuasions increasingly acknowledge the significance of stochastic influences in ecology, after decades of neglecting them. Demographic modeling now includes such influences, with some outwardly impressive developments.

For example, childhood whooping cough epidemics have been modeled in this way (Fig. 9.1), with modeled output matching the epidemiological data far better when a stochastic element is added to the model (Coulson et al. 2004).

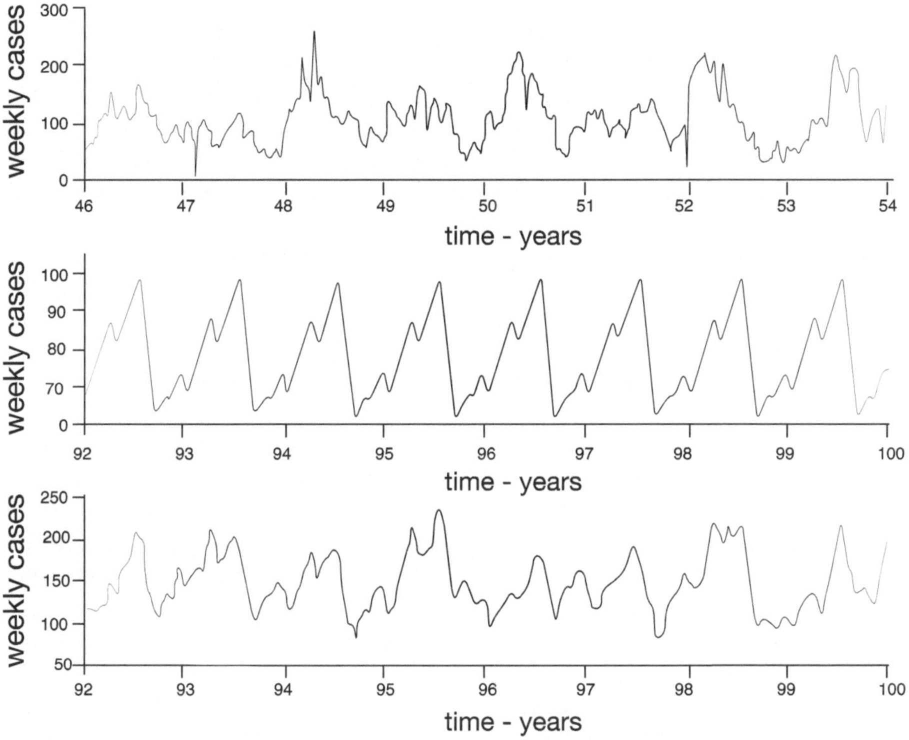

Figure 9.1 (Top) Weekly numbers of whooping cough cases in Birmingham. **(Middle)** Output from the deterministic model, containing only the "deterministic skeleton", does not match ("predict") the course of the disease. By contrast, the same model **(Bottom)** with a stochastic component matches it much more closely (reproduced from Coulson et al. (2004), with permission).

This approach, though welcome after so many decades of full focus only on density dependent deterministic processes (as explained by Coulson et al. (2004)), warrants careful consideration. Its direction comes mainly from the strong dichotomy that was set up early in ecology, at the time that the primary role assigned to deterministic processes was debated most heatedly, in the 1950's and 1980's (for history see Kingsland (1995)). Ecological processes were grouped into two categories, and the ecological influence of each tended to be seen as mutually exclusive of the other. Either density related phenomena drove the system deterministically, or weather did so stochastically and thus independently of density (e.g., Davidson and Andrewartha 1948a, b). The stochastic influence thus came to be seen in somewhat uni-dimensional terms, as a single factor that reduced the system to levels well below equilibrium or carrying capacity and tended to keep it below that level through periodic perturbation. The biological and ecological detail required to understand such systems (see Andrewartha and Birch 1954, for example) seems to have got lost in the sort of stereotyping mentioned above.

The modern demographic approach takes its lead from the historical dichotomy just outlined, but with some finer detail. Essentially, it accepts that ecological systems can be partitioned into a deterministic part that is considered to represent the essential features of the system and a stochastic part that modifies the deterministic "skeleton", as an error term (Coulson et al. 2004). The deterministic part is seen strictly in terms of demographic influences, particularly density dependent mortality and fecundity, so population equilibrium and carrying capacity are still central concepts (Coulson et al. 2004, Saether and Engen 2004, Lande et al. 2006). The stochastic part is also generalized, as it was originally, but is now partitioned into two components that influence the density related deterministic dynamics of the system. Environmental stochasticity encompasses "environmental fluctuations" driven by the weather and demographic stochasticity concerns random variation in individual fitness (Coulson et al. 2004, Saether and Engen 2004). The density independent influences had earlier been seen to determine the average level of population abundance, which was treated as relatively unimportant (Varley et al. 1973). The density dependent contribution was seen to regulate the system around its equilibrium point, although it was almost always weak when it was detectable. It was given far more prominence, though, because it was seen to explain how populations persist and do not "random walk" their way to extinction (see Cooper 2001). These developments provide an approach that, despite its apparent novelty, neither overcomes the problems identified in demographic ecology, nor fulfills the requirements specified for autecological understanding and models.

The demographic approach to stochasticity still provides a significant problem, because stochasticity is still assigned its original role—it simply disrupts the deterministic, density-governed mechanism that is believed otherwise to regulate the population deterministically. Furthermore, spatial structure and dynamics are generally ignored, but specifically so in "stochastic population theory" (Saether and Engen 2004), and it is not included in the Soay sheep models, for example, which represent an island population. Although the deterministic "skeleton" of the Soay sheep model is structured according to age and sex of the sheep, the model deals primarily with density and not individuals, as generalized by Lande et al. (2006)—"Population fluctuations are caused by demographic and environmental stochasticity, time lags due to life history, and density dependence". For analytical veracity (Chapter 8), individuals must be included and their structured lifeline must form the basis of the model.

Overall, the demographic approach is still inductive, rather than mechanistic and deductive. Whether these models have any generality remains to be seen, and has not been claimed (Coulson et al. 2004), probably because they are inductive and so cannot even be tested (see also Evans et al. (2013)). However, the framework of demographic models (when not derived from the logistic) seems to be somewhat *ad hoc* and geared more to dealing with time series population data rather than data about individuals within their ecological setting. This latter data requirement would presumably represent the "natural history" called for by so many ecologists to make ecological theory more realistic (Strong 1988, Weiner 1995, Dayton and Sala 2001, Dayton 2003). Although the fit of the curves achieved in the new demographic models looks rather impressive, the models are still steeped in concepts that are not readily quantified or are mostly undetectable in nature (carrying capacity and density dependence). Also, they are more the result of curve fitting (see Chapter 8) than of any real deductive reconstruction of the ecological processes involved. This particular problem would be exacerbated substantially if a geographic perspective were to be considered in the demographic models discussed above. In effect, therefore, these developments sidestep the major issues identified in the preceding chapters as critical to ecological understanding.

This chapter brings together the disparate contributions developed so far in the ecological literature to help ecologists deal autecologically with what is generally referred to as "abundance" in population ecology. Autecological perspectives do not see ecological systems in terms of "populations", for these are not entities with boundaries or properties (Walter 2008). Rather, autecology sets a broader context for abundance, with emphasis on the species-specific adaptations, requirements and tolerances of the organisms in question, the extent and dynamics of suitable environments, the survival and reproductive rates within localities and movement rates across localities (Hengeveld and Walter 1999, Walter and Hengeveld 2000). All of this is most appropriately seen and investigated in relation to the geographical scale that is usually defined by the species' distribution (Chapter 5), although studies need not always be so extensive spatially. Abundance is thus considered in terms of the ecological intensity of the species across localities or areas, or the distribution of its abundance, and this can be reflected visually in density prints such as those depicted in Figs. 4.8 and 4.9. The nearest autecological equivalent to "local population" is therefore the changing abundance (or ecological intensity) of the subject species within some locality of interest.

The following section introduces, with the aid of an example, the principles involved in autecology and the associated way in which ecological systems require investigation, quantification and mechanistic modeling. The emphasis is on defining the deterministic structures with which ecologists must work, and the stochastic influences that affect their operation and dynamics. Thereafter (Section 3), published observation and interpretation exemplify the way in which autecology deals with distribution, local abundance and local diversity, as approached through gradient analysis and in which the basic analytical tool is risk analysis (Section 4). Finally, an approach to the synthesis of autecological information is detailed in Section 5. Of necessity, the synthesis is illustrative rather than comprehensive, for research of this nature has been sporadic at best.

2 ADAPTIVE MECHANISMS AND THE ENVIRONMENT—DETERMINISTIC STRUCTURE AND STOCHASTIC INFLUENCES

We illustrate the approach taken in autecological investigations by considering a typical interactive sequence within the life cycle of an organism, the noctuid caterpillar pest commonly called the cotton bollworm (*Helicoverpa armigera*). It illustrates the interactions that contribute to the changing abundance of species across geographical space. Since organisms have not customarily been studied from this perspective, the evidence cited below often refers to species other than the one targeted, but it does illustrate the point and helps justify its validity.

Helicoverpa armigera moths emerge from the pupal stage in response to environmental circumstances and then move among host plants, where they periodically feed on nectar and deposit eggs. The larvae are herbivorous and tend to feed on their natal plant, although movement between plants is not uncommon (Terry et al. 1989). The different host plant species recorded for *H. armigera* attract different levels of oviposition and support different levels of developmental success, to the extent that the species to which they are primarily adapted are good in both respects. These are the primary host species (Velasco and Walter 1992, Walter and Benfield 1994, Milne and Walter 1998, Gu and Walter 1999, Milne and Walter 2000, Walter 2003, Rajapakse et al. 2006, Manners et al. 2010). Other plant species are ranked relative to the primary hosts, with some plant species hosting the insects only incidentally (Walter and Benfield 1994). Environmental conditions exert various influences on the moths' behavior and may reduce their reproductive output and shorten their life, for example when inclement weather reduces the time suitable for flight and oviposition (e.g., Shapiro 1986, Pollard 1988, Kingsolver 1989).

The issue for ecologists is how to generalize these interactions and also their consequences for the abundance of the species in the following generation. We emphasize that we could have selected any ecological process to illustrate the point, whether seed production and seedling establishment in plants, the birth and recruitment of large mammals, or the nesting of birds. Also, each such interaction needs to be set within the context of the other interactions that sustain the lifeline of the species in question (Chapter 3). In this section, we demonstrate which aspects of these particular organism-environment interactions are structured, how they are subject to stochastic influences, and the way in which ecological processes can be quantified for the purposes of developing mechanistic models.

Consider which aspects of the oviposition process outlined above can be accurately generalized in terms of an expected pattern of behavior or an average quantity. The adaptive mechanisms involved are clearly of this nature, and include the chemosensory and orientation mechanisms among others, and thus the pattern of host species affinities (which are often referred to in terms of the "preference hierarchy" (Thompson 1994)). The influence of adult nutritional and ovarian status on oviposition potential and egg deposition rates (across plant species) can also be quantified in this way, and so can the larval survival rate across plant species (Rajapakse and Walter 2007). These are all species-specific and governed by species-specific adaptive mechanisms (Walter 1988b). In autecology, this is the deterministic "structure" that relates to the organism. These behavioral and physiological mechanisms are played out within an environmental context that is also structured and dynamic, in intensity and through time. The environment is structured with respect to weather related variables, vegetation patterns, edaphic factors, and so on.

These environmental features need to be specified in relation to the life cycle tolerances and requirements of the species in question, as justified in Chapters 3 and 4, and needs to be represented spatially (Chapter 5). The local and regional environmental conditions need, therefore, to be generalized in statistical terms. For example, Australian *H. armigera* are distributed mainly down the eastern coastal and sub-coastal cropping areas of Australia (Zalucki et al. 1986, Matthews 1999), and it may be acceptable to treat them as numerically independent of those in Western Australia, where the species also occurs (although this assumption warrants independent test). Also, the mix and relative seasonal phenologies of host plant species that are available locally and geographically can be quantified spatially, as can their relative attractiveness and suitability for oviposition and larval feeding. The suitability of pupation sites, too, can be tested scientifically and specified spatially with regard to soil type and structure.

The life cycle stage that is most sensitive to environmental influence is generally that between offspring production (oviposition or birth) and recruitment to the adult population, and this is true also of *H. armigera* (Kyi et al. 1991, Zalucki et al. 2002). Other life stages may also be sensitive, again depending on species, environmental conditions and risk from biotic influences such as predation, herbivory, parasitism and the like. For example, adult penguins cannot feed while they molt, a period of high metabolic demand, and may suffer higher rates of mortality if food availability was low when they were laying down fat reserves. Further, the sensitive phase(s) of the life cycle and the source(s) of risk are likely to vary geographically (Chapters 3 and 4). For example, *Pinus sylvestris* seedlings in Spain require protection afforded by shrubs (to buffer drought, frost and physical damage, which occurs in bare areas, and to allow sufficient radiation to reach the young plants, which does not happen under trees) whereas more northern plants require sufficient warmth if they are to survive and grow (Castro et al. 2004). Sources of risk will also vary across ecological situations. Parasitism rates of herbivorous insects, for example, often vary across the host plant species used by the herbivore species within a locality (Carroll and Hoyt 1986, Leyva et al. 1991, Vinson et al. 1998, Luhring et al. 2004), and relative to temperature and thus latitude (Stireman et al. 2005). These themes recur at different points in this chapter.

Numerous aspects in the system outlined above are exposed to stochastic influences and a range of such influences may act at any particular stage of the life cycle. Furthermore, both organismal and environmental aspects of the system are altered by stochastic input. We are used to the idea of unpredictable or erratic weather change as a stochastic influence on ecological systems. In autecology, therefore, stochasticity is included as an influence of the usual course of the organisms' lifeline, and it may affect anything from direction or distance of movement to host species for oviposition to likelihood of mortality. The key is to quantify the relevant processes and variables when investigating the distribution of abundance to derive realistic estimates from the field. Indirect effects are not readily anticipated, and are almost always detected *post hoc* (e.g., Russell 2005). The specific approach taken is inevitably influenced by the question posed, the geographical extent of the relevant ecological processes, and the behavioral and physiological bases of ecological interactions, for example.

In the case of *H. armigera*, the expected (or usual) frequency distribution of eggs across host species (as determined under ideal or controlled conditions) may be modified by differential host availability or accessibility, or by weather conditions disrupting the expected course of activity (as recorded in other species (Dobkin et al. 1987, Parmesan et al. 2000)). Some moths may locate suitable hosts after migration, for example, others not. And this may be influenced by wind patterns at the time (e.g., Gregg et al. 2001). Similarly, the suitability of each host species for sustaining successful larval development can be established, again under ideal conditions. In different localities, though, those levels of suitability may be altered, by stress to the plants through aridity or because of excessive standing water (White 1969, 1970a, b), for example. In other cases, these levels may be altered by the shortened lifespan of plants of one or another species, as affected by soil type or slope perspective and associated insolation levels (Dobkin et al. 1987), or even by the differential degrees to which the different plant species attract particular parasitoids or predators (see above).

Ecological processes, as they are governed by the physiology and behavior of individuals, are thus continually distorted, disturbed, or differentially determined as they run their course. And these processes will almost inevitably produce different patterns (host use, mortality, and so on) across localities. These disruptions are imposed on the lifeline and lifetime track of the individual organisms even though they have specific properties for making their way in the environment, surviving and reproducing. Inevitably, interactions of various types may take place with other organisms and alter the course of ecological processes.

In effect, the life of individual organisms is influenced stochastically in much the same way as the ball falling down the variable Galton board

(Chapter 8, Section 5). We can thus treat the fate of individuals statistically, in probabilistic or average terms, but this needs to be done relative to the life cycle of the species of interest and relative to the risks faced by individuals at different stages of their life cycle. Although stochastic influences intrude at virtually every step of an individual's life, not all need to be quantified for inclusion in ecological models (as indicated above). Bounded by their properties, individual organisms have to cope continually with stochastic environmental uncertainty, wherever they go, which defines their risk of survival and reproduction. This is where ecological understanding, developed through appropriately designed tests (the exploration phase of research in Chapter 8), is crucial, and around which the appropriate quantification can be made (see collared dove example, Chapter 8).

The end products of the activities of organisms represent the outcomes of numerous interactions between the determinism of their adaptive mechanisms and the stochastic influences of the environment. Quantitatively, the information has to be treated collectively in frequency distributions for analytical purposes, mainly for convenience. The consequence is that the stochastic information is essentially "lost" in such models (which would otherwise have to be even more data rich). The individual-based model developed to track a highly pathogenic avian influenza virus (Savill et al. 2006) provides a good example. The aim here was to explore the impact of prophylactic vaccination on the silent spread of the pathogen between caged flocks of domestic chickens, by tracking the infection status of each bird. Each individual bird was assigned a latent period, an asymptomatic and symptomatic period, a susceptibility, infectiousness rates for virus excreted in feces or breathed out, and an antibody titer (Savill et al. 2006). These values were derived empirically, but effectively represent the end values of particular processes as affected by various stochastic influences, although the various stochastic inputs are not mentioned in this case. In effect, the stochasticity is lost in the model, but there is benefit in the information being generalized and thus simplified.

So far, we have developed the autecological perspective of ecological interactions strictly in temporal terms. In practice, this was appropriate for the avian influenza infectivity example because the spatial movement is controlled artificially. In natural systems such simple situations are not common, and the movement of organisms must be incorporated into the model. The autecological approach to spatial dynamics is developed in next, in Section 3. We emphasize that the requirements, tolerances and complex adaptations of organisms, although variable to some extent, are constant and species wide (Chapter 5). In other words, the variation that is detectable in them is relatively minor relative to the constancy in these features across the distribution of the species. This provides the basis of what we have referred to above as "the determinism of adaptive mechanisms" and which is expressed within a stochastic environment. Spatial dynamics need to be considered from this perspective.

3 INDIVIDUAL ORGANISMS, ECOLOGICAL DYNAMICS AND ECOLOGICAL INTENSITY

3.1 Individuals and Movement—Defining the Problem

Section 2 covered the structure of the mechanistic processes by means of which organisms interact with their entire environment. Here, environment includes not only abiotic features but also biotic ones like vegetation, hosts, predators and the like (Chapter 5). These "structures" of organisms are the complex properties that underpin their interactions with the environment (Chapter 6), which is dynamic spatially and temporally, with considerable variability (Chapter 4). The range of stochastic influences that impinge on ecological systems was also covered, for these divert the deterministic course that might otherwise have been taken in the lifeline of the organism in question. Note that the stochastic intrusions that influence the ecology of organisms, that is their behavior, physiology, survival, reproduction and so on, are not confined to the extreme events that are usually discussed in the ecological literature (e.g., Parmesan et al. 2000). That is, the ecological background against which individual organisms operate daily is one of flux. The environment is dynamic, at different scales of resolution (Chapter 4), and stochastic intrusions are frequent, unpredictable and generally unrelated to population density. The latter cannot, therefore, provide a sound basis for ecological generalization. These intrusions need to be considered within the context of the environmental structure that is relevant to the species of interest.

A central feature by which organisms cope with such flux, both short and long term, is movement. This point was introduced in relation to the life cycle (Chapter 3) and relative to the distributional shifts of species as a result of climate change (Chapter 4). We therefore describe approaches to incorporating movement into general ecological models. Before doing that we illustrate, with examples, a feature of movement not covered so far. At all scales of resolution, spatial and temporal, organisms move, and it is this movement, as much as the more obvious invasions into relatively uniform and extensive agricultural landscapes (collared dove, Chapter 8, Section 6) and the seasonal migrations of some species (life cycles, Chapter 3, Section 3) that warrant consideration. In most ecological systems the movement is not so obvious (e.g., Keller et al. 2001, Ward 2005), but is nevertheless an integral aspect of everyday ecological dynamics, with short and long term consequences.

Why do organisms or their offspring continually shift spatially, with all the risks that such movement seems inevitably to bring? Demographic ecology maintains that biotic influences, mainly density dependent, are the primary causes of such movement (e.g., Muller-Landau et al. 2003, Guo et al. 2005), but evidence increasingly suggests that this, although it does happen (Brunzel 2002, Nelson et al. 2002, Hanski and Meyke 2005), is not the cause in many situations. For example, the numbers of carabids that disperse by flight are similar for both sparse and numerous species, which eliminates density as a cause (den Boer 1990). Immature blue crabs (*Callinectes sapidus*) leave localities in their seagrass habitat at higher rates when densities are lower (Etherington et al. 2003). Many other organisms are also known to leave localities that are favorable to their wellbeing and reproduction, such as the milkweed beetle *Tetraopes tetraophthalmus* (e.g., Davis 1981a, 1984, 1986a). Passerine birds tend to show nesting site fidelity, but many do shift territories for the new season. Mostly this is a response to nesting failure during the previous season, but this is not inevitably so (e.g., Sedgwick 2004, Middleton et al. 2006). Note that this consideration of movement does not draw at all on the relatively vast literature on insect pests, whose movement and invasive potential is well known, even in perennial systems of cultivation.

Other aspects of movement are also not inevitably related to density. The density of immigrants into habitat fragments is not necessarily related to the size of such fragments (and population density is so only in about 50% of cases (e.g., Bowers and Matter 1997, Bender et al. 1998, Debinski and Holt 2000)). These results "demonstrate the importance of considering autecological mechanisms, such as immigration behavior" when considering the ecology of fragmented biotopes or habitats (Bowman et al. 2002). That is, demographic processes cannot account for all species or situations in dealing with the organisms in such fragments. By contrast, autecological processes can do so because autecological research focuses explicitly on individual organisms within an environmental setting, and any residual effects caused by density can readily be related to the underlying primary processes in the ecology of the organisms concerned.

Even movement that may be considered trivial, in the sense of being small scale and without migratory purpose, can result in sufficient numbers of individuals moving into new situations or localities that they alter the subsequent dynamics (e.g., Keller et al. 2001). The recolonization of seasonally available resources is a case in point, sometimes across extensive ocean, as seen in many small species of bugs (cicadellids, delphacids and aphids) (Kisimoto 1976, Rosenberg and Magor 1987, Wada et al. 1987, Kisimoto and Rosenberg 1994, Kisimoto and Sogawa 1995) and even their parasitoids (Antolin and Strong 1987, Corbett and Rosenheim 1996). In some of these cases the insects set up viable populations that are exterminated each winter, as in the planthoppers on rice in east Asia and on seagrasses on the east coast of North America. Such movements presumably influence the ecology of the species at both ends of the migratory path but it is not clear whether the seasonally occupied sites yield individuals that move to yet other parts of the system.

The movements outlined above contribute to the ongoing redistribution of individuals. This is the constant "probing" of the environment by organisms that results in the colonization of newly established habitat, and thus contributes to the spatial dynamics of the species. Movement of this nature drove the colonization of new polders by carabid beetles

(Chapter 4, Subsection 3.2) and the invasion of the collared dove (Chapter 8, Section 6). It also drives the reinvasion of crops by bugs and newly established host populations by parasitoids (previous paragraph) and underpins the recent shifts in butterfly distribution in response to climate change (Parmesan et al. 2000, Parmesan 2003). Nevertheless, organisms often remain sedentary and give the impression of permanence, thus seeming to provide justification that ecology can be treated locally, as in territorial birds. However, the appropriate empirical evidence would be needed to justify such a stance, as even long-lived sessile organisms, typified by plants, have a stage in the life cycle that is motile and which ultimately contributes to distributional shifts of their species (Fig. 4.18). Recent molecular analyses of the parentage of Mediterranean maritime pines (*Pinus pinaster*) indicated that saplings established at distances as great as 120m from their seed parent (and with an average sapling dispersal distance of 40-60m) (Gonzalez-Martinez et al. 2006). This average contrasts with that found for seed dispersal distance, which was only 12.2m, and indicates greater survival rates at greater distances from source. This, again, is a manifestation of the way in which shifting environmental conditions are tracked. Thus, in sedentary organisms too, movement is as fundamental a process as in any other living organisms (see also Truong et al. 2006), even though it may not be immediately or obviously apparent. Some plants, however, persist, although they seem to have lost their biotic dispersal agents through extinction (Janzen and Martin 1982, Snow and Walter 2007, Hall and Walter 2013). The more motile the organism, the sharper the association with particular biotopes, even in dynamic environments.

Table 9.1 lists some recent measures of movement across a broad range of taxa. When movement is targeted in the field for quantification, it is almost inevitably found to be a feature of the subject organisms' ecology, even in many short-term studies (Table 9.1), and in taxa where it is not expected (e.g. juvenile wrasse (Overholtzer-McLeod 2005)). That ecological studies readily detect movement in the subject organisms contrasts strongly with the situation relative to density dependence, which is not always detectable, even though it is so avidly sought (Dempster 1983, den Boer 1987a, 1988, Stiling 1987, Cronin and Strong 1994, Murdoch 1994, Wolda et al. 1994, Chitty 1996). That organisms effectively track shifting habitat is thus almost inevitable. They also manage to cross areas of unfavorable habitat, to some extent at least (Bonte et al. 2004, Roth et al. 2006).

There is some probability, too, that dispersing individuals will settle in unfavorable localities. Stochastic influences thus intrude, again, in the lifeline of individual organisms, and this can have significant consequences for the local distribution of abundance of a species. Organisms cannot, however, always detect that the area or locality in which they find themselves is unsuitable, and their sensory physiology and developmental processes may be positively geared to localities or hosts that are not best for their growth and reproductive development (as in the carabids beetles discussed in Chapter 6, Section 5). Parasitoids oviposit into hosts from which

Table 9.1 A non-exhaustive survey of some relatively recent studies that have quantified movement (non-directed and non-migratory, in contrast to the traditional sense of a return migration) in various organisms. Movement is clearly a persistent feature of organisms, and organisms move substantial distances (relative to traditional definitions of local population) on a regular basis. Many of the studies reported below were of short duration relative to the average life cycle duration of the subject organisms, so the distances recorded are not lifetime maxima.

ORGANISM	*EXTENT OF MOVEMENT*	*SIZE*	*AUTHORITY*
GYMNOSPERMS: Maritime pine – *Pinus pinaster*	Average distance of sapling from seed parent = 40–60 m. Maximum >120 m.	30–40 m	(Gonzalez-Martinez et al. 2006)
ARANEAE: Wolf spider – *Pardosa monticola*	Relatively more activity/movement in low-density patches (poorer habitat with lower vegetation). Cross non-suitable habitat in process.	6 mm body length	(Bonte et al. 2004)
CRUSTACEA: Blue crab – *Callinectes sapidus*	Total loss rates of marked juveniles and colonization rates by unmarked juveniles were extremely high (range = 10-57 crabs/m^2/6 h for loss and 17–51 for colonization).	2–9 mm carapace width.	(Etherington et al. 2003)
FISH: Yellowhead and bluehead wrasse - *Halichoeres garnoti* & *Thalassoma bifasciatum*	Average distance for juveniles of both species was 7 m, and about 17% of individuals left study sites, with this not related to density.	Mature at about 35mm, maximum size of males 19–25 cm	(Overholtzer-McLeod 2005)
AMPHIBIA: Spotted salamander - *Ambystoma maculatum*	44 to 467 (mean = 145) m (n = 30). Maximum distance 2x as far as earlier studies suggested.	11–25 cm	(Le Galliard et al. 2005)
Toad – *Bufo fowleri*	Long-distance movement not rare, up to 34 km. 0.15% moved at least 15 km.	50 to 80 mm	(Smith and Green 2006)
MAMMALS: New Zealand sea lion – *Phocarctos hookeri*	Unappreciated high level of dispersal by territorial adult male NZ sea lions.*	2–3.3 m long and 300 – 450 kg (males) and 1.6 - 2 m and 160 kg (females)	(Robertson et al. 2006)
Dunnart – *Sminthopsis youngsoni*	Mean of 412m/night. 2.6km by a male and 0.8 by female.	9 g	(Haythornthwaite and Dickman 2006)
Wolf – *Canis lupus*	3.4% of recorded movements (n = 298 cases) > 500 km (straight line, with some up to 4000 km actual).	20–75 kg and 1.5–2 m total body length	(Linnell et al. 2005)

*This is said to be "... contrary to the current paradigm of otariid breeding behavior, which involves prolonged, uninterrupted male territoriality based on intense male-male competition and sustained fasting" (Robertson et al. 2006).

successful emergence is precluded (Carroll and Hoyt 1986), herbivorous insects oviposit onto unsuitable host plants (Thompson 1988) and the larvae of benthic sea urchins (*Holopneustes purpurascens*) preferentially settle, to metamorphose, on algal species that are not the best hosts for their development and subsequent reproductive output, and individuals generally move as adults onto more suitable algal species (Williamson et al. 2004).

Why do organisms not simply remain, then, in an environment that is suitable to them, even optimal, especially given the risks associated with dispersal? The level of risk is indicated, for example, by only an estimated 0.6% of autumn migrants of the bird cherry-oat aphid finding hosts in southern Scotland (Ward et al. 1998). Such movement seems certainly to be associated with the risks brought by the dynamic nature of the environment, for favorable sites frequently become unfavorable and *vice versa*. Movement of individuals would thus contribute to the persistence of lineages. Empirical support is found in the ground beetles and bush crickets that vary in the degree of their wingedness. Macropterous individuals, capable of full flight, are associated with more ephemeral habitats and localities whereas the unflighted brachypterous ones are found in more persistent habitats, a relationship that also holds, to some extent, across species with different flight abilities (Darlington 1943, Lindroth 1949, Roff 1990, Simmons and Thomas 2004).

A central task for ecologists is therefore the quantification and mathematical representation of the structures and spatio-temporal dynamics outlined so far in this chapter, for these contribute to the dynamic patterns in the local distribution of abundance across (and even beyond) the distribution range of species. Mathematical models of ecological systems, to be realistic, must incorporate the ongoing movement and redistribution of organisms and also the changes to the environment, whether those changes are brought about through the development of vegetation or under the influence of climatic changes (or both, simultaneously).

3.2 Environmental Dynamics—Spatial and Temporal Variation

Historically, ecologists did grapple with ecological systems and ecological theory at much broader scales than those delimited conceptually by local concepts like population and community. The geographical perspective, developed early by the plant geographers (Cittadino 1990), was kept alive by the contributions of Gleason (1926, 1939), Gause (1932) and Bodenheimer (1958), among others (McIntosh 1985). This thrust has survived in the form of gradient analysis (Whittaker 1953, Whittaker 1967, Drury and Nisbet 1973), an approach that is expanded in this and subsequent subsections.

Gause (1932) accommodated the spatial dimension in ecology by plotting (or ordinating) field observations of the densities of various locust species against particular environmental variables (including temperature, relative humidity and evaporation rates), obtaining curves of similar shape across the species (Fig. 9.2). The curves for each species were distributed differently from one another across the different ecological variables and, unsurprisingly, he also showed that locust densities varied independently by species across space in the field, as related to habitat conditions. He accepted that local densities of each species result from different population growth rates as influenced by the microclimatic conditions within each habitat. Also, in his growth experiments with yeast (Fig. 9.3), for example, he showed that the density ultimately achieved varied with temperature (Gause 1932). This density he referred to as the asymptotic or saturating population density. He even conceptualized this approach through his plot of the asymptotic population densities achieved by individuals of a species when exposed to different environmental conditions, for he found this repeatable across species in the laboratory and field (Fig. 9.4) (Gause 1932).

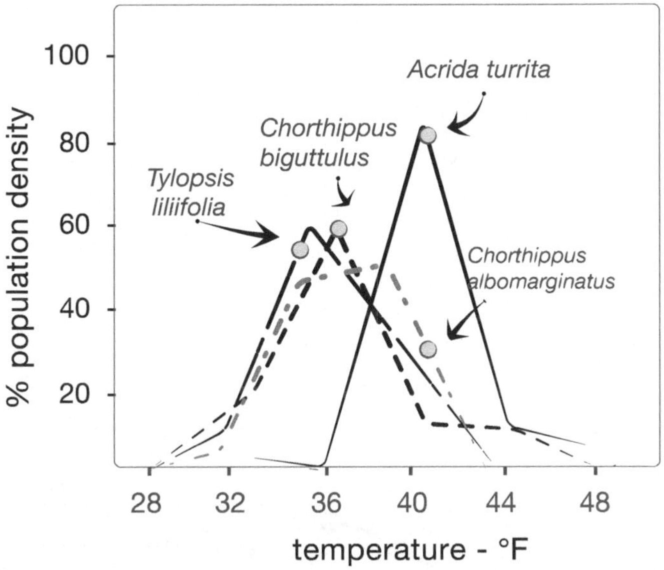

Figure 9.2 Ordination of densities of various locust species against environmental temperature. Redrawn from Gause (1932).

Following his collaboration with Raymond Pearl, however, Gause (1934) shifted the emphasis of his interpretation rather significantly, but without any justification. He now attributed the different population densities achieved at different temperatures, with specific reference to the data on yeast cells, to different values of carrying capacity (*K*) in the logistic equation, a development that has been followed consistently in demographic theory (e.g., Guo et al. 2005) (Fig. 9.5). He did this despite the knowledge that the production rate of alcohol, a waste product, and the sensitivity of yeast cells to alcohol toxicity varied with temperature (Richards 1928). This latter information, on the individual-level mechanistic

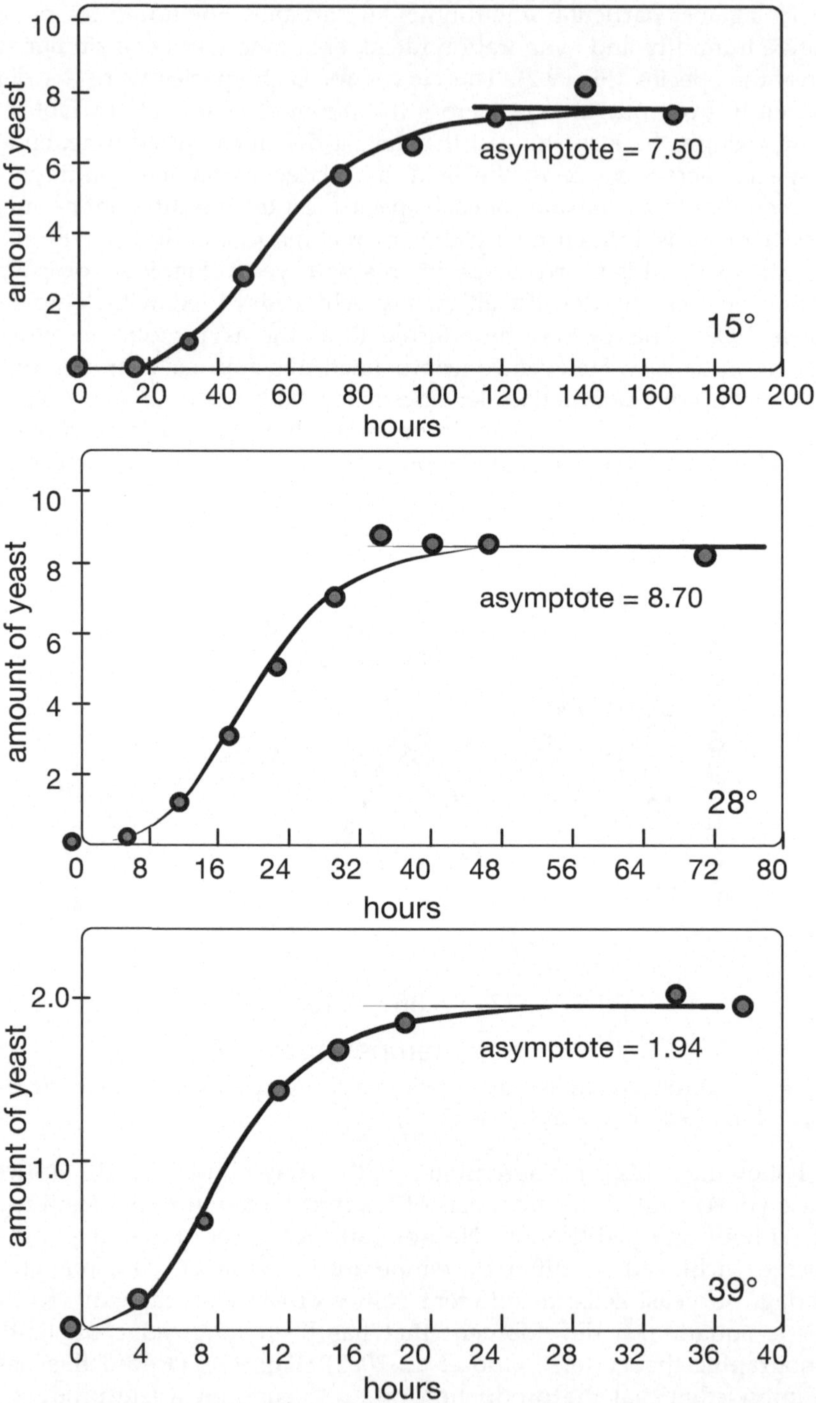

Figure 9.3 Values of *K* varying with temperature in yeast cultures. Redrawn from Gause (1932).

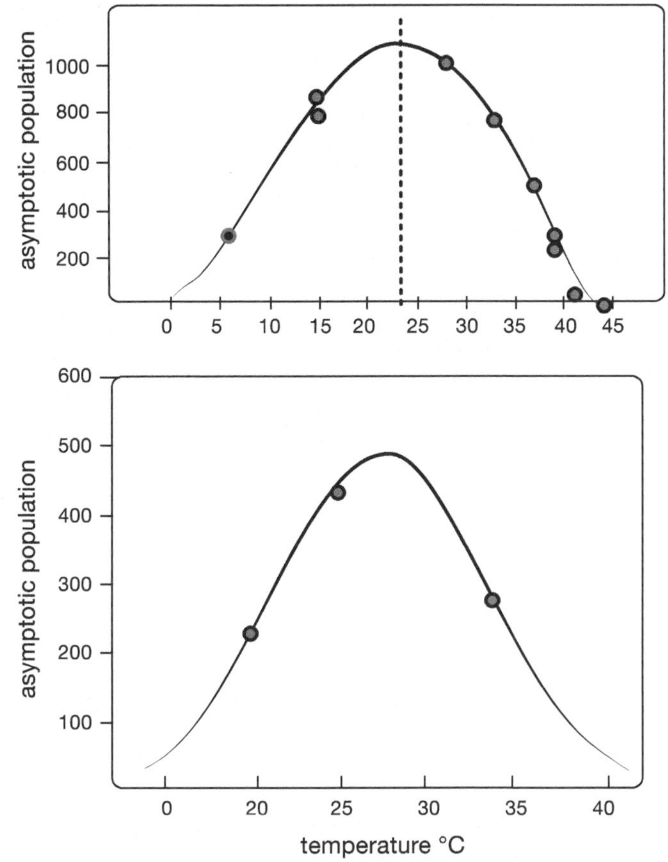

Figure 9.4 Influence of temperature on the asymptotic population level of yeast cells (**top**) and a water flea (**bottom**). Redrawn from Gause (1932).

interaction with the environment, demonstrates that population densities are, instead, determined by environmental quality, so the logistic equation (as ultimately applied to this system by Gause (1934)) misrepresents the system as suffering the gradual exhaustion of a resource and a consequent asymptotic approach to *K*. Instead, the response of the yeast cells follows an optimum curve and expresses the differences in the vigor of the individuals as a consequence of their exposure to various conditions, each of which provides a different degree of environmental match to the physiological requirements and tolerances of the organisms. The vigor of the individuals is greatest under optimal temperature conditions and, conversely, their sensitivity to alcohol the least. From the perspective of the logistic equation, therefore, *K* appeared to depend on temperature, rather than on *r*. Despite the unjustified conceptual leap from environmental influences on individuals to density compensation at the population level, the entire approach was applied, unchanged, to interpretations of competition between species (see Gause (1934) and almost any recent ecology text (Chapter 7)).

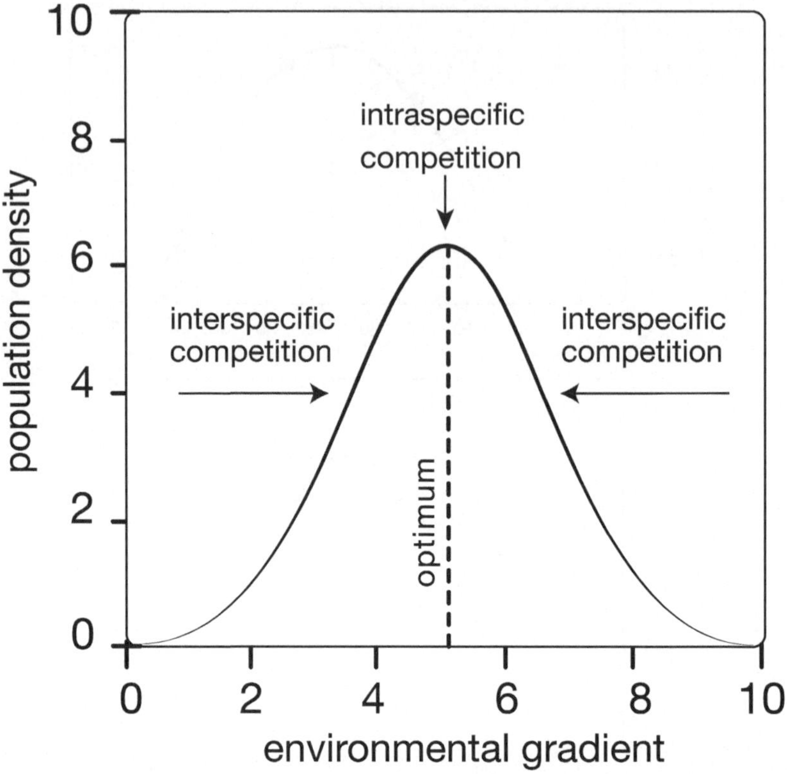

Figure 9.5 Conceptual model relating metapopulation theory to species distribution through environmental gradients. The primary ecological process is density dependence, and the optimum environment, maximum habitat carrying capacity and population density are located at the center of the species' range (in this case), which is indicated by the vertical dashed line. Reproduced from Guo (2005), with permission.

The reinterpretation of Gause's results, above, to emphasize the influence of environmental quality (relative to the species of interest), rather than carrying capacity, is supported by the observations on sea plantain plants (*Plantago maritima*) growing along a 60 m environmental gradient that ran inland from the Baltic coast in Sweden (Fig. 9.6). Near the waterfront, population turnover was high because rates of fertility, seed set and seedling establishment were high (Jerling 1985). Conversely, life expectancy of sea plantain there was relatively low because of frequent flooding. The turnover of plant individuals decreased away from the waterfront, and so did fertility, seed set and seedling establishment. Plants were also differentiated along this gradient in terms of genetically based variation in seed size, germination pattern and phenology (Jerling 1988). We return to this example as we extend interpretation of environmental dynamics, the way in which organisms respond to them and how to conduct a gradient analysis.

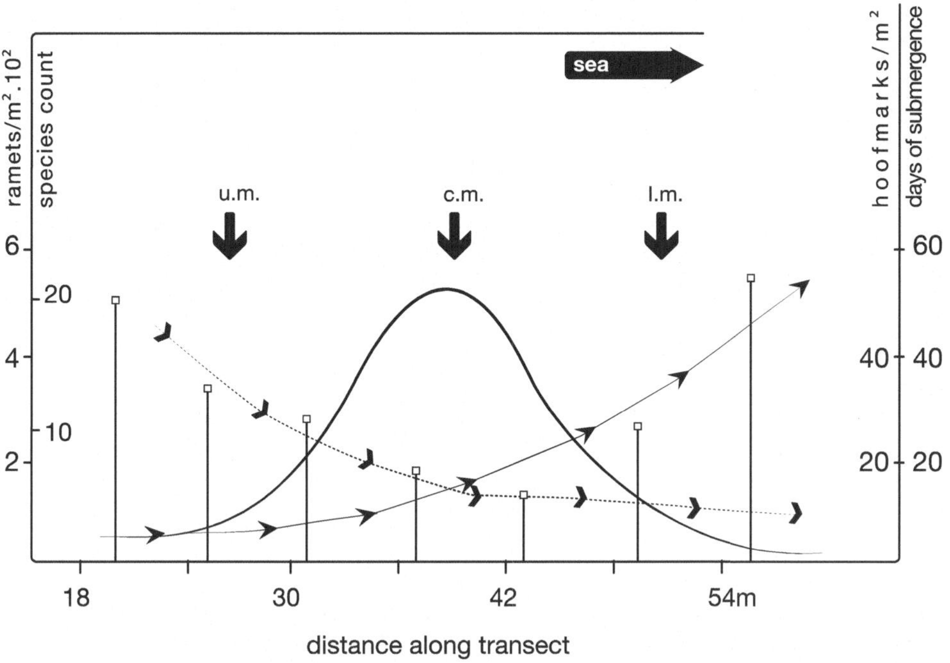

Figure 9.6 The environmental gradient along which sea plantain (*Plantago maritima*) ecology was investigated, where ❯ indicates the number of plant species, □ the relative grazing intensity (measured indirectly by cattle hoof marks), and ➤ the average number of days with submersion of vegetation. The solid curve represents the average sea plantain density, and "u.m." specifies the upper meadow, "c.m." the central meadow, and "l.m." the lower meadow, at the water's edge. Reproduced from Jerling (1988), with permission.

The central idea of ecological gradient analysis is that the performance of plant or animal individuals along an environmental gradient varies, perhaps starting with relatively poor performance at the margin of the geographical distribution of the species (or local area under investigation), gradually improving to some modal or maximal value, after which conditions deteriorate towards the opposite margin (whether local or geographic), with poorer performances as a consequence. The shape of the curve is not necessarily strongly correlated with resource availability although it is clearly not irrelevant (Alcala and Dominguez 2003). Plants provide typical examples (Fig. 9.7) but similar relationships are evident even in large mammals, as expressed in recruitment rates of deer against rainfall (Ginnett and Young 2000). Such changes are not necessarily symmetrical across the distribution of a species (Sagarin 2006, Sagarin et al. 2006), and different factors may influence the organisms' ecology in different parts of its distribution (see Section 2). Similar patterns are evident across altitudinal gradients, as in European alpine plants (Fig. 9.8; and see Scottish oats example, below) (Hengeveld 1989a).

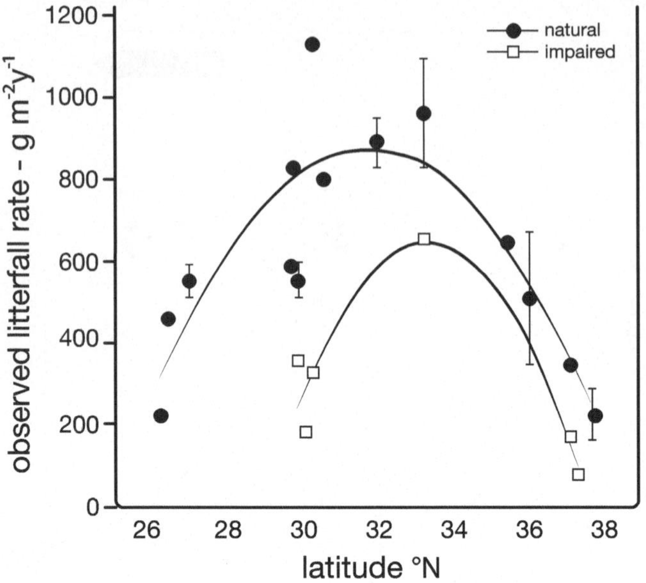

Figure 9.7 Total litterfall rates measured (as a surrogate for productivity and population density) in bald cypress swamps dominated by *Taxodium distichum* across latitude in North America. Swamps with natural hydrology (●) and impaired hydrology (□) are plotted separately. Reproduced, with permission, from Middleton and McKee (2004).

The examples above are of "direct" gradient analysis as opposed to "indirect" analyses (Whittaker 1967). In the former, the gradient in habitat conditions is roughly continuous across space, and so are the variables by which we assess the responses of the organisms concerned. The changes in realized response values are thus directly observable as we move across space and up mountains. In indirect gradient analysis, the realized response values are scattered in space and are thus independent of one another (Garrido et al. 2003). These fragments can, however, be arranged to form a regular sequence, by means of ordination techniques (e.g., Whittaker 1978, Gauch 1982). An optimum curve of performance can be derived, and from this a continuous gradient in environmental conditions can be inferred and tested, as explained below.

The different densities along environmental gradients, both direct and indirect, can be explained entirely by the quality or condition of the local habitat. Any numerically significant interchange of individuals, their new offspring or reproductive propagules that does take place across various sites or localities results in differences in local residence times, with these times being proportional to the quality of conditions within the localities (e.g., wolf spiders in inferior habitat (Bonte et al. 2004)). Differences in local densities thus reflect differences in environmental conditions for the individuals in question, in combination with their mobility, even if mobility is restricted to their time as propagules.

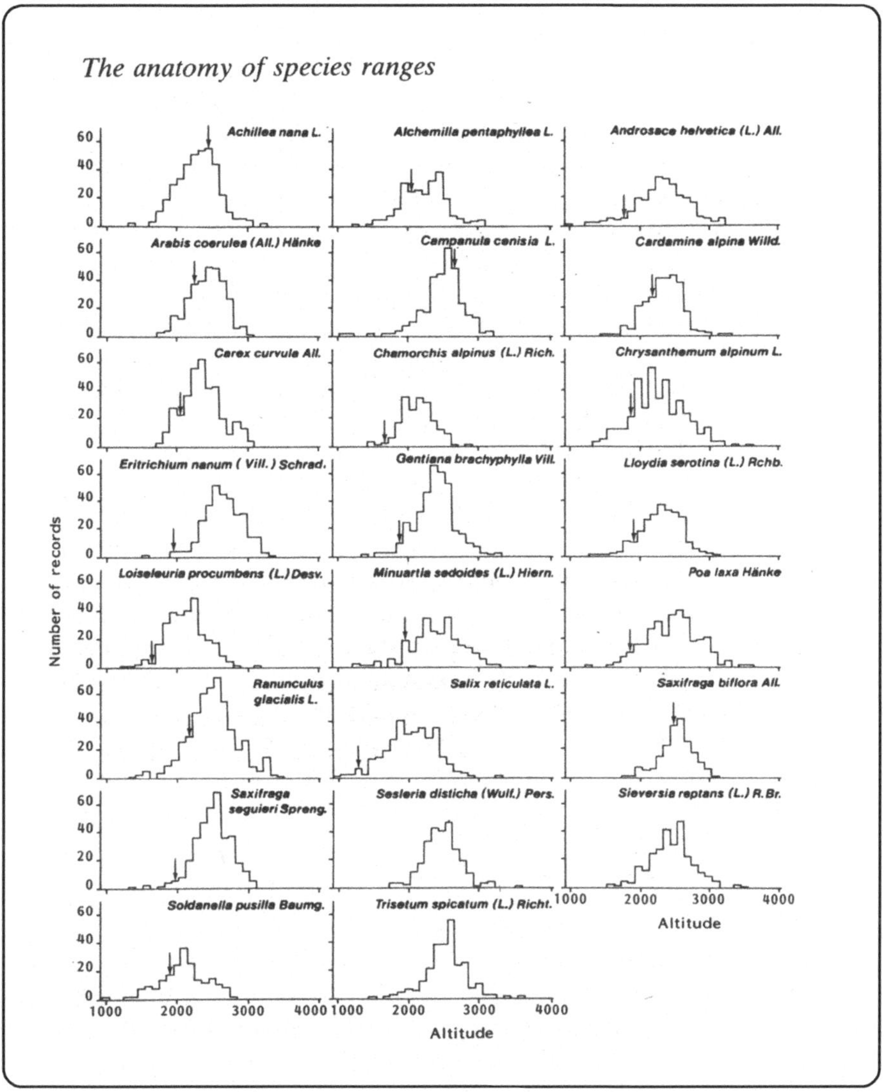

Figure 9.8 Altitudinal gradient for 23 plant species in the European Alps. The number of herbarium records has been plotted against altitude. Reproduced from Hengeveld (1989a), with permission.

Note that we have not dealt with environmental gradients as related to so-called community effects. These tend to focus on the strength of demographic interactions (competition and predation) or local (community) diversity and, although informative in many respects (e.g., the investigation by Dahlhoff et al. (2001) of the physiological status of individuals relative to a range of variables), the consequences for the survival and reproductive output of the organisms need to be included, for otherwise the observed community pattern is difficult to interpret in terms of all the other ecological processes that impact on the individuals at the same time. An autecological approach is required.

Organisms themselves also contribute substantially to the environmental dynamic, and this is usually conceptualized in terms of ecological succession. This is an area fraught with conceptual and practical difficulties, and general models have tended not to stand scrutiny, from Clements' (1916) organismic concept of vegetation succession, through deterministic models with various combinations of competition and facilitation (Connell and Slatyer 1977, McIntosh 1999) to the Markovian models (Horn 1975, 1981, van Hulst 1979) that tried to evade the problems of determinism, but also failed for their ahistorical and non-spatial structure (Facelli and Pickett 1990). Some success, albeit limited, seems to have been achieved by taking into account what has been referred to as the autecological features of a few of the most conspicuous species in the area (Billings 1952, Noble and Slatyer 1980, Stockey and Hunt 1994, Kammesheidt 2000). Again, achievements have been only modest, probably because the focus tends to fix on biotic interactions and how they are influenced by the abiotic (e.g., Pugnaire and Luque 2001), rather than on autecological aspects per se, with its focus on individual organisms and the multiple axes of differentiation of their species-specific properties. Also, expectations of our potential to predict the future states of ecological systems have probably been over-optimistic, perhaps because stochastic influences have been downplayed so much. A firmer focus on understanding is perhaps warranted, with predictive abilities to be determined after that. Such an approach may help to steer research on succession away from the over-generalization that seems to have plagued it.

Autecologically, successional systems can be seen as stochastically influenced aggregations of individuals from various species, with each having its own species-specific ecological traits, and with each species responding to a subset of the features within the locally changing environment (Walter and Hengeveld 2000). Ecological systems, seen in this way, are essentially transitory or fluid in nature across space and through time. Whilst change in such systems takes place with time, the organisms of different species may influence their environment by their presence, shading, resource utilization, and so on, with such consequences being related to the species' properties and the local context. Organisms of a particular species may even alter their local environment so much that it becomes unsuitable for individuals of their own species, perhaps the seedling stage, or for organisms of yet other species. Legumes, for example, have allowed other plants to flourish in soils once much poorer in nutrients (Vitousek et al. 1987) and moss growth within a forest has raised the water table enough to provide an environment that eliminates trees (Heinselman 1963, Drury and Nisbet 1973), so shade levels, substrate availability, edaphic factors, allelochemicals, host availability, nutrient levels, and so on are liable to change. Numerous axes of differentiation may be affected, directly and obviously, or indirectly and subtly.

Simultaneously, the altered conditions will allow organisms of yet other species to establish. Gradually, the composition of the local flora and fauna changes character, shifting with time to very different aspect, physiognomy and species composition. These are not definable stages along a continuum (Schoenly and Reid 1987, 1989) even though they may sometimes be discernible and can be categorized (although this is an exceptionally grey area, and not explanatory), they are not intrinsically stable ecological units that resist perturbation or invasion, but are dynamic and open (Drury and Nisbet 1973). They are aggregations of species that can be defined only subjectively, usually with reference to their most conspicuous elements. Further, the end point is not predetermined, and even "reverse succession" is not unusual (Drury and Nisbet 1973). Stochastic influences frequently intrude, in terms of which propagules reach the site and establish (Glenn-Lewin 1980) and the direction and nature of environmental change (Walker 1970, Drury and Nisbet 1973, Johnson and Mayeux 1992, Tanner et al. 1996, Fernandez-Gimenez and Allen-Diaz 1999).

Ecological systems may appear to stabilize to a certain extent, despite the turnover in individuals and species, and this will depend on the dynamics of the external conditions and the organisms themselves. Thus, given local conditions and the dynamics imposed on them by cyclical changes and periodic disturbances, a temporally more or less determinate state is reached. This state is established dynamically and influenced stochastically by physico-chemical conditions, including moisture levels and temperature. Biotic features also play a role, but principally in terms of the lifespan and species-specific characteristics of the plants (Gleason 1926, 1927, Drury and Nisbet 1973, Noble and Slatyer 1980). At broader temporal scales of resolution, disturbances from fire or storms, which occur with a particular frequency distribution, also exert an influence (e.g., Tanner et al. 1996). Changes in climate exert influences at yet broader scales of temporal resolution.

The importance of the spatial dynamics of organisms in relation to the temporal dynamics of a locality is also clear. For example, the arrival of individuals of a species not already in the local area under consideration, when climate or the local vegetation changes, depends on them coming from localities that lie elsewhere (see below) or, for example, they may have been there inconspicuously for protracted periods (decades or even centuries) as viable seeds (Baskin and Baskin 2006). The integration of both spatial and temporal elements into one comprehensive spatio-temporally dynamic process is thus an inevitable theoretical progression for autecology, but has yet to be achieved.

Ecological dynamics are consequently seen, in autecology, as progressions of open systems, sometimes with the appearance of steady states, defined in terms of the rate of organisms coming and going, developing, reproducing, and dying. Locally, ongoing change to species composition is an inevitable consequence. Two conceptual advances follow. One, to see ecological systems as distinct units of ecological integration, whether populations, communities or climax stages, misrepresents the dynamic processes that support and change any such steady states. Two, the open-ended and dynamic nature of ecological systems raises the issue of scale, which was dealt with in Chapter 5 in terms of the dynamics of individuals within the geographical distribution of the species. In this respect demographic ecology draws a distinction between local areas (the area of ecological interest, for whatever reason that may be) and areas external to that, which are referred to in terms of their providing the regional pool of species (the potential colonizers of the local area of interest). This bipartite division of the environment fails to accommodate the inevitably enormous differential in geographical distribution of all the species involved, and usually this is an area that incorporates quite different biogeographic source areas (as in Andalusian vegetation in Subsection 4.3). Incorporation of this differential is crucial to understanding diversity from both the ecological and evolutionary perspective, but dealing with this in terms of ecological principles is not readily achieved except autecologically (Walter and Hengeveld 2000).

Individual organisms thus constitute the units of ecological interaction, which does not deny that density-related influences are excluded. But it is typical of ecological studies that measurements of predation, competition and other such biotic processes in relation to density are extremely inconsistent in detecting the desired density dependent signal (Chapter 8, Subsection 3.2), a situation that contrasts strongly with that in which movement of organisms is quantified (Subsection 3.1), for example. The implication, simply, is that populations, communities and successional stages (particularly the climax), cannot be considered distinct ecological units in their own right. Autecological consideration of the environmental context of individual organisms introduces the mechanisms and processes that are crucial to their survival and reproduction, and thus the dimensions and features that are needed for ecological understanding.

3.3 Ecological Intensity (Relative to Concepts of Population)

The previous chapters and sections have shown that organisms are adapted to particular conditions, with some degree of tolerance or ecological plasticity. Indeed, the extent of the environmental gradient inhabited by individuals of a particular species represents the degree of environmental variation they are likely able to tolerate. A broad gradient is frequently associated with a diversity of "biotypes". Although different selection pressures undoubtedly prevail in different parts of the species' distribution (or across the gradient they inhabit), and different allele frequencies and indicators of performance are frequently detectable (Volis et al. 2002, Petru et al. 2006), the limits to gene pools (species) need to be considered carefully in this context. For example, gradient analysis on *Littorina* periwinkles has often been conducted, and studies may include more than one "type", with the results often unusual compared with those involving only one species (e.g., Pardo and Johnson 2005). *Littorina saxatilis sensu lato* remains unresolved in terms of defining species limits (Tatarenkov and Johannesson 1998, 1999), but may well comprise a species complex. Molecular results indicate some gene flow across "types" but it is limited and it does not obliterate their differential properties and ecological requirements. Although these differences have been interpreted as representing a case of divergence in progress, this seems unlikely considering the same situation prevails across a broad front, including different islands and other widely separated localities. We have no reason to believe that evolutionary processes act in such persistent and deterministic fashion across extensive geographical areas, despite claims now of "parallel speciation" (e.g., Rundle and Nosil 2005, Schluter 2009) (and see the comments on ecological speciation in Chapter 8, Subsection 3.1).

Different ecological variables almost inevitably predominate as one moves across the continuous distribution of a particular species, or across the different "fragments" that make up any disjunct distribution of a particular species (for which any gradient analysis becomes indirect because it relies on scattered fragments for its input). The usual consequence is that the densities of the species in question differ across such sites and, frequently enough, so does the structure of what is usually referred to as the "population". These measures, though they are frequently used to infer process, are merely composite measures that allow little inference about the underlying process (although this is a common approach in demographic population modeling). They are the end products of numerous processes that affect the survival rates of individuals, their reproductive success and movement, as demonstrated by the sparrowhawk data in Fig. 7.3. (Incidentally, reproductive success is often referred to as fitness, although fitness is defined in demographic terms as the expected contribution of an individual to future generations (Stearns 1992)), which is seen to be homologous to the definition of intrinsic rate of increase for a population despite the fact that "Fitness, like the intrinsic rate of increase for a population (r), can almost never be measured directly" (Mitchell and Powell 2003).

An understanding of these underlying processes gives us the information necessary to interpret changing local abundance, as seen in the sea plantain example above (Fig. 9.6). The degree of local aggregation of these plants, their seed set, and number of biotopes occupied represent the "population" attributes of this species. Such measures can also be used as the basis for predicting extinction risk, survival probability, and so on. When such attributes are represented geographically (or relative to biotope, if appropriate) they reflect the dynamic "ecological intensity" of the species within a particular area or locality and this can be measured and presented at different spatial scales (Hengeveld and Haeck 1982). This measure of ecological representation differs fundamentally from that provided by the concept of the local population in demographic ecology, for this latter is spatially confined. Table 9.2 contrasts the autecological concept of ecological intensity (with measures including abundance, density, number of biotopes occupied, amount of movement, mortality, offspring production, temporal persistence, and so on) with the various (demographic) uses of the term population.

Table 9.2 Multiple uses of the term "population", the differences and relationships between these concepts, and their contrast with the Mendelian population and the autecological measure of ecological intensity.

CONCEPT	*COMMENT*
Ecological population	The ecological population is essentially the local population. Almost invariably it is arbitrarily defined by the observer, to include a number of organisms of the same species, and which form a "more or less frequently interconnected group" that is separated "more or less clearly" from other groups of the species (Solomon 1949). See also Turchin (2003).
Metapopulation	This is a population of local populations. They have the potential to exchange individuals, a process that ultimately balances the metapopulation. Developed for species with fragmented distributions (Hanski 2004).
Locally adapted population	Has much in common with the ecological population. Any differences in the features of individuals among such populations of conspecific organisms, however slight, are seen to be significant in an evolutionary sense, although this is debatable (Walter 1999).
Mendelian population (gene pool)	Defined in terms of positive assortative mating (or who mates with whom) (see Waples and Gaggiotti 2006). All such populations of which the constituent individuals share in a common specific-mate recognition system (or, more broadly, fertilization mechanism) constitute a species (= species gene pool).
Ecological intensity	Dynamic spatio-temporal representation of a species across its distribution over time, with measures including abundance, density, number of biotopes occupied, amount of movement, mortality, offspring production, temporal persistence, and so on.

In principle, the ecological intensity of a species varies according to the suitability of the prevailing conditions for the individuals of that species, because the individuals respond somewhat directly to their immediate environmental conditions. Thus, low densities generally result from a poor matching of the requirements of the individuals of a species with the prevailing conditions in space and time, whether through low survival, unsuccessful reproduction or low rates of arrival at the locality. Sometimes, however, low densities do not accurately reflect the suitability of the local conditions, because of time lags or when secondary influences such as predation or human exploitation reduce densities of organisms temporarily and independently of ecological conditions. High mobility generally reduces such time lags so that the local ecological intensities of highly motile organisms tend to reflect differences in the suitability of local living conditions more strongly than do the ecological intensities of less motile species. Time lags can also be reduced substantially when even small refuges sustain relatively few individuals, as seen in the differential repopulation of New Zealand by fur seals and sea lions, after human persecution was stopped (Taylor et al. 1995, Lalas and Bradshaw 2001, 2003, McConkey et al. 2002).

What is significantly different in this autecological view, relative to demographic ecology theory, is the emphasis on (i) the "external" environmental factors that influence local ecological situations, (ii) the species-specificity of influences and responses (as well as the differences across species) in the environmental axes of differentiation that affect them, and (iii) these environmental variables as they impinge on individual organisms as they trace their life cycle and shift spatially relative to the conditions that prevail. Generally, these variables are not driven internally by the density dependent processes that provide the dynamic for the logistic equation and the recent extensions of demographic theory, as outlined below relative to developments in metapopulation theory in particular.

Environmental conditions change all the time, and continually disrupt any deterministic density-related processes that may be initiated. For example, in Gause's (1934) yeast growth experiments, temperature (or a correlated variable) was the spatially scattered external determining factor. In the sea plantain system in Sweden, described above, the causal pathway in determining individual densities in localities is somewhat more tangible, and illustrates better the subtleties involved in ecology. The physiological processes of the plants are clearly exposed directly to an environmental gradient, which is made up by a complex of abiotic factors, including salinity, edaphic factors, inundation frequency and duration of inundation. Thus, the statistical attributes of the local "populations", which derive from the ecological performance of the constituent individuals (and which is often referred to as their "fitness"), are determined primarily by processes that are driven from outside of the locality specified. Plant density does, however, exert an ecological influence, through shading, but is not the sole driver (nor the primary driver) of the system. Organisms thus live and reproduce under circumstances of continually varying risk, imposed externally, so ecological analysis is more realistically focused on the analysis of such risks and their environmental context. These issues are covered in Section 4.

Although local ecology is influenced by external processes that simultaneously vary at different scales of resolution, as detailed above, the local dynamics are not necessarily consistent across localities, so neither the patterns nor the processes can be generalized spatially. This perspective is entirely different from that which sees populations driven internally by local density and which has been extended now to models of ecological speciation through invoking an empty niche as the primary influence of any perceived divergence (e.g., Nosil et al. 2003). Such external influences and their variable impact render effective generalization difficult, and condemn any typological treatment of population dynamics. Spatially differential external influences were recognized early as fundamental to theory development (e.g., Dobkin et al. 1987, Hengeveld 1989b), have subsequently been increasingly demonstrated empirically (e.g., Garrido et al. 2003), and are now generally acknowledged by ecologists. Nevertheless, the relevant development of ecological theory has not taken place. Even the recent *ad hoc* adjustments to demographic ecology that have been set in place to try and accommodate this deficiency are inadequate, for the problem lies with the core of the theory. The latter sees all primary influences as density related, even under conditions of climate change (see Guo et al. 2005). This premise, appealing as it is on the surface, represents little that is consistent with what is observed in ecological systems. Yet it pervades ecological theorizing. It directs the questions asked, data collected and interpretations offered, as indicated in Chapter 2. Ultimately, the organisms and their life cycle are pushed to the periphery of ecological interpretation, rather than being central to it (and the significance of which is illustrated in Chapters 3 to 6). This has the following significant consequences for contemporary developments in demographic theory.

First is that spatially local "populations" are treated as discrete entities that can be "rescued" by immigration when their numbers decline (Lawton 1993b, Walde 1994, Hanski 1997, Guo et al. 2005, Smith and Green 2006). The problem with such a view is that populations are treated typologically, as spatially independent ecological entities that have an existence governed principally by internal density-driven demographic processes. Consequently, the reason for the local population dwindling is essentially ignored, despite warnings, and especially so in patch occupancy models where stochastic influences are generally assumed to be responsible (see Hanski 1999). If conditions locally have become unsuitable because of changes in the external factors that impose on local conditions, then such a "population" cannot in general be rescued by individuals from elsewhere, for conditions in that locality are no longer suitable for them. Similarly, "sink" as a population-level concept is typological and has no relevance to individual organisms. For individuals, it is their interaction with the environment that is significant, as well as the outcome of the interaction. Surprisingly, biotope quality has typically been left out of consideration in the rescue processes of classical metapopulation theory, although moves have been made to incorporate successional changes and their differential suitability to the species under consideration (e.g., Wilcox et al. 2006).

Metapopulation theory was originally developed to deal specifically with species that have fragmented distributions. Such metapopulations are seen within a hierarchy of scales, between local and geographical, and defined on the basis of individuals moving infrequently from one place or population to another (Hanski and Gilpin 1991, Hanski 2004). It is thus consistent with the views of evolutionary ecologists on the prevalence and inevitability of local adaptation (e.g., Hanski et al. 2004, Saccheri and Hanski 2006), but suffers from the same arbitrariness of definition. What is a local population, what a metapopulation and what should be treated geographically? Published debate reflects the dilemma well enough (e.g., Dupre and Ehrlen 2002, Freckleton and Watkinson 2002, 2003, Ehrlen and Eriksson 2003, Baguette 2004, Hanski 2004, Shreeve et al. 2004) and theory moves imperceptibly from one scale to the other, meanwhile still invoking "rescue" as a basic process (e.g., Guo et al. 2005), even in situations in which extensive areas of direct environmental gradients, as opposed to indirect (or fragmented) gradients, have been included. In direct gradients, ecological processes are continuous. Fragments cannot be specified and movement is a typical part of the process, so "rescue" is again an inappropriate conceptualisation.

Evolutionary processes have been invoked to help explain the rescue process across direct gradients. Central populations are seen to be in better habitat and are therefore at higher density and should have more genetic diversity. They can thus, it is said, provide gene flow into less fit marginal populations (Guo et al. 2005). However, the concepts of habitat tracking and ecological intensity suggest that ecological systems actually operate in the opposite way to that proposed by the rescue effect. Differences in density would be enhanced when individuals move among locations with different environmental conditions, staying longest in those locations with conditions best matching their requirements. If the "rescue" process operated as specified by theory, differences in local conditions would effectively become blurred, because they would have a sameness to them, instead of the differences and ongoing changes depicted through ecological intensity (Chapter 4). Spatial spreading of risk, which assumes a great number of independent and unidentified factors are operating on populations (den Boer 1968), if it reflected ecological reality, would also yield a similar outcome to metapopulation rescue. Here again, a statistical cancelling influence is invoked such that a regular density distribution is expected along environmental gradients because of the summation of many ecological factors, as applied also to metapopulation (Hanski 1991) and regional (geographical) scales (Brown 1984). However, the real distributions of abundances are humped (see Fig. 9.2), which suggests that only a limited number of identifiable factors are responsible for the distribution of abundance, rather than a large number of unidentifiable ones. Colonization of local areas is therefore more likely to be related to the nature of the environment than to the status (or fate) of populations or even to the balance across the populations that are seen to make up the metapopulation.

The logistic model of population dynamics is essentially static, for it implicitly excludes any spatial component. Diffusion components have been added to the logistic to incorporate movement into ecological theory and rectify the problem, but this effort to introduce spatial dynamics simply imports the problems inherent in the logistic into the proposed solution.

Conceptually, therefore, the principles underlying metapopulation dynamics theory are thus also static, for they basically follow the logistic equation (Hanski 1991, Hanski and Gilpin 1991), but at the level of populations (as typified by the "rescue effect") rather than individuals. A further problem with the logistic is that measures of ecological intensity, such as the density of individuals (*N*) or their intrinsic rate of reproduction (*r*, which is an average, not a statistical distribution), cannot be realistically isolated from their ecological context and then be treated as the cause or consequence of the other, as is inherent in the structure of the logistic equation (and thus, by extension, holds also for metapopulation theory) (Chapter 7).

Under the variable conditions that are typical of ecological systems, the parameters of the logistic can be estimated only retrospectively. They thus lose all predictive power, which helps explain why the logistic is typically applied only to data from experimental arenas, or is used in theory. It is clearly inappropriate to Jerling's (1985) sea plantain data. More specifically, no direct mechanistic link has ever been specified between individual, climatic influence, population density and carrying capacity, for example. This is true even of the most recent models (e.g. Guo et al. 2005, Lande et al. 2006), which thus tend to be descriptive rather than mechanistic. The sea plantain data also illustrate the real difficulty in defining or delineating a population when mechanistic considerations are introduced. The concept of population as a supposed unit of integration in ecological theory thus represents an awkward diversion because the relevant processes can be analyzed and explained directly only in terms of the individuals involved and their interactions with the environmental axes of differentiation that are significant to them. Indeed, the distinction between factors and processes that are supposedly internal and external to the population thus becomes irrelevant.

All group attributes that make up ecological intensity, including population density, are simply the collective outcome of processes that operate at the level of individual organisms. The parameter values that ultimately determine densities in the logistic equation are interpreted as if set by conditions that remain constant from the start of the process onwards. The course of population change is thus seen to be predetermined by those conditions and parameters. In the sea plantain example, by contrast, density is determined by factors that vary temporally, and which thus affect the plants at varying intensities through their lifetime. These processes act continuously and change dynamically, to the extent that nothing is predetermined. This latter approach is thus similar to that taken in the tomato and sweet corn studies outlined in Chapter 3, and developed further in Section 5.

Gause's (1932) approach is spatially static (Hengeveld 1990, p. 196), because each point of the optimum distribution is considered to represent the level *K* set locally by the carrying capacity that regulates population growth at that point in the gradient (Fig. 9.4). The values of *K* are therefore determined by some factor (ultimately resource availability, but through density dependent competition) varying in space, but not in time, thus forming the direct or indirect gradient of habitat conditions. Solomon (1949) referred to the process that establishes the various levels of *K* as "ecological determination". When a population drops below this level, or overshoots it, regulatory processes are thought to set in to dampen the deviation and regulate the system. Thus, in this static interpretation, two types of processes are recognized—(i) determination, which establishes the equilibrium levels at the various points along the gradient that ultimately constitute the bell-shaped optimum curve, and (ii) regulation of numbers when they deviate relative to these levels. The determining factors were thought to be largely abiotic, and the regulatory ones biotic (e.g., Varley et al. 1973), a view that persists in recent models (e.g., Coulson et al. 2004). Not surprisingly, the distinction, in practice, between processes of determination and those of regulation is not possible, as expanded below. This brings significant problems for demographic theory and its application.

The autecological perspective is quite different from that outlined above. For example, consider a species distributed along an environmental gradient. The ecological intensity of the species shifts back and forth spatially along a gradient in the field, in response to changing values of the ecological variable(s) that determine it, and which are abiotic. Locally, the population is fluctuating. Distinguishing processes of determination from those of regulation is not even possible, despite its significance to demographic theory and practice. If numbers of individuals are increasing locally, for example, the population could be increasing to a higher equilibrium level because the local determining factor has shifted, or it could signal a larger deviation from the original equilibrium level, and thus reflect regulatory processes (Solomon 1949). Effectively, linking *K* mechanistically to any abiotic influence simply cannot be achieved, and is diversionary. Such ambivalence applies equally to the fragmented, scattered parts of an indirect gradient, at least when each one is interpreted individually as a distinct population, as in demographic theory. This inability to distinguish between two processes that are thought to differ from one another in principle damages the theoretical backbone of demographic population dynamics severely. It is for these reasons that the logistic equation cannot deal with the spatially broader perspective that is required in ecological theory. A much broader perspective on the species and its environmental interactions is needed.

Naturally, local density-related (or demographic) variables may exert ecological influence, but such influences are entirely secondary in terms of the primary pattern that is set by the adaptations of the organisms as they interact with the environment. Density-related processes are therefore best interpreted in relation to an understanding of the influence of externally determined processes of habitat tracking and environmental matching that takes place at a regional and geographic scale. Without this appropriate context, studies of local diversity convey relatively little ecological insight or understanding. This is not to diminish the possible influence of predation or herbivory, for example, but to place it appropriately within a broader context.

The realignment recommended by autecology should provide novel insights for understanding, investigating and interpreting resource and pest management (see Walter (2003, pp. 235-269)). For example, high population densities of a commercial fish species can be expected to be generated when survival and reproduction are enhanced under particularly favorable conditions (e.g., Robin and Denis 1999). Nevertheless, the influence of climate in such cases is frequently downplayed, even when the appropriate research and analysis has not been conducted. Elevated population densities may be maintained even after the period of suitable environmental conditions has passed and survival rates and reproductive success have dropped somewhat. Replacement level reproduction should hold populations to levels at around that point, unless environmental conditions lower reproduction levels even further. Alternatively, high population densities may be forced down by natural enemies, as in successful biocontrol and drastic overfishing. Even if the predation pressure is lifted (as in the cessation of fishing), populations may not automatically increase again, particularly if reproductive success is still at replacement level, because the environment is not conducive to higher reproductive output and recruitment. This would explain why, after successful biological control has reduced densities of the pest to low levels, the pest remains at relatively low densities, even in the face of relatively lower natural enemy activity (Walter 2003).

3.4 Ecology and the Significance of Species

Individual organisms must adjust spatio-temporally to their ecological conditions throughout their lifeline. These adjustments may well be spatial, as in the larval movement of organisms with a pelago-benthic life cycle and the autumn and winter movements of Bewick's swans as they respond to cold, winds and food availability (Nolet et al. 2001, Beekman et al. 2002), or they may be physiological, as in the responses of perennial plants and aestivating lungfish or frogs (Kayes et al. 2009). Completion of sexual life cycles requires individual organisms to respond appropriately and synchronously, through physiological adjustment, environmental tracking and the recognition and localization of an appropriate mating partner (Chapters 5 (Section 6) and 6 (Subsection 4.4)) (Paterson 1985, 1986). The autecological interpretation thus focuses its core generalizations on the mechanistic workings of organisms in their environment and includes the requirements of sexual reproduction. Before expanding this point, we note that most eukaryotic species are sexual, at least to some extent or under certain environmental conditions. Asexuals can still be treated as outlined below, but without the requirements of sex. In any case, their genetic system generates clones, so the ecological consequences remain the same as specified below for sexual species.

Populations defined sexually, or in Mendelian terms, have traditionally been the purview of population genetics, as in the Hardy-Weinberg equation. In evolutionary biology, or population biology *sensu lato*, species are defined in population genetics terms, with cognizance of the fact that conspecific individuals may well exist in geographically separated populations, in allopatry. Individuals from such different areas, although separated in space, may well have the potential for sexual union (in future, perhaps, as climate change drives the two populations together, as in the postglacial history of many species (e.g., Hewitt 1999, 2000, Stehlik et al. 2001, Stehlik 2003, Sommer and Benecke 2005, Sommer and Nadachowski 2006)) and are thus conspecific. Such systems of disparately distributed conspecifics comprise "species gene pools" (e.g., Ayala 1982) and they represent the "field for gene recombination" (Carson 1957). The ecological significance is that the species' distribution (including consideration even of the areas that separate the populations) helps to specify the primary environmental requirements of the species, and the conditions under which they cannot persist. Species, seen in this way are not so much units (a term that is sometimes used and perhaps connotes typology) as self-defining bisexual systems (Lambert et al. 1987) whose individuals operate within the environment according to their species-specific properties or mechanisms. Again, asexuals also behave in this way, but without the requirements of sex. A good proportion of asexuals does, however, retain the capacity for sex, often when conditions are inimical to their survival. They can thus be investigated and interpreted in the same way as are sexuals (Paterson 1985, 1991, Walter 2003, Najar-Rodriguez et al. 2009).

Despite the action of the rather precise adaptations that relate organisms to their environment, individuals are influenced stochastically at virtually every point within the operation of any such mechanism. The requirements of life cycle completion and sex ensure that these dynamic systems are under stabilizing selection, because the complexities of the system and the wide range and sequential structure of the specific requirements and behaviors of the individuals mean that genetically coded variants will almost inevitably be selected out of the gene pool. This mechanistic and functional definition of species as Mendelian populations contrasts strongly with the way in which the ecological units of demographic ecology (i.e., ecological populations, communities and ecosystems) are defined (Table 9.2). Their definition can be achieved only in statistical terms, for they essentially represent stochastically generated spatially and temporally indeterminate states that can change composition, fragment and dissipate without consequence. In this sense they are arbitrarily defined, as detailed further in Section 4, whereas species (as gene pools rather than taxonomic units) exist independently of human observers and can be objectively specified through the structure of their Specific-mate Recognition System. See Walter (2003, Chapter 6) for more detail.

4 AUTECOLOGICAL ANALYSIS

4.1 Markovian Habitat Tracking, Stochastic Dynamics and Risk Analysis

The determination of the level of local densities of organisms needs to be approached from a dynamic viewpoint that deals with individual organisms. We do so on the basis of the observation that individual organisms move about within the environment according to their species-specific properties, even if they do so only as spawn, neonates or propagules. Below we outline the autecological consideration of how individual organisms relate to specific localities and how this translates into the geographical distribution of the species concerned. Although the example is an idealization and is biased toward motile organisms, it does illustrate the significant principles that relate individual organisms to the most general ecological pattern observable, that of the spatio-temporal dynamics of species.

Organisms have specific rates of dispersal and generally remain in more favorable sites longer than they do in less favorable ones, and keep moving in areas that are altogether unsuitable to their requirements and tolerances (unless they are entirely sessile, when the dynamic is somewhat slower and may cover several generations, although it follows the same principles). For modeling purposes, ecologists can thus allot individuals to the various biotopes or sites in the region according to spatially differential occupancy rates and transition probabilities. This means that ecological populations can be treated as open aggregations of individual organisms that show turnover with respect to the individuals that constitute them (Subsection 3.1). The resulting numerical dispersion (or ecological intensity) of the individuals in the region reflects the spatial pattern of the relative suitability of the various sites, with respect to both biotic and abiotic conditions. When this pattern of suitability changes, as inevitably it will, so the dispersion of ecological intensity will change accordingly.

The processes just outlined reflect the transitional, local living conditions to which individual organisms respond, and for the most part they do so independently of one another. These responses have a physiological and behavioral basis that allows the individuals to track the particular environmental conditions to which they are primarily adapted. The higher the dispersal rate of the species, the greater the rate of spatial turnover of individuals, and (in principle) the more specific and strict their environmental requirements and tolerances are likely to be. By contrast, individuals of species with lower dispersal rates are likely to be more flexible physiologically and ecologically, and the spatial distribution of their densities relative to variation in environmental conditions will be more blurred and the turnover rate lower, as seen for example in perennial plants. Local density, conceived from this spatio-temporal dynamic perspective, is not a property inherent to the population or related to the "rescue" of conspecific populations (Subsection 3.3).

It is, instead, a consequence of the activities of individual organisms, an epiphenomenon.

We first demonstrate the nature of the stochastic spatial processes we have in mind with a simple numerical example (from Ashby 1956) and then generalize this to more realistic ecological situations. An individual insect of some hypothetical type lives along a waterfront and may enter the water, shelter under stones, or lie exposed on the bank. Observations are made, at set intervals, to record its location. Double or triple counts are entered when it remains in the same situation during two or three consecutive observations. The tabulated transitions can be expressed as probabilities, with each calculated as a proportion of the total number of transitions from each of the three positions to any of the others. If large numbers of such insects live in the same area, and each behaves independently of the others, the consequences of their movements would be seen from a little distance as three "clouds" of activity, one on the bank, one in the water and one under stones. These three "populations" are readily represented by three quantities that change with time.

If the transition probabilities are consistent across all individuals, the system as a whole is determinate. It assumes a particular frequency distribution across the three sites, even if periodic disturbances change the pattern, so the system has an equilibrium state. Note the sharp contrast between the static appearance of the three spatially separated components of the system, which are unchanging at equilibrium, and the dynamics of the individual insects, which move continuously and whose transitions can be described only in terms of probabilities. These average transition probabilities represent the equilibrium values of the determinate state (or steady state) of the system. The system is steady despite the ongoing movement of all the constituent individuals or, perhaps more appropriately, because of the ongoing dynamic (Ashby 1956).

In general, the ecological intensity of a species can be viewed in terms of such a determinate state within a spatially heterogeneous environment. Individuals move from one site to the next according to transition probabilities, with these being determined by the affinity of the individuals for certain environmental or habitat conditions within the various localities that are acceptable. Naturally, the spatial dynamics of the environment may be such that a steady state is never really reached. Such situations are more likely in highly dispersive species, like aphids. Even somewhat sedentary organisms may, however, not build relatively discrete local "populations", for example when the individuals are highly dispersed in space, as in the terrestrial isopods that move about widely to feed at night and then aggregate under shelter that is set almost flush to the ground, where humidity remains high enough (den Boer 1962).

Species, through the movement of their constituent individuals, thus track the environmental conditions that suit their requirements and to which they are adapted. The rates of such movement, even if it is just the movement of propagules, are likely to vary with time, but movement of this nature is essentially an ongoing process. Markov chain models describe this process of habitat tracking in terms of residence and transition probabilities, as outlined above. Such models are of particular ecological relevance when the environmental conditions vary spatially and this spatial variation is dynamic in time. Under these circumstances, sites that fit the requirements of the individuals at one time may soon become unsuitable, and *vice versa*. The individuals concerned must leave deteriorating sites and find suitable locations elsewhere, ones that may have become suitable only recently. Sessile organisms can adjust to such change, but can do so only physiologically, through directional growth for example, or by producing motile offspring or propagules.

Organisms, in general, are usually unaware of which sites fit their needs or where suitable sites are located (and even return migrating organisms cannot be aware of future conditions). Random movement thus provides the process best suited to finding suitable localities by chance, although orientation devices provide guides when within range (as in the olfactory mechanisms used by seabirds (Chapter 3)). In general, individuals or their progeny will always be on the move, their levels of mobility being modified according to the suitability of the local conditions. Organisms move persistently, with some individuals even moving from local conditions that still appear suitable to them (Subsection 3.1), at least from our perspective. Organisms do, however, tend to reside longer in sites that are suitable to them. Consequently, numbers of individuals will build up in more suitable sites through reproduction and immigration, to the point that aggregations form and it is these that are traditionally interpreted as ecological populations. The spatial dynamism and turnover of such systems means they are essentially temporary aggregations of individuals, for which such attributes as equilibrium densities, intrinsic rate of population increase (r), and regulating mechanisms are inappropriate, even if applied over a relatively short time frame. Treating epiphenomena in this way is misleading.

The outline above demonstrates why, in autecology, ecological systems are best understood in relation to the probability of survival of individual organisms in relation to the particular environmental circumstances they encounter, that is, in terms of risk. Risk is influenced primarily by the adaptations or properties of the organisms as they function relative to the prevailing environmental conditions and by the movement of the individuals across space and over time in their tracking, as best they can, of the living conditions to which they are adapted. When they are able to track such environmental circumstances adequately, and they can bridge unfavorable stretches of terrain successfully (Hengeveld and van den Bosch 1997), they match the spatio-temporal dynamics of the conditions to which they are adapted; if not, and the mismatch persists for long enough, they die out, to the point of extinction of that species. Autecological analysis is focused on the operation and dynamics of these conspecific individuals within the structured but stochastic environment to which they are adapted. From this, ecological understanding can be framed and detailed in terms of the species-specific adaptations of the organisms concerned, the spatio-temporal dynamic structure of the environment and the risks incurred by the constituent individuals. Ecological systems are, thus, not driven demographically by density, as in the equalizing processes invoked by metapopulation models, but by environmental variables. These alternative approaches, autecological versus demographic, should not be confused, as they are qualitatively different and demand different quantification.

4.2 Dealing with Spatial Variation in Terms of Risk

In this section, we expand the perspective developed above to consider the process of environmental matching through habitat tracking within a broad geographic perspective, by means of the analysis of spatial risk. Essentially, the species' distribution provides the framework for abstraction and analysis (Chapters 4 and 5). Presently, the only robust generalization that describes the distribution of the local abundances of species is the "abundant center" model, which has been derived empirically from long time series within species (Fig. 9.9), and across many species (Fig. 9.10) (Hengeveld and Haeck 1982). Criticisms of this model (Sagarin and Gaines 2002, Sagarin 2006) seem to derive from an emphasis on local adaptation, in which organisms are seen to adapt differently to heterogeneous conditions in different part of their range. However, such local adaptation to conditions at the edge of a species distribution have not been conclusively demonstrated (Hoffmann and Blows 1994, Bridle and Vines 2007), presumably because species are mobile in a spatio-temporal sense (Chapter 4, and as detailed so far in this chapter) as well as generally being under stabilizing selection (Chapter 6, Subsection 4.4). Also, the criticisms raised against the "abundant center" model neither treat species as dynamic spatio-temporal systems nor treat the distribution of abundance statistically (Hengeveld and Haeck 1982).

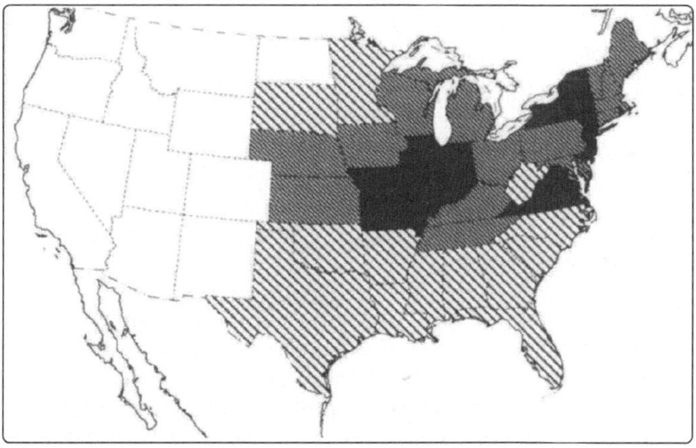

Figure 9.9 Number of outbreaks of the noctuid moth *Cirphis punctata* per state in the United States of America, over a 65 year period (1860 to 1925). Redrawn from Hengeveld (1990), with permission.

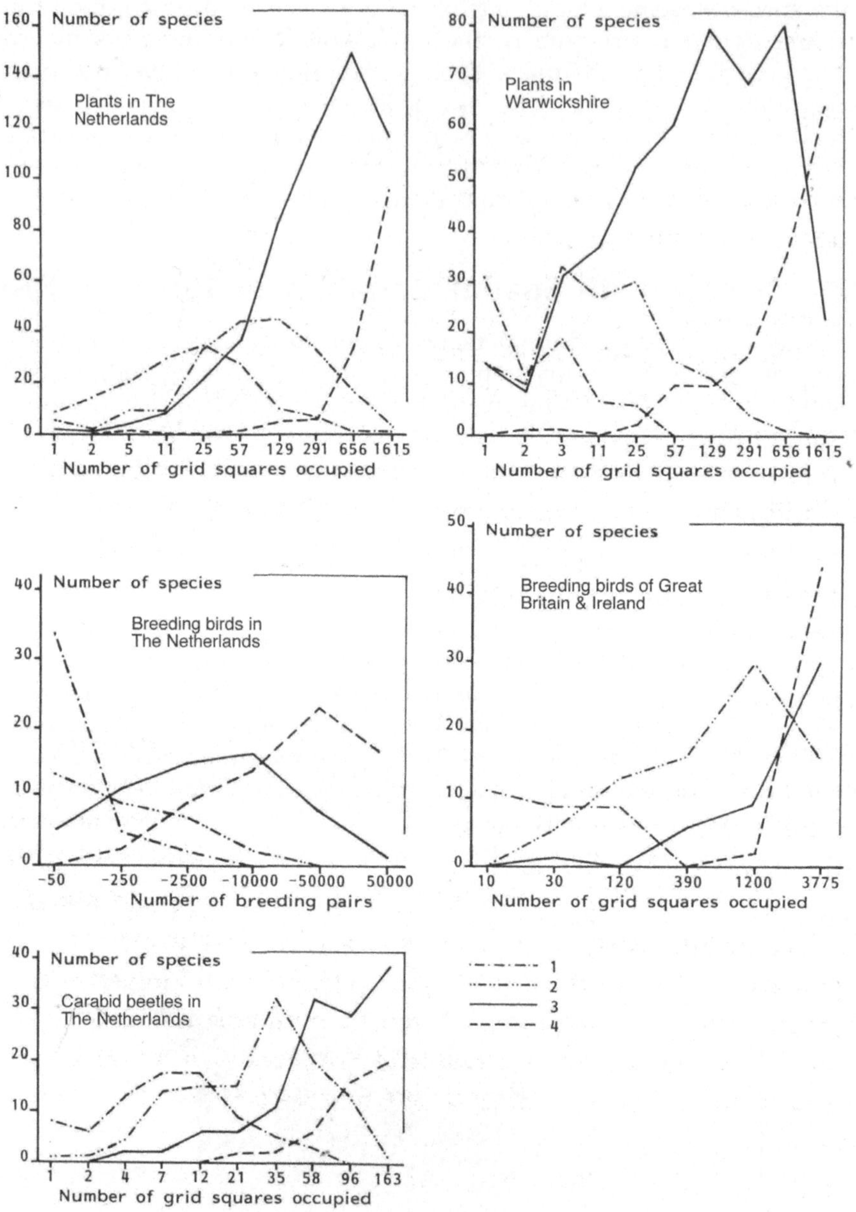

Figure 9.10 Each of these graphs plots the abundance (in terms of number of grid squares occupied by each species) of species (in different taxa and in different geographical areas, as indicated on each graph) that are at the margin of their distribution (1 in key) in the subject geographical area, relatively close to the margin of their distribution (2 in key), relatively close to the center of their distribution (3 in key) and at the center of their distribution (4 in key). Species at the center of their distribution are consistently far more abundant than those at the margin of their distribution. Reproduced from Hengeveld and Haeck (1982), with permission.

Suitable sites for individuals of a particular species generally become more widely spaced towards the margin of that species' geographical distribution, and with increasing inter-site distances their suitability will generally decrease (Hengeveld and Hemerik 2002). Moreover, with such decreasing general suitability, the more marginally suitable sites will occasionally become unsuitable for that species, with the result that it dies out locally. When the sites are randomly dispersed, and the time periods for which they are suitable are distributed randomly as well, and with both being Poisson distributed, the distances between them will be exponentially distributed. This means that the further from the distribution center, the increasingly farther apart the suitable sites, and the lower the frequency at which they will be suitable. These more marginal sites will have an exponentially decreasing chance of being found and occupied by the individuals of the species of interest, whether it is an animal or plant, as the individuals move stochastically within the environment as a whole. Also, the frequency of extinction of such local populations increases exponentially away from the center. This means, in turn, that the risk of these species dying out in any particular locality they currently inhabit increases sharply towards the range margins, because the distances between suitable sites and periods between unfavorable years increase exponentially as one leaves the center of distribution. This in itself constitutes just one component of spatial risk, the one defined by the spatio-temporal variation in ecological suitability of the environment. Such an approach to analysis can, in principle, be adapted for application to abundance distributions of any species.

A second component of risk is evident in the movement of individuals as they encounter sites with suitable conditions at various rates in different parts of their geographical distribution. As the sites become more isolated from one another, so the amount of unfavorable area between them increases. This means that the organisms spend more time in searching for suitable sites and/or in crossing areas that are unsuitable to them (in a variety of possible ways), which increases the overall distances covered, as Figure 9.11 shows. Consequently, mortality rates from exhaustion, duration of exposure to unsuitable conditions, predation, and so on, also increase. At some point, the risk negates any advantage that there originally might have been in moving across unsuitable areas. Figure 9.11 shows that at that point the curve bends down sharply, which defines the maximum extent of the unfavorable area that can be crossed, although this is (in addition to that effect) also influenced by the varying degree to which the conditions between the suitable sites are unfavorable. This curve thus represents the individuals' spatial risk, which increases towards the margins of the species' distribution in the hypothetical example above. This measure of spatial risk, together with the first spatial risk component, the one quantifying the spatio-temporal variation in ecological suitability of the environment, specifies the limits of the species' distribution. Again, we stress that this is a general principle of autecology, for it applies to all species. That is, it is applicable to each species, its adaptations, requirements and tolerances, as well as to the spatio-temporal dynamics of the environment that it must match.

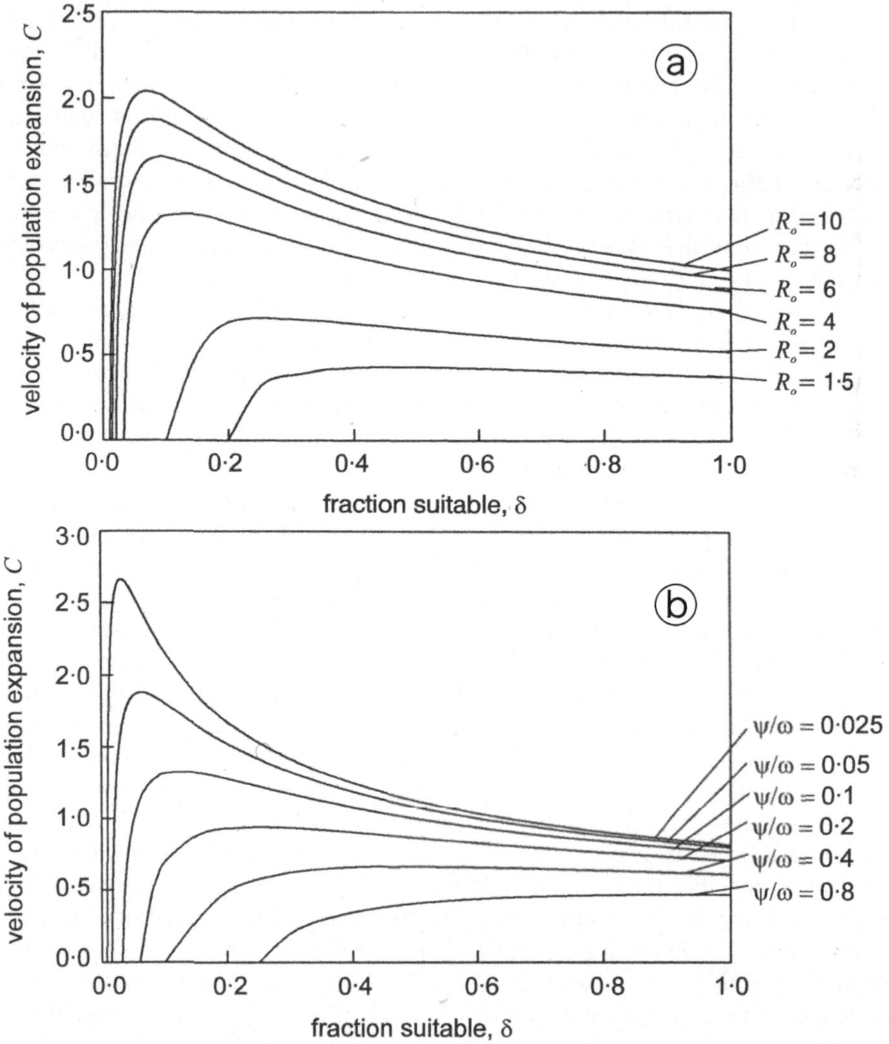

Figure 9.11 The velocity of spatial expansion of a species (*C*) as a function of the fraction of habitat that is suitable in heterogeneous conditions **(a)** for various values of R_0, where the maximum rate of expansion shifts to the left (where there are fewer favorable sites) when reproduction rates are higher. When the dispersal risk (ψ/ω) is higher **(b)** the maximum rate of expansion shifts to the right (where there are relatively more favorable sites). Reproduced from Hengeveld and van den Bosch (1997), with permission.

Under regional or geographically variable conditions, as driven externally by climate change for example, the various parameters of the stochastic distributions outlined above will change and the distributions of the species concerned will also show relatively rapid regional expansions or contractions, which we see and document as shifts in geographical distribution. Again, these spatial changes need not be symmetrical, either in terms of the shape of the species' distribution, or in terms of rates of change in different parts of that distribution or at different edges of the distribution, as illustrated by the historical shifts in distribution of beech and hemlock (Davis 1987).

4.3 Environmental Stochasticity, Adaptations and Risk Analysis

Autecological analysis requires an understanding of how the individuals of a particular species are dependent on a particular trait for survival (or achieving a specified reproductive outcome) relative to a particular environmental feature. Environmental variables (or measures) of this nature were characterized, for autecological purposes, as axes of differentiation (Chapter 5, Section 2).

Each individual organism perpetually faces risk, as indicated in the section above, mainly because environmental conditions change persistently but also because organisms move (at least at some stage of their life cycle). Organisms may thus suffer adverse circumstances, including starvation, desiccation, exhaustion, displacement or disorientation, overheating or chilling, or they may lack the temporal and spatial synchronization required for reproduction. They may therefore die through failure to maintain their energy flow, or they may fail to reproduce, through the influence of one or more ecological variables (of numerous potential influences). These insights provide the basis for quantifying, in terms of risk, the ecological implications of the interactions between organisms and the environment and they provide an autecological conceptualization of the *raison d'être* of adaptations in general.

All organisms have properties whose function is to contribute to the maintenance of a positive energy flow, in part by maintaining physiological stasis relative to the external environment and, in part, to reduce the impact of particular risks from the environment and thus enhance the probability of survival and reproduction. The adaptive mechanisms that have evolved to these ends almost invariably differ across species, but may show similarities, and they relate to the ecological circumstances at the time of speciation. These latter are the environmental conditions to which each species is primarily adapted (Chapters 5 (Section 6) and 6 (Subsection 4.3)).

Pre-adaptation is also a significant consideration with respect to the way in which organisms of particular species interact with the environment (Herrera 1992, Johnson and Mayeux 1992). For example, the flora of Andalusia is typically "Mediterranean" in its properties. That is, different species share similar properties to the extent that one can correlate the presence of various characters into so-called "character syndromes" (in the sense of Stebbins (1974)). The plants of the more ancient lineages in Andalusia acquired their "Mediterranean" features in relation to tropical environments. They were pre-adapted for an environmental match with the structure of the Mediterranean climate (Herrera 1992). In general, therefore, individuals of each species match the pattern of the temporal dynamics in non-uniform environments through the operation of their species-specific adaptations.

The quantification of risk has to take into account the composite nature of the life cycle of organisms and also the way in which environmental variables interact with one another and thus influence the ecological impact of one another. These aspects are expanded below.

Because the adaptations of organisms reduce the effects of stochasticity within the type of environment the species typically inhabits, the probability of the survival of the organisms of interest as they live, move and reproduce within the confines of their spatio-temporal environment is best quantified through estimates of risk. The expected mortality rate of individuals is a typical measure, perhaps at a particular age or stage of development, and in relation to a particular activity across different environmental circumstances (or biotopes). Reproduction can also be treated in this way. For example, the risk of failure to reproduce could be related to the temperature profiles at which a plant flowers (e.g., quantified as flowering frequency) and thus the profiles at which it fails to produce flowers and fails to reproduce. The risk of local extinction, through reproductive failure for example, can thus be estimated. Autecological research can thus focus comprehensively on the temperature profiles that are required to support the physiological processes of flower production, syngamy, seed production, seed set and seedling establishment, as introduced and illustrated in Chapter 3, Subsection 3.1 and exemplified more comprehensively in Section 5.

Individual organisms, with their species-specific properties, are not only exposed to unfavorable variables, which may kill them, but also combinations of environmental factors that provide the conditions necessary for them to function effectively, survive and reproduce. Autecology emphasizes, further, the requirements and tolerances of the organisms along their entire lifeline and thus on the adaptive mechanisms (together with their plasticity) that underpin these ecological interactions (see Chapter 3). So once the requirements of an ecologically significant trait are known, as well as the relative frequency of the environmental variable that meets this requirement, the local probability of persistence of the species can be predicted for the area under scrutiny. Autecological explanation therefore does not deal simply with chance as an average influence on the entire system. Rather, the impact of stochastic environmental influences is related directly to one or other of the species-specific environmental response systems of the individuals, so they involve behavioral, physiological, anatomical and other such properties in combination. Autecology thus relates directly, through risk analysis, to all the other biological sub-disciplines that deal with the organism itself (see Rose 1997).

Risk can be quantified in two ways, individually and statistically, both of which are relevant to autecological analysis. Risk is expressed "individually" when the risk of an individual suffering a given amount of damage, perhaps from a single cause, is estimated. Individual risks are often influenced by the sensitivity of the individuals concerned and will thus typically follow a Gaussian or normal probability distribution. Such personal or individual risk is often used in ecology, for example in ecotoxicology.

The second type of risk analysis, statistical risk, specifies risk in situations in which groups of organisms are exposed to a particular set of conditions or to particular types of events. Statistical risk estimates the relative frequency of events arising from a given cause (e.g., particular environmental circumstances) and results in a particular outcome, and can be expressed as a percentage of the total number of events. Typically, statistical risk will result in a decreasing, J-shaped risk frequency. So, for example, a plot of mortality probability (from whatever cause, perhaps predation) against age will yield such an age distribution, with many young individuals and relatively few old ones affected. Such a plot provides a half-life for the population, and this depends on environmental factors, which vary across space and over time. A similar plot has been derived for species extinction rates over geological time. Minor risks are relatively common and major risks are rare. The former are reflected in the background extinction rate. By contrast, mass extinctions, in which many species disappear simultaneously, are rare (Jablonski 1991, 1994). Raup's (1991) "kill curve" plotted the number of species' extinction rates from meteor impacts (Fig. 9.12). The risk he approximated to crater diameter, so the shape of the kill curve is determined statistically by the time between impacts for meteors of different sizes. It is thus an exponential distribution if the impacts have a Poisson distribution through time. Other processes that influence extinction risk may well be represented by other relationships, and the curves may be of somewhat different shape.

The risk faced by individuals (perhaps in relation to age) requires statistical analysis to estimate the variability of that risk. The probability of that species occurring in a particular locality can then be quantified through estimation of the relative frequency of occurrence of conditions of different intensity. For example, a high relative frequency of growing seasons with too little warmth for annual plants to set seed in a particular locality will exclude them periodically, perhaps permanently, whereas perennials can compensate by delaying seed production till the following year. Each trait is ecologically sensitive relative to one or more environmental factors, and this is expressed by a particular frequency distribution of risks, a risk distribution. Interactions between variables are common, but have been subject to surprisingly little ecological analysis (but see Haefner 1970, Bjorkman 2000, Sardella et al. 2004, Thomas et al. 2007). Organisms may be able to compensate to some extent for the ill effects of some factors, as in the perennial plants above. Different traits of an organism are likely to have different sensitivities, and therefore be exposed to different levels of risk, across localities. It is only at the center of abundance that these risks, on average, are all likely to be relatively low. Small populations trapped under adverse environmental circumstances will face high risks that are likely to generate directional selection. Extinction is probable under such circumstances, but if adaptive change matches the local environment, speciation may ultimately (and incidentally) result (Chapters 5 (Section 6), 6 (Subsections 4.3 and 4.4) and 10 (Subsection 5.2.2)).

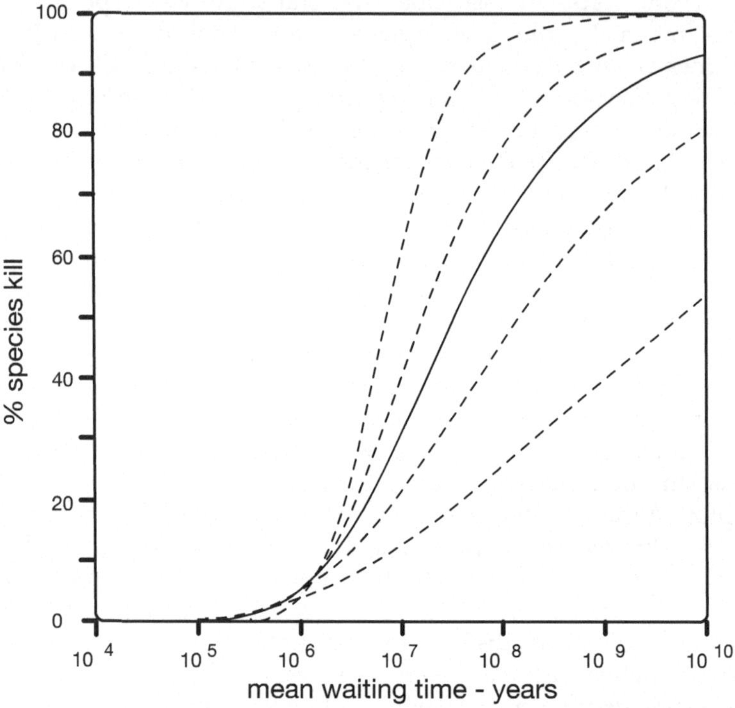

Figure 9.12 Several possible kill curves for species. For a given kill value, waiting time is the expected time between killing events of equal or greater magnitude. The solid curve provides the best fit for the Phanerozoic extinction data compiled by Raup (1991), from which the figure has been reproduced with permission.

Environmental conditions are changeable and unpredictable, to all organisms. No variable that is ecologically relevant persists indefinitely. And organisms of each species face a number of ecologically relevant variables simultaneously. For an entire suite of ecologically relevant variables to be consistent across similar time spans is that much less likely. Furthermore, such unpredictable change in environmental conditions takes place at all temporal scales, from the finest to the broadest (Fig. 9.13) (Mitchell 1976). This effectively means that the different variables will change at various scales of resolution and their effects will be superimposed interactively, which changes the environment even further and more unpredictability (Hengeveld 1997). Such variation means that the designation of ecological constants for modeling purposes, at any scale of resolution, is unrealistic and it cannot even be portrayed accurately in terms of variation around a mean value (as when environmental stochasticity is added to demographic models). The problem is that the mean value itself will inevitably vary with time; this is clearly understood when means, taken at different time scales, are compared (see Fig. 9.13). In short, no time scale exists, fine or broad, for which we can reliably take a mean value as a constant, deterministic representation (Fig. 9.13) (Hengeveld 1997), so the intensity at which organisms experience environmental variables will not remain constant over time or across space. Risk analysis deals directly with these circumstances by estimating the survival chances of conspecific individuals under field conditions at all scales jointly, as insurance models do, and therefore reflects changes in numbers over time more accurately and more realistically than any deterministic model.

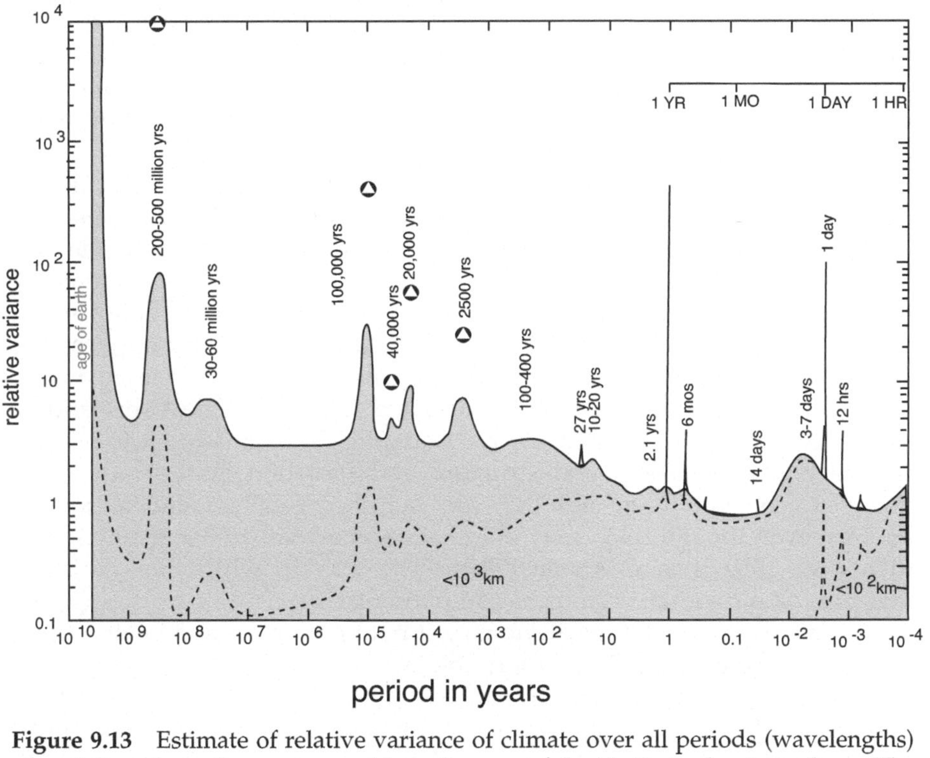

Figure 9.13 Estimate of relative variance of climate over all periods (wavelengths) of variation, from those comparable to the age of the Earth to about one hour. The stippled area represents total variance on all spatial scales of variation. Dashed curves in the lower part of the figure indicate the contributions to the total variance from processes characterized by spatial scales less than those indicated (in kilometers). Strictly periodic components of variation are represented by spikes of arbitrary width. Solid triangles indicate scaling relationship between the spikes and the amplitude of other features of the spectrum. Reproduced, with permission, from Mitchell (1976), and this reference has more detail.

With temporal variability the durations (or "waiting times" for the organisms) between comparable conditions also follow an exponential distribution. Environmental conditions also vary across space, which means their temporal dynamics differ from place to place. The distances between patches (which reflects their accessibility to organisms) also follow an exponential distribution, if Poisson distributed (Hengeveld and Hemerik 2002). Spatio-temporal variability renders the environment entirely unpredictable to animals and plants for they cannot have perfect knowledge (Rapport 1991) and so are unaware of conditions elsewhere, either in future or in terms of statistical expectations. To us, this variability is broadly predictable in determinate statistical terms of variance, skewness and so on, but is unknown to the individual organism, so that it experiences its environment as hazardous. Within limits, though, species can be adapted to environmental variation that is regular, as with seasonal cycles, because adaptations such as hibernation and dormancy can bridge particular unfavorable periods, with the match being statistical. Should these unfavorable periods extend for longer than usual, the organism may run out of resources, and if they become shorter, or the environmental cues become unreliable, their life cycle may no longer fit the temporal structure of the environment (Chapter 3, Subsection 3.3).

Environmental structure varies to a virtually unlimited extent across numerous axes of differentiation and at various scales of variation. Furthermore, evolutionary processes are undirected and contingent on such things as environmental change and which individuals ultimately make up the small populations that ultimately are forced to adapt when trapped under changing environmental circumstances. This implies that both drift and selection are likely to be important in speciation, which contrasts with the interpretations associated with "ecological speciation", as spelled out by Schluter (2009) for example. As a consequence, no two species match environmental structure and variation in the same way, not even congeneric species that are morphologically identical to one another, even though they do tend to have somewhat similar requirements (den Boer 1980, 1986b, Walter 1988a). The way in which the statistical matching of individual organisms of a particular species takes place against a structured but variable environment, at both ecological and evolutionary time scales, defines the field of autecology.

Extreme events do pose a risk and have an impact on particular organisms, even though those events may be rare. For example, even an occasional frost each spring may limit a species' geographical range (Hengeveld 1990). Furthermore, the impact of such rare events often depends on their duration, or on the associated environmental context, or on the temporal sequence in which they occur, and so on. Nevertheless, risks are seldom as dramatically obvious as this; risk analysis is not concerned exclusively with extreme values. As one moves away from a particular point of abundance within the geographical distribution of a species, for example, the frequency at which a combination of particularly favorable conditions occurs, is likely to decline at a certain rate, so that ultimately this combination becomes rare and the abundance of the organism declines commensurately (Brown 1984). In a somewhat simplified form, the frequency of the unfavorable events, such as the number of frost days in spring or extreme drops in saturation deficit, will be lowest at the geographical center of abundance of the species, and increase away from these points. Gradual changes in frequency of such events ensure that local abundances of the species concerned gradually decline. In such cases, the analysis of rare events alone is inevitably insufficient for understanding a species' ecology comprehensively; what is rare in one locality may be common elsewhere, or at another time. Notice that autecological risk analysis does not simply invoke a large number (or combination) of unidentified variables to reduce, control or balance numbers across local populations, as assumed in risk spreading theory (den Boer 1968). Autecological risk analysis is positive in always focusing on the statistical match (and thus also mismatch) of specific properties of individuals with the variation in particular elements of the environment.

Although risk can be analyzed locally, a more meaningful picture of a species' ecology emerges from a comparative analysis across localities, preferably to yield a geographical perspective across the species' distribution. At the point of highest expected abundance of the species, the overall conditions are so often favorable for the species that it may well occur in different biotopes and in large aggregations, seemingly regardless of the differences across biotopes within these localities (see Hengeveld and Haeck 1982, Hengeveld 1990, Hengeveld and Hemerik 2002). Any less favorable conditions that intrude or intervene appear to impose relatively little on the organisms when the overall conditions are generally favorable for the species.

Similarly, risks relating to biotic factors are incorporated, albeit often in conjunction with abiotic ones, including such aspects as the density and physiological states of hosts, densities of predators, and so on. Individuals exposed to adverse conditions (or particular sequences of adverse conditions) may be physiologically less vigorous under such circumstances and therefore more prone to disease, parasitism or herbivory (e.g., White 1969). Conversely, organisms under more favorable conditions are likely to be healthier, survive better and reproduce more than individuals under circumstances that are less conducive to their well-being. To understand the requirements of life for individuals of a particular species requires targeted investigation. We understand relatively little in this regard, for any species, because of the focus of population studies on common species, and these are designed to ensure density effects will be expressed as clearly as possible. Otherwise ecological studies investigate conditions at range margins and quite often these only help define minimum living conditions. Neither of these approaches reveals much about the living conditions that favor organisms of a particular species.

Essentially, the stochastic nature of the environment means that huge shifts in the distribution of abundances can be expected and data sheets are likely to include many zeros. The frequency distribution of accessibility between patches favorable to the organisms under consideration will influence whether a sufficient number of individuals reaches them on time to settle and so prolong their survival, and enhance, incidentally, the probability of the persistence of the species in question. That this spatio-temporal shifting of organisms is essentially an opportunistic process implies that average values that are taken to give meaning to concepts like population equilibrium or balance actually mislead as to what is ecologically significant, for they have no physical meaning or interpretation. Only the spatio-temporal risk distributions of particular environmental variables relative to the sensitivity and mobility of the individuals of a species have ecological relevance, for they portray what the organism experiences ecologically. Autecology thus avoids some of the misleading aspects central to demographic ecology (see Table 9.1).

4.4 Risk Distributions and Reductionist Methodology

Risk analysis in autecology is a statistical technique concerned with the consequences for survival of the environmental matching process of individual organisms and their species-specific properties. The analysis focuses on levels of variation in particular environmental variables that have been identified as being relevant and demonstrated empirically to be so. Essentially it is based on an understanding of the way in which a few variables in the environment, identified empirically, impact on the survival and reproduction of individual organisms (Chapter 8). The risk that an organism does not start or complete one or other biological process, such as germination, hatching, finding suitable food, or reproducing, expresses the outcome of a mechanistic process as it is influenced stochastically. Various such outcomes result in a particular chance distribution for the range of values realized by the biological process concerned. These outcomes together contribute to the risk distribution that represents the probability of the particular outcome from that process. The species' probability of persistence in the locality under consideration is reflected by the parameters of this distribution. Any change in the parameter values of this distribution of survival chances will sooner or later be reflected in the ecological response of the species in question. Its local abundance will change, perhaps even to the point of it disappearing from the locality.

Risk analysis is based on reductionist methodology. Changes in environmental variables and their expected influences are preferably considered one by one, although not necessarily so. When more than one process simultaneously affects the survival or performance of organisms, the effects of changes in parameter values of individual processes can easily be swamped by the joint risk of all variables together (e.g., Kooijman et al. 1989). Or their interactive effects may enhance the ecological process under consideration. Changes in such parameter values may be insignificantly small relative to the effects of the other ecologically relevant variables combined, or they may be compensated for by changes in values of other independent parameters. This swamping effect is known in preventive medicine, for example, where remedial measures are often taken for individual relief only, for they cannot reduce the risk to the population as a whole (Rose 1992). Similarly, the use of seat belts in traffic is advisable for personal benefit, yet hardly affects the overall mortality rate in traffic accidents, and does not reduce overall population mortality.

The swamping effect outlined above impacts ecological investigation and interpretation substantially. The ecological effects of the values of one parameter or process must be weighed relative to the actions of all other ecological variables together. The anticipated effects of processes that are given special prominence in demographic ecology, mainly competition and predation, should also be evaluated in this way, but this has not been done in justification of the ecological significance of these processes. Their effects are, in general, likely to be swamped by the impacts of the other ecologically relevant variables. Had this procedure been followed in the early development of ecological theory, competition in particular may never have been assigned such a central role in the theory of ecology and evolution as we now see.

5 AUTECOLOGICAL SYNTHESIS—EXAMPLES OF DYNAMIC ECOLOGICAL MATCHING

Species responses are not only physiologically determined, but they are also individualistic. Plants, in particular, have been investigated from this autecological perspective. We thus see that plant species grow optimally in environments in which the seasonal temperature profile, for example, matches their various stage-dependent temperature requirements closely. The width of the ranges of survival or optimal performance implies that temperatures can vary to some extent without doing serious harm to the plants. However, the narrower this range, the higher the risk of inimical effects from relatively extreme environmental fluctuation. The responses of the plants will differ across biotopes because the various factors have different weights, and therefore interact differently in each biotope. This is complicated further by variation across seasons and across different years. The multiple natures of the variables across space and through time generate ever-new patterns in space. All this can be expressed in terms of differing risks, and these impose different rates of mortality and reproduction.

Autecological research has been conducted on few organisms and then only on particular aspects of what is required to understand their spatio-temporal dynamics. The most completely worked example of risk analysis focuses on the altitudinal distribution of oats and its dynamics in Scotland (Parry 1978, Parry and Carter 1985), but this picture is far from complete because the frequency distribution of ecological sensitivities among individual oat plants was not included and neither were the changes in such sensitivities over their lifetime. Before detailing the oats example we therefore first give examples of such data to provide a more complete picture of the stochastic background to ecological situations in general.

We begin with the response of bluebells, *Hyacinthoides non-scripta,* to temperature and rainfall in terms of the seasonal trends in these variables, for this explains the geographical distribution of this species in western Eurasia. This study considers only two specific stages in the life cycle. To extend this perspective to the entire life cycle, we return to the responses of tomato and sweet corn to temperature throughout their lifetime (see Chapter 3), and the variation in their responses among individuals. This consideration is then extended further, with the example of oats in the Scottish mountains, to link the analysis of ecological risk with that of ecological gradients. We know of no species or system that has been subject to as comprehensive an autecological analysis as is reconstructed here from studies on different systems, but this synthesis helps to illustrate the advantages of autecological analysis and the insights to be gained from it. We note further the other aspects, covered in different parts of this book, that would help expand the understanding of such ecological systems.

5.1 Bluebell Life Cycle and Seasonal Structure of the Environment

Figure 9.14 shows the geographical distribution of bluebells in the western part of Europe (Thompson and Cox 1978). In this region, Atlantic climatic conditions prevail, which means that temperatures are relatively low and their seasonal variation moderate, whereas precipitation is high. Figure 9.15 shows the seasonal variation of temperature and rainfall in the British Isles, indicating a general decline from moderate temperatures in early August to lower ones in October. This decline is expressed as a 30-year mean value with its variance, and also as mean soil temperature. The histogram shows, too, the seasonal distribution of rainfall. This pattern of seasonal fluctuation accords with the ecological requirements that sustain seed ripening and germination of the bluebell, as given in Table 9.3. In spring, the moisture level required for the ripening of viable seed is low, whereas it is high for the germination of those seeds in autumn. In contrast, for seed ripening, the plant requires relatively high spring temperatures and low ones for seedling germination later in the year.

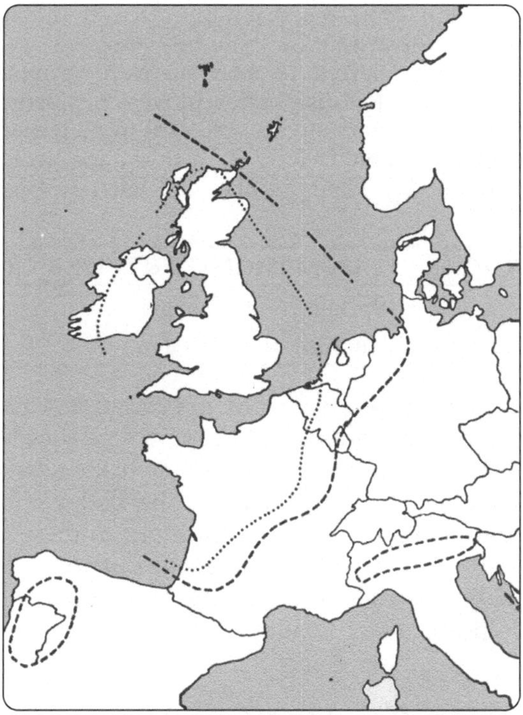

Figure 9.14 Distribution of bluebell in western Eurasia. Redrawn from Thompson and Cox (1978).

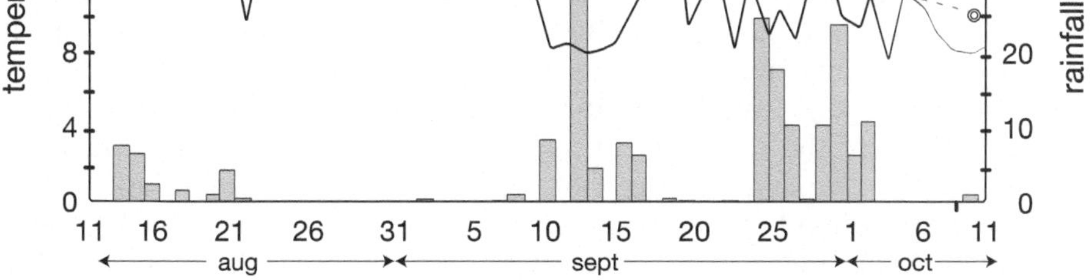

Figure 9.15 Weather conditions during August, September and October within shaded woodland where bluebells grow. Daily precipitation (mm) shown by histograms. Daily maximum and minimum temperature limits contained within shaded area. Circles indicate means of weekly soil surface temperatures. Closed symbol represents the mean monthly temperature at a weather station near the study area and at the same altitude. Redrawn from Thompson and Cox (1978).

Table 9.3 Two biological processes that take place at different stages of the bluebell's (*Hyacinthoides non-scripta*) life cycle are tabulated against two seasonal climatic variables that are relevant to their successful completion, to indicate why the bluebell has a restricted distribution in western Eurasia (Thompson and Cox 1978). Note that other factors are also relevant to understanding the ecology of this species, including edaphic factors, mycorrhizal and light requirements and the timing of their availability (Merryweather and Fitter 1995a, b, 1996, Gonzalez Sierra et al. 1996).

BIOLOGICAL PROCESS	*SEASON*	*MOISTURE LEVELS*	*TEMPERATURE*
Seed ripening	Spring	Low	> 26°C
Seed germination	Autumn	High	< 11°C

The data in Table 9.3 provide a ready appreciation for why the bluebell is restricted to those environments in Europe where the changing seasonal requirements of two parts of its life cycle match the seasonal variation in the two relevant environmental variables. The better and more frequent this match, the more optimal its living conditions in these respects, and the more abundant and widespread the bluebell. But this example, as noted above, includes only two adaptive features, which needs to be expanded for a full autecological understanding of this species.

5.2 Variation in Growing Conditions During an Organism's Lifetime

The responses of tomato plants to thermal conditions through their lifetime was assembled by Wang (1963), from the literature, as outlined in Chapter 3. The variation in responses of seedling, pre-reproductive, flowering, and fruiting plants is plotted in Fig. 3.19, which reveals three important points. (1) For each stage, there exists a range of optimal temperatures, and this lies between a range of temperatures that support lesser growth or production, and outside of this is a range where temperatures are lethal. (2) Through the lifetime of an individual, its relationship to temperature is neither constant, nor gradually decreasing or increasing, because it varies continually. Specifically, its response shows a clear dip during the vegetative stage, and then it rises and finally slowly drops again. (3) The width of the thermal optimal response, as measured against the y-axis, remains the same throughout the life of a tomato plant. Notice that these clearly defined, coherent patterns originate from observations and experiments conducted independently by different investigators. Also, the main performance indicator is yield, but the approach is informative as to autecological method and interpretation.

A comparison of similar data from a different crop species, sweet corn yields revealing ecological differences relative to the three criteria specified in relation to the tomato data (Fig. 3.20). Again, a "central area" of optimal temperatures is identifiable relative to the lifetime performance of the plants. In contrast to tomato, however, sweet corn has a temperature range that remains more or less constant. The width of the response range of sweet corn is similar to that in tomato in not being constant through the lifetime of the individuals. Initially it is broad, but it then decreases for the rest of the plants' life. Also, tomato plants have a broader temperature range for optimal growth and reproduction through most of their life than do sweet corn plants. At the same time, though, tomato requires slightly lower temperatures than does sweet corn. From these observations, one would predict that the wild type of sweet corn lived under more constant and more narrowly defined conditions than did wild type tomato.

Rainfall is also ecologically significant to these plants. The maximum rainfall during tomato flowering, for example, as measured across four weeks over several consecutive years (Fig. 3.22), influences yield, and again it follows some optimum relationship. However, annual tomato yield relative to rainfall does not follow the fitted curves closely. Rather, they are scattered rather broadly, which implies that rainfall is not the only variable determining survival and reproduction statistics in these plants. Temperature is also significant ecologically, as detailed above, but other environmental variables are possibly also influential. Statistical estimates of the relevant variables and the way in which they operate are required for combination into an analytical model that describes the overall response mechanism. See Chapter 8 for justification of such an approach to modeling and how it can be achieved. In short, a species' response is multivariate, but not in the sense of several variables operating independently. Rather, these variables are interactive (see above), and their influences must be weighed relative to one another in a mechanistic model. The model should reflect realistically the way in which the variables operate and interact with one another in relation to the various processes that underpin the persistence and reproduction of the organism. The seasonally shifting responses are thus statistically defined in terms of risk, risk distributions, survival probability, and so on.

The considerations above deal with the requirements of the organisms through their life cycle, and the variance in these requirements with which they can cope. They do not include the ecological risk faced by those individuals through their life cycle. Risks of this nature vary from year to year. That is, the variation in the temporal pattern of living conditions is unique each year. Fig. 3.21 shows the decrease in tomato yield relative to the number of days with temperatures higher than the upper temperature that supports optimal yield during flowering, 33.3°C. The greater the deviation of temperature into the sub-lethal parts of the response range, the lower the yield. Also, the longer the spell of high temperatures lasts, the steeper the regression, and lower the yield. Thus, the extent of the deviation from the upper optimum temperature represents individual risk whereas the frequency of occurrence of such a temperature, weighed by period length, represents the statistical risk (Subsection 4.3). Figure 3.21 thus integrates these two types of risk for tomato plants.

The performance of tomato plants is independent across years (Fig. 3.23), because this species is annual in its life cycle. In one year this species may yield well because of favorable temperatures whereas the following year's performance may be disastrous, because of heat or cold. Yet other ecological influences may intrude. Rainfall may be too low or too high for reproduction and so on. The risk of survival and reproduction thus varies substantially and independently across growing seasons because of a few specific, identifiable ecological factors that operate relative to different species-specific physiological properties.

Figure 3.23 shows that the yield of 1948 (top left hand data point) was clearly aberrant, with the crop failing because of too much rain, despite temperatures being close to optimal. This suggests that the data in Figs. 3.21 and 3.22 should be integrated for a more complete picture of the ecology of tomato plants under cultivation. Fig. 3.23 thus shows areas of equal yields delimited by yield isopleths as a function of both the frequency of occurrence of temperatures that are too high on the x-axis and the amount of rainfall on the y-axis. The yield of 1948 is now in the same range as the yields of 1949 and 1952, each of which experienced less excessive rainfall but higher temperatures. Yields across these years were similar to one another, but for different ecological reasons. In this way, one can forecast the yield from information on temperature and rainfall during the flowering period. Also, local regions can be delimited and chosen for tomato production, based on the frequency distribution of temperatures and rainfall in the recent past. Yield expectation can thus be estimated based on autecological understanding.

Biotic and edaphic variables such as insects, diseases and poor agricultural practices have also been considered (Wang 1963). The correlation between these variables and temperature and rainfall proved to be high, so any deviations from the correlation were considered to represent influence from these other variables and that they were, therefore, of only minor importance. Consequently they could be dropped from further analysis. This procedure of stripping minor causes of variation from those that are the principal determinants of the process under investigation, starting from a complete set of observations, is a good approach to revealing the relevance of potentially influential factors in autecological studies.

5.3 Risk Analysis in Scottish Oats

The yield of oats grown in Scotland has been analyzed in terms of risk (Parry 1978, Parry and Carter 1985), primarily in relation to annual climatic variation. In these studies, the specific requirements of the plants were not considered in the way outlined in the previous subsection for other crops, and neither was the variation in physiological responses between individual plants or the seasonal structure of the physiological requirements and responses of oats. Despite this simplification, however, the methodology of risk analysis is clearly revealed by the oats study and, through this, the spatial responses of oats can be tracked across several centuries. This example thus extends the illustration of how autecological investigations and analyses need to be structured.

The analysis of oats production focused on the minimum number of growing degree-days (GDD), with GDD being a measure of the thermal energy that each plant needs during the growing season for it to grow from seed and into a mature plant that yields viable seeds. Not only is the GDD requirement species-specific, but so too is the lower temperature level above which physiological processes operate. The required number of GDDs increases by 30 for every 100 mm of precipitation above a minimum of 200 during the vegetation period, at least for barley in Scandinavia (Bergthorsson 1985). In the case of oats, the minimum number of GDDs required for seed production is 970, and the minimum temperature at which thermal energy induces a physiological response is 4.4°C.

With the figures above, the chance of oats not reproducing within any locality can be calculated, simply by estimating the proportion of years when this minimum GDD requirement is not met. This figure defines, directly, the risk of oat plants failing to reproduce. Furthermore, the risks can be calculated for different altitudes, by incorporating the decrease in temperature with every 100 m increase in altitude, the lapse rate, for the particular latitude and geographical setting concerned. Isolines can thus be mapped across localities with different levels of agricultural risk. The approach to risk analysis outlined above can be taken with respect to historical periods that experienced different climatic conditions. The history of oats cultivation in a mountainous area like Scotland can thus be reconstructed and, in turn, be tested with historical data on the dates of occupation or abandonment of fields at various altitudes. Figure 9.16 shows how high-altitude oats cultivation during the Early Middle Ages retreated to lower altitudinal levels in subsequent centuries, the Little Ice Age, as expressed by the abandonment of settlements as temperatures dropped. Oats cultivation at higher altitude was resumed when temperatures increased in the 20th century.

An ecologically more revealing approach is to calculate the extinction risk of oats, given its minimum requirement for GDDs, and so on, as a function of the lapse rate for each of the two historical time periods separately. These two curves reveal not only a shift in risk across the two historical periods (Fig. 9.17, left), but also shows the equivalent curve that plots those times when inimical conditions span, by chance, two consecutive years in succession (Fig. 9.17, right). The conditions in each such two-year period need not be identical to one another, but need only be poor for the growth of oat plants. Such coincidence is ecologically relevant, for the seed bank of wild species reduces at specific rates, and may thus be exhausted when unsuitable conditions continue for too many successive generations.

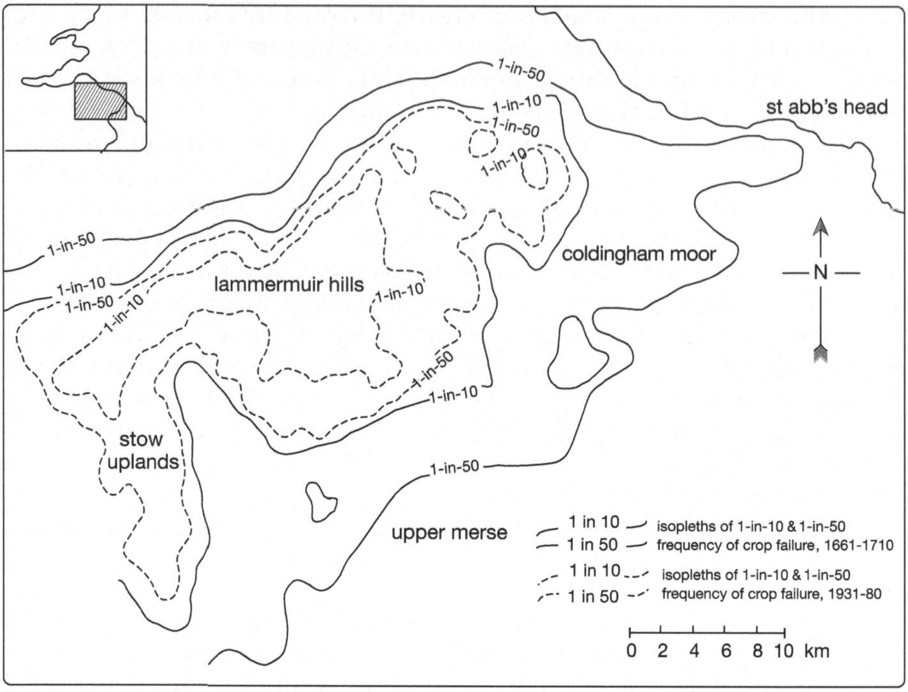

Figure 9.16 Map to show how oats cultivation was shifted altitudinally between the cool (1661 - 1710) and warm (1931 - 1980) periods investigated, according to the risk of crop failure (expressed as a frequency). Reproduced from Parry and Carter (1985), with permission.

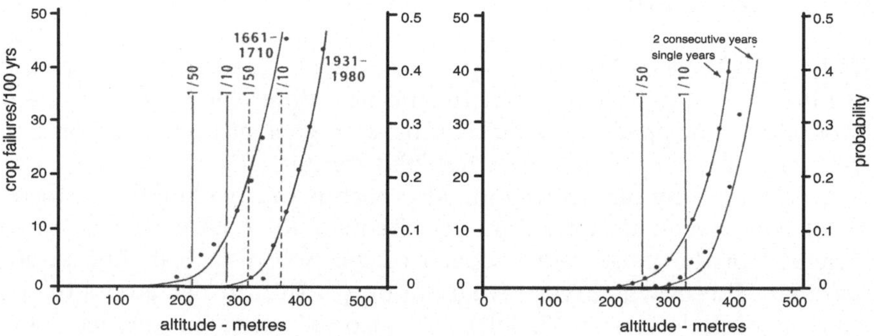

Figure 9.17 Risk distribution of oats as the number of crop failures per century as a function of altitude on a Scottish mountain. The curves represent the expected risk distribution and the dots represent observations. The left hand diagram includes risk distributions for two historical periods that were climatically different from one another. The right hand graph shows the probability of a crop failure in a single year and also that for failure in two consecutive years. Reproduced, with permission, from Parry and Carter (1985).

Figure 9.17 also shows that the risk curve is not straight, but flattens at lower altitudes. A similar plot across latitude would show the risk values increasing again, with the species again failing to reproduce, but now because of higher temperatures towards lower latitudes. The risk of extinction plot is the reverse of the species' survival probability, so the trough can be inverted to interchange the two probabilities. The resulting curve now resembles an optimal response curve in relation to temperature or precipitation, as in the curves for tomato and sweet corn (Figs. 3.19 and 3.20). In the case of oats, however, the curve does not represent a purely physiological response, for it also contains the stochastic element of variable conditions over the years. As an inverted curve of extinction risk, this probability distribution of survival and reproduction is based on a physiologically relevant energy measure, GDD. It thus also resembles curves generated by gradient analysis in vegetation research.

5.4 Conclusion

The synthesis presented in this section, as a whole, reveals how autecological research can explain the spatio-temporal dynamics of any species. The principles are clear, and so is the autecological generalization. Specific cases, though, demand the appropriate data. This aspect of autecology has been seen as a fatal weakness, and a justification for the search for the overarching generalities sought in population ecology, community ecology and macroecology. The autecological information presented in this section helps, though, to illustrate why such demographic generalizations, as desirable as they may be, are unlikely to hold for natural systems. Ecologists must work within the realities of the subject material rather than pursue idealizations of nature.

6 THE DEVELOPMENT OF AUTECOLOGY—CENTRAL QUESTIONS

The central questions in autecology focus on the way in which the adaptations of individual organisms influence their probability of survival in a spatially and temporally varying environment. Answers are sought with respect to how the properties and requirements of individuals match the fluctuating environmental conditions to which they are exposed throughout their life. Quantification of the precision of that match is crucial for autecological understanding. The organismal context for such understanding is the way in which the properties and requirements of individuals change with age, as illustrated for plants in the preceding section (and see also the examples in Chapter 3, Section 3).

To what degree do the properties and requirements of organisms match the shorter or longer seasons as latitude increases, and how do they cope with summer drought or extreme winter cold? How easily can individuals track the conditions to which they are adapted, across space, when conditions change? We know that even sessile organisms, like plants, can track such change, but we remain uncertain about plants with seed dispersal limited mostly to the seeds dropping to the base of the parent plant, presumably because they have lost their biotic vectors through extinction (e.g., Snow and Walter 2007, Hall and Walter 2013). If they cannot track suitable conditions do they have properties that compensate for such a shortfall in mobility? A crucial reality for ecology is the species-specificity of the adaptations and requirements that are ecologically relevant, and the general observation that these features are species-wide (Rajapakse and Walter 2007, Popple et al. 2008) (and see Fig. 5.1). This is the ecological realization of the punctuated equilibrium pattern that is evident in the fossil record, and which is now widely accepted as representative (Eldredge et al. 2005).

Ecological questions posed from the perspective just outlined enrich ecological understanding, because the demographic questions and interpretations that currently dominate ecology focus on patterns that are the numerical consequences of autecological matching. Demographic ecology thus downplays the significance of the adaptive mechanisms that make up organisms and make up the mechanistic focus of evolutionary theory, at least that theory associated with the recognition concept of species (Walter et al. 1984, Paterson 1986, Hulley et al. 1988a, Walter and Paterson 1994, Hengeveld and Walter 1999, Walter and Hengeveld 2000). We are not saying that no progress has been made in ecology. Substantial progress has been made, but this has almost invariably come from insights developed independently of demographic theory (Walter and Paterson 1994). This point is developed in the following chapter where the two approaches to ecology are contrasted and placed in the context of all the other biological sub-disciplines.

Chapter 10

Contrasting the Ecological Paradigms—Principles, Related Sub-disciplines, General Laws and Prediction

1 INTRODUCTION

The earlier chapters developed the argument that ecology currently has two contrasting approaches that are mutually exclusive. They are paradigmatic in that each defines a set of coherent, interrelated theories that are based on mutually exclusive foundational assumptions, or "hard cores" as Imre Lakatos called them (Kuhn 1962, Chalmers 2003).

Naturally, the description above of the situation in ecology is somewhat clinical, for the lines demarcating the two paradigms are blurred by current interpretation and practice. A contributing problem is that autecology theory has, until now, lacked clear articulation and so has remained inadequately defined. Autecological research and information is therefore almost invariably and misleadingly portrayed as an adjunct to demographic theory (in ways outlined by Walter (2013)) rather than as a valid alternative approach to ecology (Hengeveld and Walter 1999, Walter and Hengeveld 2000). It has even been placed, inexplicably, within the ecological hierarchy (Beeby 1993, Pickett et al. 2007). Autecology is thus readily portrayed as descriptive natural history that is subservient to demographic ecology theory and research, even if simultaneously it is seen to have some practical use (e.g., Lawton 1993a, Kareiva 1994, Murdoch and Briggs 1996).

The structured theory for autecology developed in Chapters 2 to 9 does allow a more accurate contrast between the two paradigms to be developed, the goal of this chapter. Before expanding on how this is done, we consider the ambivalence that is evident in any general inspection of ecological research and interpretation. Although demographic theory may work well enough in some circumstances, most of the recent advances in understanding ecology have been developed independently of demographic theory. These include insights about the shifting distributions of species in response to climate change and the prevalence of species stability over time and across space (Chapter 5, Section 3). Although developments of this nature challenge the core of demographic ecology, they have simply been expropriated and annexed to demographic theory. In some cases adjunct sub-disciplines have been developed in such form that the new observations and interpretations no longer challenge the core assumptions of demographic theory. Spatial ecology and landscape ecology seem to have developed in this way and tend to focus on the associated technology more than on organisms and their interactions with the environment, as pointed out by Mitchell and Powell (2003). In other cases, the very pattern that presents the challenge is explained in terms of demographic processes. For example, species stability has even been interpreted in terms of local, rapid adaptation that somehow "averages out" into stability over time (e.g., Eldredge et al. 2005), despite the unusual number of unlikely steps and added assumptions, and despite the straightforward alternative interpretation offered by the Recognition Concept of species and its associated interpretation of adaptation (Chapters 5 (Section 6) and 6 (Subsections 4.3 and 4.4)).

Most current ecological research is conducted under the guidance of demographic theory, either to extend that theory or simply in conjunction with it. Much of the former aims at the verification of the importance of particular processes, primarily density dependence and competition, or of particular outcomes driven by such processes. This is particularly evident in the search for examples of character displacement, local adaptation and sympatric speciation (which is often dealt with under the unjustifiably expansive cover of "ecological speciation" (Chapter 8, Subsection 3.1)). Such observations and research, which adopt demographic principles, are readily assigned to their "home" paradigm. So, too, is the search for consistent patterns across species. Robust patterns of this nature have proved elusive (e.g., Greig-Smith 1986) and yet are still sought, even over regional or global spatial scales (e.g., Ricklefs 1987, Lawton 1999, Blackburn and Gaston 2001, 2006). Presumably the aim is to verify the significance given to biotic factors in demographic ecology and its evolutionary arm, evolutionary ecology.

Other research is structured from influences outside of demographic theory, as in biological sub-disciplines somewhat removed from the premises of that theory (e.g., paleontology and applied fields in which particular species are targeted for detailed scrutiny (Walter 2003)). Such research is more difficult to categorize because the questions that underpin it often do not relate directly to demographic theory, and the data and interpretation are commonly at odds with the principles of demographic ecology. Paradoxically, demographically influenced interpretations are often appended, however inappropriate. Some of this research is referred to as autecology or natural history, both of which are widely considered to be anecdotal and removed from theory, but nevertheless useful. The formalization of autecology through Chapters 3 to 10 should help remove this ambivalence about the status of autecology. A more realistic assessment of the overall structure of ecological theory and the place of different pieces of ecological research becomes possible, and the way in which we analyze this is outlined in Section 2.

Ecologists might wish to steer a middle course and treat the two paradigms as complementary, two extremes of a continuum, or as not reflecting anything fundamentally different within ecology. Indeed, the paradigmatic discrepancies in ecological theory and practice are frequently handled in this way. To illustrate how such a compromise approach fails, we represent in tabular form the nature and relative structures of demographic ecology and autecology as we developed them earlier (Hengeveld and Walter 1999, Walter and Hengeveld 2000), to contrast the principles and internal logic of each paradigm and thus demonstrate that the differences in approach are fundamental and irreconcilable (Section 3, Table 10.1). Such a stark comparison allows a paradigm-based re-examination of community and ecosystem ecology, and scrutiny of the widely accepted view of ecology being hierarchical (Section 4).

Section 5 extends the comparison, begun above, across the two paradigms, to represent the way in which each of them is related logically to other sub-disciplines in biology (Table 10.2). An appreciation of these latter relationships is critical to understanding that autecology has a different intellectual support structure, within biology, from that of demographic ecology. Significantly, these sub-disciplines are themselves open to alternative interpretation, and these alternatives align differentially with the alternative ecological paradigms. The link between ecology and behavior provides a clear illustration, as follows.

Autecology has mechanistic links to ethology, which developed with a strong focus on the complex nature of adaptations and how they relate to the environment. A good example is the Specific-mate Recognition System (a subset of the Fertilization Mechanism) by means of which individuals of motile species locate mates, interact with them, and achieve syngamy. These are strongly structured catenary sequences, with each intermediate step serving a particular function within a particular environmental or behavioral context (Fig. 3.12). The overall outcome tends to be consistent, given particular circumstances. Adaptive change to such systems is seen to take place only when forced, in small populations under strong selection from the environment (Chapters 5 (Section 6) and 6 (Subsections 4.3 and 4.4)).

By contrast, demographic ecology has direct logical ties to behavioral ecology, which focuses on optimization and the efficiency of particular expressions of behavior, and which is seen to be driven demographically by competitive selection. Even the recognition system, as it pertains to sexual species, has been dealt with in this way, and interpreted in terms of preference (e.g., Sherman et al. 1997), rather than the strictly defined process of recognition (Paterson 1982b, 1985). Such conflation of strict alternatives not only misleads about the structure of ecological thought, but also obscures vision with regard to extending theory and understanding, and thus improving the practice of ecology. Consequently, ecology may appear far more factual to many practitioners than it indeed is. This chapter argues against the view that the direction of current ecology is the best available.

We believe that an appreciation of the alternative interpretations and relationships among the biological sub-disciplines should help substantially in evaluating the mutual independence of the two ecological paradigms. All of this emphasizes that autecology is not intended to be seen simply as a modular replacement for demographic ecology within the current arrangement of biological sub-disciplines. Instead, the acceptance of autecology as an alternative invites re-examination of the relationships among sub-disciplines across the breadth of biology. Fortunately, such analyses have already been conducted within most other biological sub-disciplines and the alternatives available have been well articulated, for example in species theory (Paterson 1993c), molecular genetics (Rose 1997) and developmental genetics (Carroll et al. 2005a).

The confluence of these theoretical developments is revealing of the historical development of autecology. Andrewartha and Birch's (1954) book has been highly influential and is still frequently cited. It did not manage, though, to dent the dominance of demographic ecology. Two particular issues in autecology seem to have played a role, the evolutionary underpinning of ecological theory (Orians 1962) and the question of whether autecology was to be pitched at individual organisms or at species. These problems can now be resolved (Section 6) before we illustrate how the difference between the two paradigms runs even deeper than detailed in Sections 2 to 5. Section 7 considers how each paradigm and its expectations relate to general perceptions of ecology having "laws of nature" and thus predictive ability (Table 10.3). Indeed, the distinction between the two paradigms should help resolve some of the topical issues of the nature of ecology as a science, and this point is expanded later.

2 COHERENCE IN ECOLOGICAL THEORY

Before examining the two ecological paradigms in detail, we explain how we have approached the construction and description of each paradigm, for this is necessarily a process of simplification. Basically, we dismantle current ecological practice and theory into the two paradigms, to extract the principles that drive each of them. This is a process of constructing generalizations. Surprisingly, no coherent set of ecological principles has ever been explicitly stated (as pointed out by Cooper (2001) in relation to the concept of population equilibrium), which we take as reflecting four points.

1) A common set of basic principles, which we have labeled demographic principles (Hengeveld and Walter 1999, Walter and Hengeveld 2000), is relatively widely accepted, albeit generally tacitly so (Cooper 2001, Cuddington 2001). This situation exists despite the widespread confusion about what constitutes a principle in a discipline that is as fragmented and multipartite as ecology is today.
2) Modern ecology accepts hierarchical levels within the discipline (population, community, and ecosystem), even though no clear mechanistic or logical connection is evident between the proposed hierarchical levels (Chapter 6, Subsection 3.2.3).
3) There seems to be general satisfaction with the ongoing *ad hoc* expansion of ecological theory to accommodate conflicting evidence and interpretation as it arises (Lawton 1999, Paine 2002). This is done without any adjustment to the underlying ecological principles even though a substantial volume of observational evidence conflicts with the foundational statements of demographic ecology (e.g., Walter and Paterson 1994, 1995).
4) The most widely-appreciated "dissidents" in ecology (Andrewartha and Birch 1954, 1984, Murray 1979, Simberloff 1980, 1984, Strong and Simberloff 1981, Peters 1991, White 1993, 2001, den Boer and Reddingius 1996, Drury 1998, Murray 1999) have not articulated a comprehensive alternative set of principles and associated methodology to provide a way around the problems they have identified over the years.

The situation encapsulated in the four points above raises the issue of the development of generalizations about each ecological paradigm in this chapter. Because of the extreme fragmentation of ecology and ecological opinion (see Graham et al. 2002), apparent exceptions and deviating opinions are likely to be ever-present. Counter-arguments may therefore be raised immediately against the validity of any particular generalization that is made about ecological theory as this chapter is developed. We therefore request patience so the argument we present can be examined in its totality. Any interim distractions can readily be assigned later to a place in one or other of the alternative structures we outline. The commentary so far in this chapter implies that autecology is not altogether new, and that autecological research has been practiced for some time. The problem, however, is that autecology has had neither a generalized description nor a formalized structure, until now. Its strength in explaining ecological phenomena could therefore not be appreciated and its full potential not achieved.

Central to our approach is that a paradigm is definable in terms of fundamental principles that are common to the set of theories that represent the paradigm. This is the hard core of the paradigm or research program (Chalmers 2003, p. 131). Our aim is to state these principles as starkly as possible, and then to trace the supplementary statements that inevitably follow and which form the various theoretical constructs that shape the paradigm and the research conducted in its name. Such sequences of statements thus exhibit the paradigms as conceptual models, and detail the coherence and logical relationships between the ecological theories that constitute each paradigm.

Making robust generalizations in science (and about science) is always a difficult process, because different individuals almost invariably hold different views to some extent. The structure of ecology, as a science, coupled with the complexity of natural systems, combine to hinder efforts at extracting principles. Nevertheless, generalization must be possible for otherwise ecologists would have to admit to science without any fundamentals in place. Some may feel that the advocacy of pluralism in ecology breaks the need for general principles. But this is not so, for the diversity within ecological pluralism must inevitably be bound at a deeper level by common statements of principle (Walter and Paterson 1994, Walter 1995, 2008). These deeper principles represent the foundational statements from which theoretical developments and practice are launched, and these are the generalizations we seek.

3 BASIC APPROACH AND PRINCIPLES OF EACH PARADIGM

The starting point of each ecological paradigm is defined in relation to the most fundamental statements that can be made about each; these are their foundational statements and they relate to such issues as the primary patterns to be explained in ecological systems, the features of ecological systems that are given central prominence, and so on. Table 10.1 outlines the different perspectives that each ecological paradigm emphasizes. Before detailing these specifically for each paradigm, we emphasize that these are the most basic thoughts or assumptions about ecological systems (Part A of Table 10.1) and thus represent the most basic principles involved in each of the overall views of ecological systems. They effectively direct thought and questions. They thus direct what data are emphasized and what information is downplayed through its being cut from consideration (the "Bronowski cut" (Walter 2003, pp. 33 and 223)). In effect, these principles

Table 10.1 Contrast between the ways in which the two current paradigms in ecology (demographic ecology and autecology) perceive, investigate and interpret ecological systems. Note that the generalized statements below derive principally from the research questions and interpretations of ecologists, and thus reflect the way in which ecologists conceptualize ecological systems (**Part A**) and work with them (**Part B**). Part A thus represents the principles of each paradigm.

	SYSTEM OF ECOLOGICAL THOUGHT (OR ECOLOGICAL PARADIGM)	
	DEMOGRAPHIC ECOLOGY	*AUTECOLOGY*
A. Perceptions of ecological systems		
Primary pattern to be explained	Population equilibrium (persistence) and community structure	Geographical spatio-temporal dynamics of species
Central feature of system	Resource availability and density dependent influences (demographic forces)	Individual organisms and their species-specific adaptations relative to their relevant axes of environmental differentiation
Environment—primary components	Biotic (demographic structure) and abiotic (disruptive)	Structured physico-chemically (climate, soils, etc.) and biotically (hosts, vegetation, etc.)
Organism-environment interaction	Density and coexistence relative to resource availability	Environmental requirements and tolerances of individuals relative to dynamic spatio-temporal structure of environment
Nature of ecological system	Equilibrium and homogeneous	Heterogeneous and dynamic open systems ("non-equilibrium" is misleading)*
Level of explanation	Supra-organismal	Individual organism
B. Approach to ecological analysis		
Variables	Carrying capacity, r_{max}, density dependence, optimal fitness, limiting similarity, empty niche	Nutrient intake, energy loss, energy partitioning, physico-chemical factors, tolerances, environmental axes, survival rate, reproductive rate
Climate—temporal variation	Reduced to average value	Incorporates scales of variation
Stochastic (chance) influences	Eliminated or amalgamated into a single error term (reduced by ecologist)	Organisms must cope with chance processes, whose influence is reduced by species-specific adaptations
Spatio-temporal scale of interpretation	Locality and time specific, with phylogeny factored out	Species-wide, potentially global and relative to phylogeny

* Perceived disruptions to equilibrium result in ecological systems often being portrayed as non-equilibrium. This does not represent an alternative conception of ecological systems to the demographic one (Walter 2008).

dictate how ecologists approach their investigation (Part B of Table 10.1) through the ecological variables that are given significance, how stochastic factors are treated, and which scales of variation are seen as important. That we are dealing with fundamentally different "ways of seeing" (Hughes and Lambert 1984) is clear from the comparison in Table 10.1. The reasons for the differences in emphasis across the two paradigms are also explored, as this helps to explain their distinctness and why they are mutually exclusive.

3.1 Demographic Ecology

Demographic ecology is concerned principally with processes considered to reduce the amplitude of variation of population numbers around the mean or equilibrium population density that is assumed to be typical for each population, or even species. The equilibrium is usually treated as static and is usually characterized by the mean population density although the estimates are invariably made at different times. Algebraically this is a problem because different "types" of objects are effectively sampled, with these being different individuals that are likely to have been subjected to quite different combinations of ecological influences. Averaging them to derive a "higher order" quantity seems to be inappropriate.

Even if one accepts that this algebraic obstacle is not debilitating, additional problems are evident. By definition, the observed numbers deviate from the mean, and these deviations are interpreted as fluctuations around equilibrium and therefore considered to be representative of a continuous range of fluctuation around the mean. Outbreaks or extinctions are considered to lie outside this range, and are treated as exceptional disruptions to equilibrium conditions (Pimm 1991). Non-equilibrium systems are therefore interesting only for that reason: "Why are they not in equilibrium"? Clearly, the idea of non-equilibrium ecology needs to be treated carefully (Rohde 2005). If ecology is "non-equilibrium", it does not need non-equilibrium theory, it needs an alternative theory that is based on the primary ecological influences that affect organisms (Walter 1995, 2008), as stressed throughout this book.

The concepts emphasized in demographic ecology, namely population equilibrium and persistence, coexistence, density dependence and so on (Part A, Table 10.1), are not readily related to the geographical distribution of organisms. Recent efforts to incorporate spatial elements into demographic theory (Guo et al. 2005) are as awkward and unrealistic as earlier ones. Specifically, island biogeography (e.g., MacArthur and Wilson 1967, MacArthur 1972b) and metapopulation ecology (Gilpin and Hanski 1991, Hanski 2004) tend to deal with the spatial element almost entirely in terms of island (or patch) size and distance from source. A further significant problem with these treatments is that the mechanistic connections between the variables that are seen as significant and the conceptual constructs are not explicit, mainly because the demographic principles are not amenable to such treatment (Chapter 7).

Usually, therefore, studies are carried out on populations and communities in arbitrarily restricted local areas. The parameter values for reproduction and mortality tend to be treated as population characteristics that can be generalized (e.g., *r* (Chapter 7)). Demographic ecology is thus concerned with supra-organismal phenomena or processes. That is, populations with particular quantitative characteristics are treated as basic units or entities. Within this context, natural selection implicitly involves group selection, even among several species simultaneously (Hengeveld 1988b, Walter 1991, 1995). Invoking group selection, even if it is done implicitly, means that demographic ecology effectively uses the parameter values of populations as species-wide properties (e.g., in the evolution of niche characteristics), and considers these values to be subject to natural selection. Extension of these aspects to regional levels (e.g., Ricklefs 1987, Ricklefs 1989, Mora et al. 2003, Munguia 2004, He et al. 2005) is even less appropriate. And since environmental conditions do vary spatially, demographic studies are usually carried out locally and over the short-term, to render environmental conditions as constant and uniform as possible. Demographic conditions are thus time and locality specific, and the characteristics of organisms are interpreted to be consistent with these local and temporally constrained strictures. A significant consequence is that ecological systems in all localities are seen to drive the adaptive process somewhat ineluctably and differentially across localities. The geographical mosaic of coevolution (Thompson 2005) is a typical example.

The difficulty for the approach to ecology and adaptation just outlined is the observation that the physiological, behavioral and other properties of individual organisms are clearly characteristic of the species as a whole, given a small amount of intraspecific variation (e.g., Walter et al. 1984, Paterson 1993d, Rajapakse and Walter 2007, Popple et al. 2008) (Chapter 5, Section 3). Even evolutionary ecologists, who represent the evolutionary extension of demographic ecology, now accept that species are stable, but still try to interpret the adaptations of their study organisms in terms of local, spatio-temporally static processes (e.g., Eldredge et al. 2005).

Demographic ecology does not consider that variation in the rates of reproduction or mortality could be a consequence of the interactions between local environmental conditions and the complex adaptive mechanisms of the individual organisms that make up the populations under consideration (as detailed in Chapter 9). And it does not consider how those adaptations could have evolved in relation to the demographic processes that are given priority in demographic theory. The problem is that the adaptations of organisms are not related directly to demographic processes like density dependence. Indeed, such a connection cannot be made. For instance, how can selection for efficiency in resource use relative to other species really translate into the transformation of physiological and other mechanisms (Walter and Donaldson 1994)? No answer follows directly from the demographic principles or equations. Yet, differences within or between species in physiology, morphology and behavior are often obviously relevant ecologically, as illustrated so clearly in the Adelie penguin example of Chapter 3, for example. Understanding these properties, their evolutionary origins and, especially, their ecological consequences represents the field of autecology.

3.2 Autecology

Autecology provides an alternative perspective of the living world to that just outlined, and poses different questions about the distribution of abundance of organisms. What are the primary factors that influence the survival of individual organisms and, as a consequence, the spatio-temporal dynamics and persistence of the species they constitute? The focus is firmly on the properties that enable individual organisms to survive through their life cycle within the context of a dynamic and heterogeneous (and therefore unpredictable) environment, and to reproduce within that context (Chapters 3 and 4). It is the nature of the environment and thus the organism-environment interaction that gives these issues their significance. In autecology, the population density and community structure (in demographic ecology terms) within any specific locality are set aside, as issues of secondary or epiphenomenal importance. They can be explained (in terms of ecological intensity and local species diversity, including rare species) once the fundamentals of the components of the system as a whole are understood. The explicit connection between individual organisms and the environment provides a direct logical link to explaining the spatio-temporal dynamics of species (Chapter 9).

The most fundamental process in relation to living systems remaining operational within an environmental setting is related to the energy flow through the individual, in conjunction with material (or mass) flow (Kooijman 2010). What are the requirements of organisms for achieving this end, and what properties enable them to do so? These issues extend to questions of how these various properties that make up individuals are combined in their operations (Chapter 5), and how they interact to yield a functional organism that can maintain itself (Haukioja 1982) within an environmental context that is temporally dynamic and spatially heterogeneous (Hengeveld and Walter 1999, Walter and Hengeveld 2000)? How do organisms cope with the vagaries of the environment, which, although structured, is dynamic and strongly stochastic in nature? Besides energy sources, organisms interact with elements of the temperature profile of a locality (extremes, durations, diel and seasonal dynamics, and so on (Chapters 3 and 4)), as well as the physico-chemical and biotic factors that continuously affect the various operations of the organism through the various phases of its existence. These factors include humidity, edaphic conditions, vegetation quality and structure, host availability, host quality and so on (see Appendix). Such influences are well known to ecologists but the subtle variation (across species' life stages and across species) in the ways in which they operate is not widely appreciated, and especially not in relation to the spatio-temporal dynamics of the environments that organisms inhabit. In any case, the adaptations of organisms that relate to their ecological requirements and tolerances are species-specific, highly idiosyncratic, and often very ancient "tools" for maintaining their energy flow.

As long as an organism's energy flow is maintained, the individual can survive. When one or more of its attributes fail, energy will ultimately cease to flow and the individual will inevitably die. This energy flow follows general physical principles and applies to all organisms (Kooijman 1986, 2010), and this was so from the earliest stages in the evolution of life (Hengeveld and Fedonkin 2007). This does not translate into general rules for the structure and nature of adaptations or the properties of organisms, for these are idiosyncratic, and for this reason beyond prediction. Also, adaptations are not acquired as deterministic refinements relative to ecological continua (e.g., the proposed *r* & *K* selection and specialist-generalist continua (Walter 2003, pp. 222 to 224, Finlay-Doney and Walter 2012a)), although such comparative studies are often conducted across species and along such continua. At a particular stage of the life cycle, reproductive organs develop (or mature) and reproductive behavior is initiated. Reproduction thus correlates with a particular compartmentalization and level of energy flow within the individual organism, and usually within the context of particular environmental conditions, and is the inevitable outcome of internally developing and evolving conditions, as influenced by environmental circumstances.

Autecology stresses that organisms exist within a variable physico-chemical environment, and that organisms cope mainly through two types of adjustment. On the one hand, phenotypic adjustments to biochemical, developmental, physiological and behavioral mechanisms, in response to environmental circumstances, are widely appreciated, often as "reaction norms" (e.g., Stearns 1992, Gilbert and Epel 2009). The resulting organismal flexibility is relatively small in amplitude. Organisms also adjust their position geographically, and this is as true of sessile organisms, which have highly motile offspring or propagules, as it is of motile ones. Individual organisms move across various distances and track their usual environmental conditions by dispersal movements and habitat tracking (Davis 1976, 1986b, Coope 1978, 1979, Huntley and Webb 1989, Huntley 1991, Parmesan et al. 1999) (Chapters 4 and 9).

Habitat and environmental tracking effectively renders the individuals' external living conditions as constant as possible. Any associated changes in reproductive and mortality rates are best seen as consequences (or effects) of the changing organism-environment interaction, rather than driving forces in the competitive evolutionary adjustment process envisaged under the demographic paradigm. Fitness, as measured under such circumstances, is a consequence of the actions of organisms, not a cause for their actions (as portrayed in demographic ecology). In any case, movement away from deteriorating conditions will work against local adaptation. That this interpretation is realistic is illustrated by the properties of individual organisms being species-wide (see above).

The autecological view does not deny evolutionary adjustment. Adaptation takes place when small populations are trapped for extended periods under environmental conditions that differ from those to which the organisms are primarily adapted. Spatial adjustment is denied and conditions may deteriorate as climate changes (Paterson 1985, 1986, Walter and Paterson 1994). Such contingent circumstances may not be uncommon, but they are not observed as a regular process in ecological systems. The best possibilities, apart from island systems (which do not necessarily reflect well the geographical elements of continental and oceanic systems), may be seen in relictual populations, as in the locally distributed subspecies of some butterflies (Walter and Zalucki 1999). Although genetic differences are often readily detected across localities, which may be as restricted as individual trees hosting herbivorous insects (e.g., Boecklen and Mopper 1998, Mopper and Strauss 1998), ecologically meaningful change to major mechanisms is not.

Physiologists, starting with Claude Bernard, have long appreciated the central significance of organisms buffering their "internal milieu" against changing outside conditions. This, ultimately, is the explanation of why the various forms of adjustment outlined above are necessary. Various adaptive mechanisms are known to operate to this end, which is the context for explaining spatial adjustment; it is a supplementary mechanism in maintaining internal constancy within the organism. The ecology of an organism thus represents the way in which it structures its energy flow relative to various environmentally induced losses and maintains itself to the point(s) of reproduction (Kooijman 2010). The mechanism of energy extraction and conservation evolved some four billion years ago and has been basic to all organisms ever since (Kooijman and Hengeveld 2005). For the earliest energy conservation mechanisms to remain functional under variable and changing conditions, biological mechanisms evolved to protect the basic process and then to improve it. Thus, the initial energy gain from inorganic reactions was essential to the operation of biological membranes (Hengeveld and Fedonkin 2004, 2007).

The original chemo-litho-autotrophic basis for existence gave rise to systems that replaced mineral energy sources with photosynthetic (photo-autotrophic) energy gains, such as in cyanobacteria. Improved storage systems could then be added to organisms, one for storing the energy gained from the environment as a buffer under variable feeding conditions (chemical storage) (Kooijman 2010) and another for stabilizing the internal processes within the system (genetic material) (Haukioja 1982, Rose 1997). These aspects cannot be related directly to the principles of the demographic ecology paradigm, which again provides a strong contrast with autecology. Autecology sees these storage processes as significant ecological adaptations that evolved within the context of the environment prevailing at the time of their acquisition.

The subsequent diversification of adaptations was broad, and centered around the provision of mechanisms for the temporal stabilization of environmental conditions. Organisms thus evolved effective self-stabilization mechanisms to compensate for the absence of uniform spatio-temporal environmental conditions. They could thereby hold their internal conditions within much narrower limits than those imposed stochastically from without. This compensatory adaptive adjustment to the dynamics of the environment is imperative for the within-generation survival and across-generation persistence of living systems and represents the essence of living entities (Haukioja 1982). Phenotypic, behavioral and spatial adjustments serve the same purpose but at the ecological scale. Autecology centers on the various types of adjustments that safeguard all organisms' requirements for maintaining the function of its central life process, its internal energy flow, and for producing the following generation. Thus, autecology is primarily concerned with the processes that enable the survival of individual organisms and their production of offspring. Fitness, though, is not seen as a cause, but a consequence.

The points raised above extend to a broader scale, to questions of how one set of properties within an organism influences its other properties. Arthur (2004) refers to this as co-adaptation within the organism, and sees it as a significant but neglected aspect, revived only recently by evolutionary developmentalists. For instance, how does mobility or a sessile way-of-life impose on the biochemical, physiological, anatomical and behavioral mechanisms that constitute the other adaptive mechanisms that make up the organism, and how does it affect its requirements and means of fulfilling these? Such influences are likely to be interactive or reciprocal, because organisms are functional wholes. For example, what demands do any of these features make on hearing or vision? How do the properties form complexes of co-adapted adaptations, which are referred to as complex adaptations or character complexes (Chapter 5)? If one trait undergoes evolutionary change, what consequence does that have for the others, and how does that affect interactions with the environment and thus the survival process within the ecological setting?

When conditions in a particular region become favorable for the existence of individuals of a particular species, relative residence time increases (Hengeveld and Hemerik 2002) and the survival of individuals and their reproductive output is enhanced (e.g., White 1969, 1970a, b, 1993). Numbers increase and more individuals are supported than previously, and there are more of them per unit of biotope than are present in an ecologically less favorable region. Individuals move to more favorable regions without prior knowledge of their existence. Such random movement is well known and may involve vast distances (see Chapter 9, Subsection 3.1). Individuals even leave from favorable localities, but on average the influx will usually outnumber the efflux. Inevitably, such favorable conditions are reversed at some point, a process that takes place at intervals, whether regularly or irregularly. Even when numbers are very high, when density-related processes are expected to play a role, the primary influence of the environment cannot be ignored as that is what drove the dynamic in the first place, so an autecological approach is still warranted (Chapter 9).

3.3 Section Summary

The issues raised in this section illustrate the fundamental and mutually exclusive difference between autecology and demographic ecology. Whereas demographic ecology is concerned primarily with processes external to the organisms, in the form of demographic influences that are seen to maintain equilibrium, autecology focuses intimately on processes that are internal to the organism and how these, in turn, relate to the outside environment (which does not exclude the biotic). These are the processes that result in organisms as a whole matching with the environment, as defined in relation to its spatio-temporal dynamics. External factors such as predation and competition may modify the pattern, but these are secondary influences (Andrewartha and Birch 1984). They are the ones, nevertheless, that are assigned primary importance under the demographic paradigm.

4 COMMUNITIES, ECOSYSTEMS AND HIERARCHY —PARADIGMATIC PERCEPTIONS

Community ecology is an integral part of demographic ecology, for it is underpinned by the same principles that support population ecology. Autecology, by contrast, deals with the subject of local diversity in a different way. Each species is seen to present its own ecological problem as to why it occurs in the area of interest and the densities it reaches there. Similarly, ecosystem ecology is perceived differently by the two approaches to ecology. From the demographic perspective it is seen simply as part of the ecological hierarchy, and the main representative of the abiotic environment. The autecological perspective, on the other hand, sees it as a related sub-discipline in biology, one that looks at ecological systems from a different perspective and scale, and which informs about the nature and scale of various features of the environment. This is particularly true, for example, of such aspects as biogeochemical cycles and vegetation type. The place of food chain theory is, however, not so certain, a point expanded below.

4.1 Community Ecology

Community ecology goes beyond single-species population dynamics to the broader scope of several interacting species. The context, however, remains strongly focused on those biotic interactions that can be interpreted with respect to density, namely predation, parasitism and competition (e.g., Kingsolver 1989, Dahlhoff et al. 2001, Pugnaire and Luque 2001).

Population dynamics and community ecology are thus two sides of the same demographic coin, as illustrated by their common reliance on competitive resource use. The extension of the logistic population model to cope with more than one species in the Lotka-Volterra models (as explained by Pianka 1994, for example) demonstrates this point beyond doubt. The basic questions addressed by these two areas of demographic ecology are concerned with the factors and processes that affect numerical fluctuations in equilibrium and the balance achieved among species within the community context.

Field experimental results demonstrate that the environmental context is crucial to the outcomes of such interactions (e.g., Buss 1990, Dahlhoff et al. 2001). They thus support, indirectly, the autecological view that a broad perspective of environment must be included in ecological theory. These biotic processes, in other words, need to be investigated and understood in much broader terms that relate to the ecological intensity of the organisms in question, and relative to all the processes governing ecological intensity. Focusing only on the outcomes of the biotic interactions conveys relatively little about the ecological system as a whole, or about the ecology of the organisms within that system. In other words, the appropriate deductive methods are not used to assess the relative contribution of these variables relative to all the others that influence the system in question (Chapter 8, Section 4). Nevertheless, this is the approach most widely practiced in multiple species field studies in demographic community ecology.

Communities and guilds are explicitly defined on the basis of common resource use, with the "member" species judged *a priori* to have at least the potential for interacting competitively (see Walter 1988a). This *potential* is considered to provide the full justification for studying processes that might ameliorate competitive effects, whether anticipated or real. The effects of the competitive interactions that take place between the species are seen to modify the carrying capacity (K) of that environment, and thus the mean population level of the species concerned. Since the interactions are investigated and interpreted in relation to a common resource that is shared, or partitioned, among species, questions of balance are inevitable. These questions are now phrased in terms of community structure, assembly rules and the like (e.g., Leibold et al. 2004), and they generate such questions as how many species can coexist, or live in a state of balance, on that resource (e.g., Leibold et al. 2004, Zhang et al. 2004, Inouye 2005, Urban et al. 2008). These are issues that are concerned with the way in which particular species are perceived to function within the framework of a multi-species community, which thus verges on the typological (Hengeveld 1988b, Walter 1995).

Autecological theory emphasizes that environmental matching is not only a species related attribute (see Chapter 7), but that each species responds to different subsets and ranges of the physico-chemical variables that make up the environmental circumstances of any particular locality. The organisms involved also respond in this way to the different subsets of biotic components within that environment. Because of this ecological idiosyncrasy, the individuals of a species usually find themselves in different biotopes and in different floral and faunal compositions. Quite simply, the co-occurrence of species within localities is explicable in relation to that area suiting each of the different species at that time, each in its own way. The idiosyncrasy of species is thus readily appreciated, as is the point that "communities" do not exist. They have neither a history nor a future of their own (Fig. 3.1) (Walter et al. 1984, Hulley et al. 1988a, b, Walter 1988a, 1995, Walter and Paterson 1995).

4.2 Ecosystem Ecology

Ecosystem theory and practice holds an ambivalent relationship with ecology, and this stems from ecosystem ecology having at least two emphases, each of which coincides with one of the ecological paradigms.

On the one hand, ecosystem theory interprets ecological systems as physico-chemical systems, with energy as the binding force. Ecosystems are thus considered to have properties that are not those of its parts, but which result from its internal dynamics (see McIntosh 1981). In this way, organisms are effectively discarded altogether, and are essentially collapsed into operational collectives that occupy positions on different trophic levels (Golley 1993, p. 5). It is irrelevant which species converts the energy, so long as that role is performed. Species can thus readily replace one another within the context of the ecosystem dynamics. Externally induced changes in the numbers of organisms are not considered either; for it is implicit that energy conversion will continue. The emphasis here is on food web structure and dynamics, a perspective that is strongly related to demographic ecology theory (e.g., Pimm 1991, Pimm et al. 1991), but which has delivered little of theoretical or practical significance (O'Neill 2001).

Ecosystem ecology has another focus, one that encompasses such aspects as biogeochemical cycles and which, therefore, informs on how the environment of organisms is constituted and explains its underlying dynamics. Although autecology clearly relies strongly on such information, ecosystem ecology of this persuasion differs fundamentally from the autecological approach to understanding the spatio-temporal dynamics of organisms. Ecosystem ecology does, however, constitute an important part of autecological interpretation in helping to explain the dynamics of the environment inhabited by organisms.

In demographic ecology, ecosystems are currently seen as part of the ecological hierarchy (e.g., Price and Waldbauer 1994, Pickett et al. 2007). However, the history of ecosystem ecology shows it was not developed in this way. In other words, demographic ecology has co-opted the ecosystem concept, and turned it to demographic purposes, apparently to negate the early criticism that abiotic processes were excluded from demographic ecology. Ecosystem theory had originally been developed as an alternative to the community ecology of the day. Indeed, it was developed as a reductionist alternative, as described by Worster (1977, pp. 301-302). The impetus for an alternative came from Tansley's (1935) criticism of the super-organism concept of community development. He suggested the concept of community be dropped and he expressed the need for a reductionist alternative. His introduction of ecosystem was thus a new, alternative view of organization in natural systems.

The development of ecosystem theory was not favored by community ecologists at the time. They considered it took too much focus from organisms and the evolutionary process that was being incorporated more fully into the ecology of the day. Indeed, the development of evolutionary ecology in the 1960's was seen as a remedy for this situation (Golley 1993, p. 5). The two, however, tackle rather different issues, as outlined above, and we suggest ecosystem ecology should be seen in the broader context provided by autecology.

Autecology sees ecosystem ecology as addressing questions that are quite different from the question of the spatio-temporal distribution of abundance that is addressed in ecology *per se*. But the issues of ecosystem ecology do have relevance to questions that are primarily ecological in nature. Information on biomes, vegetation structure, geochemical cycles and so on helps autecologists answer their primary question of the geographical spatio-temporal dynamics of their subject species. The research of Vitousek (2006) shows how minerals are provided to vegetation of different age on the Hawaiian Islands from different primary sources as the systems have aged. Ecosystem theory and practice thus cover the range of spatio-temporal scales covered in autecology and, frequently enough, ecosystems research intersects with autecology in much the same way as biochemistry and physiology interdigitate with autecology. Autecology thus sees ecosystem ecology as a subject area with a different perspective on natural systems (i.e., that concerned with energy and nutrient cycles). It addresses different questions from those raised in autecology, and does not represent a hierarchical level of organization in ecological systems.

4.3 Why Ecological Systems are not Hierarchical

Ecological systems are frequently portrayed as having hierarchical structure. The "levels of organization" within ecological systems run upwards from the individual, to the population level, to metapopulations, through communities, and the ecosystem level, and then via biomes to the global biosphere. Before considering ecological hierarchies in more detail, we clarify some serious misconceptions about the validity of hierarchies in biological reasoning, as detailed by Rose (1997). In interpreting the workings of individual organisms, biologists frequently refer to various levels of organization, and these range from the molecular, through the biochemical, physiological, and behavioral to the whole organism. However, neither the organism itself, nor the various processes involved actually operate according to different rules within these different "levels". Nevertheless, these levels of perception are usually represented as levels of organization, with each subsequent level in the hierarchy seen to embody new properties or processes that are not evident at the level below.

The processes seen by biologists to operate at the various "levels" outlined above are all parts of one process; the ion channels open (biochemistry), nerve action potentials are generated (physiology), the frog jumps (behavior), it does so for some reason (evolutionary biology), and it may thus survive to reproduce (ecology). This operation, as a whole, allows the organism, as a living system, to function effectively. The categorization into levels of organization is a question of interpretation, and is misleading. These levels simply represent levels of perception. For our own convenience, we impose order on the various ways of asking questions, and we derive molecular, biochemical, physiological, behavioral, evolutionary and ecological answers. The distinction is therefore one of epistemology or method, but we aim at a single ontology or interpretation of the world (Rose 1997). Hierarchies in biology are thus artificial, and are specifiable only in relation to our viewpoint or relative to the types of questions we ask and the scale of resolution of our perceptions and instruments. In ecological systems, therefore, the processes that are significant ecologically exert their influences on the individual organisms, and they do so in combination and simultaneously.

The idea of ecological hierarchies with different levels of organization is therefore not valid, for it is conspecific individuals, with their species-specific adaptations, operating within the spatio-temporal variation of that subset of environmental variables that is significant to their existence and reproduction. "Populations" and "communities" are thus not levels of organization; they are simply statistical epiphenomena. Ecosystems, too, are not levels of ecological organization because, conceptually, they are reductionist abstractions of the ecological system cast in terms of energy and material flow. As outlined above, an understanding of the dynamic biogeochemical processes that influence ecosystem structure helps us define the environmental dynamics to which organisms respond.

Significantly, species are usually excluded from ecological hierarchies (see Walter 2003), for the demographic principles almost invariably associated with such hierarchies effectively remove them from any realistic consideration. The particular properties and behaviors of individual organisms and species, those that are considered "idiosyncratic" in terms of demographic ecology expectations, are the features that are effectively eliminated. These, as we have argued throughout this book, are the features that influence and dictate the ecology of organisms.

5 RELATIONSHIPS OF THE TWO PARADIGMS TO OTHER SUB-DISCIPLINES

A crucial aspect for any sub-discipline in biology, including ecology, is that its subject material and results can be directly related to the subject matter of the other sub-disciplines that support it. Biochemistry, for instance, deals with the subject matter covered by physiology, but from a different perspective, or scale of perception. These two sub-disciplines do not reflect different levels within an organizational hierarchy (see previous section), for the process with which each one deals is contemporaneous and within the same organ (Rose 1997). Each sub-discipline represents a different problem area within biology, and poses different questions about the same subject. The interpretations thus generated within each sub-discipline must be mechanistically related across the sub-disciplines. The subject matter and interpretations within ecology should also be directly referable to the other sub-disciplines within biology. Here we specify how this is achieved by each of the ecological paradigms, and do so with reference to the summary in Table 10.2.

5.1 Demographic Ecology

5.1.1 Evolutionary and behavioral ecology

Demographic ecology is explicitly and logically well integrated into evolutionary theory, but it is a particular interpretation within evolutionary biology with which it is associated. The area of overlap between these disciplines is termed "evolutionary ecology" and this tends to focus strongly on measures of fitness and optimization and is peculiarly demographic in constitution. Natural selection is not considered to act on individuals as a whole, but rather on specific traits. Different studies on a species (even ones conducted concurrently) can claim significant directional selection on several traits simultaneously, an aspect of evolutionary ecology that is puzzling, given the observed stability of species (Chapter 5).

The traits given particular significance are those that relate to resource use within the demographic context and those related to choice of mate (i.e., sexual selection). They are dealt with mainly in behavioral terms, and in relation to reproductive output, which, in relation to conspecific organisms in other experimental or observational treatments, is accepted as a measure of future fitness. Behavioral ecology thus represents the executive arm of evolutionary ecology, in a way that physiological ecology cannot. The more advantageous traits are those that are considered to increase the efficiency of resource acquisition or the maximization of offspring production. In turn, each improvement in such efficiency is assumed to be coded by an individual and independent allele, and each such allele codes for an increase in efficiency in a quantitative way. Such genes have been postulated, but never isolated.

The demographic circumstances of any population in question are assumed to generate selection pressures continuously. Organisms are considered to have their ecologically significant traits scrutinized by natural selection, and the organizing concept is the assumption that reproductive rates are under selection pressure for maximizing fitness. Incidentally,

Table 10.2 Relationship between the premises of each ecological paradigm (demographic ecology and autecology) and various features of organisms, measures of ecological success, nature of evolutionary change, interactions between organisms, and related biological sub-disciplines.

	DEMOGRAPHIC ECOLOGY	*AUTECOLOGY*
Organismal features:		
Important traits of organisms	Aspects that increase efficiency of resource acquisition	Complex adaptations that relate the organism to its usual environment
Genetic and developmental systems	Small, quantitative differences coded by postulated alleles. Development seen as malleable	Genetics and development interrelated, with strict requirements on timing of processes. Complexity imposes cross-generational stability
Measuring ecological success:		
Natural selection	Driven by intraspecific competition, and based on relative reproductive success	Non-competitive, based primarily on mortality (the failure of individuals)
Measure of ecological success	Measured fitness benefits assumed to project to future generations	Survival and reproductive output at time of assessment
Nature of evolutionary change:		
Consequence of selection	Instantaneous, locality-specific optimization, with divergence into biotypes, races and then species	Stabilizing selection, so species are stable and adaptations are distributed species-wide
Tempo of evolutionary change	Phyletic gradualism, ongoing adaptation in local populations, seen sometimes to be halted by various constraints	Stasis usual, with adaptation sporadic and forced when small population is trapped in adverse environment (speciation may result)
Interacting organisms:		
Prey (host) and consumer	Co-evolutionary "arms race"	Sequential evolution
Competing species	Local adaptive adjustment in resource use because interspecific competition seen to be strong	Interspecific competition spatio-temporally sporadic, so evolutionary effects not expected

	DEMOGRAPHIC ECOLOGY	*AUTECOLOGY*
Related sub-disciplines:		
Evolutionary process and speciation	Evolutionary ecology - adaptation relative to efficient resource use across species (e.g., sympatric speciation reflects efficient capture of unexploited resources), protected by reproductive isolation	Recognition concept of species - adaptation incidental and contingent on strong environmental influences imposed on small allopatric populations. Species delimited by system of recognition among potential mating partners
Behavior	Behavioral ecology - emphasis on efficiency and optimization	Ethology - emphasis on complex behavioral mechanisms within environmental context
Biochemistry and physiology	Indirectly related in terms of proximate mechanisms that are selected for increased ecological efficiency	Ecophysiology, with emphasis on tolerances and mechanisms in relation to environmental context
Ecosystem ecology	A hierarchical level in ecology that introduces abiotic conditions and the perceived holism of food webs	An epistemological adjunct to autecology that deals with biogeochemical cycles and their consequences, and helps interpret environmental structure and dynamics

"success" and "efficiency" thus have a logical place within this competitive framework, and provide a strong contrast with the "sufficient" and "effective" of autecology. Selection is thus seen to be competitive by nature, and adaptation should be ongoing, unless bounded by constraints. Even complex adaptations are assumed to be built up gradually by this process, as indicated by the co-evolutionary "arms race", "Red Queen" and "genetic handshaking" metaphors (e.g., Ridley 1994, Thompson 2005).

Selection should thus lead to instantaneous change within a locality, depending on the local species composition. Clearly this implies that the demographic forces across different localities, and thus across the different populations that make up a species, could not be the same. Within species, divergence of populations is expected to be the norm, and within trophic levels competing species are seen to adjust to one another locally in relation to their resource use and thus reduce the intensity of competition. This latter is seen to facilitate local coexistence. Down trophic levels, an ongoing co-evolutionary adjustment between predator and prey, or herbivore and host plant, is seen to take place.

5.1.2 Local ongoing adaptation versus species-wide adaptations

The central place in evolutionary ecology of ongoing adaptation to local conditions generates a serious problem, because complex adaptations, such as those associated with human locomotion (and the others covered in Chapter 3, Section 2), tend to be species-wide. Evolutionary ecology does not provide a valid explanation for this large-scale spatio-temporal uniformity in the adaptive mechanisms of species, with only *ad hoc* modifications to interpretations of local adaptation within species being offered (e.g., Thompson 1994, 2005, Eldredge et al. 2005). So, we have the disjunction between demographically driven evolutionary change being ongoing and locality-specific, on the one hand, and the most significant adaptations of organisms being species-wide and stable through time on the other. What demands explanation is the way in which the slight adaptive improvements that are considered to arise at intervals in different populations of the species are supposed to infiltrate the entire species' distribution. Any explanation must also include the undoubted fact that the "selective environment" will differ across local populations of a species. A real difficulty is how such changes could be co-ordinated temporally and spatially, so that the build up of any complex adaptation is in the appropriate sequence across geographical space. Coordination at such scales is simply not feasible. That it should happen and can happen is simply accepted in evolutionary ecology, albeit tacitly. More recently, the idea of parallel speciation has been introduced (e.g., Nosil et al. 2002, Schluter 2009) although alternative explanations for the observed patterns are seldom seriously considered. Perhaps telling in this regard is the effort expended on research to verify that natural selection drives local adaptation.

The relationship between population-specific optimization of fitness and the evolution of a new species presents further problems. The first problem relates to the logic associated with the defining characteristics of species, namely isolating mechanisms. The isolating mechanism of any species is specifiable only in relation to areas of overlap with the species against which isolation was evolved, because it is an *ad hoc* mechanism for this purpose, as recognized by Ernst Mayr (Paterson 1978). By contrast, the behavioral and physiological components of these mechanisms that lead to mating and fertilization (but which are termed "isolating mechanisms") are, again, species-wide in distribution (e.g., Henderson and Lambert 1982, Popple et al. 2008). These components make up part of the complex mechanism that brings members of the opposite sex together for mating, the Specific-Mate Recognition System (Paterson 1985). These adaptations should therefore be viewed positively as an inter-individual mechanism rather than as a group centered mechanism to keep gene pools "pure". So, once more, we see evolutionary interpretations cast in local terms, whereas the characteristics involved are strictly species-wide in distribution. In empirical terms, the search for robust examples of local character displacement (whether ecological or reproductive) has been as long and hard as it has been fruitless. Most claimed examples have eventually fallen with further critical investigation (e.g., Grant 1972, 1975, Cherrill and James 1987, Saloniemi 1993).

The same argument holds for characteristics presumed to be ecologically significant in a community context. "Communities" or species aggregations are local, but the characteristics of the organisms are species-wide in distribution (Walter et al. 1984, Walter 1991) (and see Huey (1979) for an example). These difficulties are further compounded in evolutionary ecology by another logical disjunction, for species are also treated typologically in community ecology (Hengeveld 1988b, Walter and Paterson 1995) because species are seen as units that evolve to fill empty niches. This postulated process has assumed such significant proportions that the niche, which remains unspecifiable in natural systems, has even been elevated to a property of species (Mayr 1982) (but see Hengeveld (1988b)). Character displacement in feeding structures is believed to enhance differences between populations (often seen as incipient species) and to reduce competition by minimizing niche overlap as they evolve. Belief in the importance of this process also provides the *raison d'être* for the recently revitalized acceptance of sympatric speciation (often now in the form of "ecological speciation"). Again, putative cases are usually studied locally (Hereward et al. 2013) although the different types are likely to have geographical distributions that are substantially independent of one another (Popple et al. 2008), implying the study areas are within areas of secondary contact.

Species are thus treated as adaptive devices that evolve to improve the efficiency of resource use (Paterson 1978). This view is central to such concepts as adaptive radiation, as originally developed, although its influence is often hidden now (see analyses of Walter (1988b) and Masters and Rayner (1993)). Even critics of the adaptationist program make statements that are implicitly based on the belief that radiations are held back until mass extinctions produce empty niches that can be filled by the next radiation (e.g., Gould 1982b), so its influence is strong. The precise evolutionary processes that are responsible remain unspecified, with the belief in local adaptation still intact. One of the major problems is that sympatric speciation has to be invoked (Masters and Rayner 1993), and this elevates the logical inconsistency between competitive selection driving evolution, which is necessarily sympatric, and the allopatric model of speciation. This also helps, perhaps, to explain the current popularity of sympatric speciation. Logically, it is required, but empirically it remains weak because tests almost invariably represent efforts aimed at verification, because possible alternatives (e.g., that the study includes two cryptic species that are largely independent in their geographical distribution (Popple et al. 2008)) are virtually never canvassed.

That evolutionary ecologists see selection as an optimizing process that increases fitness through increased efficiency of resource use raises further considerations. How do "efficiency" and "optimization" relate to the adaptive mechanisms of organisms, such as those detailed in Chapters 3 to 5? The minimum requirement, logically, is a direct mechanistic (or causal) link. Again, evolutionary ecology falters because efficiency cannot be directly related to the behavioral processes as ethologists study them. Efficiency, in evolutionary terms, is relevant only to the concept of optimization, as it has been developed in behavioral and evolutionary ecology. For this reason, a strict division is maintained between evolutionary processes, which are referred to as ultimate mechanisms, and the behavioral processes that are called proximate mechanisms (Mayr 1961, Ariew 2003). Use of this distinction in this particular way papers over the disjunction between the material side of behavior—the proximate mechanism—and the abstraction that represents "efficiency"—the ultimate mechanism (Walter 2013). Sometimes, efficiency is linked to behavior more directly, as in the organisms responding to more specific sensory information and thus becoming more efficient, but the connection is still not mechanistic.

The competitive process was originally seen, in relation to evolutionary change, to be generated and maintained by a discrepancy between the demand for resources and the amount of resource available, under conditions of population growth. This is the "Malthusian discrepancy". The discrepancy derives from the rift between a geometrically increasing demand and a linearly increasing supply (Darwin 1859). However, this is not an accurate portrayal of what takes place in natural systems (Himmelfarb 1959), and cannot drive the relative evolution of organisms on different trophic levels. The supply, such as prey, for example, also increases geometrically, although this is likely to be at a rate different from the predators' rate of increase, and this need not even be at a lower rate. In short, the competitive argument in modern evolutionary ecology is flawed in several significant ways.

5.2 Autecology

5.2.1 Complex adaptations or character complexes

Organisms, from the autecological perspective, are made up of sets of interrelated mechanisms. These have been called complex adaptations (Paterson 1985) and character complexes (Hengeveld 1981), and these include the environmental response systems of organisms. "Complex" refers to the coordinated operation of interrelated mechanisms of various types (structural, behavioral, physiological and so on) to perform particular tasks (Frazzetta 1975, Hengeveld 1981) and also refers to the coordination of a sequence of subsidiary functional steps that together, acting in sequence, achieve a particular end (Paterson 1985). Both of these aspects are likely to come into play together in considering the function of any particular complex adaptation. Each step or aspect of such complexes would be relatively insignificant were it to operate in isolation (Paterson 1985). Examples of complex adaptations include the catenary sequence of events that (i) lead an insect herbivore to its usual host plant (Wapshere 1989, Marohasy 1998) (Fig. 3.11), (ii) result in predatory carabids capturing their collembolan prey (Hengeveld 1981), (iii) enable arboreal pythons to strike prey within a tree canopy, hold onto it and ingest it (Frazzetta 1975), and (iv) maximize the chances of males and females reaching the point of fertilization and syngamy (Paterson 1985) (Fig. 3.12).

Each species is unique in the specifics of its complex adaptations and in the mix of components that make up character complexes. Species are thus given special prominence in autecology, as integrated systems with objectively definable limits, but they are not treated typologically (Walter 1995, Hengeveld and Walter 1999, Walter and Hengeveld 2000). Although species are considered as idiosyncratic with regard to their ecology or, at least, with regard to the expectations of demographic ecology, autecology can still develop generalizations to cover all species. These include a general concept of species, how organisms interact with the environment and the spatio-temporal consequences of those interactions, for example. The form of the generalization is clearly different from the generalizations hoped for in demographic ecology (e.g., Price 2002), and is covered in Chapter 6 (Subsection 4.5) and Section 7 of this chapter.

Genes code in a rather complex way for aspects of the morphological, physiological and behavioral systems of organisms, although we know most about those responsible for form (Carroll et al. 2005a, Carroll 2008b). That is, the systems that influence the way in which organisms respond to their environment are likely to be under genetic control that is not straightforward. The intricacy of the processes involved in the development of each organism through each of its life stages imposes an infinitely more "rigid" structure on the organism and its development than anticipated in evolutionary ecology (Rose 1997, Arthur 2004). The major consequence of the functional requirements of organisms in their usual environment is that any alteration to function is more likely to be deleterious than advantageous. With subsidiary functional steps involved in complex adaptations, the selective consequences are compounded. This evidently imposes strong limits on change when organisms are in the environment to which they are primarily adapted, are free to track suitable environments, and form part of a relatively extensive network of individuals.

Physiological changes coded by allelic substitutions have been recorded in nature, including resistance to malaria in human populations and to chemical pesticides in a range of organisms. These typically involve structural changes to particular proteins, in some cases after duplication of the gene (Devonshire and Sawicki 1979). However, adaptive change in even these relatively simple coding systems evidently needs intense selection pressure. Malaria and insecticides kill, and it is this aspect that drives the alleles involved through populations that are reduced, sometimes dramatically. Some such genes have, by contrast, been stable through billions of years, for example those derived with eukaryogenesis from the Archaea (e.g., Carroll 2008a). That evolutionary change is not the ongoing, everyday process that is envisaged in evolutionary ecology can be gauged by the common observation that artificially selected organisms that return to the wild tend to revert to their wild type, and this takes place in a similar way in different localities (e.g., Sol 2008). Natural selection may screen persistently, but under everyday conditions it is not driving the directional change anticipated in evolutionary ecology.

Evolutionary change that affects form mainly involves the regulatory genes that control development (Carroll et al. 2005a, 2008, Carroll 2008b). These regulate the body-building and body-patterning genes (Chapter 5, Section 3). They regulate the expression of each, and do so by targeting specific genes in different localities within the organism and at different times through the life of the organism, by switching them on and off as required. Coordination is clearly the key, for transcription factors typically regulate scores to hundreds of individual target genes and change at this level is likely to be disruptive. Such mosaic pleiotropy means that most proteins regulating development (e.g., sonic hedgehog) participate in multiple, independent developmental processes and the formation and patterning of morphologically disparate body structures. They therefore function independently in different cell types, germ layers, body parts, and developmental stages of an individual organism. This means that mutations that alter the function or activity of such proteins are likely to have widespread and potentially many negative effects on development, survival and reproductive output, which again implies that genetically based change is most likely to be selected out (all from Carroll 2008b). The genetic systems of organisms can be highly variable (e.g., Gilbert and Epel 2009), but to a large extent this variation may exist to maintain function, and allow different responses to changing environments (plasticity).

Thus, the mechanisms of maintenance and reproduction of organisms are complex and restricted in how they can change in an evolutionary sense, by their mode of formation and the strong constraints imposed by the functional demands on the adaptive mechanisms involved in the organism-environment interaction. This process takes place in similar ways across the geographical distribution of the species, since individuals track suitable environmental conditions. Further stabilization is conferred in sexual species (i.e., most eukaryotic species) by the requirements of sex. The mechanisms that coordinate potential mating partners in time, and bring them together in a heterogeneous world, are complex adaptations that must act consistently if individuals are to contribute to the following generation (Paterson 1985, 1986). Behavioral and physiological changes are thus likely to stand in the way of individuals contributing to the following generation. So even those genetical systems that generate variability, that is sexual systems, are stabilized under field conditions.

In summary, autecology is built on the premise that species are predominantly under stabilizing selection after speciation, and this is consistent with the Recognition Concept of species in this regard. In corollary, evolutionary change is seen as remarkable, something that remains difficult to explain. By contrast, in evolutionary ecology, which is logically strongly related to the premises of demographic ecology, adaptive change is seen as an everyday occurrence. Existence itself is consequently treated as unremarkable, under the demographic perspective, and without need of explanation, as pointed out by Haukioja (1982). This view of living systems seems unable to consider that successful survival across generations demands the construction of offspring that are functional equivalents of the adults that produced them (Haukioja 1982), and that stability across geographical space and over time should be commonplace.

5.2.2 Adaptation and speciation within geographical context

The primary means by which individual organisms can adjust to a continually changing and heterogeneous environment is through phenotypic plasticity or through movement to more suitable localities. Organisms thus maintain themselves and may eventually reproduce. They cannot do so to optimize their future fitness—the world is too variable for this and the future is unknown (Rapport 1991). Genetic adaptation is not possible at the fine scale required, except for relatively simple allelic substitutions. Such genetic changes that have been demonstrated (e.g., chromosome inversions relative to environmental temperature and alleles that confer resistance to malaria) result from environmentally related selection pressures, and the process need not be cast in terms of optimization relative to other individuals.

Species, as groups of individuals, track suitable environmental conditions. Such habitat tracking regularly extends to continental or oceanic limits, and may even be global. Even sessile organisms track environmental change, through the motility of their seeds or larvae (Ford 1982, Huntley 1991) (Chapter 4). Such movement is risky, even for organisms that can navigate and control their course. Mortality is high, which again works against fine levels of optimization evolving. Through environmental tracking, the different individuals of a species tend to be exposed to similar conditions to one another, which is referred to as their constancy in factor space (Chapter 6, Section 5). The overall outcome is the observable dynamics of species across geographical space and over time. The local ecology of organisms can thus be understood only in relation to the broader context of their species-specific properties, their environmental requirements and broader geographical distribution, and the history of climate and environmental change.

The stability of species is a consequence of at least five different processes (Chapter 5, Section 3). These include (1) the complexity of their developmental processes (Rose 1997, Arthur 2004), (2) the imperative that the species-specific physiological and behavioral properties of individuals match their environmental requirements, (3) the synchronization across individuals of their life cycle, including sexual maturation, with changing environmental conditions, (4) the environmental matching of individuals with their species-specific habitat requirements through habitat recognition, and (5) the complex procedure of recognition and localization of appropriate mating partners, followed by the subsequent achievement of syngamy. The stability of complex adaptations is the norm in nature (Chapter 5, Section 3), and such adaptations may persist unchanged through numerous speciation events (Hengeveld 1981), because of their persistent functional role.

Genetic variation is widely sought and often recorded in the literature in evidence of the adaptability of species. Such demonstrations perhaps often fall under the banner of Chamberlin's (1897) "precipitate explanation", for the expectation that the demonstration of variation is sufficient evidence of adaptation in progress seems to be seen to reduce the need for critical test of the entire set of propositions. For example, some claims of intraspecific variation involve abstract characters that are unlikely to represent accurately the functional aspect really involved. Mate choice would be a good example, for non-functional aspects of behavior are undoubtedly included in the measures presented (e.g., Butlin 1995). The functional aspect in the interaction of potential mating partners is undoubtedly recognition, which is neurally more simple, is more likely to be operating within the environmental contexts inhabited by organisms, and is also more appropriate at the chemical level and so holds also for plants and so on (Clarke and Knox 1978, Paterson 1993b, Radutoiu et al. 2003). Interpretation of variation in such systems requires functional understanding and care (Paterson 1993b). Further, recorded intraspecific variation may also be artificial when, for example, two or more species are erroneously conflated as one (Paterson 1991, 1993a), and this is likely the case in many contemporary ecological (and sympatric) speciation studies (Walter 2003, Popple et al. 2008). Furthermore, the demonstrations of adaptive change tend to involve relatively minor changes, as in insecticide resistance.

The above considerations suggest that the evolution of a new species requires special circumstances, ones that destabilize the co-ordination within the system (Paterson 1985, 1986). This demands relatively small numbers of individuals to be prohibited from adjusting in a spatial fashion to changed conditions, as happened for example to the organisms inhabiting the Olympic Peninsula of Washington state during the glacials (Pielou 1991). In addition, the environment in the area of geographic isolation would have to shift in its structure or dynamics for directional selection to be set up across all individuals in the population, and thus to force adaptive change. Selection in this context is not competitive and is not necessarily acting strongly on differential reproductive output among individuals. Rather, which individuals survive, and which die, which ones produce offspring that survive, and which ones do not, decide outcomes in such circumstances. This is the so-called "hard selection" advocated originally by Alfred Russel Wallace (Wallace 1968, pp. 427 ff.) and it acts on the basis of whether the physiological and behavioral mechanisms of the individual organisms enable survival through the entire life cycle in the locality in which they are forced to stay.

5.2.3 Evolutionary contingency and allopatric speciation

The evolution of a new species is a consequence of the interplay between variability and selection on the one hand, and the contingency of where and when a relatively small group of individuals is trapped and destabilized in isolation of other conspecifics. Also open to contingency is the direction and extent of environmental change to which the small isolate is exposed, and whether the lineage(s) involved can survive the environmental pressures for long enough that adaptation actually takes place. Such adaptive change may proceed so far that a new species is forged incidentally. In other cases adaptive change may stop at some point before such an extreme is reached. The latter explains different subspecies in different areas, and hybrid zones that persist across extensive areas for long periods. The Recognition Concept does not give special evolutionary significance to hybridization, for it cannot drive further evolutionary change through reinforcement (Paterson 1978, Spencer et al. 1986, 1987).

Species themselves are therefore not adaptive devices that fit into a community or that evolve relative to other species on an "ecological continuum" (e.g., $r - K$ selection and generalist-specialist continuum) (Finlay-Doney and Walter 2012a). Species are, instead, the consequences of change forced on a small group of individuals that has been spatially restricted by external conditions (Paterson 1985). Should the conditions to which the small isolate has adapted become more widespread, the newly-adapted species will spread geographically into areas of suitable habitat that are accessible to the movement of the individual organisms. As individuals move to new areas, they carry the features of their parents, and this process explains the species-wide distribution of species' characteristics. A further point of difference from the demographic approach is that greater taxonomic similarity among species implies they have greater ecological similarity to one another than to more distantly related species. This is contrary to Hardin's (1960) competitive exclusion principle, for related species are more prone to co-occur because they are more similar in their ecological requirements (Lindroth 1949, den Boer 1980, Hengeveld 1985a).

Individuals, and thus species, proceed in time through a haphazard, directionless and opportunistic process of matching the environmental dynamics within their ambit, and which also are directionless (Fig. 9.14). In a sense, they shift in relation to ever-changing conditions, the variation of which can only be described in statistical terms that are, of course, unknown to the individual organisms that are exposed to them (Hengeveld and Walter 1999, Walter and Hengeveld 2000). Ecologically, they can be specified for a particular species in statistical terms and relative to (i) survival and reproduction rates under particular environmental conditions, (ii) migration distances between natal and nesting sites, and (iii) the processes that determine or influence the meeting chances of potential mating partners, and so on (see Chapter 9). These processes are the result of stochastic influences on individual organisms as they track suitable environmental conditions, coordinate specific (usually seasonal) mating times, recognize and localize potential mating partners that signal and respond appropriately, and so on. All of these properties vary between species, at least to some extent. This means that the environmental components to which they are exposed and, in turn, to which they respond, represent a small subset of the entire spectrum of components of the environment (or axes of differentiation) as a whole.

As a result of the way in which adaptation, and thus species-formation, takes place, the responses to the environment of each species are independent of one another (except for obligate associations across species (e.g., obligate pollination mutualisms (Pellmyr and Thompson 1992, Terry et al. 2007)) and therefore specific. Each species varies at its own spatial and temporal scale, and when its constituent individuals meet other conditions, they respond in their own idiosyncratic, species-specific ways and rates. The development of specific properties that match with particular environmental conditions is one of the principal driving forces in the origin of species. The historical origin of mechanisms relevant to the ecology of organisms, as described in Chapter 3, Subsection 2.6 for the origins of humans, the species we know best in this regard, should therefore replace the relatively insignificant search for allelic substitutions, behavioral optimization, and minor adaptive alterations that is current in attempts to verify the evolutionary ecology perspective on evolutionary change.

6 RESOLVING THE AMBIVALENCE IN THE TERM AUTECOLOGY

The relationship between species and individuals outlined above, and justified through the rest of this book, overcomes a problem associated with the term autecology. The autecology label has been applied indiscriminately to individual species and to individual organisms, without any clarification of the relationship between them or of the structure of autecology.

However, early plant autecology was strongly associated with physiology and was often interpreted as the physiological study of individual plants (Hagen 1986, Cittadino 1990) or animals (Allee et al. 1949, pp. 48 and 227). Others applied it equally, but without clarification, to both individuals and species (Chapman 1931). The term "autecology" was therefore rejected by some and, ironically, was not used by Andrewartha and Birch (1954, 1984), who have become most strongly associated with this approach to ecology.

We have shown, however, that the term relates to individuals and species simultaneously, as the two are inextricably linked to one another; the individual is the operative unit in an ecological landscape, and it derives from a sexual union from within the species' gene pool. Individuals thus derive their species-specific characteristics from their sexual origin within a particular species. In turn, the specific sexual behavior of an individual ties it to a larger group of organisms, in principle the entire species gene pool. The species, as a genetical nexus among individuals, is definable in terms of the intersexual communication and behavior that takes place between male and female organisms. To all biologists, the concept of "species" has further utility, for each can be referred to as an abstract unit to convey information about the adaptations carried by its constituent individuals, and thus their requirements and tolerances. "Species" can thus be used as shorthand to transmit a body of ecological information.

Why Andrewartha and Birch (1954, 1984) did not detect the ecological relevance of this duality in understanding organisms in nature is unclear. They did stress that the physiology and behavior of the animal must be investigated (Andrewartha and Birch 1954, p. 10), but provided little general direction beyond that, for they did not base their ecological interpretation upon the concept of adaptive mechanisms as opposed to the adaptive process. Although they are associated with the autecological tradition of the 20th century, they were seen mostly as taking a strong stance against the ecological importance attributed to density dependence and standing for the notion of density independence. However, the latter is incorrect (Andrewartha and Birch 1954, p. 19), but the persistent and general emphasis in ecology on populations and density, and the associated concern for population persistence may well have been diversionary, for Andrewartha and Birch (1984) ultimately turned to the group-related idea of "spreading of risk" and its attendant problems (Hengeveld 1989b).

Autecology can be developed only when the strong methodological links to local populations, density related phenomena, and questions of population persistence are broken (Walter 2008). These arguments explain why we have moved away from the local population as a unit of study, towards individuals, which are potentially wide-ranging spatially. The approach advocated by autecology illustrates why populations and communities do not have the special meaning given to them under demographic ecology. They are epiphenomena, and the organisms seen to comprise such entities are better understood by an examination of the organisms themselves (Walter 2008).

A further difficulty for Andrewartha and Birch (1954) was the disjunction between their approach to ecology, which implicitly treated species as stable and as having species-wide properties, and the evolutionary theory of the day (Orians 1962). Andrewartha and Birch (1954) were seen to be keeping ecology apart from evolution, but this apparent disjunction can be reconciled through the association of autecology with the processes of adaptation and speciation developed through the recognition concept of species (Chapters 5 (Section 6) and 6 (Section 4)) (Walter 2008), but which was not available until the 1980s. The autecology of Andrewartha and Birch (1954) is indeed inconsistent with the precepts of evolutionary ecology, which was still nascent at that time. And specifically it clashed strongly with the core assumption of evolutionary ecology, that selection pressures are developed through density dependent competitive processes, to the point that adaptive local change is inevitable and ongoing, in much the same way as adaptive optimization is seen to take place (see Walter 2008). All of this suggests that Andrewartha and Birch (1954) could, perhaps, be portrayed as having been prescient in their treatment of species, as pertains to stability and so on. They did, however, accept that evolutionary change is an ongoing process and did try to relate ecological investigation to the understanding of change in this way (Walter 2008).

The advantage originally presented by the demographic paradigm, and still widely sought, is a set of principles that would negate the idiosyncratic aspects of species. Density dependence, if it worked in nature as postulated, would be very useful. But experience shows that no such process is reliable enough for the desired purpose. Ultimately, ecologists must work within the constraints of the subject material (Walter 2003, p. 233). Autecological theory has been developed to do this, and has the evolutionary backing of the recognition concept of species as well as the strengthening empirical evidence of the stability of species.

7 SCIENTIFIC LAWS, PREDICTION AND THE TWO ECOLOGICAL PARADIGMS

Each of the ecological paradigms can be related, differentially, to particular perceptions of the nature of science and scientific laws. These perspectives are reflected in the somewhat different ambitions of each paradigm in terms of understanding and prediction (Walter 2013). These relationships are spelled out in Table 10.3, and expanded upon below. At the outset, we clarify the two ways in which the term prediction is used in science.

Prediction in reference to scientific theories relates, on the one hand, to testing theory. Outcomes from the scientific manipulation of systems, through experimentation or intervention in the laboratory or field, are predicted on the basis of theory. The scientific test is then run, and the outcome matched with prediction to judge the implications of the research. Prediction has another meaning, that of forecasting future states in natural

Table 10.3 Relationships of the two ecological paradigms (demographic ecology and autecology) to scientific method, the nature of ecological interpretation and ecological prediction.

EXTERNAL REFERENCE	*DEMOGRAPHY*	*AUTECOLOGY*
Existence of external laws	Yes, population stability and community structure imposed by external laws (e.g., competitive exclusion principle)	No such external laws; models of organismal properties and processes are developed (e.g., Recognition Concept of species)
Heuristic laws	External laws used as basis for interpretation	Used to guide questions, investigation and interpretation
Ecological systems	Hierarchical organization and interpretation	Not hierarchical, for ecology is a methodological "level" within biology, as a whole, and all such methods contribute to a single world view (ontology)
"Prediction" of spatio-temporal conditions by organisms (and natural selection)	Assumed possible, as in rescue effect, risk spreading and adoption of strategies to optimize future fitness	No, spatio-temporal heterogeneity
Prediction of future states by ecologists	Assumed possible for population density, community structure and establishment of introduced species, with laws sought to this end	No specific predictions because of environmental stochasticity and species-specific, idiosyncratic ecological properties of organisms (limited by weather prediction capabilities)
Role of mathematical models	Guide observation and prediction of future states of ecological systems	Reconstruct ecological processes, for explanatory purposes and short-term prediction

systems, as practiced in relation to short term local weather changes and the movement of planets, moons and comets in our own solar system. Naturally, the two aspects may coincide in a particular set of observations. Furthermore, predictions of both types are possible under demographic theory and autecological theory. How this is achieved needs to be considered for each of the two approaches to ecological understanding, as do the limitations on forecasting future states of ecological systems.

At the heart of science we find "laws". These are surprisingly difficult to define, because they lie at the center of the contentious issue of what, exactly, defines science (Chalmers 2003, pp. 213ff, Colyvan and Ginzburg 2003, Cooper 2003, p. 112). We do not contribute to this debate, but do accept that core assumptions about the nature and dynamics of ecological systems are reflected in statements that are law-like, and useful in that regard (e.g., Chalmers 2003, Colyvan and Ginzburg 2003, Cooper 2003). They represent understanding and are primarily heuristic in nature. They suggest to ecologists how to look at systems and, thus, how to investigate them, for they are representative of the underlying principles that represent the subject matter of ecology.

In ecology, the term "law" is said to be used in two ways (O'Hara 2005). Correlative laws reflect observed regularities or patterns. The example given by O'Hara (2005) is Rapaport's rule, which has the diversity of species increasing as one approaches the equator. Predictably, the Cameroon supports a high diversity of species because it is on the equator. Although predictive in this rather limited way, such laws do not offer much in terms of ecological interpretation. The primary patterns in nature seem to be more significant in this regard. Causal laws, on the other hand, deal explicitly with cause and are intended to encapsulate the mechanisms or processes that give rise to observed regularities. The Malthusian law is said to be a good example (O'Hara 2005).

We examine each ecological paradigm from this background perspective.

7.1 Demographic Ecology and Laws Aimed at Prediction

The way in which demographic ecologists use laws relates not so much to organisms and their properties, nor to observable pattern, but rather to expectations of how natural systems work. The competitive exclusion principle, for example, is law-like in the way it generates expectations (Walter 1988a, Weber 1999), and so is the balance of nature argument, with its emphasis on population stability. These are both used in law-like fashion, although neither was developed from a strong observational basis in pattern justification (Chapter 8, Section 3). Take, for example, Hardin's (1960) competitive exclusion principle, which represents a clear constraint that species are expected to impose on one another with regard to resource utilization. The principle was developed *a priori* (Simberloff 1980, Walter 1988a) and, like a law in society, is considered to be imposed as a force on the species concerned (Walter 2013). Individuals must thus follow suit, but this is implicit in the argument, and never discussed. Significantly, this law does not follow as a logical development from the properties and requirements of the individual organisms that make up the species concerned, nor from their behavior in relation to the ecological context. These same points relate equally to arguments about population equilibrium, although recent developments need to be taken into account (as detailed below, in this section).

The ecological influences detailed above are not specified to be operating on the organisms directly, at least not on their known behavioral, physiological and biochemical systems. Rather, the influence acts on groups of organisms (abstracted as populations and communities). In other words, the influence on individuals is indirect and is seen to act on species and population parameters to the extent that the influence on individual organisms remains unspecified. These laws do not relate directly to individual organisms themselves, for they are supra-organismal (Table 10.1).

The treatment of species as just outlined is therefore typically typological (Hengeveld 1988b, Walter 1993a). Law-like influences, external to individual organisms, are thus seen to underlie processes in demographic ecology (Table 10.3), although acceptance of this approach may be implicit more than explicit. In any case, expectations are thus imposed on the organisms, rather than the generalization being derived from the organisms themselves, so these laws are used heuristically as well. Ecological systems and, by implication, the organisms that are part of those systems, are expected to conform to those laws. An understanding of the laws themselves is expected to confer better predictive power of future states of ecological systems, and the search for periodic table equivalents for ecology stands testimony to the power of these ideas (e.g., Keddy 1994).

The treatment of ecological systems as hierarchies organized around ecological laws that have a global quality supports the popular view that ecology (and evolutionary ecology) can and should become predictive in the sense of ecologists being able to predict future states in ecological systems. For applied ecologists, especially conservationists, who often focus on restoration and balance, and pest managers, who introduce natural enemy species for biological control of pests, this is an appealing possibility. The search for the appropriate laws is thus an ongoing process for predicting future population trends. Our actual predictive abilities are likely to be far more limited than expected under the demographic perspective, and are dealt with in the following subsection.

The law associated with the balance of nature has been modified in response to criticisms of the original logistic law that represented ecological equilibrium (see Kingsland 1995). The more recent emphasis on the Malthusian law of population growth (Colyvan and Ginzburg 2003, Turchin 2003, O'Hara 2005) seems to be a retraction from the equilibrium aspect of population dynamics and a retraction to the process seen to generate the anticipated pressure from population density. This process is given law status because it is believed to be universally true in the sense that population growth would be exponential (although geometrical in organisms with discrete generation times) if all external ecological influences were removed. The reason this particular law is accepted so widely by ecologists may reflect the way it relates directly to individual organisms, albeit somewhat indirectly, and because it derives from observation that is readily verifiable. Individual organisms produce far more offspring than will ever survive. Notice that this law does not derive from a pattern that is observable in nature, whether that is a pattern of population growth or population stability, for example. Rather, it represents the pattern one would expect from the reproductive capacity of organisms, a point strongly emphasized by Malthus and Darwin, and then by those with an interest in ecological and evolutionary processes ever since.

Although the Malthusian law generates legitimate questions, it does seem to fail as a general, causal law. The approach is undermined, yet again, by species not being ecologically equivalent to one another, and by the vast diversity of axes of differentiation (Chapter 5) with which different species and all their different life stages can interact with their specific environment. The Malthusian law therefore conveys too little to be of much practical use. It is as true as the equation $N = b - d + i - e$, and as devoid of insight, for description cannot substitute for interpretation. Can ecological laws, in other words, be realistically expected if they treat individuals as being independent of their species-specific properties? The problem is that the individuals that make up a species do not react in the same way as those of another species to given ecological circumstances. All else is not equal in ecological systems, and we need to look further.

In short, the focus of study in demographic ecology is on the external laws expected to influence the ecology of organisms and this helps explain why the idiosyncrasy of species is seen to be a problem. The following subsection explains how autecology is different; the focus is on extracting generalities from observations of organisms so as to organize our knowledge of natural systems.

7.2 Autecology and Laws Aimed at Interpretation

Autecology has a perception of ecological systems that is fundamentally different from that of demographic ecology, because its emphasis is focused firmly on individual organisms within the context of a capricious external environment that is simultaneously physico-chemical and biotic in composition, rather than bipartite in this regard. Of primary importance are the adaptations that govern the primary responses of organisms to external conditions. Ecological interest thus focuses on the lifeline and lifetime track of organisms and the ecological consequences of these features as they relate to the distribution and local abundance of particular species. In this context, organisms are not constrained at all by external laws of nature, of the sort advocated under demographic ecology. The emphasis is thus transferred to the properties of organisms within the environmental context.

Autecological interpretation does develop generalized laws in the form of models, but these relate to organisms and their species-specific properties within the context of the environment they usually occupy. Such models are similar in spirit to that underpinning the Recognition Concept of species, for example, and provide a heuristic for guiding questions, investigation and interpretation (Table 10.3). This concept of species is based upon the behavioral and physiological properties of the sexes, the actions of which bring appropriate mating partners together for fertilization and syngamy within the usual habitat of the species (Paterson 1985). The organisms are not seen to be conforming, as a group, to external influences, as in the competitive exclusion principle or relative to a proposed population equilibrium. Rather, the general interpretation is developed from a functional analysis of what the organisms themselves do, as influenced by the adaptations they carry and the external conditions they face at the time. Such general statements or laws thus guide investigation (e.g., Fernando and Walter 1997, Rafter and Walter 2013, Rafter et al. 2013). The functional interpretation stems from the underlying pattern observed in nature, that of organisms being grouped distinctly into species in which like mates with like (Chapter 8, Section 3).

Laws in autecology therefore comprise generalized statements that express expectations as to how organisms interact with specific subsets of the entire environment. Again, the functional interpretation derives from the underlying pattern of the spatio-temporal dynamics of species in relation to the dynamic structure of the environment. The details of such interactions will be species-specific because species' properties are acquired in relation to very specific, local environmental circumstances at the time of speciation (Chapters 5 (Section 6) and 6 (Section 4)). Law-like statements of this nature do not allow extrapolation of explanation to different situations and across all species, as expected of the external laws of demographic ecology. Rather, they suggest ways in which to approach interpretation and investigation of particular species and their ecology (Walter 1995). This approach generates the alternative possible explanations required to test ecological interpretation (e.g., Finlay-Doney and Walter 2012a). Laws in autecological theory do not place organisms within a demographic context that is constrained by other organisms, as seen with niche and carrying capacity.

The most basic prediction we can derive from autecology is that the ecological properties of individuals that are potentially part of the same gene pool in nature should be stable across space and over time, which provides a basis for extrapolating information and making predictions about aspects of the ecology of these organisms, (i) from one place of derivation to another, (ii) from place of derivation to place of release (if we are transplanting biocontrol agents or organisms of conservation significance), (iii) from one place of introduction to another (again biocontrol agents), or (iv) from one time to another. However, the population consequences of these properties and the interactions they mediate will vary with locality and time, as environmental conditions vary. In other words, the population consequences of a natural enemy's activity in its area of endemicity cannot necessarily be expected to be mirrored in areas where it is introduced (from Walter 2003, p. 268).

In short, then, the nature of adaptations implies that the basic pattern will hold consistently, with the most basic patterns being that organisms are "organized" into gene pools (species) and these species show particular spatio-temporal responses in their distribution to changes in environmental structure and dynamics. Nevertheless, ancillary patterns above this level will also be discernible, but will not necessarily hold universally. For example, visual signals are restricted to diurnal organisms, but those animals belonging to groups with bioluminescence can signal at night, and diurnal organisms also use non-visual modes in their interactions (Paterson 1991). In ecological terms, the most basic pattern is the dynamic spatio-temporal distribution of species, from which various ancillary correlations can be identified, such as those related to patterns of resource use in organisms considered to be generalists. From there, the mechanisms that account for such patterns can be identified and investigated (e.g., Rajapakse et al. 2006, Rajapakse and Walter 2007, Finlay-Doney and Walter 2012b, Rafter and Walter 2013).

An appreciation of the distinction between demographic ecology and autecology will help to reassess the nature of ecological laws and their place in ecology as a science (Walter 2013).

8 CONCLUSIONS

Each of the two paradigms in ecology represents a logical, coherent flow of interpretation and this implies that each needs to be examined as a whole (Tables 10.1–10.3). Components are not free to be removed and replaced willfully, and the entire structure of each does not stand or fall on such realignments. For demographic ecology resource utilization, as played out by the demographic flux, represents the primary mechanism by which the primary patterns of equilibrium or ecological balance are considered to be attained. The autecological equivalent represents the maintenance of individual budgets and flow of energy and matter, by means of the spatio-temporal flux within individuals in their spatio-temporal dynamic matching of environmental conditions. Each of these sets of logic leads to its own distinct interpretation of evolutionary change.

Ecologists seem to have spent little effort in extracting the underlying principles of each paradigm (Cooper 2001, Cuddington 2001). Instead, criticisms and new complexities raised in relation to demographic ecology theory have most frequently been accommodated by incorporating adjunct theories to what was already available. In many ways, this has screened the underlying demographic ecology principles from scrutiny, and protected them. This seems simultaneously to have generated the feeling that ecology was adjusting to changed perceptions of natural systems (e.g., Graham and Dayton 2002, Paine 2002). Such *ad hoc* adjustment seems to explain the frustrations of McIntosh (1985), that a "New Ecology" emerges every 10 years or so, yet nothing actually changes fundamentally. That such an approach has apparently become routine is unfortunate, as it has inhibited understanding of the discipline, and has stood in the way of further strengthening and developing ecology as a science.

The two ecological approaches, demographic ecology and autecology, differ fundamentally from one another, and are based on mutually exclusive views of the constitution of organisms, the environment they inhabit, and the organism-environment interaction. The two approaches are, therefore, irreconcilable, in much the same way as the "molecular-centered" paradigm (of Dawkins (1989) for example) differs from the "organism-centered" paradigm, as justified by Rose (1997). Indeed, the autecological approach and its view of ecological understanding is consistent with the organism-centered view favored by Rose (1997), for it (1) is also materialistic (rather than idealistic), (2) sees the environment as presenting a challenge to living systems, (3) is designed to cope with the observable species-specificity seen among organisms, and (4) recognizes that contingency must form an appropriate part of the explanation of what we observe within the environmental context. The emphasis on the exclusive pre-eminence of DNA in the molecular paradigm, which represents a strongly deterministic outlook (Rose 1997), is reminiscent of the deterministic twist given to virtually all aspects of demographic ecology.

Because autecology is an organism-centered discipline, it concentrates on the probability of survival of individual organisms as a result of their ecological operations under variable conditions. It thus provides ecology with a direct connection to the other biological disciplines for the first time (Table 10.2), by placing the genetic, biochemical, physiological, anatomical, and behavioral properties of organisms into their environmental context. And these properties can only be understood through the way in which they function internally, that is within the organism, and respond externally, that is to the environment, but through the organism as a whole. The properties of the organisms may change in the way they function, under the influence of strong directional selection under particular specific circumstances of space and time, and this perspective connects autecology directly with the evolutionary process as envisaged under the recognition concept of species.

In closing, we point out that autecology is frequently taken as some sort of elaborate, qualitative natural history, without any recognition that it forms a coherent alternative to the currently widely accepted and quantitatively developed demographic approach. Autecology, though, does not only quantify ecological processes, it also retains the individuality and ecological idiosyncrasy of species and their component individuals in the generalization it provides of such systems. The challenge in autecology is therefore to find ways and means of quantification and modeling, to reformulate much of the research focus in ecology, and develop a more coherent relationship with other sub-disciplines in biology. New perceptions and questions will inevitably follow.

Appendix

AXES OF ENVIRONMENTAL DIFFERENTIATION

Axes of environmental differentiation are physico-chemical and biotic variables that are ecologically relevant to the survival and reproduction of organisms. We often measure them and represent them indirectly and in various ways (Spomer 1973). Any particular variable may affect different organisms in numerous ways and the brief summary list below, which has anecdotal information on a range of species, is included to illustrate this. In some cases the organisms concerned and the biological properties concerned are also given. The different measures may be specific to species, or relevant to several species. Exposure of the organisms concerned to the circumstances listed below is said to impact on them to the extent that their spatio-temporal dynamics are affected. Note that (i) ecologists have studied such influences over a long period (see Bodenheimer 1938, 1958, Billings 1952, 1974, Daubenmire 1974), particularly prior to the advent of evolutionary ecology in the 1950s, and (ii) in very few of these cases has the underlying physiological mechanism actually been investigated and determined.

Temperature

Annual mean temperature (Fang and Lechowicz 2006)

Seasonal temperatures (Huntley and Webb 1989)

Air temperature (Kingsolver 1989)

Temperature extremes (Criddle et al. 1994)

Mean northern May temperature (laying date for tree swallows) (Dunn and Winkler 1999)

Mean temperature for warmest month (Pigott 1975, Fang and Lechowicz 2006)

Mean temperature for coldest month (Pigott 1975)

Mean maximum temperature for warmest month (Pigott 1975)

Monthly mean temperature (Fang and Lechowicz 2006)

Duration of growing season with favourable temperatures, or growing season warmth (Criddle et al. 1994, Fang and Lechowicz 2006)

Rates and magnitudes of daily changes in temperature (Criddle et al. 1994)

Number summer days with night temperatures greater than threshold (moth flight) (Battisti et al. 2006)

Maximum air temperature & temperature sum (sphecid moth foraging) (Strohm and Linsenmair 1998)

More than 20-24 days below freezing during winter and more than nine consecutive days below freezing (armadillos) (Taulman and Robbins 1996)

Number of winter days with air temperature remaining below 0°C all day (benthic marine predatory polychaetes) (Beukema et al. 2000)

Cumulative winter temperature (Fang and Lechowicz 2006)

Average winter temperature (grass growth) (Bergthorsson 1985)

Cumulative days of frost through winter (Cawthorne and Marchant 1980)

Number of frost-free days in March (Gray 1997)

July temperature (Gray 1997)

Winter minimum temperature (Kincer 1928)

Continentality index (Fang and Lechowicz 2006)

Winter-spring ice extent (for Atlantic cod recruitment), which seems to be influenced by water temperature, ice conditions, and volume and extent of the cold intermediate layer in the water column that overlies the coastal shelf (Taggart et al. 1994)

Growing season temperature sum (or accumulated temperature) (Parry and Carter 1985)

Unusually high (northern) winter temperatures (plus high temperatures in July to September) (Schwerdtfeger 1932, cited by Ito (1961))

Cool early summer (June and July) temperatures (Miyashita 1955; cited by Ito (1961))

Water temperature and wind (for barnacles, see Ito (1961))

Months in which temperature is greater than 0°C and less than 30°C (Fang and Lechowicz 2006)

Solar radiation (Kingsolver 1989)

Rainfall

Annual precipitation (Fang and Lechowicz 2006)

Evapotranspiration relative to precipitation (Bartlein et al. 1986, Fang and Lechowicz 2006)

Monthly precipitation (Fang and Lechowicz 2006)

Rainfall events (sporadic or at a particular time, as in autumn rain for octopi) (Ito 1961)

Rainfall when salmon fry are still in rivers (see Ito (1961, p. 66))

March-July precipitation totals (northern hemisphere) (Ginnett and Young 2000)

Rainfall in May (Gray 1997)

Periodicity/Frequency of rainfall (Davidson and Andrewartha 1948a, Ito 1961)

Positive to drought (locusts, as soil moisture seems to affect eggs) (Ma 1958 cited by Ito (1961))

Rainfall during dry season. Positive effects of higher rainfall (tropical herbivorous insects and large herbivorous mammals) (Wolda 1978b, Fryxell 1987).

Positive effects of low rainfall (van der Laan 1959; cited by Ito (1961))

Rainfall in previous breeding season (tiny carnivorous marsupials) (Lada et al. 2013)

Deep snow and freezing rain that freezes over vegetation rendering it unavailable to large herbivores (Vors and Boyce 2009)

Relative Humidity (RH)

RH June/July current (northern) summer (Kincer 1928)

RH July/August previous summer (Kincer 1928)

Cloud Cover

Percentage cloudiness previous (northern) June/September (Kincer 1928)

Cloud cover relative to fog drip (plant recruitment) (Fischer et al. 2009)

Aquatic Conditions

Depth-averaged (0-50m) summer salinity (Taggart et al. 1994)

Oceanic currents and disturbance of them (by wind) (for cod and other marine fish, see Ito (1961, p. 66))

Depth-averaged summer salinity measurement (for Atlantic cod) (see Taggart et al. (1994)).

Sea ice cover and laying date (guillemots) (Gaston et al. 2005)

Edaphic Factors

Soil texture (Jha and Singh 1990)

Indirect Influences

Timing of budburst and/or early leaf growth of host plant (Butterfield and Coulson 1997)

Shading (Preece 1997)

Combinations of Factors

Naturally, combinations of factors mentioned above may be significant ecologically (Gray 1997). For example, both summer and winter temperatures may be important in determining the distributional ranges of species (e.g., Iversen (1944) on *Hedera, Viscum* and *Ilex*) and physiological and growth response studies have shown the importance of freezing tolerance, growing season length requirements, drought intolerance and critical conditions for completion of particular stages of the life cycle of beech (Huntley et al. 1989, p. 556).

Correlated Indices or Proxy Measures

For example, winter minimum temperature may be highly correlated with monthly mean temperature for January (e.g. Huntley et al. (1989) on *Fagus* distribution). Mean temperatures of January and July are frequently used in northern hemisphere studies as measures of temperature minima and maxima, respectively. Also, mean annual precipitation may be used in this way. The use of correlated indices is likely to obscure which variable(s) are actually influencing the ecology of the organisms in question.

Subjectivity Inherent in Measures

Because we tend to use indirect measures, subjectivity will be a problem. To illustrate, a "frost day" has been defined in terms of there being frozen ground at 0900 h at more than six coastal sites (Cawthorne and Marchant 1980). Clearly this is indicative and other measures may be more appropriate.

Glossary

A LEXICON FOR AUTECOLOGY

Note that not all terms are strictly autecological—various terms that are used more generally or that are particular to demographic ecology are also included for their nuances relative to autecological theory.

Abiotic forcing External influences, such as climate, may drive ecological systems and influence the evolution of the organisms concerned, and such drivers of change are interpreted in terms of abiotic forcing, perhaps to emphasise that the expected biotic influences (like density dependence) were not responsible.

Adaptation Adaptation is a term that covers the process of adaptation as well as the mechanism that evolves. Each is expanded below, in relation to demographic ecology and autecology.

Adaptive mechanism An adaptation is an evolved device or mechanism that functions to support the survival or reproduction of individual organisms. The phrase "adaptive mechanism" is therefore a tautology, but is retained to emphasize that it is not the process under consideration, because the emphasis in evolutionary ecology is on the process of adaptation rather than the mechanism.

Adaptive process The adaptive process in evolutionary ecology (which is intimately associated with the logic and processes of demographic ecology) is seen to be driven by optimizing selection. Autecology, instead, sees the adaptive process in relation to the organism-environment interaction ("environment" encompasses the biotic (see below), such as vegetation, hosts, and so on) so is not seen in the competitive terms of individuals being more "efficient" than others in the production of offspring.

Advection term of spatial diffusion The movement of organisms of a particular species within a particular area proceeds from a particular source (or sources) and tends in a particular direction (or directions). This is represented by the advection term of spatial diffusion, for modeling purposes.

Arrhenius temperature coefficient See "Temperature coefficient", to which this phrase is frequently abbreviated.

Axiom See "foundation statement".

Basic assumption See "foundation statement".

Biogeochemical cycles The pathways followed by chemicals (often nutrients) through various components that make up ecological systems, including organisms, water, soil and air. The scales at which such pathways (or cycles) operate are often very large relative to the scale at which ecological systems are often investigated.

Biological species concept Species were traditionally treated typologically as fixed, invariant entities. With the development of evolutionary biology, with its emphasis on population level processes through population genetics, a different approach was required. Sex combines the genes from two parents, so sexual behavior and fertilization provided the criterion for interpreting species' limits. The biological species concept, developed to a large extent by Mayr (1963), emphasized the reproductive isolation of species although mating behavior was also seen as crucial to fertilization being achieved. Reproductive behavior was interpreted almost solely in terms of it being a mechanism (or adaptation) to isolate species. The conflation inherent in this concept was disentangled by Paterson (1985) who relabeled the biological species concept the isolation concept (because of its almost sole concern with reproductive isolation and isolation mechanisms) and developed the alternative recognition concept of species. The latter focuses on reproductive behavior and associated processes as adaptations that mediate the sexual interactions of individuals, and sees the reproductive isolation of groups as a by-product. The focus of the isolation concept on the isolation of groups stands in contrast to the individual selection that provides the cornerstone of Darwinian evolution.

Biotic In general, "biotic" in biology refers to environmental influences that derive from living organisms, and includes non-living organic matter (Andrewartha and Birch 1954, p. 16). It provides a contrast with chemical and physical factors. "Biotic" is used in this sense in autecology. In demographic ecology, "biotic" refers to density dependent regulatory influences.

Biotope A biotope occupied by individuals of a species is the place where a species is found within a particular part of its geographical range, and is usually characterized by its physical appearance (e.g., open woodland or slow-flowing stream). See Habitat for further clarification.

By-product See epiphenomenon.

Chance See stochastic.

Chaos Deterministic systems that are highly sensitive to even very slight differences in initial conditions are chaotic. The outcomes are so widely variable they are effectively unpredictable.

Co-evolution The co-adaptation of two species, with adaptive change in one eliciting an adaptive response in the other. Several finer definitions than this are available, but at the core of co-evolutionary theory is that the process is ongoing. The broad use of the term means that the meaning intended needs to be spelt out at each use of the term.

Coexistence This is an interpretation of how species that have similar diet (or have similarities in the use of other resources) are considered to live together in a locality without one of them driving the other to extinction through competition for the particular resource in question (i.e., through competitive exclusion). Such explanations lie at the core of community ecology.

Community In ecology a community is a suit of species that use the same resource and are, in theory at least, potential competitors (and are sometimes said, more vaguely, to have the potential to interact).

Contact distribution Frequency distribution of distances covered by conspecific organisms when they move from a particular starting point (e.g., their natal site). Typically it is *J*-shaped as in Fig. 7.4.

Demographic ecology Demographic ecology encompasses what is referred to as population ecology and community ecology, each of which is seen to deal with a particular level of the ecological hierarchy. Because the primary influence on populations and communities is considered to be the demographic influences of density (and thus competition within or between species), this perspective on ecological systems is referred to collectively as demographic ecology. Naturally, other influences are acknowledged, but the underlying theory emphasizes density effects.

Demography Demography is the study of the statistics of populations, to determine what variables affect population change.

Dispersal kernel The dispersal kernel is the mathematical representation of the rate of movement of the individuals of the subject species in the geographical area of interest, and thus its rate of spread through that area.

Divergent selection or divergent natural selection This is natural selection that leads to the divergence of conspecific populations (whether in sympatry or allopatry) and is commonly associated with ecological speciation. Use of the term "divergent" implies that the selection is acting principally to produce a new species (expressly or tacitly to specialize on unused resources (= fill an empty niche)).

Ecological determination The process in ecology in which the general level of abundance of a population is set (Solomon 1949). Abiotic factors are usually considered to do this, whereas population regulation around that level of abundance is seen to be achieved by density-dependent processes such as competition and predation.

Ecological fitting This term was introduced initially to indicate the existence of a species in a locality through pre-adaptation to the local conditions (rather than through adaptive change in response to species seen to be competitors, so as to facilitate coexistence) (Janzen 1985). More recently it has been used to indicate the fitting of a species into an empty niche within a community (Agosta 2006, Agosta and Klemens 2008). The latter is a demographic concept and should be compared with the autecological equivalent of "environmental matching".

Ecological hierarchy Ecological systems are often portrayed as being hierarchical, starting with individual organisms and proceeding higher to populations, communities, ecosystems and thence to larger scales. Autecological theory does not acknowledge such divisions.

Ecological idiosyncrasy Community ecologists have long sought consistent patterns, in nature, across species. However, the ecology of species has been frustrating in this regard. Consistency in community structure and succession is almost invariably disrupted by the species-specificity inherent in the ecology of organisms. The ecology of species is therefore said to be idiosyncratic, although this is not explanatory of their ecology at all. See entry on individualism of species.

Ecological intensity Measure of performance of a species relative to the two-dimensional response surface that represents its geographical distribution. It is measured in terms of the density of organisms (of a particular species) per unit of biotope, by the number of biotopes occupied within a geographical region, number of offspring produced, and so on. Although partly equivalent, numerically, to local abundance and local population density, it is unlike these concepts in that it does not see organisms confined within subjectively specified boundaries within which density dependent effects are seen to operate.

Ecological measure of species stability See species stability.

Ecological process A process in ecology is simply any action or series of actions that take place in the environment (or an experimental arena). The term is a neutral one in that it does not imply a particular end. Ecological processes may be mediated or influenced by the action of adaptations (or mechanisms (see below)) of the organisms involved in such actions and care is needed to differentiate the functional aspects of such mechanisms from any incidental consequences. For example, a filter feeder may discard sand particles into the water around it, cloud the water and thus render that area unsuitable for organisms that require a lot of light. The negative effect on the latter organisms is incidental to the process that incidentally alters the environment.

Ecological risk The number of failures of a species (e.g., to establish, reproduce) within a certain area or region relative to the total number of potential generations.

Ecological system The local diversity of flora and fauna and the dynamic that underpins it. In autecology, the scale is defined in terms of locality where the species of interest (e.g., herbivorous insect, host plant and natural enemies) live/s, and the geographical distribution of that/those species. The climatic and biogeochemical cycles that are relevant to understanding the system are likely to be of far larger scale. In demographic ecology, ecological systems are seen as comprising a hierarchy of scales, from local, metapopulation, regional to ecosystem.

Ecotone An ecotone is the spatial transition from one ecological situation to another, for example from one vegetation type to another, or one biotope to another.

Empiricism Broadly, empiricism is the approach to knowledge that emphasizes sensory experience, as opposed to rationalism, in which reason is the main means for extending knowledge. Empiricism more narrowly characterizes the inductive method pioneered by Francis Bacon, because natural laws were expected to become evident from objectively collected data.

Energy budget (maintenance of) Organisms survive through maintenance of the flow of energy through them.

Engineer species Organisms of numerous species, as they interact with their environment, impact on their surrounds and change them. Numerous conspecifics in the same area may thus change the local environment. Such organisms have been called "engineer species". This metaphor treats species typologically (see below) in that it implies a constant and even inevitable "role" for the species concerned. However, the impact of these organisms is likely to vary through their life cycle and in relation to the influence of different environmental conditions, so the typology is unwarranted. The environmental impact of organisms is better described simply as such, and they are readily integrated into interpretations of biogeochemical cycles and environmental dynamics.

Environment In demographic ecology the environment of organisms is conceptualized as being bipartite, with an abiotic component that imposes stochasticity through adverse weather conditions on the population or community of interest and disrupts the influence of the biotic component that operates primarily through deterministic density effects. Autecology, by contrast, conceptualizes the environment of organisms in terms of the species-specific requirements and tolerances of organisms. See other entries beginning with "Environment".

Environmental axes of differentiation The environment is multi-faceted in the way it impacts on organisms. Organisms, in turn, have requirements of the environment. Each environmental variable, such as temperature, humidity and nutrients, can impact organisms in a variety of ways. Temperature, for instance, may impact on organisms in relation to the provision of heat over the growing season, extremes, or even the timing of frost events early in the spring. Each such measure is an axis of differentiation to which the species of interest may be adapted. The array of such environmental features (or axes of differentiation) that impact on organisms is potentially limitless.

Environmental dynamics The environment of organisms goes through various changes, some cyclical (e.g., daily, seasonally or even longer term climatic cycles) and others not (as in climate change). It is also impacted by biogeochemical cycles and the activities of organisms (whether conspecific or not). These changes represent the environmental dynamics that organisms face.

Environmental matching Organisms persist in a locality in which the environment (defined autecologically rather than demographically (see above)) provides for their requirements and does not exceed their tolerances. That is, the organisms match the local environmental conditions, with these being the ones to which they are primarily adapted. They do so through the process of habitat tracking.

Environmental response system The combined mechanisms (see mechanism below) by which organisms respond to their environment. They include the integrated actions of genetical, biochemical, physiological, behavioral and morphological mechanisms. Organisms, by means of these systems, are able to match their environmental requirements.

Environmental structure The environment of organisms is dynamic, but nevertheless can be characterized in terms of its structure, as represented for example by such features as seasonal temperature profiles to indicate the distribution through time of the provision of heat or the frequency of days with temperatures above a particular level that might impact negatively on the organisms in question.

Environmental tracking All organisms can move at some stage of their life cycle, if only passively as seeds or spores. Organisms thus move into suitable areas away from where they were born or developed. Through such movement they colonize areas that are newly suitable with respect to their requirements and tolerances. Individual organisms thus cope with environmental heterogeneity and changed environmental circumstances by environmental tracking in a spatio-temporal sense. Also, they may adjust developmentally, physiologically or behaviorally. This is the primary determinant of the observed spatio-temporal intensity of occurrence of organisms. See habitat tracking.

Epiphenomenon An epiphenomenon is a side-effect of a primary cause. Interpreting an epiphenomenon relies on understanding the primary cause. Studying the epiphenomenon in itself can be extremely misleading.

Equilibrium in ecology Populations are considered, in demographic ecology, to be in a state of equilibrium. Fluctuations in population density are considered to lead to influences (higher or lower levels of competition) that return the population to equilibrium (or a state of balance).

Evolutionary stability See species stability.

First principle See "foundation statement".

Fitness Fitness is a central concept in evolutionary ecology, where reproductive output ("fitness") is assumed to be optimized relative to environmental circumstances, but with particular emphasis on outcompeting other organisms. Indeed, intraspecific competition is seen as a central driving force in evolutionary ecology (Brady 1982, Paterson 2005). Fitness is defined in demographic terms as the expected contribution of an individual to future generations (Stearns 1992), which is seen to be homologous to the definition of intrinsic rate of increase for a population despite the fact that "Fitness, like the intrinsic rate of increase for a population (r), can almost never be measured directly" (Mitchell and Powell 2003). In autecology, by contrast, reproductive output is seen to reflect the ecological "performance" of individuals of a particular species under given circumstances, and is not taken as an indication of the direction of evolutionary change. Rather, performance is seen to reflect how adaptive mechanisms operate within particular ecological contexts, as in the performance of an herbivorous insect species across different host species.

Foundation(al) statement All interpretations are based on premises (or prior beliefs) about how the system in question works. These premises are also known as (basic) assumptions, axioms, first principles or foundation(al) statements, and these play the crucial role of guiding questions and interpretations. They are almost invariable strongly held views and often are tacit or poorly expressed.

Habitat In autecology, the habitat of a particular species is the set of environmental conditions that satisfy the requirements of the individual organisms that make up that species (i.e., its environmental requirements). This is, therefore, an abstraction that is centered on the elements of the environment that the organism requires for its persistence and reproduction. A species may occupy multiple biotopes if all of them satisfy its environmental requirements.

Habitat recognition Many organisms interact with their environment through their neurosensory system. They respond to environmental information and are attracted, settle, and so on, in a process of habitat recognition. Even single-celled organisms and pelago-benthic larval forms respond in this way.

Habitat tracking The process (in environmental tracking) by which organisms that have sensory capabilities, use sensory information from the environment (about vegetation structure, substrate chemistry, and so on) to detect and localize suitable habitat conditions. One can thus talk in terms of habitat recognition.

Holocene This is the current geological period, having started about 12,000 years ago, when the Pleistocene ended. See Pleistocene and Quaternary.

Hydrophilous That which relies on water, so hydrophilous species are associated with moist environments.

Idiosyncrasy of species See Ecological idiosyncrasy.

Individualism of species The phrase introduced by H.A. Gleason (1926) to indicate that the ecology of species does not fit the expectations of community ecologists and does not fit into a predetermined community structure or successional process. That is, the ecology of species is not tied into that of other species unless there is a particularly strong association between them, as in use of hosts, pollinator mutualisms, and so on. See Ecological idiosyncrasy.

Intensity, Intensity of occurrence See "ecological intensity".

Isolation concept of species See "biological species concept".

Isopoll maps Maps that plot relative pollen frequency, usually through time, and from which changes in plant species distribution and vegetation structure can be derived.

Life cycle Each individual organism develops through various stages until it reaches reproductive competence, and this is generalized or conceptualized into the life cycle of the species of which it is a "member". The life cycle is thus a metaphor that expresses how a species is perpetuated. The experience of individual organisms in nature is conceptualized better in terms of its actual lifeline or ecological trajectory.

Lifeline (see lifetime track) The lifeline of an organism is the particular track (or trajectory) that the individual takes, defined spatio-temporally, through its entire life. It thus includes successive reproductive seasons in iteroparous organisms, and can be thought of as the lifetime ecological trajectory (realized or predicted) of the individual.

Lifetime track The spatial displacement of an individual organism from the place of its birth or initial development, through any foraging areas, breeding areas and so on, perhaps linked by migration routes, to the point of its death (see Drake et al. 1995, Dingle 1996, Walter 2003).

Locality Many detailed ecological studies are conducted within a particular locality. The locality is identified as of interest for various reasons (a species may be common or rare there, intense competition may be expected there, and so on) and may be defined in various ways and at various scales, often for convenience. That is, the extent of a locality is almost invariably defined arbitrarily for convenience. Autecology specifies that the locality should be seen within the context of the species' distribution because organisms are not restricted to such places.

Mechanism In evolutionary biology the term "mechanism" has a particular meaning, and infers that one is looking at an adaptation, something that evolved to serve a particular function. In the ecological literature the word mechanism is often used when process (see ecological process, above) would be more accurate.

Mesophilous Refers to organisms that require temperatures considered moderate, roughly between 25 and 40°C.

Metapopulation A population of populations—developed for those species with a patchy distribution. Each such population is presumed to follow, in demographic ecology (logistic equation) terms, the usual dynamics of local populations but to be linked by movement among patches.

Multivoltine A species (or particular population of a species) that produces more than one new generation in a year. See univoltine.

MyBP Million years before present.

Net reproductive rate (R_0) Mean number of female offspring produced by a cohort of parent females, through their lifetime, and surviving to reproductive age.

Operational environment This is the immediate environment of an organism of a particular species defined in terms of the interactions that take place between those individuals and the environment. These are the interactions in which mass or energy are exchanged (see Spomer 1973). It is thus defined in relation to its species-specific requirements and tolerances, and these may be biotic or abiotic for in autecology the environment of organisms is not seen as bipartite.

Optimizing selection This is natural selection operating strongly on the basis of competition between organisms, to the extent that "efficiency" (in locating hosts, for example, or harvesting resources) is positively selected. The connection between efficiency and the physiological and behavioral mechanisms of the organisms is left unspecified. See fitness.

Optimum curve Non-linear, humped response curve of a species relative to a certain environmental factor.

Palynology In ecology this is usually the study of pollen (including fossil pollen) in the environment. More generally, it is the study of any small particles.

Patch occupancy model A particular approach to modeling the population dynamics of a metapopulation, which is one that comprises several smaller, local populations. Patch occupancy models are based on colonization rates of empty patches. Although empty, these patches are presumed to have lost their inhabitants through stochastic processes, so the patch is available for immediate recolonization. Models the "equilibrium values" of patch occupancy. See also metapopulation.

Permissive range The range of values, for any physico-chemical factor (e.g. temperature, humidity), between which survival and reproduction of organisms of a particular species are supported (Criddle et al. 1994).

Pleistocene The geological period lasting from about two and a half million years ago until about 12,000 years ago, when the Holocene started. Together these two periods comprise the Quaternary period (see below).

Population biology Population biology represents the theory and investigation of Mendelian (or sexual) populations. It provides the link between species (and thus species theory) as seen as "species gene pools", population genetics (which again deals explicitly with Mendelian populations) and ecology. Note that "population" here is defined in reference to gene flow and the potential for gene flow through organisms that share in the common behaviors and so on that lead to fertilization. This definition stands in strong contrast to the "population" (strictly, the "local population") of demographic ecology (and its sub-discipline population ecology). The different meanings of the word population are readily conflated.

Population ecology Population ecology encompasses population dynamics and metapopulation dynamics theory and investigation. Population, here, is defined arbitrarily, usually for convenience but sometimes in relation to the relative influence of migration relative to other processes (see Chapter 7, Section 4).

Population equilibrium See Equilibrium in ecology.

Pre-adaptation This is a key concept in understanding how evolution takes place and in interpreting many aspects of ecology, so the concept has two applications. Essentially, a pre-adaptation is a feature of an organism that (i) has been seized upon by natural selection and turned to a different function, as in the rigid fins of crossopterygian fish that were the precursors for the two sets of limbs of terrestrial vertebrates, and (ii) allows an organism to make a "new" environmental match because its properties incidentally allow it to do so, as in cliff-dwelling peregrine falcons living and nesting in high-rise cities.

Prediction The term "prediction" has a duality in science and these two uses seem to be conflated in many statements about science. The more mundane sense of prediction is to "forecast", as in forecasting a future state of a system. Weather forecasting is a good example. In the epistemological sense, by contrast, a prediction, developed on the basis of an understanding of the system under consideration (e.g., verbal model, mathematical model), is made as to what to expect from an experimental manipulation of the system or a set of observations of the system. It refers to expected or calculated values according to some mechanistic model. This tests scientific understanding. It is true that a prediction of the first type could be used in scientific tests.

Process See ecological process.

Q_{10} Change in rate of a reaction in response to a 10°C change in temperature.

Quaternary ecology Ecology of the Quaternary period. Many fossils of this period are representative of organism still alive today. Comprises the Holocene plus Pleistocene.

R_0 See Net reproductive rate.

Random Distribution according to a particular chance process.

Recognition concept of species Explained under the Biological species concept.

Refugium Relatively small area in which individuals of a species occur within a larger region unsuitable to them. Implicitly the larger area was once inhabited by them and has since become unsuitable, or the larger area is now inhabited after a period spent in a refugium, or both of these.

Rescue effect Process in metapopulation ecology of numerical compensation of a relatively small population by individuals of a large one.

Spatial adjustment Process of a group of individuals matching, in space, the environmental conditions to which they are adapted.

Spatial diffusion The spatial dispersion of particles according to a stochastic process, starting from one point.

Spatial dynamics The dynamics of individuals across space typically through the processes of habitat tracking, habitat matching, environmental matching, and so on.

Spatial stability See species stability.

Spatio-temporal dynamics Each species covers a particular geographic distribution, which changes with time. This change is simply a consequence of the shifting, across space and over time, of individual organisms. Individuals move in the normal course of events although the rate is likely to increase (in many species) when their surrounding environment deteriorates relative to their species-specific requirements and tolerances. "Populations" of a species thus wax and wane and localities that are newly suitable to the species are colonized. The distribution of the species may change as a consequence.

Species specificity Properties of individuals of a species that are consistent throughout the entire species. Use of the phrase infers that those properties are different from equivalent properties in closely related species and influence their behavior, environmental interactions, and so on, differently from that in the other species.

Species stability The idea of species remaining stable (i.e., unchanging) through time is best known as punctuated equilibrium. For species to remain stable through time, they would have to remain unchanging (stable) across space, which is the spatial equivalent of punctuated equilibrium.

Spreading of risk The view that numerous ecological influences, of all types, operate on populations with the net effect that their collective influence cancels out extremes of density and thus generates a pattern that resembles one of population stability (den Boer 1968, Reddingius 1971, Andrewartha and Birch 1984).

Stochastic A stochastic influence is one that affects an outcome in random (or chance) fashion, to the extent that the outcomes of even deterministic processes are most accurately described probabilistically.

Surface response analysis Statistical distribution of a species' ecological intensity of occurrence across geographical space. Statistically, it shows the highest intensities to occur in the centre of a species' range with a tapering towards the margins.

Teleology The idea that a process (e.g., evolutionary change) or an end-product (e.g., the production of a new species) is achieved for a purpose. Physics and chemistry eliminated teleology relatively early in their history. Darwin freed biology of what might be termed naïve teleology, but the adaptive process does produce mechanisms that lead to a particular end-point (or goal). This particular form of teleology (which has attracted alternative names, including teleonomy (Mayr 1974)), signals the way in which the biological sciences differ from the physical sciences (Rosenberg 1985).

Temperature coefficient The rate of change in a biochemical reaction in response to an increase in temperature. The Q_{10} value (rate of increase with temperature change of +10°C) is typically 2–3 across the range of temperatures to which the enzymes of interest are adapted. Provides a practical means for comparing the temperature responses of different life stages or species, for example.

Typology Emphasis on constancy and typical values. The overall treatment excludes a realistic consideration of the components. For example, species seen as having a "place" among other species in a community disregards the individuals that make up the species and how they interact with the environment.

Univoltine A species (or particular population of a species) that produces only one new generation in a year. See multivoltine.

Way-of-life Term to indicate the general type of existence of particular organisms, for example pelagic birds (life above deep ocean waters), sessile (attached to a substrate), free living (as opposed to parasitic) or nomadic (if it tracks conditions that change rather rapidly over the short term).

Xerophilous Pertains to organisms that live in hot, dry environments.

References

Aarssen LW. 1997. On the progress of ecology. Oikos 80: 177-178.

Abbot P, Abe J, Alcock J, Alizon S, Alpedrinha JAC, Andersson M, Andre JB, van Baalen M, Balloux F, Balshine S, Barton N, Beukeboom LW, Biernaskie JM, Bilde T, Borgia G, Breed M, Brown S, Bshary R, Buckling A, Burley NT, Burton-Chellew MN, Cant MA, Chapuisat M, Charnov EL, Clutton-Brock T, Cockburn A, Cole BJ, Colegrave N, Cosmides L, Couzin ID, Coyne JA, Creel S, Crespi B, Curry RL, Dall SRX, Day T, Dickinson JL, Dugatkin LA, El Mouden C, Emlen ST, Evans J, Ferriere R, Field J, Foitzik S, Foster K, Foster WA, Fox CW, Gadau J, Gandon S, Gardner A, Gardner MG, Getty T, Goodisman MAD, Grafen A, Grosberg R, Grozinger CM, Gouyon PH, Gwynne D, Harvey PH, Hatchwell BJ, Heinze J, Helantera H, Helms KR, Hill K, Jiricny N, Johnstone RA, Kacelnik A, Kiers ET, Kokko H, Komdeur J, Korb J, Kronauer D, Kummerli R, Lehmann L, Linksvayer TA, Lion S, Lyon B, Marshall JAR, McElreath R, Michalakis Y, Michod RE, Mock D, Monnin T, Montgomerie R, Moore AJ, Mueller UG, Noe R, Okasha S, Pamilo P, Parker GA, Pedersen JS, Pen I, Pfennig D, Queller DC, Rankin DJ, Reece SE, Reeve HK, Reuter M, Roberts G, Robson SKA, Roze D, Rousset F, Rueppell O, Sachs JL, Santorelli L, Schmid-Hempel P, Schwarz MP, Scott-Phillips T, Shellmann-Sherman J, Sherman PW, Shuker DM, Smith J, Spagna JC, Strassmann B, Suarez AV, Sundstrom L, Taborsky M, Taylor P, Thompson G, Tooby J, Tsutsui ND, Tsuji K, Turillazzi S, Ubeda F, Vargo EL, Voelkl B, Wenseleers T, West SA, West-Eberhard MJ, Westneat DF, Wiernasz DC, Wild G, Wrangham R, Young AJ, Zeh DW, Zeh JA, Zink A. 2011. Inclusive fitness theory and eusociality. Nature 471: E1-E4.

Abele LG, Simberloff DS, Strong DR, Thistle AB. 1984. Preface. Pages vii-vix in Strong DR, Simberloff D, Abele LG, Thistle AB, eds. Ecological Communities: Conceptual Issues and the Evidence. Princeton: Princeton University Press.

Adams MA. 2000. Contrasts between *Eucalyptus* subgenera. Funct Ecol 14: 654-654.

Agosta SJ. 2006. On ecological fitting, plant-insect associations, herbivore host shifts, and host plant selection. Oikos 114: 556-565.

Agosta SJ, Klemens JA. 2008. Ecological fitting by phenotypically flexible genotypes: implications for species associations, community assembly and evolution. Ecol Lett 11: 1123-1134.

Ainley DG. 2002. The Adélie Penguin: Bellwether of Climate Change. New York: Columbia University Press.

Ainley DG, Ballard G. 2012. Non-consumptive factors affecting foraging patterns in Antarctic penguins: A review and synthesis. Polar Biol 35: 1-13.

Alcala RE, Dominguez CA. 2003. Patterns of prey capture and prey availability among populations of the carnivorous plant *Pinguicula moranensis* (Lentibulariaceae) along an environmental gradient. Am J Bot 90: 1341-1348.

Alexander RM. 1999. Energy for Animal Life. Oxford: Oxford University Press.

Allee WC, Emerson AE, Park O, Park T, Schmidt KP. 1949. Principles of Animal Ecology. Philadelphia: W.B. Saunders.

Allen JRM, Brandt U, Brauer A, Hubberten H-W, Huntley B, Keller J, Kraml M, Mackensen A, Mingram J, Negendank JFW, Nowaczyk NR, Oberhansli H, Watts WA, Wulf S, Zolitschka B. 1999. Rapid environmental changes in southern Europe during the last glacial period. Nature 400: 740-743.

Anderson ML. 1987. A re-appraisal of method: the way of confrontation. S Afr J Sci 83: 89-93.

Andrewartha HG. 1984. Ecology at the crossroads. Aust J Ecol 9: 1-3.

Andrewartha HG, Birch LC. 1954. The Distribution and Abundance of Animals. Chicago: University of Chicago Press.

Andrewartha HG, Birch LC. 1984. The Ecological Web: More on the Distribution and Abundance of Animals. Chicago: University of Chicago Press.

Anekonda TS, Criddle RS, Bacca M, Hansen LD. 1999. Contrasting adaptation of two *Eucalyptus* subgenera is related to differences in respiratory metabolism. Funct Ecol 13: 675-682.

Anekonda TS, Criddle RS, Hansen LD. 2000. Contrasts between *Eucalyptus* subgenera—Response. Funct Ecol 14: 654-656.

Angert AL, Schemske DW. 2005. The evolution of species' distributions: Reciprocal transplants across the elevation ranges of *Mimulus cardinalis* and *M. lewisii*. Evolution 59: 1671-1684.

Angus RB. 1983. Evolutionary stability since the Pleistocene illustrated by reproductive compatibility between Swedish and Spanish *Helophorus lapponicus* Thomson (Coleoptera, Hydrophilidae). Biol J Linn Soc 19: 17-25.

Antolin MF, Strong DR. 1987. Long-distance dispersal by a parasitoid (*Anagrus delicatus*, Mymaridae) and its host. Oecologia 73: 288-292.

Ariew A. 2003. Ernst Mayr's 'ultimate/proximate' distinction reconsidered and reconstructed. Biology & Philosophy 18: 553-565.

Arizaga J, Campos F, Alonso D. 2006. Variations in wing morphology among subspecies might reflect different migration distances in Bluethroat. Ornis Fenn 83: 162-169.

Arsuaga JL, Martinez I. 2006. The Chosen Species. The Long March of Human Evolution. Oxford: Blackwell Publishing.

Arthur W. 2004. Biased Embryos and Evolution. Cambridge: Cambridge University Press.

Ashby WR. 1956. An Introduction to Cybernetics. New York: Wiley.

Ashworth AC. 1997. The response of beetles to Quaternary climate changes. Pages 119-127 in Huntley B, Cramer W, Morgan AV, Prentice HC, Allen JRM, eds. Past and Future Rapid Environmental Changes: The Spatial and Evolutionary Responses of Terrestrial Biota. Berlin: Springer.

Atkinson TC, Briffa KR, Coope GR. 1987. Seasonal temperatures in Britain during the past 22,000 years, reconstructed using beetle remains. Nature 325: 587-592.

Ayala FJ. 1982. Population and Evolutionary Genetics: A Primer. Menlo Park, CA: Benjamin/Cummings.

Baguette M. 2004. The classical metapopulation theory and the real, natural world: A critical appraisal. Basic Appl Ecol 5: 213-224.

Baker RR. 1978. The Evolutionary Ecology of Animal Migration. London: Hodder & Stoughton.

Baldwin MW, Winkler H, Organ CL, Helm B. 2010. Wing pointedness associated with migratory distance in common-garden and comparative studies of stonechats (*Saxicola torquata*). J Evol Biol 23: 1050-1063.

Balter M. 2009. A Neandertal primer. Science 323: 870-870.

Bang BG. 1966. Olfactory apparatus of tubenosed birds. Acta Anat 65: 391-415.

Barrett G, Silcocks A, Barrry S, Cunningham R, Poulter R. 2003. The New Atlas of Australian Birds. Hawthorn East: Royal Australasian Ornithologists Union.

Bartlein PJ, Prentice IC, Webb T. 1986. Climatic response surfaces from pollen data for some eastern North American taxa. J Biogeogr 13: 35-57.

Baskin CC, Baskin JM. 2006. The natural history of soil seed banks of arable land. Weed Sci 54: 549-557.

Battisti A, Stastny M, Buffo E, Larsson S. 2006. A rapid altitudinal range expansion in the pine processionary moth produced by the 2003 climatic anomaly. Global Change Biol 12: 662-671.

Baur B, Baur A. 1993. Climatic warming due to thermal radiation from an urban area as possible cause for the local extinction of a land snail. J Appl Ecol 30: 333-340.

Beeby A. 1993. Applying Ecology. London: Chapman & Hall.

Beekman JH, Nolet BA, Klaassen M. 2002. Skipping swans: Fuelling rates and wind conditions determine differential use of migratory stopover sites of Bewick's Swans *Cygnus bewickii*. Ardea 90: 437-460.

Begon M, Townsend CR, Harper JJ. 2006. Ecology. From Individuals to Ecosystems. 4th ed. Oxford: Blackwell Publishing.

Bender DJ, Contreras TA, Fahrig L. 1998. Habitat loss and population decline: A meta-analysis of the patch size effect. Ecology 79: 517-533.

Bennett KD. 1997. Evolution and Ecology: The Pace of Life. Cambridge: Cambridge University Press.

Bensch S, Andersson T, Akesson S. 1999. Morphological and molecular variation across a migratory divide in willow warblers, *Phylloscopus trochilus*. Evolution 53: 1925-1935.

Bensch S, Helbig AJ, Salomon M, Siebold I. 2002. Amplified fragment length polymorphism analysis identifies hybrids between two subspecies of warblers. Mol Ecol 11: 473-481.

Benson SR, Croll DA, Marinovic BB, Chavez FP, Harvey JT. 2002. Changes in the cetacean assemblage of a coastal upwelling ecosystem during El Niño 1997-98 and La Niña 1999. Prog Oceanogr 54: 279-291.

Berger J. 1992. Facilitation of reproductive synchrony by gestation adjustment in gregarious mammals—A new hypothesis. Ecology 73: 323-329.

Berger TD, Trinkaus E. 1995. Patterns of trauma among the Neandertals. J Archaeol Sci 22: 841-852.

Bergthorsson P. 1985. Sensitivity of Icelandic agriculture to climatic variations. Clim Change 7: 111-127.

Berry RJ. 1989. Ecology: where genes and geography meet. J Anim Ecol 58: 733-759.

Beukema JJ, Essink K, Dekker R. 2000. Long-term observations on the dynamics of three species of polychaetes living on tidal flats of the Wadden Sea: The role of weather and predator-prey interactions. J Anim Ecol 69: 31-44.

Bever JD, Schultz PA, Pringle A, Morton JB. 2001. Arbuscular mycorrhizal fungi: More diverse than meets the eye, and the ecological tale of why. Bioscience 51: 923-931.

Bickford D, Lohman DJ, Sodhi NS, Ng PKL, Meier R, Winker K, Ingram KK, Das I. 2007. Cryptic species as a window on diversity and conservation. Trends Ecol Evol 22: 148-155.

Billings WD. 1952. The environmental complex in relation to plant growth and distribution. Q Rev Biol 27: 251-265.

Billings WD. 1974. Environment: concept and reality. Pages 9-35 in Strain BR, Billings WD, eds. Vegetation and Environment. The Hague: Junk.

Birch LC. 1944. The effect of temperature and dryness on the survival of the eggs of *Calandra oryzae* L. (small strain) and *Rhyzopertha dominica* Fab. (Coleoptera). Aust J Exp Biol Med. 22: 265-269.

Birch LC. 1948. The intrinsic rate of natural increase of an insect population. J Anim Ecol 17: 15-26.

Bjorkman C. 2000. Interactive effects of host resistance and drought stress on the performance of a gall-making aphid living on Norway spruce. Oecologia 123: 223-231.

Blackburn TM, Gaston KJ. 2001. Linking patterns in macroecology. J Anim Ecol 70: 338-352.

Blackburn TM, Gaston KJ. 2006. There's more to macroecology than meets the eye. Glob Ecol Biogeogr 15: 537-540.

Blanquart F, Kaltz O, Nuismer SL, Gandon S. 2013. A practical guide to measuring local adaptation. 16: 1195-1205.

Boavida LC, Vieira AM, Becker JD, Feijo JA. 2005. Gametophyte interaction and sexual reproduction: How plants make a zygote. Int J Dev Biol 49: 615-632.

Bodenheimer FS. 1938. Problems of Animal Ecology. London: Oxford University Press.

Bodenheimer FS. 1958. Animal Ecology To-day. Den Haag: Dr W. Junk.

Boecklen WJ, Mopper S. 1998. Local adaptation in specialist herbivores: theory and evidence. Pages 64-88 in Mopper S, Strauss SY, eds. Genetic Structure and Local Adaptation in Natural Insect Populations: Effects of Ecology, Life History, and Behavior. New York: Chapman & Hall.

Bonner JT. 1993. Life Cycles: Reflections of an Evolutionary Biologist. Princeton, NJ: Princeton University Press.

Bonte D, Lens L, Maelfait JP. 2004. Lack of homeward orientation and increased mobility result in high emigration rates from low-quality fragments in a dune wolf spider. J Anim Ecol 73: 643-650.

Botkin DB. 2001. The naturalness of biological invasions. West N Am Nat 61: 261-266.

Bowers MA, Matter SF. 1997. Landscape ecology of mammals: Relationships between density and patch size. J Mammal 78: 999-1013.

Bowman J, Cappuccino N, Fahrig L. 2002. Patch size and population density: the effect of immigration behavior. Conserv Ecol 6(1): 9.

Bowyer RT, van Ballenberghe V, Kie JG. 1998. Timing and synchrony of parturition in Alaskan moose: Long-term versus proximal effects of climate. J Mammal 79: 1332-1344.

Bowyer RT, van Ballenberghe V, Kie JG, Maier JAK. 1999. Birth-site selection by Alaskan moose: Maternal strategies for coping with a risky environment. J Mammal 80: 1070-1083.

Bowyer RT, van Ballenberghe V, Kie JG. 2003. Moose (*Alces alces*). Pages 931-964 in Feldhamer GA, Thompson BC, Chapman JA, eds. Wild

Mammals of North America Biology, Management, and Conservation 2nd ed. Baltimore: Johns Hopkins University Press.

Bradshaw CJA. 2008. Having your water and drinking it too: Resource limitation modifies density regulation. J Anim Ecol 77: 1-4.

Brady RH. 1982. Dogma and doubt. Biol J Linn Soc 17: 79-96.

Brakefield PM. 2011. Evo-devo and accounting for Darwin's endless forms. Philos Trans R Soc Lond B Biol Sci 366: 2069-2075.

Brakefield PM, Gates J, Keys D, Kesbeke F, Wijngaarden PJ, Monteiro A, French V, Carroll SB. 1996. Development, plasticity and evolution of butterfly eyespot patterns. Nature 384: 236-242.

Branch GM. 1985. Competition: Its role in ecology and evolution in intertidal communities. Pages 97-104. in Vrba ES, ed. Species and Speciation Transvaal Museum Monograph No 4. Pretoria: Transvaal Museum.

Brewis IA, Wong CH. 1999. Gamete recognition: Sperm proteins that interact with the egg zona pellucida. Rev Reprod 4: 135-142.

Bridle JR, Vines TH. 2007. Limits to evolution at range margins: when and why does adaptation fail? Trends Ecol Evol 22: 140-147.

British Trust for Ornithology. 2012. BirdTrends 2012: trends in numbers, breeding success and survival for UK breeding birds. http://www.bto.org/about-birds/birdtrends/2012. Accessed Jan. 2013.

Brown JH. 1984. On the relationship between abundance and distribution of species. Am Nat 124: 255-279.

Brown JH. 1995. Macroecology. Chicago: University of Chicago Press.

Brown JH, Kodric-Brown A. 1977. Turnover rates in insular biogeography: Effect of immigration on extinction. Ecology 58: 445-449.

Brunzel S. 2002. Experimental density-related emigration in the cranberry fritillary *Boloria aquilonaris*. J Insect Behav 15: 739-750.

Burkhardt RW. 2005. Patterns of Behavior. Konrad Lorenz, Niko Tinbergen, and the Founding of Ethology. Chicago: University of Chicago Press.

Burroughs WJ. 2003. Weather Cycles. Real or Imaginary? Cambridge: Cambridge University Press.

Burton JF. 1995. Birds and Climate Change. London: Christopher Helm.

Bush GL. 1994. Sympatric speciation in animals: New wine in old bottles. Trends Ecol Evol 9: 285—287.

Buss LW. 1990. Competition within and between encrusting clonal invertebrates. Trends Ecol Evol 5: 352-356.

Butlin R. 1995. Genetic variation in mating signals and responses. Pages 327-366 in Lambert DM, Spencer HG, eds. Speciation and the Recognition Concept Theory and Application. Baltimore: Johns Hopkins University Press.

Butterfield JEL, Coulson JC. 1997. Terrestrial invertebrates and climate change: physiological and life-cycle adaptations. Pages 401-412 in Huntley B, Cramer W, Morgan AV, Prentice HC, Allen JRM, eds. Past and Future Rapid Environmental Changes: The Spatial and Evolutionary Responses of Terrestrial Biota. Berlin: Springer.

Byrne DN, Blackmer JL. 1996. Examination of short-range migration by *Bemisia*. Pages 17-28 in Gerling D, Mayer RT, eds. *Bemisia* 1995: Taxonomy, Biology, Damage, Control and Management. Andover: Intercept.

Cain ML, Bowman WD, Hacker SD. 2011. Ecology, 2nd ed. Sunderland, Mass.: Sinauer Associates.

Cameron RD, Smith WT, Fancy SG, Gerhart KL, White RG. 1993. Calving success of female caribou in relation to body-weight. Can J Zool 71: 480-486.

Campos PF, Willerslev E, Sher A, Orlando L, Axelsson E, Tikhonov A, Aaris-Sorensen K, Greenwood AD, Kahlke RD, Kosintsev P, Krakhmalnaya T, Kuznetsova T, Lemey P, MacPhee R, Norris CA, Shepherd K, Suchard MA, Zazula GD, Shapiro B, Gilbert MTP. 2010. Ancient DNA analyses exclude humans as the driving force behind late Pleistocene musk ox (*Ovibos moschatus*) population dynamics. Proc Natl Acad Sci USA 107: 5675-5680.

Carroll DP, Hoyt SC. 1986. Hosts and habitats of parasitoids (Hymenoptera: Aphidiidae) implicated in biological control of apple aphid (Homoptera: Aphididae). Environ Entomol 15: 1171-1178.

Carroll SB. 2005. Evolution at two levels: On genes and form. PLoS Biol 3: 1159-1166.

Carroll SB. 2008a. The Making of the Fittest. DNA and the Ultimate Forensic Record of Evolution. London: Quercus.

Carroll SB. 2008b. Evo-devo and an expanding evolutionary synthesis: A genetic theory of morphological evolution. Cell 134: 25-36.

Carroll SB, Grenier JK, Weatherbee SD. 2005a. From DNA to Diversity. Molecular Genetics and the Evolution of Animal Design. Oxford: Blackwell Publishing.

Carroll SB, Prud'homme B, Gompel N. 2008. Regulating evolution. Sci Am 298: 60-67.

Carroll SP, Loye JE, Dingle H, Mathieson M, Famula TR, Zalucki MP. 2005b. And the beak shall inherit—evolution in response to invasion. Ecol Lett 8: 944-951.

Carson HL. 1957. The species as a field for gene recombination. Pages 23-38 in Mayr E, ed. The Species Problem. Washington, D.C.: American Association for the Advancement of Science, Publication No. 50.

Carson HL. 1987. Colonization and speciation. Pages 187-206 in Gray AJ, Crawley MJ, Edwards PJ, eds. Colonization, Succession and Stability. Oxford: Blackwell.

Castro J, Zamora R, Hodar JA, Gomez JM. 2004. Seedling establishment of a boreal tree species (*Pinus sylvestris*) at its southernmost distribution limit: consequences of being in a marginal Mediterranean habitat. J Ecol 92: 266-277.

Caswell H. 2001. Matrix Population Models: Construction, Analysis, and Interpretation, 2nd ed. Sunderland MA: Sinauer Associates.

Caughley G. 1994. Directions in conservation biology. J Anim Ecol 63: 215-244.

Caviedes CN, Fik TJ. 1992. The Peru-Chile eastern Pacific fisheries and climatic oscillation. Pages 355-375 in Glantz MH, ed. Climatic Variability, Climate Change, and Fisheries. Cambridge: Cambridge University Press.

Cawthorne RA, Marchant JH. 1980. The effects of the 1978/79 winter on British bird populations. Bird Study 27: 163-172.

Chalmers AF. 1999. What is This Thing Called Science? An Assessment of the Nature and Status of Science and Its Methods. St Lucia: University of Queensland Press.

Chalmers AF. 2003. What is This Thing Called Science? 3rd ed. St Lucia: University of Queensland Press.

Chamberlin TC. 1897. Studies for students: The method of multiple working hypotheses. J Geol 5: 837-848.

Chapleau F, Johansen PH, Williamson M. 1988. The distinction between pattern and process in evolutionary biology: the use and abuse of the term 'strategy'. Oikos: 136-138.

Chapman RN. 1931. Animal Ecology with Special Reference to Insects. New York: McGraw-Hill.

Chapman T, Arnqvist G, Bangham J, Rowe L. 2003. Sexual conflict. Trends Ecol Evol 18: 41-47.

Cherrill AJ, James R. 1987. Character displacement in *Hydrobia*. Oecologia 71: 618-623.

Chitty D. 1996. Do Lemmings Commit Suicide? Beautiful Hypotheses and Ugly Facts. New York: Oxford University Press.

Chown SL, Nicolson SW. 2004. Insect Physiological Ecology. Mechanisms and Patterns. Oxford: Oxford University Press.

Cittadino E. 1990. Nature as the Laboratory. Darwinian Plant Ecology in the German Empire, 1880-1900. Cambridge: Cambridge University Press.

Claridge MF, den Hollander J, Morgan JC. 1985. The status of weed-associated populations of the brown planthopper, *Nilaparvata lugens* (Stål)—host race or biological species? Zool J Linn Soc 84: 77-90.

Clark LR, Geier PW, Hughes RD, Morris RF. 1967. The Ecology of Insect Populations in Theory and Practice. London: Methuen.

Clarke AE, Knox RB. 1978. Cell recognition in flowering plants. Q Rev Biol 53: 3-28.

Clarke AR, Walter GH. 1995. "Strains" and the classical biological control of insect pests. Can J Zool 73: 1777-1790.

Clements FE. 1916. Plant Succession: An Analysis of the Development of Vegetation. Washington, DC: Publication No. 242, Carnegie Institution of Washington.

Cohen JE. 1995. How Many People Can the Earth Support? New York: W.W. Norton.

Collins JP. 1986. Evolutionary ecology and the use of natural selection in ecological theory. J Hist Biol 19: 257-288.

Colwell RK. 1984. What's new? Community ecology discovers biology. Pages 387-396 in Price PW, Slobodchikoff CN, Gaud WS, eds. A New Ecology Novel Approaches to Interactive Systems. New York: John Wiley.

Colwell RK. 1986. Community biology and sexual selection: lessons from hummingbird flower mites. Pages 406-424 in Diamond J, Case TJ, eds. Community Ecology. New York: Harper and Row.

Colyvan M, Ginzburg LR. 2003. Laws of nature and laws of ecology. Oikos 101: 649-653.

Connell JH. 1978. Diversity in tropical rain forests and coral reefs. Science 199: 1302-1310.

Connell JH. 1979. Tropical rain forests and coral reefs as open non-equilibrium systems. Pages 141-163 in Anderson RM, Turner BD, Taylor LR, eds. Population Dynamics 20th Symposium of The British Ecological Society. London: Blackwell Scientific Publications.

Connell JH. 1980. Diversity and the coevolution of competitors, or the ghost of competition past. Oikos 35: 131-138.

Connell JH, Slatyer RO. 1977. Mechanisms of succession in natural communities and their role in community stability and organization. Am Nat 111: 1119-1144.

Connor EF, Simberloff D. 1979. The assembly of species communities: chance or competition. Ecology 60: 1132-1140.

Coope GR. 1978. Constancy of insect species versus inconstancy of Quaternary environments. Pages 176-186 in Mound LA, Waloff N, eds. Diversity of Insect Faunas. Oxford: Blackwell.

Coope GR. 1979. Late Cenozoic fossil Coleoptera: evolution, biogeography, and ecology. Annu Rev Ecol Syst 10: 247-267.

Coope GR. 1987a. The response of late Quaternary insect communities to sudden climatic changes. Pages 421-438 in Gee JHR, Giller PS, eds. Organization of Communities Past and Present. Oxford: Blackwell.

Coope GR. 1987b. Fossil beetle assemblages as evidence for sudden and intense climatic changes in the British Isles during the last 45,000 years. Pages 147-150 in Berger WH, Labeyrie LD, eds. Abrupt Climatic Change. Dordrecht: D. Reidel.

Coope GR. 1994. The response of insect faunas to glacial-interglacial climatic fluctuations. Philos Trans R Soc Lond B Biol Sci 344: 19-26.

Coope GR. 2004. Several million years of stability among insect species because of, or in spite of, Ice Age climatic instability? Philos Trans R Soc Lond B Biol Sci 359: 209-214.

Cooper G. 2001. Must there be a balance of nature? BiolPhilos 16: 481-506.

Cooper GJ. 2003. The Science of the Struggle for Existence. Cambridge: Cambridge University Press.

Corbett A, Rosenheim JA. 1996. Quantifying movement of a minute parasitoid, *Anagrus epos* (Hymenoptera: Mymaridae), using fluorescent dust marking and recapture. Biol Control 6: 35-44.

Coulson T, Godfray HCJ. 2007. Single-species dynamics. Pages 17-34 in May RM, McLean AR, eds. Theoretical Ecology: Principles and Applications. Oxford: Oxford University Press.

Coulson T, Catchpole EA, Albon SD, Morgan BJT, Pemberton JM, Clutton-Brock TH, Crawley MJ, Grenfell BT. 2001. Age, sex, density, winter weather, and population crashes in Soay sheep. Science 292: 1528-1531.

Coulson T, Rohani P, Pascual M. 2004. Skeletons, noise and population growth: the end of an old debate? Trends Ecol Evol 19: 359-364.

Cramp S, (Ed.). 1992. The Birds of the Western Palearctic Vol. VI Oxford: Oxford University Press.

Criddle RS, Hopkin MS, McArthur ED, Hansen LD. 1994. Plant distribution and the temperature coefficient of metabolism. Plant Cell Environ 17: 233-243.

Criddle RS, Breidenbach RW, Fontana AJ, Henry JM, Smith BN, Hansen LD. 1996. Plant respiration responses to climate determine geographic distribution. Russ J Plant Physiol 43: 698-704.

Criddle RS, Smith BN, Hansen LD. 1997. A respiration based description of plant growth rate responses to temperature. Planta 201: 441-445.

Cronin H. 1991. The Ant and the Peacock. Altruism and Sexual Selection from Darwin to Today. Cambridge: Cambridge University Press.

Cronin JT, Strong DR. 1994. Parasitoid interactions and their contribution to the stabilization of Auchenorrhyncha populations. Pages 400-428 in Denno RF, Perfect TJ, eds. Planthoppers: Their Ecology and Management. New York: Chapman & Hall.

Crozier LG. 2004. Field transplants reveal summer constraints on a butterfly range expansion. Oecologia 141: 148-157.

Cuddington K. 2001. The "balance of nature" metaphor and equilibrium in population ecology. BiolPhilos. 16: 463-479.

Cunningham JP. 2012. Can mechanism help explain insect host choice? J Evol Biol 25: 244-251.

Curio E. 1976. The Ethology of Predation. Berlin: Springer-Verlag.

Dahlhoff EP, Buckley BA, Menge BA. 2001. Physiology of the rocky intertidal predator *Nucella ostrina* along an environmental stress gradient. Ecology 82: 2816-2829.

Darlington PJ. 1943. Carabidae of mountains and islands: data on the evolution of isolated faunas, and on atrophy of wings. Ecol Monogr 13: 37-61.

Darwin C. 1859. On the Origin of Species by Means of Natural Selection, or the Preservation of Favoured Races in the Struggle for Life. Facsimile reprint, 1964, Harvard University Press, Cambridge, Mass. London: John Murray.

Daubenmire RF. 1974. Plants and Environment: A Textbook of Plant Autecology. 3rd ed. London: John Wiley & Sons.

Davidson J, Andrewartha HG. 1948a. The influence of rainfall, evaporation and atmospheric temperature on fluctuations in the size of a natural population of *Thrips imaginis* (Thysanoptera). J Anim Ecol 17: 200-222.

Davidson J, Andrewartha HG. 1948b. Annual trends in a natural population of *Thrips imaginis* (Thysanoptera). J Anim Ecol 17: 193-199.

Davies AP, Ceballo FA, Walter GH. 2004. Is the biocontrol potential of *Coccidoxenoides perminutus*, a mealybug parasitoid, limited by climatic or nutritional factors? Biol Control 31: 181-188.

Davis MA. 1981a. The flight capacity of dispersing milkweed beetles, *Tetraopes tetraopthalmus*. Ann Entomol Soc Am 74: 385-386.

Davis MA. 1984. The flight and migration ecology of the red milkweed beetle (*Tetraopes tetraophthalmus*). Ecology 65: 230-234.

Davis MA. 1986a. Geographic patterns in the flight ability of a monophagous beetle. Oecologia 69: 407-412.

Davis MB. 1976. Pleistocene biogeography of temperate deciduous forests. Geoscience and Man 13: 13-26.

Davis MB. 1981b. Quaternary history and the stability of forest communities. Pages 132-153 in West DC, Shugart HH, Botkin DB, eds. Forest Succession. New York: Springer.

Davis MB. 1986b. Climatic instability, time lags, and community disequilibrium. Pages 269-284 in Diamond JM, Case TJ, eds. Community Ecology. New York: Harper & Row.

Davis MB. 1987. Invasions of forest communities during the Holocene: beech and hemlock in the Great Lakes region. Pages 373-393 in Gray AJ, Crawley MJ, Edwards PJ, eds. Colonization, Succession and Stability. Oxford: Blackwell.

Dawkins R. 1989. The Selfish Gene. Oxford: Oxford University Press.

Dawson MN, Jacobs DK. 2001. Molecular evidence for cryptic species of *Aurelia aurita* (Cnidaria, Scyphozoa). Biol Bull 200: 92-96.

Day WH. 1994. Estimating mortality caused by parasites and diseases of insects—comparisons of the dissection and rearing methods. Environ Entomol 23: 543-550.

Dayton PK. 2003. The importance of the natural sciences to conservation. Am Nat 162: 1-13.

Dayton PK, Sala E. 2001. Natural history: the sense of wonder, creativity and progress in ecology. Sci Mar 65: 199-206.

de Vargas C, Norris R, Zaninetti L, Gibb SW, Pawlowski J. 1999. Molecular evidence of cryptic speciation in planktonic foraminifers and their relation to oceanic provinces. Proc Natl Acad Sci USA 96: 2864-2868.

DeBach P. 1969. Uniparental, sibling and semi-species in relation to taxonomy and biological control. Isr J Entomol 4: 11-28.

Debinski DM, Holt RD. 2000. A survey and overview of habitat fragmentation experiments. Conserv Biol 14: 342-355.

Dempster JP. 1982. The ecology of the Cinnabar moth, *Tyria jacobaeae* L. (Lepidoptera: Arctiidae). Adv Ecol Res 12: 1-36.

Dempster JP. 1983. The natural control of populations of butterflies and moths. Biol Rev 58: 461-481.

Dempster JP, Pollard E. 1986. Spatial heterogeneity, stochasticity and the detection of density dependence in animal populations. Oikos 46: 413-416.

den Boer PJ. 1962. The ecological significance of activity patterns in the woodlouse *Porcellio scaber* Latr (Isopoda). Arch Neerl Zool 14: 283-409.

den Boer PJ. 1968. Spreading of risk and stabilization of animal numbers. Acta Biotheor 18: 165-194.

den Boer PJ. 1970. On the significance of dispersal power for populations of carabid beetles (Coleoptera, Carabidae). Oecologia 4: 1-28.

den Boer PJ. 1979a. Populations of carabid beetles and individual behaviour. General aspects. Pages 145-149 in den Boer PJ, Thiele HU, Weber F, eds. On the Evolution of Behaviour in Carabid Beetles Miscellaneous Papers, Agricultural University Wageningen, The Netherlands, vol. 18. Wageningen: Veenman & Zonen.

den Boer PJ. 1979b. The individual behaviour and population dynamics of some carabid beetles of forests. Pages 151-166 in den Boer PJ, Thiele HU, Weber F, eds. On the Evolution of Behaviour in Carabid Beetles. Zoological Institute of the University of Cologne: H. Veenman and B.V. Zonen, Wageningen.

den Boer PJ. 1980. Exclusion or coexistence and the taxonomic or ecological relationship between species. Neth J Zool 30: 278-306.

den Boer PJ. 1986a. Density dependence and the stabilization of animal numbers. 1. The winter moth. Oecologia 69: 507-512.

den Boer PJ. 1986b. The present status of the competitive exclusion principle. Trends Ecol Evol 1: 25-28.

den Boer PJ. 1987a. The density dependence debate: detecting density dependence. Trends Ecol Evol 2: 77-78.

den Boer PJ. 1987b. Density dependence and the stabilization of animal numbers. 2. The pine looper. Neth J Zool 37: 220-237.

den Boer PJ. 1988. Density dependence and the stabilization of animal numbers. 3. The winter moth reconsidered. Oecologia 75: 161-168.

den Boer PJ. 1990. The survival value of dispersal in terrestrial arthropods. Biol Conserv 54: 175-192.

den Boer PJ. 1999. Natural selection or the non-survival of the non-fit. Acta Biotheor 47: 83-97.

den Boer PJ, Reddingius J. 1996. Regulation and Stabilization Paradigms in Population Ecology. London: Chapman and Hall.

Desender K, Dufrene M, Maelfait J-P. 1994. Long term dynamics of carabid beetles in Belgium: a preliminary analysis on the influence of changing climate and land use by means of a database covering more than a century. Pages 247-252 in Desender K, Dufrene M, Loreau M, Luff ML, Maelfait J-P, eds. Carabid Beetles Ecology and Evolution. Dordrecht: Kluwer Academic Publishers.

Devonshire AL, Sawicki RM. 1979. Insecticide resistant *Mysus persicae* as an example of evolution by gene duplication. Nature 280: 140-141.

Dijksterhuis EJ. 1969. The Mechanization of the World Picture. New York: Oxford University Press.

Dingle H. 1996. Migration: The Biology of Life on the Move. New York: Oxford University Press.

Dobkin DS, Olivieri I, Ehrlich PR. 1987. Rainfall and the interaction of microclimate with larval resources in the population dynamics of checkerspot butterflies (*Euphydryas editha*) inhabiting serpentine grassland. Oecologia 71: 161-166.

Drake VA, Gatehouse AG, Farrow RA. 1995. Insect migration: a holistic conceptual model. Pages 427-457 in Drake VA, Gatehouse AG, eds. Insect Migration. Cambridge: Cambridge University Press.

Drew RAI, Lloyd AC. 1987. Relationship of fruit-flies (Diptera, Tephritidae) and their bacteria to host plants. Ann Entomol Soc Am 80: 629-636.

Drew RAI, Courtice AC, Teakle DS. 1983. Bacteria as a natural source of food for adult fruit-flies (Diptera, Tephritidae). Oecologia 60: 279-284.

Drury WH. 1998. Chance and Change. Ecology for Conservationists. Berkeley: University of California Press.

Drury WH, Nisbet ICT. 1973. Succession. J Arnold Arboretum 54: 331-368.

Duffy DC, Rudolf A. 1986. Cape cormorant mass mortality during a cold-water event off Namibia. Ostrich 57: 247-248.

Duffy JE. 1996. Species boundaries, specialization, and the radiation of sponge-dwelling alpheid shrimp. Biol J Linn Soc 58: 307-324.

Dumas C, Knox RB, Gaude T. 1984. Pollen-pistil recognition: new concepts from electron microscopy and cytochemistry. Int Rev Cytol 90: 239-272.

Dunn PO, Winkler DW. 1999. Climate change has affected the breeding date of tree swallows throughout North America. Proc R Soc Lond B Biol Sci. 266: 2487-2490.

Dupre C, Ehrlen J. 2002. Habitat configuration, species traits and plant distributions. J Ecol 90: 796-805.

Dusenbery DB. 1992. Sensory Ecology. How Organisms Acquire and Respond to Information. New York: W.H. Freeman.

Eber S, Brandl R. 1996. Metapopulation dynamics of the tephritid fly *Urophora cardi*: an evaluation of incidence-function model assumptions with field data. J Anim Ecol 65: 621-630.

Edney EB. 1982. The truth about saturation deficiency—an historical perspective. J Exp Zool 222: 205-214.

Ehrlen J, Eriksson O. 2003. Large-scale spatial dynamics of plants: a response to Freckleton & Watkinson. J Ecol 91: 316-320.

Ehrlich PR. 2004. Preface. Pages vii—x in Ehrlich PR, Hanski I, eds. On the Wings of Checkerspots: A Model System for Population Biology. Oxford: Oxford University Press.

Ehrlich PR, Hanski I. 2004. Checkerspot research. Background and origins. Pages 3-16 in Ehrlich PR, Hanski I, eds. On the Wings of Checkerspots: A Model System for Population Biology. Oxford: Oxford University Press.

Ehrlich PR, Hanski I, Boggs CL. 2004. What have we learned? Pages 3-16 in Ehrlich PR, Hanski I, eds. On the Wings of Checkerspots: A Model System for Population Biology. Oxford: Oxford University Press.

Eiseley L. 1958. Darwin's Century. Evolution and the Men who Discovered It. New York: Doubleday.

Eldredge N. 1971. The allopatric model and phylogeny in Paleozoic invertebrates. Evolution 25: 156-167.

Eldredge N. 1985. Unfinished Synthesis. New York: Oxford University Press.

Eldredge N, Gould SJ. 1972. Punctuated equilibria: an alternative to phyletic gradualism. Pages 82-115 in Schopf TJM, ed. Models in Paleontology. San Francisco: Freeman, Cooper.

Eldredge N, Thompson JN, Brakefield PM, Gavrilets S, Jablonski D, Jackson JBC, Lenski RE, Lieberman BS, McPeek MA, Miller W. 2005. The dynamics of evolutionary stasis. Paleobiology 31: 133-145.

Elias SA. 1994. Quaternary Insects and their Environments. Washington: Smithsonian Institution Press.

Elias SA. 2010. Advances in Quaternary Entomology. Elsevier: Amsterdam.

Elkinton JS, Buonaccorsi JP, Bellows TS, Vandriesche RG. 1992. Marginal attack rate, k-values and density dependence in the analysis of contemporaneous mortality factors. Res Popul Ecol 34: 29-44.

Ellegård A. 1958. Darwin and the General Reader. The Reception of Darwin's Theory of Evolution in the British Periodical Press, 1859-1872. Göteborg: Göteborg University Press.

Elsasser WM. 1987. Reflections on a Theory of Organisms. Holism in Biology. Baltimore: Johns Hopkins University Press.

Endler JA. 1995. Multiple-trait coevolution and environmental gradients in guppies. Trends Ecol Evol 10: 22-29.

Eremaea Birds. 2013. Available at: http://www.eremaea.com. Accessed Aug. 2013.

Etherington LL, Eggleston DB, Stockhausen WT. 2003. Partitioning loss rates of early juvenile blue crabs from seagrass habitats into mortality and emigration. Bull Mar Sci 72: 371-391.

Evans MR, Grimm V, Johst K, Knuuttila T, de Langhe R, Lessells CM, Merz M, O'Malley MA, Orzack SH, Weisberg M, Wilkinson DJ, Wolkenhauer O, Benton TG. 2013. Do simple models lead to generality in ecology? Trends Ecol Evol 28: 578-583.

Ezard THG, Bullock JM, Dalgleish HJ, Millon A, Pelletier F, Ozgul A, Koons DN. 2010. Matrix models for a changeable world: the importance of transient dynamics in population management. J Appl Ecol 47: 515-523.

Facelli JM, Pickett STA. 1990. Markovian chains and the role of history in succession. Trends Ecol Evol 5: 27-30.

Fang JY, Lechowicz MJ. 2006. Climatic limits for the present distribution of beech (*Fagus* L.) species in the world. J Biogeogr 33: 1804-1819.

Feller W. 1940. On the logistic law of growth and its empirical verifications in biology. Acta Biotheor 5: 51-66.

Fernandez-Gimenez ME, Allen-Diaz B. 1999. Testing a non-equilibrium model of rangeland vegetation dynamics in Mongolia. J Appl Ecol 36: 871-885.

Fernando LCP, Walter GH. 1997. Species status of two host-associated populations of *Aphytis lingnanensis* (Hymenoptera: Aphelinidae) in citrus. Bull Entomol Res 87: 137-144.

Ferris PJ, Pavlovic C, Fabry S, Goodenough UW. 1997. Rapid evolution of sex-related genes in *Chlamydomonas*. Proc Natl Acad Sci USA 94: 8634-8639.

Finlay-Doney M, Walter GH. 2012a. The conceptual and practical implications of interpreting diet breadth mechanistically in generalist predatory insects. Biol J Linn Soc 107: 737-763.

Finlay-Doney M, Walter GH. 2012b. Behavioral responses to specific prey and host plant species by a generalist predatory coccinellid (*Cryptolaemus montrouzieri* Mulsant). Biol Control 63: 270-278.

Fischer DT, Still CJ, Williams AP. 2009. Significance of summer fog and overcast for drought stress and ecological functioning of coastal California endemic plant species. J Biogeogr 36: 783-799.

Fisher RA. 1948. Statistical Methods for Research Workers. 10th ed. Edinburgh: Oliver & Boyd.

Flenley JR. 1979. The Equatorial Rain Forest: A Geological History. London: Butterworths.

Foottit RG. 1997. Recognition of parthenogenetic insect species. Pages 291-307 in Claridge MF, Dawah HA, Wilson MR, eds. Species: the Units of Biodiversity. London: Chapman & Hall.

Ford J. 1974. Concepts of subspecies and hybrid zones, and their application in Australian ornithology. Emu 74: 113-123.

Ford MJ. 1982. The Changing Climate: Responses of the Natural Fauna and Flora. London: George Allen & Unwin.

Frazzetta TH. 1975. Complex Adaptations in Evolving Populations. Sunderland: Sinauer.

Freckleton RP, Watkinson AR. 2002. Large-scale spatial dynamics of plants: metapopulations, regional ensembles and patchy populations. J Ecol 90: 419-434.

Freckleton RP, Watkinson AR. 2003. Are all plant populations metapopulations? J Ecol 91: 321-324.

Fryxell JM. 1987. Food limitation and demography of a migratory antelope, the white-eared kob. Oecologia 72: 83-91.

Garrido B, Hampe A, Maranon T, Arroyo J. 2003. Regional differences in land use affect population performance of the threatened insectivorous plant *Drosophyllum lusitanicum* (Droseraceae). Divers Distrib 9: 335-350.

Gaston AJ, Gilchrist HG, Hipfner JM. 2005. Climate change, ice conditions and reproduction in an Arctic nesting marine bird: Brunnich's guillemot (*Uria lomvia* L.). J Anim Ecol 74: 832-841.

Gaston KJ. 2003. The how and why of biodiversity. Nature 421: 900-901.

Gatehouse AG. 1987. Migration and low population density in armyworm (Lepidoptera: Noctuidae) life histories. Insect Sci Applic 8: 573-580.

Gause GF. 1932. Ecology of populations. Q Rev Biol 7: 27-46.

Gause GF. 1934. The Struggle for Existence (1969 reprint). New York: Hafner.

Gauch HG. 1982. Multivariate Analysis in Community Ecology. Cambridge: Cambridge University Press.

Geiser F, Pavey CR. 2007. Basking and torpor in a rock-dwelling desert marsupial: survival strategies in a resource-poor environment. J Comp Physiol B 177: 885-892.

Ghiselin MT. 1974. The Economy of Nature and the Evolution of Sex. Berkeley: University of California Press.

Giannini NP, Bertelli S. 2004. Phylogeny of extant penguins based on integumentary and breeding characters. Auk 121: 422-434.

Gilbert SF, Epel D. 2009. Ecological Developmental Biology: Integrating Epigenetics, Medicine, and Evolution. Sunderland, Mass: Sinauer.

Gilpin M, Hanski I, (Eds). 1991. Metapopulation Dynamics: Empirical and Theoretical Investigations. London: Academic Press.

Ginnett TF, Young ELB. 2000. Stochastic recruitment in white-tailed deer along an environmental gradient. J Wildl Manag 64: 713-720.

Ginzburg L, Colyvan M. 2004. Ecological Orbits. How Planets Move and Populations Grow. Oxford: Oxford University Press.

Gleason HA. 1917. The structure and development of the plant association. Bull Torrey Bot Club 44: 463-481.

Gleason HA. 1926. The individualistic concept of the plant association. Bull Torrey Bot Club 53: 7-26.

Gleason HA. 1927. Further views on the succession concept. Ecology 8: 299-326.

Gleason HA. 1939. The individualistic concept of the plant association. Am Midl Nat 21: 92-110.

Glenn-Lewin DC. 1980. The individualistic nature of plant community development. Vegetatio 43: 141-146.

Godfray HCJ, Waage JK. 1991. Predictive modelling in biological control: the mango mealy bug (*Rastrococcus invadens*) and its parasitoids. J Appl Ecol 28: 434-453.

Golley FB. 1993. A History of the Ecosystem Concept in Ecology. More than the Sum of the Parts. New Haven: Yale University Press.

Gonzalez-Martinez SC, Burczyk J, Nathan R, Nanos N, Gil L, Alia R. 2006. Effective gene dispersal and female reproductive success in Mediterranean maritime pine (*Pinus pinaster* Aiton). Mol Ecol 15: 4577-4588.

Gonzalez Sierra G, Penas Merino A, Alonso-Herrero E. 1996. Phenology of *Hyacinthoides non-scripta* (L.) Chouard, *Melittis melissophyllum* L. and *Symphytum tuberosum* L. in two deciduous forests in the Cantabrian mountains, Northwest Spain. Vegetatio 122: 69-82.

Gotelli NJ, Graves GR. 1996. Null Models in Ecology. Washington: Smithsonian Institute.

Gould JL. 1982a. Ethology: The Mechanism and Evolution of Behavior. New York: W.W. Norton & Co.

Gould SJ. 1982b. Free to be extinct. Nat Hist 91: 12-16.

Graham MH, Dayton PK. 2002. On the evolution of ecological ideas: Paradigms and scientific progress. Ecology 83: 1481-1489.

Graham MH, Dayton PK, Hixon MA. 2002. Paradigms in ecology: Past, present, and future. Ecology 83: 1479-1480.

Graham RW. 1986. Response of mammalian communities to environmental changes during the Late Quaternary. Pages 300-313 in Diamond J, Case TJ, eds. Community Ecol. New York: Harper & Row.

Grant PR. 1972. Convergent and divergent character displacement. Biol J Linn Soc 4: 39-68.

Grant PR. 1975. The classical case of character displacement. Pages 237-337 in Dobzhansky T, Hecht MK, Steere WC, eds. Evolutionary Biology, vol. 8. New York: Plenum.

Grant PR, Grant BR. 2002. Unpredictable evolution in a 30-year study of Darwin's finches. Science 296: 707-711.

Grant WE, Swannack TM. 2008. Ecological Modeling: A Common-Sense Approach to Theory and Practice. Oxford: Blackwell Publishing.

Gray AJ. 1997. Climate change and the reproductive biology of higher plants. Pages 371-380 in Huntley B, Cramer W, Morgan AV, Prentice HC, Allen JRM, eds. Past and Future Rapid Environmental Changes: The Spatial and Evolutionary Responses of Terrestrial Biota. Berlin: Springer.

Grayson DK. 2007. Deciphering North American Pleistocene extinctions. J Anthropol Res 63: 185-213.

Greenville AC, Wardle GM, Dickman CR. 2012. Extreme climatic events drive mammal irruptions: regression analysis of 100-year trends in desert rainfall and temperature. Ecol Evol 2: 2645-2658.

Gregg PC, Del Socorro AP, Rochester WA. 2001. Field test of a model of migration of moths (Lepidoptera: Noctuidae) in inland Australia. Aust J Entomol 40: 249-256.

Greig-Smith P. 1986. Chaos or order—organization. Pages 19-29 in Kikkawa J, Anderson DJ, eds. Community Ecology: Pattern and Process. Oxford: Blackwell Scientific Publications.

Gribbin J. 1990. Hothouse Earth. The Greenhouse Effect and Gaia. London: Bantam Press.

Grimm V. 1996. A down-to-earth assessment of stability concepts in ecology: dreams, demands, and the real problems. Senckenb Marit 27: 215-226.

Grimm V, Wissel C. 1997. Babel, or the ecological stability discussions: an inventory and analysis of terminology and a guide for avoiding confusion. Oecologia 109: 323-334.

Grimm V, Railsback SF. 2005. Individual-based Modeling and Ecology. Princeton: Princeton University Press.

Groves CP, Leslie DM. 2011. Order Artiodactyla. Family Bovidae (hollow-horned ruminants). Pages 444-779 in Wilson DE, Mittermeier RA, eds. Handbook of the Mammals of the World Vol 2 Hoofed Mammals. Barcelona: Lynx Edicions.

Gu H, Walter GH. 1999. Is the common sowthistle (*Sonchus oleraceus*) a primary host plant of the cotton bollworm, *Helicoverpa armigera* (Lep., Noctuidae)? Oviposition and larval performance. J Appl Entomol 123: 99-105.

Gunn A, Adamczewski J. 2003. Muskox (*Ovibos moschatus*). Pages 1076-1094 in Feldhamer GA, Thompson BC, Chapman JA, eds. Wild Mammals of North America Biology, Management, and Conservation 2nd ed. Baltimore: Johns Hopkins University Press.

Guo QF, Taper M, Schoenberger M, Brandle J. 2005. Spatial-temporal population dynamics across species range: from centre to margin. Oikos 108: 47-57.

Guthrie RD. 1990. Frozen Fauna of the Mammoth Steppe. The Story of Blue Babe. Chicago: University of Chicago Press.

Guthrie RD. 2006. New carbon dates link climatic change with human colonization and Pleistocene extinctions. Nature 441: 207-209.

Gyaneshwar P, Hirsch AM, Moulin L, Chen WM, Elliott GN, Bontemps C, Estrada-de los Santos P, Gross E, dos Reis FB, Sprent JI, Young JPW, James EK. 2011. Legume-nodulating Betaproteobacteria: diversity, host range, and future prospects. Mol Plant Microbe In 24: 1276-1288.

Haccou P, Meelis E. 1994. Statistical Analysis of Behavioural Data: an Approach Based on Time-structured Models. Oxford: Oxford University Press.

Hadid Y, Tzur S, Pavlicek T, Sumbera R, Skliba J, Lovy M, Fragman-Sapir O, Beiles A, Arieli R, Raz S, Nevo E. 2013. Possible incipient sympatric ecological speciation in blind mole rats (*Spalax*). Proc Natl Acad Sci USA 110: 2587-2592.

Haefner PA. 1970. Effect of low dissolved oxygen concentrations on temperature-salinity tolerance of sand shrimp, *Crangon septemspinosa* Say. Physiol Zool 43: 30-37.

Hagen JB. 1986. Ecologists and taxonomists: divergent traditions in twentieth-century plant geography. J Hist Biol 19: 197-214.

Hall CAS. 1988. An assessment of several of the historically most influential theoretical models used in ecology and of the data provided in their support. Ecol Model 43: 5-31.

Hall JA, Walter GH. 2013. Seed dispersal of the Australian cycad *Macrozamia miquelii* (Zamiaceae): are cycads megafauna-dispersed "grove-forming" plants? Am J Bot 100: 1127-1136.

Hall MI, Ross CF. 2007. Eye shape and activity pattern in birds. J Zool 271: 437-444.

Hamilton WD. 1967. Extraordinary sex ratios. Science 156: 477-488.

Hanski I. 1982. Dynamics of regional distribution: the core and satellite hypothesis. Oikos 38: 210-221.

Hanski I. 1991. Single-species metapopulation dynamics: Concepts, models and observations. Biol J Linn Soc 42: 17-38.

Hanski I. 1997. Predictive and practical metapopulation models: the incidence function approach. Pages 21-45 in Tilman D, Kareiva P, eds. Spatial Ecology The Role of Space in Population Dynamics and Interspecific Interactions. Princeton: Princeton University Press.

Hanski I. 1998. Spatial structure and dynamics of insect populations. Pages 3-27 in Dempster JP, McLean IFG, eds. Insect Populations in Theory and in Practice 19th Symposium of the Royal Entomological Society. Dordrecht: Kluwer Academic Publishers.

Hanski I. 1999. Habitat connectivity, habitat continuity, and metapopulations in dynamic landscapes. Oikos 87: 209-219.

Hanski I. 2004. Metapopulation theory, its use and misuse. Basic Appl Ecol 5: 225-229.

Hanski I, Gilpin M. 1991. Metapopulation dynamics: brief history and conceptual domain. Biol J Linn Soc 42: 3-16.

Hanski I, Meyke E. 2005. Large-scale dynamics of the Glanville fritillary butterfly: landscape structure, population processes, and weather. Ann Zool Fenn 42: 379-395.

Hanski I, Hellmann JJ, Boggs CL, McLaughlin JF. 2004. Checkerspots as a model system in population biology. Pages 245-263 in Ehrlich PR,

Hanski I, eds. On the Wings of Checkerspots: A Model System for Population Biology. Oxford: Oxford University Press.

Hardin G. 1960. The competitive exclusion principle. Science 131: 1292-1297.

Hare J. 1997. The wild Bactrian camel *Camelus bactrianus* ferus in China: The need for urgent action. Oryx 31: 45-48.

Harvey PH, Colwell RK, Silvertown JW, May RM. 1983. Null models in ecology. Annu Rev Ecol Syst 14: 189-211.

Harwood JD, Sunderland KD, Symondson WOC. 2001. Living where the food is: web location by linyphiid spiders in relation to prey availability in winter wheat. J Anim Ecol 38: 88-99.

Hassell MP, May RM. 1985. From individual behaviour to population dynamics. Pages 3-32 in Sibly RM, Smith RH, eds. Behavioural Ecology: Ecological Consequences of Adaptive Behaviour 25th Symposium of the British Ecological Society, Reading, 1984. Oxford: Blackwell Scientific Publications.

Hassell MP, Godfray HCJ. 1992. The population biology of insect parasitoids. Pages 265-292 in Crawley MJ, ed. Natural Enemies: The Population Biology of Predators, Parasites and Diseases. Oxford: Blackwell Scientific Publications.

Hassell MP, Southwood TRE, Reader PM. 1987. The dynamics of the viburnum whitefly (*Aleurotrachelus jelinekii*): a case study of population regulation. J Anim Ecol 56: 283-300.

Haukioja E. 1982. Are individuals really subordinated to genes? A theory of living entities. J Theor Biol 99: 357-375.

Hausdorf B. 2011. Progress toward a general species concept. Evolution 65: 923-931.

Haythornthwaite AS, Dickman CR. 2006. Long-distance movements by a small carnivorous marsupial: how *Sminthopsis youngsoni* (Marsupialia: Dasyuridae) uses habitat in an Australian sandridge desert. J Zool 270: 543-549.

He FL, Gaston KJ, Connor EF, Srivastava DS. 2005. The local-regional relationship: Immigration, extinction, and scale. Ecology 86: 360-365.

Hebets EA, Papaj DR. 2005. Complex signal function: developing a framework of testable hypotheses. Behav Ecol Sociobiol 57: 197-214.

Heck KL. 1976. Some critical considerations of the theory of species packing. Evol Theory Rev 1: 247-258.

Heck KL. 1980. Competitive exclusion or competitive delusion? Paleobiology 6: 241-242.

Heinselman ML. 1963. Forest sites, bog processes, and peatland types in glacial Lake Agassiz region, Minnesota. Ecol Monogr 33: 327-374.

Helbig AJ, Seibold I, Martens J, Wink M. 1995. Genetic differentiation and phylogenetic relationships of Bonelli warbler *Phylloscopus bonelli* and green warbler *P. nitidus*. J Avian Biol 26: 139-153.

Hellmann JJ, Weiss SB, McLaughlin JF, Ehrlich PR, Murphy DD, Launer AE. 2004. Structure and dynamics of *Euphydryas editha* populations. Pages 34-62 in Ehrlich PR, Hanski I, eds. On the Wings of Checkerspots: A Model System for Population Biology. Oxford: Oxford University Press.

Helmuth B, Kingsolver JG, Carrington E. 2005. Biophysics, physiological ecology, and climate change: Does mechanism matter? Annu Rev Physiol 67: 177-201.

Hemerik L, Busstra C, Mols P. 2004. Predicting the temperature-dependent natural population expansion of the western corn rootworm, *Diabrotica virgifera*. Entomol Exp Appl 111: 59-69.

Henderson NR, Lambert DM. 1982. No significant deviation from random mating of worldwide populations of *Drosophila melanogaster*. Nature 300: 437-440.

Hengeveld R. 1979. The analysis of spatial patterns of some ground beetles (Col. Carabidae). Pages 333-346 in Cormack RM, Ord JK, eds. Spatial And Temporal Analysis in Ecology, vol. 8. Fairland, Maryland: International Co-operative Publishing House.

Hengeveld R. 1981. The evolutionary relevance of feeding habits of ground beetles (Coleoptera, Carabidae). Entomol Scand Suppl 15: 305-315.

Hengeveld R. 1985a. Dynamics of Dutch beetle species during the twentieth century (Coleoptera, Carabidae). J Biogeogr 12: 389-411.

Hengeveld R. 1985b. On the explanation of the elevation effect by a dynamic interpretation of species distribution along altitudinal gradients. Blumea 30: 353-361.

Hengeveld R. 1988a. Mechanisms of biological invasions. J Biogeogr 15: 819-828.

Hengeveld R. 1988b. Mayr's ecological species criterion. Syst Zool 37: 47-55.

Hengeveld R. 1989a. Dynamics of Biological Invasions. London: Chapman & Hall.

Hengeveld R. 1989b. Caught in an ecological web. Oikos 54: 15-22.

Hengeveld R. 1990. Dynamic Biogeography. Cambridge: Cambridge University Press.

Hengeveld R. 1992. Potential and limitations of predicting invasion rates. Fla Entomol 75: 60-72.

Hengeveld R. 1993. Ecological biogeography. Progr Phys Geogr 17: 448-460.

Hengeveld R. 1997. Impact of biogeography on a population-biological paradigm shift. J Biogeogr 24: 541-547.

Hengeveld R, Hogeweg P. 1979. Cluster analysis of the distribution patterns of Dutch carabid species (Col.). Pages 65-86 in Orloci L, Rao CR, Stiteler WM, eds. Multivariate Methods in Ecological Work. Fairland, Maryland: International Co-operative Publishing House.

Hengeveld R, Haeck J. 1981. The distribution of abundance. 2. Models and implications. P K Ned Akad C Biol C 84: 257-284.

Hengeveld R, Haeck J. 1982. The distribution of abundance. 1. Measurements. J Biogeogr 9: 303-316.

Hengeveld R, van den Bosch F. 1991. The expansion velocity of the collared dove (*Streptopelia decaocto*) population in Europe. Ardea 79: 67-72.

Hengeveld R, van den Bosch F. 1997. Invading into an ecologically non-uniform area. Pages 217-225 in Huntley B, Cramer W, Morgan AV, Prentice HC, Allen JRM, eds. Past and Future Rapid Environmental Changes. Berlin: Springer.

Hengeveld R, Walter GH. 1999. The two coexisting ecological paradigms. Acta Biotheor 47: 141-170.

Hengeveld R, Hemerik L. 2002. Biogeography and dispersal. Pages 303-324 in Bullock JM, Kenward RE, Hails RS, eds. Dispersal Ecology 42nd Symposium of the British Ecological Society. Oxford: Blackwell Science.

Hengeveld R, Fedonkin MA. 2004. Causes and consequences of eukaryotization through mutualistic endosymbiosis and compartmentalization. Acta Biotheor 52: 105-154.

Hengeveld R, Fedonkin MA. 2007. Bootstrapping the energy flow in the beginning of life. Acta Biotheor 55: 181-226.

Hepworth SR, Valverde F, Ravenscroft D, Mouradov A, Coupland G. 2002. Antagonistic regulation of flowering-time gene SOC1 by CONSTANS and FLC via separate promoter motifs. EMBO Journal 21: 4327-4337.

Hereward JP, Walter GH, DeBarro PJ, Lowe AJ, Riginos C. 2013. Gene flow in the green mirid, *Creontiades dilutus* (Hemiptera: Miridae), across arid and agricultural environments with different host plant species. Ecol Evol 3: 807-821.

Herrera CM. 1992. Historical effects and sorting processes as explanations for contemporary ecological patterns: character syndromes in Mediterranean woody plants. Am Nat 140: 421-446.

Hewitt G. 2000. The genetic legacy of the Quaternary ice ages. Nature 405: 907-913.

Hewitt GM. 1999. Post-glacial re-colonization of European biota. Biol J Linn Soc 68: 87-112.

Higgins RW, Ruff CB. 2011. The effects of distal limb segment shortening on locomotor efficiency in sloped terrain: implications for Neandertal locomotor behavior. Am J Phys Anthropol 146: 336-345.

Highnam KC, Hill L. 1977. The Comparative Endocrinology of the Invertebrates. 2nd ed. London: Edward Arnold.

Hillis DM. 1988. Systematics of the *Rana pipiens* complex: puzzle and paradigm. Ann Rev Ecol Syst 19: 39-63.

Himmelfarb G. 1959. Darwin and the Darwinian Revolution. New York: Anchor Books.

Hockett B, Haws JA. 2003. Nutritional ecology and the human demography of Neandertal extinction. Pages 21-34. Symposium on the Extinction of the European Neandertals during Isotope Stage 3 held at the 16th INQUA Congress. Reno, NV.

Hoekstra HE, Coyne JA. 2007. The locus of evolution: Evo devo and the genetics of adaptation. Evolution 61: 995-1016.

Hoffmann AA, Blows MW. 1994. Species borders: ecological and evolutionary perspectives. Trends Ecol Evol 9: 223-227.

Hölldobler B, Wilson EO. 1990. The Ants. Cambridge, Massachusetts: Belknap Press of Harvard University Press.

Holliday NJ. 1977. Population ecology of winter moth (*Operophtera brumata*) on apple in relation to larval dispersal and time of bud burst. J Appl Ecol 14: 803-813.

Holzner W, Immonen R. 1982. Europe: an overview. Pages 203-226 in Holzner W, Numata M, eds. Biology and Ecology of Weeds. The Hague: Junk.

Horn HS. 1975. Forest succession. Sci Am 232: 90-98.

Horn HS. 1981. Succession. Pages 253-271 in May RM, ed. Theoretical Ecology: Principles and Applications (2nd ed). Oxford: Blackwell.

Huber F. 1985. Approaches to insect behavior of interest to both neurobiologists and behavioral ecologists. Fla Entomol 68: 52-78.

Huber F. 1990. Cricket neuroethology: Neuronal basis of intraspecific acoustic communication. Adv Stud Behav 19: 299-356.

Huey RB. 1979. Parapatry and niche complementarity of Peruvian desert geckos (*Phyllodactylus*): the ambiguous role of competition. Oecologia 38: 249-259.

Huffaker CB, Berryman A, Turchin P. 1999. Dynamics and regulation of insect populations. Pages 269-312 in Huffaker CB, Gutierrez AP, eds. Ecological Entomology. New York: John Wiley.

Hughes AJ, Lambert DM. 1984. Functionalism, structuralism and "ways of seeing". J Theor Biol 111: 787-800.

Hulley PE, Walter GH, Craig AJFK. 1988a. Interspecific competition and community structure, II. The recognition concept of species. Riv Biol/Biol Forum 81: 261-283.

Hulley PE, Walter GH, Craig AJFK. 1988b. Interspecific competition and community structure, I: Shortcomings of the competition paradigm. Riv Biol/Biol Forum 81: 57-71.

Huntley B. 1991. How plants respond to climate change: migration rates, individualism and the consequences for plant communities. Ann Bot 67: 15-22.

Huntley B, Birks HJB. 1983. Past and Present Pollen Maps for Europe 0-13,000 Years Ago. Cambridge: Cambridge University Press.

Huntley B, Webb T. 1989. Migration: species' response to climatic variations caused by changes in the earth's orbit. J Biogeogr 16: 5-19.

Huntley B, Bartlein PJ, Prentice IC. 1989. Climatic control of the distribution and abundance of beech (*Fagus* L.) in Europe and North America. J Biogeogr 16: 551-560.

Hutchinson GE. 1978. An Introduction to Population Ecology. New Haven: Yale University Press.

Inouye BD. 2005. Scaling up from local competition to regional coexistence across two scales of spatial heterogeneity: insect larvae in the fruits of *Apeiba membranacea*. Oecologia 145: 188-196.

Irwin DE, Alstrom P, Olsson U, Benowitz-Fredericks ZM. 2001. Cryptic species in the genus *Phylloscopus* (Old World leaf warblers). Ibis 143: 233-247.

Isaacs R, Byrne DN. 1998. Aerial distribution, flight behaviour and eggload: their inter-relationship during dispersal by the sweetpotato whitefly. J Anim Ecol 67: 741-750.

Ito Y. 1961. Factors that affect the fluctuations of animal numbers, with special reference to insect outbreaks. Bull Natl Inst Agric Sci Ser C (Plant Pathol Entomol) 13: 57-89.

Iversen J. 1944. *Viscum, Hedera* and *Ilex* as climatic indicators. Geol Foren Stock For 66: 463-483.

Jablonski D. 1991. Extinctions: a paleontological perspective. Science 253: 754-757.

Jablonski D. 1994. Extinctions in the fossil record. Philos Trans R Soc Lond B Biol Sci 344: 11-17.

Jagers op Akkerhuis GAJM. 2008. Analysing hierarchy in the organization of biological and physical systems. Biol Rev 83: 1-12.

Janzen D. 1985. On ecological fitting. Oikos 45: 308-310.

Janzen DH, Martin PS. 1982. Neotropical anachronisms—the fruits the gomphotheres ate. Science 215: 19-27.

Jasny BR, Chin G, Chong L, Vignieri S. 2011. Again, and again, and again .... Science 334: 1225-1225.

Jenkins G. 1974. A note on climatic cycles and the rise of Chinggis Khan. Cent Asiat J 18: 217-226.

Jeong S, Rebeiz M, Andolfatto P, Werner T, True J, Carroll SB. 2008. The evolution of gene regulation underlies a morphological difference between two *Drosophila* sister species. Cell 132: 783-793.

Jerling L. 1985. Population dynamics of *Plantago maritima* along a distributional gradient on a Baltic seashore meadow. Vegetatio 61: 155-161.

Jerling L. 1988. Genetic differentiation in fitness related characters in *Plantago maritima* along a distributional gradient. Oikos 53: 341-350.

Jha CS, Singh JS. 1990. Composition of dry tropical forest in relation to soil texture. J Veg Sci 1: 609-614.

Johnson CG. 1969. Migration and Dispersal of Insects by Flight. London: Methuen.

Johnson HB, Mayeux HS. 1992. A view on species additions and deletions and the balance of nature. J Range Manag 45: 322-333.

Jolly D, Taylor D, Marchant R, Hamilton A, Bonnefille R, Buchet G, Riollet G. 1997. Vegetation dynamics in central Africa since 18,000 yr BP: pollen records from the interlacustrine highlands of Burundi, Rwanda and western Uganda. J Biogeogr 24: 495-512.

Kammesheidt L. 2000. Some autecological characteristics of early to late successional tree species in Venezuela. Acta Oecol 21: 37-48.

Karasov WH, Martinez del Rio C. 2007. Physiological Ecology. How Animals Process Energy, Nutrients, and Toxins. Princeton: Princeton University Press.

Kareiva P. 1994. Ecological theory and endangered species. Ecology 75: 583.

Kayes SM, Cramp RL, Franklin CE. 2009. Metabolic depression during aestivation in *Cyclorana alboguttata*. Comp Biochem Physiol Part A Mol Integr Physiol 154: 557-563.

Keddy P. 1994. Applications of the Hertzsprung-Russell star chart to ecology: reflections on the 21st birthday of Geographical Ecology. Trends Ecol Evol 9: 231-234.

Keech MA, Bowyer RT, Ver Hoef JM, Boertje RD, Dale BW, Stephenson TR. 2000. Life-history consequences of maternal condition in Alaskan moose. J Wildl Manag 64: 450-462.

Keller LF, Jeffery KJ, Arcese P, Beaumont MA, Hochachka WM, Smith JNM, Bruford MW. 2001. Immigration and the ephemerality of a natural population bottleneck: evidence from molecular markers. Proc R Soc Lond B Biol Sci 268: 1387-1394.

Keyfitz N. 1968. Introduction to the Mathematics of Population. Reading, Mass: Addison and Wesley.

Keyfitz N, Flieger W. 1968. World Population. An Analysis of Vital Data. Chicago: University of Chicago Press.

Kikkawa J. 1977. Ecological paradoxes. Aust J Ecol 2: 121-136.

Kiltie RA. 1984. Seasonality, gestation time, and large mammal extinctions. Pages 299-314 in Martin PS, Klein RG, eds. Quaternary Extinctions. Tucson: University of Arizona Press.

Kincer JB. 1928. Weather and the cotton boll weevil. Mon Weather Rev 56: 301-304.

Kingsland SE. 1995. Modeling Nature. Episodes in the History of Population Ecology. 2nd ed. Chicago: University of Chicago Press.

Kingsolver JG. 1989. Weather and the population-dynamics of insects—integrating physiological and population ecology. Physiol Zool 62: 314-334.

Kingsolver JG, Woods HA, Buckley LB, Potter KA, MacLean HJ, Higgins JK. 2011. Complex life cycles and the responses of insects to climate change. Integr Comp Biol 51: 719-732.

Kishimoto-Yamada K, Itioka T. 2013. Seasonality in phytophagous scarabaeid (Melolonthinae and Rutelinae) abundances in an 'aseasonal' Bornean rainforest. Insect Conserv Divers 6: 179-188.

Kisimoto R. 1976. Synoptic weather conditions inducing long-distance immigration of planthoppers, *Sogatella furcifera* Horvath and *Nilaparvata lugens* (Stål). Ecol Entomol 1: 95-109.

Kisimoto R, Rosenberg LJ. 1994. Long-distance migration in delphacid planthoppers. Pages 302-322 in Denno RF, Perfect TJ, eds. Planthoppers Their Ecology and Management. New York: Chapman & Hall.

Kisimoto R, Sogawa K. 1995. Migration of the brown planthopper Nilaparvata lugens and the white-backed planthopper Sogatella furcifera in East Asia: the role of weather and climate. Pages 67-91 in Drake VA, Gatehouse AG, eds. Insect Migration. Cambridge: Cambridge University Press.

Klowden MJ. 2007. Physiological Systems in Insects. 2nd edition. Burlington, MA: Academic Press.

Koch PL, Barnosky AD. 2006. Late Quaternary extinctions: State of the debate. Annu Rev Ecol Evol Syst vol. 37: 215-250.

Kohn M. 2004. A Reason for Everything. Natural Selection and the English Imagination. London: Faber & Faber.

Kokko H, Jennions MD. 2010. Behavioral ecology: the natural history of evolutionary theory. Pages 269-290 in Bell MA, Futuyma DJ, Eanes WF, Levinton JS, eds. Evolution Since Darwin: The First 150 Years. Sunderland, Mass.: Sinauer Associates.

Kooijman S, Vanderhoeven N, Vanderwerf DC. 1989. Population consequences of a physiological model for individuals. Funct Ecol 3: 325-336.

Kooijman SALM. 1979. A spatial model explaining species-abundance curves. Pages 333-347 in Ord J, Patil G, Taillie C, eds. Statistical Distributions in Ecological Work. New York: Pergamon Press.

Kooijman SALM. 1986. Population dynamics on basis of budgets. Pages 266-297 in Metz JAJ, Diekmann O, eds. Lecture Notes in Biomathematics 68 The Dynamics of Physiologically Structured Populations. Berlin: Springer-Verlag.

Kooijman SALM. 2010. Dynamic Energy Budget Theory for Metabolic Organisation, 3rd ed. Cambridge: Cambridge University Press.

Kooijman SALM, Hengeveld R. 2005. The symbiontic nature of metabolic evolution. Pages 159-202 in Reydon TAC, Hemerik L, eds. Current Themes in Theoretical Biology: A Dutch perspective. Dordrecht: Springer.

Kooyman GL. 2002. Evolutionary and ecological aspects of some Antarctic and sub-Antarctic penguin distributions. Oecologia 130: 485-495.

Krebs CJ. 2008. The Ecological World View. Collingwood: CSIRO Publishing.

Krebs JR, Davies NB, (Eds). 1997. Behavioural Ecology: an Evolutionary Approach, 4th ed Oxford: Blackwell Science.

Kuhn TS. 1962. The Structure of Scientific Revolutions. Chicago: Chicago University Press.

Kyi A, Zalucki MP, Titmarsh IJ. 1991. An experimental study of early stage survival of *Helicoverpa armigera* (Lepidoptera: Noctuidae) on cotton. Bull Entomol Res 81: 263-271.

Lack D. 1954. The Natural Regulation of Animal Numbers. Oxford: Oxford University Press.

Lada H, Thomson JR, Cunningham SC, MacNally R. 2013. Rainfall in prior breeding seasons influences population size of a small marsupial. Austral Ecol 38: 581-591.

Lakatos I. 1970. Falsification and the methodology of scientific research programmes. Pages 91-196 in Lakatos I, Musgrave A, eds. Criticism and the Growth of Knowledge. Cambridge: Cambridge University Press.

Lalas C, Bradshaw CJA. 2001. Folklore and chimerical numbers: review of a millennium of interaction between fur seals and humans in the New Zealand region. NZ J Mar Freshw Res 35: 477-497.

Lalas C, Bradshaw CJA. 2003. Expectations for population growth at new breeding locations for the vulnerable New Zealand sea lion (*Phocarctos hookeri*) using a simulation model. Biol Conserv 114: 67-78.

Lambers H, Chapin FS, Pons TL. 1998. Plant Physiological Ecology. New York: Springer.

Lambert DM, Henderson NR. 1986. The stability of the specific-mate recognition system of *Drosophila melanogaster*. Behav Genet 16: 369-373.

Lambert DM, Michaux B, White CS. 1987. Are species self-defining? Syst Zool 36: 196-205.

Lande R, Engen S, Saether BE, Coulson T. 2006. Estimating density dependence from time series of population age structure. Am Nat 168: 76-87.

Latto J, Hassell MP. 1988. Generalist predators and the importance of spatial density dependence. Oecologia 77: 375-377.

Lawton JH. 1993a. On the behaviour of autecologists and the crisis of extinction. Oikos 67: 3-5.

Lawton JH. 1993b. Range, population abundance and conservation. Trends Ecol Evol 8: 409-413.

Lawton JH. 1999. Are there general laws in ecology? Oikos 84: 177-192.

Lawton JH, MacGarvin M. 1986. The organization of herbivore communities. Pages 163-186. in Kikkawa J, Anderson DJ, eds. Community Ecology: Pattern and Process. London: Blackwell Scientific Publications.

Le Galliard JF, Ferriere R, Clobert J. 2005. Effect of patch occupancy on immigration in the common lizard. J Anim Ecol 74: 241-249.

Leduc RG, Robertson KM, Pitman RL. 2008. Mitochondrial sequence divergence among Antarctic killer whale ecotypes is consistent with multiple species. Biol Lett 4: 426-429.

Leibold MA, Holyoak M, Mouquet N, Amarasekare P, Chase JM, Hoopes MF, Holt RD, Shurin JB, Law R, Tilman D, Loreau M, Gonzalez A. 2004. The metacommunity concept: a framework for multi-scale community ecology. Ecol Lett 7: 601-613.

Lensink R. 1997. Range expansion of raptors in Britain and the Netherlands since the 1960s: testing an individual-based diffusion model. J Anim Ecol 66: 811-826.

Levins R. 1970. Extinction. Pages 77-107 in Gerstenhaber M, ed. Some Mathematical Problems in Biology. Providence, R.I.: American Mathematical Society.

Lewin R. 2005. Human Evolution. An Illustrated Introduction. Oxford: Blackwell Publishing.

Leyva JL, Browning HW, Gilstrap FE. 1991. Effect of host fruit species, size, and color on parasitization of *Anastrepha ludens* (Diptera: Tephritidae) by *Diachasmimorpha longicaudata* (Hymenoptera: Braconidae). Environ Entomol 20: 1469-1474.

Limpens E, Bisseling T. 2003. Signaling in symbiosis. Curr Opin Plant Biol 6: 343-350.

Limpens E, Mirabella R, Fedorova E, Franken C, Franssen H, Bisseling T, Geurts R. 2005. Formation of organelle-like N_2-fixing symbiosomes in legume root nodules is controlled by DMI2. Proc Natl Acad Sci USA 102: 10375-10380.

Lindroth CH. 1949. Die Fennoskandischen Carabiden. Vol. 3. Allgemeiner Teil. Stockholm: Kungliga Vetenskaps—och Vitterhets-Samhället i Göteborg.

Linnell JDC, Broseth H, Solberg EJ, Brainerd SM. 2005. The origins of the southern Scandinavian wolf *Canis lupus* population: potential for natural immigration in relation to dispersal distances, geography and Baltic ice. Wildl Biol 11: 383-391.

Liss M, Kirk DL, Beyser K, Fabry S. 1997. Intron sequences provide a tool for high-resolution phylogenetic analysis of volvocine algae. Curr Genet 31: 214-227.

Lockwood R, Swaddle JP, Rayner JMV. 1998. Avian wingtip shape reconsidered: wingtip shape indices and morphological adaptations to migration. J Avian Biol 29: 273-292.

Lokkers C, Jones RE. 1999. The cabbage white, *Pieris rapae* (Pieridae). Pages 153-172 in Kitching RL, Scheermeyer E, Jones RE, Pierce NE, eds. Monographs on Australian Lepidoptera Volume 6 Biology of Australian Butterflies. Collingwood: CSIRO Publishing.

Lomnicki A. 1988. Population Ecology of Individuals. Princeton: Princeton University Press.

Lorenz EN. 1993. The Essence of Chaos. Seattle: University of Washington Press.

Lough JM, Wigley TML, Palutikof JP. 1983. Climate and climate impact scenarios for Europe in a warmer world. J Climate Appl Meteor 22: 1673-1684.

Lovejoy CO. 2005a. The natural history of human gait and posture—Part 1. Gait Posture. 21: 95-112.

Lovejoy CO. 2005b. The natural history of human gait and posture—Part 2. Gait Posture. 21: 113-124.

Lovejoy CO. 2007. The natural history of human gait and posture—Part 3. Gait Posture. 25: 325-341.

Lovejoy CO, McCollum MA. 2010. Spinopelvic pathways to bipedality: why no hominids ever relied on a bent-hip-bent-knee gait. Philos Trans R Soc Lond B Biol Sci 365: 3289-3299.

Lovtrup S. 1984. The eternal battle against empiricism. Riv Biol/Biol Forum 77: 183-209.

Loxdale HD, Lushai G. 1999. Slaves of the environment: the movement of herbivorous insects in relation to their ecology and genotype. Philos Trans R Soc Lond B Biol Sci 354: 1479-1495.

Loxdale HD, Lushai G, Harvey JA. 2011. The evolutionary improbability of 'generalism' in nature, with special reference to insects. Biol J Linn Soc 103: 1-18.

Luhring KA, Millar JG, Paine TD, Reed D, Christiansen H. 2004. Ovipositional preferences and progeny development of the egg parasitoid *Avetianella longoi*: factors mediating replacement of one species by a congener in a shared habitat. Biol Control 30: 382-391.

Luzhang R, Lixun Z, Naifa L, Zuhao H, Muli M, Kerong X. 2005. Distribution and population status of the wild bactrian camel (*Camelus bactrianus ferus*) in Gansu Province, China. J Camel Pract Res 12: 59-63.

Mabey R. 1995. The Oxford Book of Nature Writing. Oxford: Oxford University Press.

MacArthur RH. 1972a. Coexistence of species. Pages 253-259 in Behnke J, ed. Challenging Biological Problems. Oxford: Oxford University Press.

MacArthur RH. 1972b. Geographical Ecology: Patterns in the Distribution of Species. New York: Harper & Row.

MacArthur RH, Wilson EO. 1967. The Theory of Island Biogeography. Princeton: Princeton University Press.

Madsen T, Shine R. 1999. Rainfall and rats: climatically-driven dynamics of a tropical rodent population. Aust J Ecol 24: 80-89.

Magee B. 1973. Popper. London: Fontana/Collins.

Malausa T, Dalecky A, Ponsard S, Audiot P, Streiff R, Chaval Y, Bourguet D. 2007. Genetic structure and gene flow in French populations of two *Ostrinia* taxa: host races or sibling species? Mol Ecol 16: 4210-4222.

Mallet J. 1995. A species definition for the modern synthesis. Trends Ecol Evol 10: 294-299.

Mallet J. 2005. Hybridization as an invasion of the genome. Trends Ecol Evol 20: 229-237.

Mallet J. 2008. Hybridization, ecological races and the nature of species: empirical evidence for the ease of speciation. Philos Trans R Soc Lond B Biol Sci 363: 2971-2986.

Mandelbrot BB. 1977. Fractals: Form, Chance, and Dimension. San Francisco: W.H. Freeman.

Mangel M. 2008. Uncertainty in ecology: a retrospective and prospective. Bull Br Ecol Soc 39: 25-28.

Manners AG. 2008. The ecological principles that underpin host testing for weed biological control—a case study with the lantana sap-sucking bug, *Acanophora compressa* Walker (Hemiptera: Membracidae). Unpublished PhD thesis. Brisbane: The University of Queensland.

Manners AG, Walter GH. 2009. Multiple host use by a sap-sucking membracid: population consequences of nymphal development on primary and secondary host plant species. Arthropod Plant Interact 3: 87-98.

Manners AG, Palmer WA, Dhileepan K, Hastwell GT, Walter GH. 2010. Characterising insect plant host relationships facilitates understanding multiple host use. Arthropod Plant Interact 4: 7-17.

Manning A, Dawkins MS. 2012. An Introduction to Animal Behaviour. 6th ed. Cambridge: Cambridge University Press.

Marchant JH, Hudson R, Carter SP, Whittington P. 1990. Population Trends in British Breeding Birds. Tring, Hertfordshire: British Trust for Ornithology.

Marchetti K, Price T, Richman A. 1995. Correlates of wing morphology with foraging behavior and migration distance in the genus *Phylloscopus*. J Avian Biol 26: 177-181.

Marohasy J. 1996. Host shifts in biological weed control: real problems, semantic difficulties or poor science? Int J Pest Manag 42: 71-75.

Marohasy J. 1998. The design and interpretation of host-specificity tests for weed biological control with particular reference to insect behaviour. Biocontrol News Info 19: 13-20.

Martínez I. 1992. Order Sphenisciformes. Family Spheniscidae (Penguins). Pages 140-160 in del Hoyo J, Elliot A, Sargatal J, eds. Handbook of the Birds of the World Vol. 1. Ostrich to Ducks. Barcelona: Lynx Edicions.

MathsIsFun.com. 2010. Quincunx. Available at: http://www.mathsisfun.com/data/quincunx.html. Accessed Jan. 2013.

Masters JC, Rayner RJ. 1993. Competition and macroevolution: the ghost of competition yet to come. Biol J Linn Soc 49: 87-98.

Mathieu PP, Sutton RT, Dong B, Collins M. 2004. Predictability of winter climate over the North Atlantic European region during ENSO events. J Clim 17: 1953-1974.

Matthews M. 1999. Heliothine Moths of Australia. A Guide to Pest Bollworms and Related Noctuid Groups. Collingwood: CSIRO Publishing.

May RM. 1976. Simple mathematical models with very complicated dynamics. Nature 261: 459-467.

Mayhew PJ. 2006. Discovering Evolutionary Ecology: Bringing Together Ecology and Evolution. Oxford: Oxford University Press.

Maynard Smith J. 1981. Macroevolution. Nature 289: 13-14.

Mayr E. 1961. Cause and effect in biology. Science 134: 1501-1506.

Mayr E. 1963. Animal Species and Evolution. Cambridge: Harvard University Press.

Mayr E. 1974. Teleological and teleonomic: a new analysis. Philos Sci 14: 91-117.

Mayr E. 1982. The Growth of Biological Thought. Diversity, Evolution, and Inheritance. Cambridge: Harvard University Press.

McConkey S, Heinrich S, Lalas C, McConnell H, McNalley N. 2002. Pattern of immigration of New Zealand sea lions *Phocarctos hookeri* to Otago, New Zealand: implications for the management of the species. Aust Mammal 24: 107-116.

McIntosh RP. 1981. Succession and ecological theory. Pages 10-23 in West DC, Shugart HH, Botkin DB, eds. Forest Succession Concepts and Application. New York: Springer-Verlag.

McIntosh RP. 1985. The Background of Ecology: Concept and Theory. Cambridge: Cambridge University Press.

McIntosh RP. 1999. The succession of succession: a lexical chronology. Bull Ecol Soc Am 80: 256-265.

McKie R. 2000. Ape Man. The Story of Human Evolution. London: BBC Worldwide.

McNab BK. 2002. The Physiological Ecology of Vertebrates. A View from Energetics. Ithaca: Cornell University Press.

McNeill WH. 1990. Population and Politics since 1750. Charlottesville: University of Virginia Press.

McPhaden MJ, Yu X. 1999. Equatorial waves and the 1997-98 El Niño. Geophys Res Lett 26: 2961-2964.

Merila J, Sheldon BC, Kruuk LEB. 2001. Explaining stasis: microevolutionary studies in natural populations. Genetica 112: 199-222.

Merryweather J, Fitter A. 1995a. Phosphorus and carbon budgets: Mycorrhizal contribution in *Hyacinthoides non-scripta* (L.) Chouard ex Rothm. under natural conditions. New Phytol 129: 619-627.

Merryweather J, Fitter A. 1995b. Arbuscular mycorrhiza and phosphorus as controlling factors in the life history of *Hyacinthoides non-scripta* (L.) Chouard ex Rothm. New Phytol 129: 629-636.

Merryweather J, Fitter A. 1996. Phosphorus nutrition of an obligately mycorrhizal plant treated with the fungicide benomyl in the field. New Phytol 132: 307-311.

Middleton BA, McKee KL. 2004. Use of a latitudinal gradient in bald cypress (*Taxodium distichum*) production to examine physiological controls of biotic boundaries and potential responses to environmental change. Glob Ecol Biogeogr 13: 247-258.

Middleton HA, Morrissey CA, Green DJ. 2006. Breeding territory fidelity in a partial migrant, the American dipper *Cinclus mexicanus*. J Avian Biol 37: 169-178.

Miller FL. 2003. Caribou (*Rangifer tarandus*). Pages 965-997 in Feldhamer GA, Thompson BC, Chapman JA, eds. Wild Mammals of North America Biology, Management, and Conservation 2nd ed. Baltimore: Johns Hopkins University Press.

Miller KA, Fluharty DL. 1992. El Niño and variability in the northeastern Pacific salmon fishery: implications for coping with climate change. Pages 49-88 in Glantz MH, ed. Climate Variability, Climate Change, and Fisheries. Cambridge: Cambridge University Press.

Miller KV, Muller LI, Demarais S. 2003. White-tailed deer (*Odocoileus virginianus*). Pages 906-930 in Feldhamer GA, Thompson BC, Chapman JA, eds. Wild Mammals of North America Biology, Management, and Conservation 2nd ed. Baltimore: Johns Hopkins University Press.

Milne M, Walter GH. 1998. Host species and plant part specificity of the polyphagous onion thrips, *Thrips tabaci* (Thysanoptera: Thripidae), in an Australian cotton-growing area. Aust J Entomol 37: 115-119.

Milne M, Walter GH. 2000. Feeding and breeding across host plants within a locality by the widespread thrips *Frankliniella schultzei*, and the invasive potential of polyphagous herbivores. Divers Distrib 6: 243-257.

Milne M, Walter GH, Milne JR. 2007. Mating behavior and species status of host-associated populations of the polyphagous thrips, *Frankliniella schultzei*. J Insect Behav 20: 331-346.

Mitchell JM. 1976. Overview of climatic variability and its causal mechanisms. Quat Res 6: 481-493.

Mitchell MS, Powell RA. 2003. Linking fitness landscapes with the behavior and distribution of animals. Pages 93-124 in Bissonette JA, Storch I, eds. Landscape Ecology and Resource Management Linking Theory with Practice. Washington, D.C.: Island Press.

Moilanen A. 2000. The equilibrium assumption in estimating the parameters of metapopulation models. J Anim Ecol 69: 143-153.

Moore JA. 1957. An embryologist's view of the species concept. Pages 325-338 in Mayr E, ed. The Species Problem, vol. 50. Washington: The American Association for the Advancement of Science.

Moore JA. 1975. *Rana pipiens*—the changing paradigm. Am Zool 15: 837-849.

Moore TE, Huber F, Weber T, Klein U, Bock C. 1993. Interaction between visual and phonotactic orientation durning flight in *Magicicada cassini* (Homoptera: Cicadidae). Gt Lakes Entomol 26: 199-221.

Mopper S, Strauss SY, (Eds). 1998. Genetic Structure and Local Adaptation in Natural Insect Populations: Effects of Ecology, Life History, and Behavior. New York: Chapman & Hall.

Mora C, Chittaro PM, Sale PF, Kritzer JP, Ludsin SA. 2003. Patterns and processes in reef fish diversity. Nature 421: 933-936.

Morin PA, Archer FI, Foote AD, Vilstrup J, Allen EE, Wade P, Durban J, Parsons K, Pitman R, Li L, Bouffard P, Nielsen SCA, Rasmussen M, Willerslev E, Gilbert MTP, Harkins T. 2010. Complete mitochondrial genome phylogeographic analysis of killer whales (Orcinus orca) indicates multiple species. Genome Res 20: 908-916.

Morris D. 1970. The function and causation of courtship ceremonies. Pages 128-152 in Morris D, ed. Patterns of Reproductive Behaviour: Collected Papers by Desmond Morris. London: Jonathan Cape.

Morris RFE. 1963. The dynamics of epidemic spruce budworm populations. Mem Entomol Soc Can 31: 1-332.

Morton ES. 1975. Ecological sources of selection on avian sounds. Am Nat 109: 17-34.

Morton SR, Smith DMS, Dickman CR, Dunkerley DL, Friedel MH, McAllister RRJ, Reid JRW, Roshier DA, Smith MA, Walsh FJ, Wardle GM, Watson IW, Westoby M. 2011. A fresh framework for the ecology of arid Australia. J Arid Environ 75: 313-329.

Mouquet N, Moore JL, Loreau M. 2002. Plant species richness and community productivity: why the mechanism that promotes coexistence matters. Ecol Lett 5: 56-65.

Moyes CD, Schulte PM. 2006. Principles of Animal Physiology. San Francisco: Pearson Education.

Muller-Landau HC, Levin SA, Keymer JE. 2003. Theoretical perspectives on evolution of long-distance dispersal and the example of specialized pests. Ecology 84: 1957-1967.

Munguia P. 2004. Successional patterns on pen shell communities at local and regional scales. J Anim Ecol 73: 64-74.

Murdoch WW. 1994. Population regulation in theory and practice. Ecology 75: 271-287.

Murdoch WW, Walde SJ. 1989. Analysis of insect population dynamics. Pages 113-141 in Whittaker JB, ed. Toward a More Exact Ecology. Oxford: Blackwell.

Murdoch WW, Briggs CJ. 1996. Theory for biological control: recent developments. Ecology 77: 2001-2013.

Murphy DD, Wahlberg N, Hanski I, Ehrlich PR. 2004. Introducing checkerspots. Taxonomy and ecology. Pages 17-33 in Ehrlich PR, Hanski I, eds. On the Wings of Checkerspots: A Model System for Population Biology. Oxford: Oxford University Press.

Murray BG. 1979. Population Dynamics: Alternative Models. New York: Academic Press.

Murray BG. 1999. Can the population regulation controversy be buried and forgotten? Oikos 84: 148-152.

Najar-Rodriguez AJ, McGraw EA, Hull CD, Mensah RK, Walter GH. 2009. The ecological differentiation of asexual lineages of cotton aphids: alate behaviour, sensory physiology, and differential host associations. Biol J Linn Soc 97: 503-519.

Nelson AR, Johnson CL, Matter WJ, Mannan RW. 2002. Tests of emigration in small mammals under experimental conditions. Can J Zool 80: 2056-2060.

Neufeld A. 1987. How far do you trust your colleagues? New Sci 15 January 1987: 59.

Neumann DR. 2001. Seasonal movements of short-beaked common dolphins (*Delphinus delphis*) in the north-western Bay of Plenty, New Zealand: influence of sea surface temperature and El Niño/La Niña. N Z J Mar Freshw Res 35: 371-374.

Nevitt G. 1999a. Foraging by seabirds on an olfactory landscape. Am Sci 87: 46-53.

Nevitt G. 1999b. Olfactory foraging in Antarctic seabirds: a species-specific attraction to krill odors. Mar Ecol Prog Ser 177: 235-241.

Nevitt G, Reid K, Trathan P. 2004. Testing olfactory foraging strategies in an Antarctic seabird assemblage. J Exp Biol 207: 3537-3544.

Nevitt GA. 2000. Olfactory foraging by Antarctic procellariiform seabirds: life at high Reynolds numbers. Biol Bull 198: 245-253.

Nevitt GA, Haberman K. 2003. Behavioral attraction of Leach's storm-petrels (*Oceanodroma leucorhoa*) to dimethyl sulfide. J Exp Biol 206: 1497-1501.

Nevitt GA, Bonadonna F. 2005. Sensitivity to dimethyl sulphide suggests a mechanism for olfactory navigation by seabirds. Biol Lett 1: 303-305.

Nikolskiy PA, Sulerzhitsky LD, Pitulko VV. 2011. Last straw versus Blitzkrieg overkill: Climate-driven changes in the Arctic Siberian mammoth population and the Late Pleistocene extinction problem. Quaternary Sci Rev 30: 2309-2328.

Noble IR, Slatyer RO. 1980. The use of vital attributes to predict successional changes in plant communities subject to recurrent disturbances. Vegetatio 43: 5-21.

Nolet BA, Andreev VA, Clausen P, Poot MJM, Wessel EGJ. 2001. Significance of the White Sea as a stopover for Bewick's Swans *Cygnus columbianus bewickii* in spring. Ibis 143: 63-71.

Noor MAF. 2002. Is the biological species concept showing its age? Trends Ecol Evol 17: 153-154.

Nosil P, Crespi BJ, Sandoval CP. 2002. Host-plant adaptation drives the parallel evolution of reproductive isolation. Nature 417: 440-443.

Nosil P, Crespi BJ, Sandoval CP. 2003. Reproductive isolation driven by the combined effects of ecological adaptation and reinforcement. Proc R Soc Lond B Biol Sci 270: 1911-1918.

O'Hara RB. 2005. The anarchist's guide to ecological theory. Or, we don't need no stinkin' laws. Oikos 110: 390-393.

O'Neill RV. 2001. Is it time to bury the ecosystem concept? (With full military honors of course!). Ecology 82: 3275-3284.

Oatley TB, Earle RA, Prins AJ. 1989. The diet and foraging behaviour of the ground woodpecker. Ostrich 60: 75-84.

Oldroyd H. 1964. The Natural History of Flies. London: Weidenfeldt & Nicolson.

Oppenheimer S. 2003. Out of Eden. The Peopling of the World. London: Robinson.

Orians GH. 1962. Natural selection and ecological theory. Am Nat 96: 257-263.

Overholtzer-McLeod KL. 2005. Post-settlement emigration affects mortality estimates for two Bahamian wrasses. Coral Reefs 24: 283-291.

Packert M, Martens J, Eck S, Nazarenko AA, Valchuk OP, Petri B, Veith M. 2005. The great tit (*Parus major*)—a misclassified ring species. Biol J Linn Soc 86: 153-174.

Padgham M. 2004. Reverberation and frequency attenuation in forests-implications for acoustic communication in animals. J Acoust Soc Am 115: 402-410.

Paine RT. 2002. Advances in ecological understanding: by Kuhnian revolution or conceptual evolution? Ecology 83: 1553-1559.

Pardo LM, Johnson LE. 2005. Explaining variation in life-history traits: growth rate, size, and fecundity in a marine snail across an environmental gradient lacking predators. Mar Ecol Prog Ser 296: 229-239.

Parmesan C. 2003. Butterflies as bio-indicators for climate change impacts. Pages 541-560 in Boggs CL, Watt WB, Ehrlich PR, eds. Evolution and Ecology Taking Flight: Butterflies as Model Systems. Chicago: University of Chicago Press.

Parmesan C. 2006. Ecological and evolutionary responses to recent climate change. Annu Rev Ecol Evol Syst 37: 637-669.

Parmesan C, Yohe G. 2003. A globally coherent fingerprint of climate change impacts across natural systems. Nature 421: 37-42.

Parmesan C, Ryrholm N, Stefanescu C, Hill JK, Thomas CD, Descimon H, Huntley B, Kaila L, Kullberg J, Tammaru T, Tennent WJ, Thomas JA, Warren, M. 1999. Poleward shifts in geographical ranges of butterfly species associated with regional warming. Nature 399: 579-583.

Parmesan C, Root TL, Willig MR. 2000. Impacts of extreme weather and climate on terrestrial biota. Bull Am Meteorol Soc 81: 443-450.

Parniske M, Downie JA. 2003. Plant biology—Locks, keys and symbioses. Nature 425: 569-570.

Parry ML. 1978. Climatic Change, Agriculture and Settlement. Folkestone: Wm Dawson.

Parry ML, Carter TR. 1985. The effect of climatic variations on agricultural risk. Clim Change 7: 95-110.

Paterson HEH. 1956. Status of the two forms of house fly occurring in South Africa. Nature 178: 928-929.

Paterson HEH. 1964. Population genetic studies in areas of overlap of two subspecies of *Musca domestica* L. Pages 244-254 in Davis DHS, ed. Ecological Studies in Southern Africa. The Hague: W. Junk.

Paterson HEH. 1978. More evidence against speciation by reinforcement. S Afr J Sci 74: 369-371.

Paterson HEH. 1980. A comment on "mate recognition systems". Evolution 34: 330-331.

Paterson HEH. 1981. The continuing search for the unknown and the unknowable: a critique of contemporary ideas on speciation. S Afr J Sci 77: 113-119.

Paterson HEH. 1982a. Darwin and the origin of species. S Afr J Sci 78: 272-275.

Paterson HEH. 1982b. Perspectives on speciation by reinforcement. S Afr J Sci 78: 53-57.

Paterson HEH. 1985. The recognition concept of species. Pages 21-29 in Vrba ES, ed. Species and Speciation. Pretoria: Transvaal Museum.

Paterson HEH. 1986. Environment and species. S Afr J Sci 82: 62-65.

Paterson HEH. 1988. On defining species in terms of sterility: problems and alternatives. Pac Sci 42: 65-71.

Paterson HEH. 1989. Updating the evolutionary synthesis. Riv Biol-Biol Forum 82: 371-375.

Paterson HEH. 1991. The recognition of cryptic species among economically important insects. Pages 1-10 in Zalucki MP, ed. *Heliothis*: Research Methods and Prospects. New York: Springer-Verlag.

Paterson HEH. 1993a. Botha de Meillon and the *Anopheles gambiae* complex. Pages 39-46 in Coetzee M, ed. Entomologist Extraordinary: A Festschrift in Honour of Botha de Meillon. Johannesburg: South African Institute for Medical Research.

Paterson HEH. 1993b. Animal species and sexual selection. Pages 209-228 in Lees DR, Edwards D, eds. Evolutionary Patterns and Processes. London: Academic Press for the Linnean Society of London.

Paterson HEH. 1993c. Evolution and the Recognition Concept of Species: Collected Writings. Baltimore: Johns Hopkins University Press.

Paterson HEH. 1993d. Variation and the Specific-Mate Recognition System. Pages 209-227 in Bateson PPG, ed. Perspectives in Ethology, Volume 10; Behavior and Evolution. New York: Plenum Press.

Paterson HEH. 2005. The competitive Darwin. Paleobiology 31: 56-76.

Peach WJ, Denny M, Cotton PA, Hill IF, Gruar D, Barritt D, Impey A, Mallord J. 2004. Habitat selection by song thrushes in stable and declining farmland populations. J Appl Ecol 41: 275-293.

Pellmyr O, Thompson JN. 1992. Multiple occurences of mutualism in the yucca moth lineage. Proc Natl Acad Sci USA 89: 2927-2929.

Pennington JT. 1985. The ecology of fertilization of echinoid eggs—the consequences of sperm dilution, adult aggregation, and synchronous spawning. Biol Bull 169: 417-430.

Perez-Tris J, Ramirez A, Telleria JL. 2003. Are Iberian chiffchaffs *Phylloscopus (collybita) brehmii* long-distance migrants? An analysis of flight-related morphology. Bird Study 50: 146-152.

Perry GLW. 2002. Landscapes, space and equilibrium: shifting viewpoints. Progr Phys Geogr 26: 339-359.

Peters RH. 1991. A Critique for Ecology. Cambridge: Cambridge University Press.

Peterson I. 1993. Newton's Clock. Chaos in the Solar System. New York: W.H. Freeman.

Petit RJ, Excoffier L. 2009. Gene flow and species delimitation. Trends Ecol Evol 24: 386-393.

Petru M, Tielborger K, Belkin R, Sternberg M, Jeltsch F. 2006. Life history variation in an annual plant under two opposing environmental constraints along an aridity gradient. Ecography 29: 66-74.

Pettigrew JD, Frost BJ. 1985. A tactile fovea in the Scolopacidae? Brain Behav Evol 26: 185-195.

Pianka ER. 1994. Evolutionary Ecology. 5th ed. New York: HarperCollins College Publishers.

Pickett STA, White PS. 1985. The Ecology of Natural Disturbances and Patch Dynamics. Orlando: Academic Press.

Pickett STA, Kolasa J, Jones CG. 2007. Ecological Understanding. The Nature of Theory and the Theory of Nature. 2nd ed. London: Academic Press.

Pielou EC. 1969. An Introduction to Mathematical Ecology. New York: Wiley-Interscience.

Pielou EC. 1974. Population and Community Ecology. Principles and Methods. New York: Gordon & Breach.

Pielou EC. 1975. Ecological Diversity. New York: John Wiley.

Pielou EC. 1977. Mathematical Ecology. London: John Wiley.

Pielou EC. 1991. After the Ice Age. The Return of Life to Glaciated North America. Chicago: University of Chicago Press.

Pierce GJ, Ollason JG. 1987. Eight reasons why optimal foraging is a complete waste of time. Oikos 49: 111-117.

Piersma T. 1996. Family Scolopacidae (sandpipers, snipes and phalaropes). Pages 444-487 in del Hoyo J, Elliot A, Sargatal J, eds. Handbook of the Birds of the World Vol 3 Hoatzin to Auks. Barcelona: Lynx Edicions.

Piersma T, van Aelst R, Kurk K, Berkhoudt H, Maas LRM. 1998. A new pressure sensory mechanism for prey detection in birds: the use of principles of seabed dynamics? Proc R Soc Lond B Biol Sci 265: 1377-1383.

Pigott CD. 1974. The response of plants to climate and climatic change. Pages 32-44 in Perring F, ed. The Flora of a Changing Britain. Faringdon: E.W. Classey.

Pigott CD. 1975. Experimental studies on the influence of climate on the geographical distribution of plants. Weather 30: 82-90.

Pigott CD. 1992. Are the distributions of species determined by failure to set seed? Pages 203-216 in Marshall C, Grace J, eds. Fruit and Seed Production. Cambridge: Cambridge University Press.

Pilkey OH, Pilkey-Jarvis L. 2007. Useless Arithmetic: Why Environmental Scientists Can't Predict the Future. New York: Columbia University Press.

Pimm SL. 1991. The Balance of Nature? Ecological Issues in the Conservation of Species and Communities. Chicago: University of Chicago Press.

Pimm SL, Lawton JH, Cohen JE. 1991. Food web patterns and their consequences. Nature 350: 669-674.

Pitman RL, Ensor P. 2003. Three forms of killer whales (*Orcinus orca*) in Antarctic waters. J Cetacean Res Manag 5: 131-139.

Pitman RL, Perryman WL, Leroi D, Eilers E. 2007. A dwarf form of killer whale in Antarctica. J Mammal 88: 43-48.

Pitman RL, Durban JW, Greenfelder M, Guinet C, Jorgensen M, Olson PA, Plana J, Tixier P, Towers JR. 2011. Observations of a distinctive morphotype of killer whale (*Orcinus orca*), type D, from subantarctic waters. Polar Biol 34: 303-306.

Plotkin H. 1994. Darwin Machines and the Nature of Knowledge. Concerning Adaptations, Instinct and the Evolution of Intelligence. London: Penguin Books.

Pollard E. 1988. Temperature, rainfall and butterfly numbers. J Appl Ecol 25: 819-828.

Poole RW. 1974. An Introduction to Quantitative Ecology. New York: McGraw-Hill.

Popper KR. 1959. The Logic of Scientific Discovery. London: Hutchinson.

Popper KR. 1963. Conjectures and Refutations. London: Routledge & Kegan Paul.

Popper KR. 1983. Realism and the Aim of Science. From the Postscript to the Logic of Scientific Discovery (1956). London: Hutchinson.

Popple LW, Walter GH. 2010. A spatial analysis of the ecology and morphology of cicadas in the *Pauropsalta annulata* species complex (Hemiptera: Cicadidae). Biol J Linn Soc 101: 553-565.

Popple LW, Walter GH, Raghu S. 2008. The structure of calling songs in the cicada *Pauropsalta annulata* Goding and Froggatt (Hemiptera: Cicadidae): evidence of diverging populations? Evol Ecol 22: 203-215.

Porch N, Elias S. 2000. Quaternary beetles: a review and issues for Australian studies. Aust J Entomol 39: 1-9.

Potts MD. 2003. Drought in a Bornean everwet rain forest. J Ecol 91: 467-474.

Pough FH. 1989. Organismal performance and Darwinian fitness—approaches and interpretations. Physiol Zool 62: 199-236.

Preece PB. 1971a. Contributions to biology of mulga. I. Flowering. Aust J Bot 19: 21-38.

Preece PB. 1971b. Contributions to biology of mulga. II. Germination. Aust J Bot 19: 39-49.

Preece RC. 1997. The spatial response of non-marine Mollusca to past climate changes. Pages 163-177 in Huntley B, Cramer W, Morgan AV,

Prentice HC, Allen JRM, eds. Past and Future Rapid Environmental Changes: The Spatial and Evolutionary Responses of Terrestrial Biota. Berlin: Springer.

Price GJ. 2013. Quaternary. Pages 545-577 in Jell P, ed. The Geology of Queensland. Brisbane: Geological Society of Queensland.

Price PW. 1987. The role of natural enemies in insect populations. Pages 287-312 in Barbosa P, ed. Insect Outbreaks. New York: Academic Press.

Price PW. 1996. Empirical research and factually based theory: what are their roles in entomology? Am Entomol 42: 209-214.

Price PW. 2002. Species interactions and the evolution of biodiversity. Pages 3-25 in Herrera CM, Pellmyr O, eds. Plant-Animal Interactions An Evolutionary Approach. Oxford: Blackwell Science.

Price PW, Waldbauer GP. 1994. Ecological aspects of pest management. Pages 35-72 in Metcalf RL, Luckman WH, eds. Introduction to Insect Pest Management. New York: John Wiley.

Price PW, Denno RF, Eubanks MD, Finke DL, Kaplan I. 2011. Insect Ecology: Behavior, Populations and Communities. Cambridge: Cambridge University Press.

Price T, Lovette IJ, Bermingham E, Gibbs HL, Richman AD. 2000. The imprint of history on communities of North American and Asian warblers. Am Nat 156: 354-367.

Pugnaire FI, Luque MT. 2001. Changes in plant interactions along a gradient of environmental stress. Oikos 93: 42-49.

Radchuk V, Turlure C, Schtickzelle N. 2013. Each life stage matters: the importance of assessing the response to climate change over the complete life cycle in butterflies. J Anim Ecol 82: 275-285.

Radutoiu S, Madsen LH, Madsen EB, Felle HH, Umehara Y, Gronlund M, Sato S, Nakamura Y, Tabata S, Sandal N, Stougaard J. 2003. Plant recognition of symbiotic bacteria requires two LysM receptor-like kinases. Nature 425: 585-592.

Rafter MA, Walter GH. 2013. Mate recognition in the South African citrus thrips *Scirtothrips aurantii* (Faure) and cross-mating tests with populations from Australia and South Africa. Journal of Insect Behavior, DOI 101007/s10905-013-9391-7.

Rafter MA, Hereward JP, Walter GH. 2013b. Species limits, quarantine risk and the intrigue of a polyphagous invasive pest with highly restricted host relationships in its area of invasion. Evol Appl 6: 1195-1207.

Raghu S. 2003. Spatial and temporal partitioning of behaviour by adult dacines: direct evidence for methyl eugenol as a mate rendezvous cue for *Bactrocera cacuminata*. Physiol Entomol 28: 175-184.

Raghu S, Clarke AR, Bradley J. 2002. Microbial mediation of fruit fly—host plant interactions: is the host plant the "centre of activity"? Oikos 97: 319-328.

Railsback LB. 1990. Method of multiple working hypotheses—a chimera—comment. Geology 18: 917-918.

Railsback LB. 2004. T.C. Chamberlin's "Method of Multiple Working Hypotheses": An encapsulation for modern students (see http://www.gly.uga.edu/railsback/railsback_chamberlin.html). Houston Geol Soc Bull 47: 68-69.

Rainey RC. 1989. Migration and Meteorology. Flight Behaviour and the Atmospheric Environment of Locusts and other Migrant Pests. Oxford: Clarendon Press.

Rajapakse CNK, Walter GH. 2007. Polyphagy and primary host plants: oviposition preference versus larval performance in the lepidopteran pest *Helicoverpa armigera*. Arthropod Plant Interact 1: 17-26.

Rajapakse CNK, Walter GH, Moore CJ, Hull CD, Cribb BW. 2006. Host recognition by a polyphagous lepidopteran (*Helicoverpa armigera*): primary host plants, host produced volatiles and neurosensory stimulation. Physiol Entomol 31: 270-277.

Rapport DJ. 1991. Myths in the foundations of economics and ecology. Biol J Linn Soc 44: 185-202.

Raup DM. 1991. A kill curve for Phanerozoic marine species. Paleobiology 17: 37-48.

Raup DM, Sepkoski JJ. 1982. Mass extinctions in the marine fossil record. Science 215: 1501-1503.

Raup DM, Gould SJ, Schopf TJM, Simberloff DS. 1973. Stochastic models of phylogeny and the evolution of diversity. J Geol 81: 525-542.

Raven PH, Evert RF, Eichhorn SE. 1999. Biology of Plants, 6th ed. New York: W.H. Freeman.

Read J, Hill RS. 1985. Dynamics of *Nothofagus*-dominated rainforest on mainland Australia and lowland Tasmania. Vegetatio 63: 67-78.

Reading RP, Mix H, Lhagvasuren B, Blumer ES. 1999. Status of wild Bactrian camels and other large ungulates in south-western Mongolia. Oryx 33: 247-255.

Ready E. 2013. Neandertal foraging during the late Mousterian in the Pyrenees: new insights based on faunal remains from Gatzarria Cave. J Archaeol Sci 40: 1568-1578.

Reddingius J. 1971. Gambling for existence. Acta Biotheor 20, Supplement 1: 1-208.

Reddingius J, Den Boer PJ. 1970. Simulation experiments illustrating stabilization of animal numbers by spreading of risk. Oecologia 5: 240-284.

Remy W, Taylor TN, Hass H, Kerp H. 1994. Four hundred-million-year-old vesicular arbuscular Mycorrhizae. Proc Natl Acad Sci USA 91: 11841-11843.

Reynolds HW, Gates CC, Glaholt RD. 2003. Bison (*Bison bison*). Pages 1009-1060 in Feldhamer GA, Thompson BC, Chapman JA, eds. Wild Mammals of North America Biology, Management, and Conservation 2nd ed. Baltimore: Johns Hopkins University Press.

Richards OW. 1928. Potentially unlimited multiplication of yeast with constant environment, and the limiting of growth by changing environment. J Gen Physiol 11: 525-538.

Ricklefs RE. 1987. Community diversity: relative roles of local and regional processes. Science 235: 167-171.

Ricklefs RE. 1989. Speciation and diversity: the integration of local and regional processes. Pages 599-622 in Otte D, Endler JA, eds. Speciation and Its Consequences. Sunderland: Sinauer.

Ricklefs RE. 2008. The Economy of Nature. 6th ed. New York: W.H. Freeman.

Ridley M. 1994. The Red Queen: Sex and the Evolution of Human Nature. Harmondsworth: Penguin.

Ritchie PA, Millar CD, Gibb GC, Baroni C, Lambert DM. 2004. Ancient DNA enables timing of the Pleistocene origin and Holocene expansion of two Adelie penguin lineages in Antarctica. Mol Biol Evol 21: 240-248.

Robertson BC, Chilvers BL, Duignan PJ, Wilkinson IS, Gemmell NJ. 2006. Dispersal of breeding, adult male *Phocarctos hookeri*: Implications for disease transmission, population management and species recovery. Biol Conserv 127: 227-236.

Robin J-P, Denis V. 1999. Squid stock fluctuations and water temperature: temporal analysis of English Channel Loliginidae. J Appl Ecol 36: 101-110.

Robinson RA, Siriwardena GM, Crick HQP. 2005. Size and trends of the House Sparrow *Passer domesticus* population in Great Britain. Ibis 147: 552-562.

Rockwell LR. 2006. Introduction to Population Ecology. Oxford: Blackwell Publishing.

Rodwell MJ, Drevillon M, Frankignoul C, Hurrell JW, Pohlmann H, Stendel M, Sutton RT. 2004. North Atlantic forcing of climate and its uncertainty from a multi-model experiment. QJR Meteorol Soc 130: 2013-2032.

Roff DA. 1990. The evolution of flightlessness in insects. Ecol Monogr 60: 389-421.

Rohde K. 1979. A critical evaluation of intrinsic and extrinsic factors responsible for niche restriction in parasites. Am Nat 114: 648-671.

Rohde K. 2005. Nonequilibrium Ecology. Cambridge: Cambridge University Press.

Rohde K, (Ed.). 2013. The Balance of Nature and Human Impact. Cambridge: Cambridge University Press.

Rombola E, Marschoff E, Coria N. 2003. Comparative study of the effects of the late pack-ice break-off on chinstrap and Adelie penguins' diet and reproductive success at Laurie Island, South Orkney Islands, Antarctica. Polar Biol 26: 41-48.

Rose G. 1992. The Strategy of Preventive Medicine. Oxford: Oxford University Press.

Rose MR, Mueller LD. 2006. Evolution and Ecology of the Organism. Upper Saddle River, New Jersey: Pearson Prentice Hall.

Rose S. 1997. Lifelines. Biology, Freedom, Determinism. London: Penguin Books.

Rosenberg A. 1985. The Structure of Biological Science. Cambridge: Cambridge University Press.

Rosenberg LJ, Magor JI. 1987. Predicting windborne displacements of the brown planthopper, *Nilaparvata lugens* from synoptic weather data. I. Long-distance displacements in the North-east monsoon. J Anim Ecol 56: 39-51.

Roth D, Roland J, Roslin T. 2006. Parasitoids on the loose—experimental lack of support of the parasitoid movement hypothesis. Oikos 115: 277-285.

Roughgarden J. 1998. Primer of Ecological Theory. Upper Saddle River, New Jersey: Prentice Hall.

Ruelle D. 1991. Chance and Chaos. Princeton: Princeton University Press.

Rule S, Brook BW, Haberle SG, Turney CSM, Kershaw AP, Johnson CN. 2012. The aftermath of megafaunal extinction: ecosystem transformation in Pleistocene Australia. Science 335: 1483-1486.

Rundle HD, Nosil P. 2005. Ecological speciation. Ecol Lett 8: 336-352.

Russell D. 2005. Striper Wars. An American Fish Story. Washington, DC: Island Press, Shearwater Books.

Rutberg AT. 1984. Birth synchrony in American bison (*Bison bison*)—response to predation or season? J Mammal 65: 418-423.

Saccheri I, Hanski I. 2006. Natural selection and population dynamics. Trends Ecol Evol 21: 341-347.

Saccheri IJ, Boggs CL, Hanski I, Ehrlich PR. 2004. Genetics of checkerspot populations. Pages 199-218 in Ehrlich PR, Hanski I, eds. On the Wings of Checkerspots: A Model System for Population Biology. Oxford: Oxford University Press.

Saether BE, Engen S. 2004. Stochastic population theory faces reality in the laboratory. Trends Ecol Evol 19: 351-353.

Sagarin RD. 2006. Recent studies improve understanding of population dynamics across species ranges. Oikos 115: 386-388.

Sagarin RD, Gaines SD. 2002. The 'abundant centre' distribution: to what extent is it a biogeographical rule? Ecol Lett 5: 137-147.

Sagarin RD, Gaines SD, Gaylord B. 2006. Moving beyond assumptions to understand abundance distributions across the ranges of species. Trends Ecol Evol 21: 524-530.

Sale PF. 1988. Perception, pattern, chance and the structure of reef fish communities. Environ Biol Fishes 21: 3-15.

Salomon M, Voisin JF, Bried J. 2003. On the taxonomic status and denomination of the Iberian chiffchaffs. Ibis 145: 87-97.

Saloniemi I. 1993. An environmental explanation for the character displacement pattern in *Hydrobia* snails. Oikos 67: 75-80.

Salt G. 1968. The resistance of insect parasitoids to the defense reactions of their hosts. Biol Rev 43: 200-232.

Sardella BA, Cooper J, Gonzalez RJ, Brauner CJ. 2004. The effect of temperature on juvenile Mozambique tilapia hybrids (*Oreochromis mossambicus* × *O. urolepis homorum*) exposed to full-strength and hypersaline seawater. Comp Biochem Physiol Part A Mol Integr Physiol 137: 621-629.

Sardella BA, Kultz D, Cech JJ, Brauner CJ. 2008. Salinity-dependent changes in Na^+/K^{+-}ATPase content of mitochondria-rich cells contribute to differences in thermal tolerance of Mozambique tilapia. J Comp Physiol B Biochem Syst Environ Physiol 178: 249-256.

Savill NJ, St Rose SG, Keeling MJ, Woolhouse MEJ. 2006. Silent spread of H5N1 in vaccinated poultry. Nature 442: 757.

Sayaboc PD. 1994. Investigation of the behaviour and host relationships of Anisopteromalus calandrae (Howard) (Hymenoptera: Pteromalidae), a parasitoid of stored grain beetles. Unpublished MAgrSci thesis. 127 pp. Brisbane: The University of Queensland.

Schluter D. 2009. Evidence for ecological speciation and its alternative. Science 323: 737-741.

Schmidt DN, Thierstein HR, Bollmann J, Schiebel R. 2004. Abiotic forcing of plankton evolution in the Cenozoic. Science 303: 207-210.

Schmidt-Nielsen K. 1990. Animal Physiology: Adaptations and Environment. Cambridge: Cambridge University Press.

Schoener TW. 1986a. Mechanistic approaches to community ecology: a new reductionism. Am Zool 26: 81-106.

Schoener TW. 1986b. Overview: kinds of ecological communities—ecology becomes pluralistic. Pages 467-479 in Diamond JM, Case TJ, eds. Community Ecology. New York: Harper & Row.

Schoenly K, Reid W. 1987. Dynamics of heterotrophic succession in carrion arthropod assemblages: discrete series or a continuum of change. Oecologia 73: 192-202.

Schoenly KG, Reid W. 1989. Dynamics of heterotrophic succession in carrion revisited. A reply to Boulton and Lake (1988). Oecologia 79: 140-142.

Schonswetter P, Popp M, Brochmann C. 2006. Rare arctic-alpine plants of the European Alps have different immigration histories: the snow bed species *Minuartia biflora* and *Ranunculus pygmaeus*. Mol Ecol 15: 709-720.

Schowalter TD. 2000. Insect Ecology. An Ecosystem Approach. San Diego: Academic Press.

Schulze E-D, Beck E, Müller-Hohenstein K. 2005. Plant Ecology. Heidelberg: Springer.

Schultz TD. 1998. The utilization of patchy thermal microhabitats by the ectothermic insect predator, *Cicindela sexguttata*. Ecol Entomol 23: 444-450.

Schwartz CC, Hundertmark KJ. 1993. Reproductive characteristics of Alaskan moose. J Wildl Manag 57: 454-468.

Schwerdtfeger W, (Ed.). 1976. Climate of Central and South America. New York: Elsevier.

Sedgwick JA. 2004. Site fidelity, territory fidelity, and natal philopatry in willow flycatchers (*Empidonax traillii*). Auk 121: 1103-1121.

Shapiro AM. 1986. r-K selection at various taxonomic levels in the pierine butterflies of North and South America. Pages 135-152 in Taylor F, Karban R, eds. The Evolution of Insect Life Cycles. New York: Springer-Verlag.

Sharp GD. 1992. Climate change, the Indian Ocean tuna fishery, and empiricism. Pages 378-416 in Glantz MH, ed. Climate Variability, Climate Change and Fisheries. Cambridge: Cambridge University Press.

Shel'deshova GG. 1967. Ecological factors determining distribution of the codling moth *Laspeyresia pomonella* L. (Lepidoptera, Tortricidae) in the northern and southern hemispheres. Entomol Rev 46: 349-361.

Shepherd LD, Millar CD, Ballard G, Ainley DG, Wilson PR, Haynes GD, Baroni C, Lambert DM. 2005. Microevolution and mega-icebergs in the Antarctic. Proc Natl Acad Sci USA 102: 16717-16722.

Sherman PW, Reeve HK, Pfennig DW. 1997. Recognition systems. Pages 69-96 in Krebs JR, Davies NB, eds. Behavioural Ecology: An Evolutionary Approach. Oxford: Blackwell Science.

Shreeve TG, Dennis RLH, Van Dyck H. 2004. Resources, habitats and metapopulations—whither reality? Oikos 106: 404-408.

Simberloff D. 1980. A succession of paradigms in ecology: essentialism to materialism and probabilism. Pages 63-99 in Saarinen E, ed. Conceptual Issues in Ecology. Dordrecht: D. Reidel.

Simberloff D. 1982. The status of competition theory in ecology. Ann Zool Fenn 19: 241-253.

Simberloff D. 1983. Competition theory, hypothesis testing, and other community ecological buzzwords. Am Nat 122: 626-635.

Simberloff D. 1984. The great god of competition. The Sciences 24: 17-22.

Simberloff D. 2004. Community ecology: Is it time to move on? Am Nat 163: 787-799.

Simmons AD, Thomas CD. 2004. Changes in dispersal during species' range expansions. Am Nat 164: 378-395.

Sinclair ARE. 1977. The African Buffalo: A Study of Resource Limitation of Populations. Chicago: University of Chicago Press.

Sinclair M, Solemdal P. 1988. The development of "population thinking" in fisheries biology between 1878 and 1930. Aquat Living Resour 1: 189-213.

Singer MC, Ehrlich PR. 1979. Population dynamics of the checkerspot butterfly *Euphydryas editha*. Fortschr Zool 25: 53-60.

Smith ADM, Maelzer DA. 1986. Aggregation of parasitoids and density-independence of parasitism in field populations of the wasp *Aphytis melinus* and its host, the red scale *Aonidiella aurantii*. Ecol Entomol 11: 425-434.

Smith JJB. 1985. Feeding mechanisms. Pages 33-85 in Kerkut GA, Gilbert LI, eds. Comprehensive Insect Physiology, Biochemistry and Pharmacology Vol 4 Regulation: Digestion, Nutrition, Excretion. Oxford: Pergamon Press.

Smith MA, Green DM. 2006. Sex, isolation and fidelity: unbiased long-distance dispersal in a terrestrial amphibian. Ecography 29: 649-658.

Smith TM, Smith RL. 2012. Elements of Ecology, 8th ed. Boston: Benjamin Cummings.

Smithers RHN. 1983. The Mammals of the Southern African Subregion. Pretoria: University of Pretoria.

Snow EL, Walter GH. 2007. Large seeds, extinct vectors and contemporary ecology: testing dispersal in a locally distributed cycad, *Macrozamia lucida* (Cycadales). Aust J Bot 55: 592-600.

Sol D. 2008. Artificial selection, naturalization, and fitness: Darwin's pigeons revisited. Biol J Linn Soc 93: 657-665.

Solomon ME. 1949. The natural control of animal populations. J Anim Ecol 18: 1-35.

Sommer RS, Benecke N. 2005. The recolonization of Europe by brown bears *Ursus arctos* Linnaeus, 1758 after the Last Glacial Maximum. Mammal Rev 35: 156-164.

Sommer RS, Nadachowski A. 2006. Glacial refugia of mammals in Europe: evidence from fossil records. Mammal Rev 36: 251-265.

Southwood TRE. 1977. Habitat, the templet for ecological strategies? J Anim Ecol 46: 337-365.

Southwood TRE, Comins HN. 1976. A synoptic population model. J Anim Ecol 45: 949-965.

Southwood TRE, Hassell MP, Reader PM, Rogers DJ. 1989. Population dynamics of the viburnum whitefly (*Aleurotrachelus jelinekii*). J Anim Ecol 58: 921-942.

Spencer HG, McArdle BH, Lambert DM. 1986. A theoretical investigation of speciation by reinforcement. Am Nat 128: 241-262.

Spencer HG, Lambert DM, McArdle BH. 1987. Reinforcement, species, and speciation: a reply to Butlin. Am Nat 130: 958-962.

Spomer GG. 1973. The concept of "interaction" and "operational environment" in environmental analysis. Ecology 54: 200-204.

Spoor F, Wood B, Zonneveld F. 1994. Implications of early hominid labyrinthine morphology for evolution of human bipedal locomotion. Nature 369: 645-648.

Spoor F, Wood B, Zonneveld F. 1996. Evidence for a link between human semicircular canal size and bipedal behaviour. J Hum Evol 30: 183-187.

Spoor F, Hublin JJ, Braun M, Zonneveld F. 2003. The bony labyrinth of Neanderthals. J Hum Evol 44: 141-165.

State of New Jersey, Department of Environmental Protection, Division of Fish & Wildlife. 1996-2005. Biology of the white-tailed deer *Odocoileus virginianus*. Available at: http://www.state.nj.us/dep/fgw/deerbiol.htm. Accessed Aug. 2013.

Stearns SC. 1992. The Evolution of Life Histories. Oxford: Oxford University Press.

Stebbins GL. 1974. Flowering Plants. Evolution above the Species Level. Cambridge, MA: Harvard University Press.

Stehlik I. 2003. Resistance or emigration? Response of alpine plants to the ice ages. Taxon 52: 499-510.

Stehlik I, Schneller JJ, Bachmann K. 2001. Resistance or emigration: response of the high-alpine plant *Eritrichium nanum* (L.) Gaudin to the ice age within the Central Alps. Mol Ecol 10: 357-370.

Steinberg PD, De Nys R, Kjelleberg S. 2002. Chemical cues for surface colonization. J Chem Ecol 28: 1935-1951.

Stenseth NC, Chan KS, Tavecchia G, Coulson T, Mysterud A, Clutton-Brock T, Grenfell B. 2004. Modelling non-additive and nonlinear signals from climatic noise in ecological time series: Soay sheep as an example. Proc R Soc Lond B Biol Sci 271: 1985-1993.

Stiling P. 1988. Density-dependent processes and key factors in insect populations. J Anim Ecol 57: 581-593.

Stiling PD. 1987. The frequency of density dependence in insect host-parasitoid systems. Ecology 68: 844-856.

Stireman JO, Dyer LA, Janzen DH, Singer MS, Li JT, Marquis RJ, Ricklefs RE, Gentry GL, Hallwachs W, Coley PD, Barone JA, Greeney HF, Connahs H, Barbosa P, Morais HC, Diniz IR. 2005. Climatic unpredictability and parasitism of caterpillars: Implications of global warming. Proc Natl Acad Sci USA 102: 17384-17387.

Stockey A, Hunt R. 1994. Predicting secondary succession in wetland mesocosms on the basis of autecological information on seed and seedlings. J Appl Ecol 31: 543-559.

Strohm E, Linsenmair KE. 1998. Temperature dependence of provisioning behaviour and investment allocation in the European beewolf *Philanthus triangulum* F. Ecol Entomol 23: 330-339.

Strong DR. 1988. Parasitoid theory: from aggregation to dispersal. Trends Ecol Evol 3: 277-280.

Strong DR, Simberloff DS. 1981. Straining at gnats and swallowing ratios: character displacement. Evolution 35: 810-812.

Strong DR, Szyska LA, Simberloff DS. 1979. Tests of community-wide character displacement against null hypotheses. Evolution 33: 897-913.

Summers HA, Smith BN, Hansen LD. 2009. Comparison of respiratory and growth characteristics of two co-occurring shrubs from a cold desert, *Coleogyne ramosissima* (blackbrush) and *Atriplex confertifolia* (shadscale). J Arid Environ 73: 1-6.

Sundell J, Huitu O, Henttonen H, Kaikusalo A, Korpimaki E, Pietiainen H, Saurola P, Hanski I. 2004. Large-scale spatial dynamics of vole populations in Finland revealed by the breeding success of vole-eating avian predators. J Anim Ecol 73: 167-178.

Sung SB, Amasino RM. 2004. Vernalization and epigenetics: how plants remember winter. Curr Opin Plant Biol 7: 4-10.

Sutherland WJ. 1996. From Individual Behaviour to Population Ecology. Oxford: Oxford University Press.

Sutton RT, Hodson DLR. 2005. Atlantic Ocean forcing of North American and European summer climate. Science 309: 115-118.

Sutton RT, Mathieu PP, Collins M, Dong B. 2005. Predictability of winter climate over the North Atlantic European region during ENSO events—Reply. J Clim 18: 2773-2774.

Svensson E, Hedenstrom A. 1999. A phylogenetic analysis of the evolution of moult strategies in Western Palearctic warblers (Aves: Sylviidae). Biol J Linn Soc 67: 263-276.

Swanson RL, Williamson JE, De Nys R, Kumar N, Bucknall MP, Steinberg PD. 2004. Induction of settlement of larvae of the sea urchin *Holopneustes purpurascens* by histamine from a host alga. Biol Bull 206: 161-172.

Swanson RL, de Nys R, Huggett MJ, Green JK, Steinberg PD. 2006. In situ quantification of a natural settlement cue and recruitment of the Australian sea urchin *Holopneustes purpurascens*. Mar Ecol Prog Ser 314: 1-14.

Sylvan R. 1994. Illusion and illogic in evolution. Riv Biol-Biol Forum 87: 191-221.

Taggart C.T, Anderson J, Bishop C, Colbourne E, Hutchings J, Lilly G, Morgan J, Murphy E, Myers R, Rose G, Shelton P. 1994. Overview of cod stocks, biology, and environment in the Northwest Atlantic Region of Newfoundland, with emphasis on northern cod. ICES Mar Sci Symp 198: 140-157.

Tagu D, Lapeyrie F, Martin F. 2002. The ectomycorrhizal symbiosis: genetics and development. Plant Soil 244: 97-105.

Tanner JE, Hughes TP, Connell JH. 1996. The role of history in community dynamics: a modelling approach. Ecology 77: 108-117.

Tansley AG. 1935. The use and abuse of vegetational concepts and terms. Ecology 16: 284-307.

Taplin LE, Grigg GC. 1981. Salt-glands in the tongue of the estuarine crocodile *Crocodylus porosus*. Science 212: 1045-1047.

Tatarenkov A, Johannesson K. 1998. Evidence of a reproductive barrier between two forms of the marine periwinkle *Littorina fabalis* (Gastrapoda). Biol J Linn Soc 63: 349-365.

Tatarenkov A, Johannesson K. 1999. Micro- and macrogeographic allozyme variation in *Littorina fabalis;* do sheltered and exposed forms hybridize? Biol J Linn Soc 67: 199-212.

Taulman JF, Robbins LW. 1996. Recent range expansion and distributional limits of the nine-banded armadillo (*Dasypus novemcinctus*) in the United States. J Biogeogr 23: 635-648.

Taylor KJ, Lowe AJ, Hunter RJ, Ridgway T, Gresshoff PM, Rossetto M. 2005. Genetic diversity and regional identity in the Australian remnant *Nothofagus moorei*. Aust J Bot 53: 437-444.

Taylor LL, Leake JR, Quirk J, Hardy K, Banwart SA, Beerling DJ. 2009. Biological weathering and the long-term carbon cycle: integrating mycorrhizal evolution and function into the current paradigm. Geobiol 7: 171-191.

Taylor LR. 1986. Synoptic dynamics, migration and the Rothamsted insect survey. J Anim Ecol 55: 1-38.

Taylor LR, Taylor RAJ. 1983. Insect migration as a paradigm for survival by movement. Pages 181-214 in Swingland IR, Greenwood PJ, eds. The Ecology of Animal Movement. Oxford: Clarendon Press.

Taylor RH, Barton KJ, Wilson PR, Thomas BW, Karl BJ. 1995. Population status and breeding of New Zealand fur seals (*Arctocephalus forsteri*) in the Nelson-Northern Marlborough region, 1991-94. N Z J Mar Freshw Res 29: 223-234.

Terry I, Bradley JR, van Duyn JW. 1989. Establishment of early instar *Heliothis zea* on soybeans. Entomol Exp Appl 51: 233-240.

Terry I, Walter GH, Moore C, Roemer R, Hull C. 2007. Odor-mediated push-pull pollination in cycads. Science 318: 70-70.

Thiery A. 1997. Horizontal distribution and abundance of cysts of several large branchiopods in temporary pool and ditch sediments. Hydrobiologia 359: 177-189.

Thomas CD, Singer MC. 1998. Scale-dependent evolution of specialization in a checkerspot butterfly: from individuals to metapopulations and ecotypes. Pages 343-374 in Mopper S, Strauss SY, eds. Genetic Structure and Local Adaptation in Natural Insect Populations: Effects of Ecology, Life History, and Behavior. New York: Chapman & Hall.

Thomas PB, Morris EC, Auld TD. 2007. Response surfaces for the combined effects of heat shock and smoke on germination of 16 species forming soil seed banks in south-east Australia. Austral Ecol 32: 605-616.

Thompson JN. 1988. Evolutionary ecology of the relationship between oviposition preference and performance of offspring in phytophagous insects. Entomol Exp Appl 47: 3-14.

Thompson JN. 1994. The Coevolutionary Process. Chicago: University of Chicago Press.

Thompson JN. 2005. The Geographic Mosaic of Coevolution. Chicago: University of Chicago Press.

Thompson JN. 2009a. The coevolving web of life. Am Nat 173: 125-140.

Thompson JN. 2009b. Which ecologically important traits are most likely to evolve rapidly? Oikos 118: 1281-1283.

Thompson PA, Cox SA. 1978. Germination of the bluebell (*Hyacinthoides non-scripta* (L.) Chouard) in relation to distribution and habitat. Ann Bot 42: 51-62.

Thygerson T, Harris JM, Smith BN, Hansen LD, Pendleton RL, Booth DT. 2002. Metabolic response to temperature for six populations of winterfat (*Eurotia lanata*). Thermochim Acta 394: 211-217.

Tiainen J, Hanski IK. 1985. Wing shape variation of Finnish and Central European willow warbler *Phylloscopus trochilus* and chiffchaffs *P. collybita*. Ibis 127: 365-371.

Tilman D, Lehman CL, Kareiva P. 1997. Population dynamics in spatial habitats. Pages 3-20 in Tilman D, Kareiva P, eds. Spatial Ecology: The Role of Space in Population Dynamics and Interspecific Interactions. Princeton: Princeton University Press.

Townsend CR, Begon M, Harper JL. 2008. Essentials of Ecology. 3rd ed. Oxford: Blackwell Publishing.

Tregenza T, Wedell N, Chapman T. 2006. Introduction. Sexual conflict: a new paradigm? Philos Trans R Soc Lond B Biol Sci 361: 229-234.

Trueman CNG, Field JH, Dortch J, Charles B, Wroe S. 2005. Prolonged coexistence of humans and megafauna in Pleistocene Australia. Proc Natl Acad Sci USA 102: 8381-8385.

Truong C, Palme AE, Felber F. 2006. Recent invasion of the mountain birch *Betula pubescens* ssp. *tortuosa* above the treeline due to climate change: genetic and ecological study in northern Sweden. J Evol Biol 20: 369-380.

Tulgat R, Schaller GB. 1992. Status and distribution of wild bactrian camels *Camelus bactrianus ferus*. Biol Conserv 62: 11-19.

Turchin P. 1999. Population regulation: a synthetic view. Oikos 84: 153-159.

Turchin P. 2001. Does population ecology have general laws? Oikos 94: 17-26.

Turchin P. 2003. Complex Population Dynamics: A Theoretical/Empirical Synthesis. Princeton, NJ: Princeton University Press.

Turin H, den Boer PJ. 1988. Changes in the distribution of carabid beetles in The Netherlands since 1880. II. Isolation of habitats and long-term time trends in the occurrence of carabid species with different powers of dispersal (Coleoptera, Carabidae). Biol Conserv 44: 179-200.

Turin H, Haeck J, Hengeveld R. 1977. Atlas of the Carabid Beetles of The Netherlands. Amsterdam: North-Holland Publishing Company.

Turner A, Paterson HEH. 1991. Species and speciation: evolutionary tempo and mode in the fossil record reconsidered. Geobios 24: 761-769.

Tyndale-Biscoe H. 1973. Life of Marsupials. London: Edward Arnold.

Tyndale-Biscoe H. 2005. Life of Marsupials. Collingwood: CSIRO publishing.

Urban MC, Leibold MA, Amarasekare P, De Meester L, Gomulkiewicz R, Hochberg ME, Klausmeier CA, Loeuille N, de Mazancourt C, Norberg J, Pantel JH, Strauss SY, Vellend M, Wade MJ. 2008. The evolutionary ecology of metacommunities. Trends Ecol Evol 23: 311-317.

Valverde F, Mouradov A, Soppe W, Ravenscroft D, Samach A, Coupland G. 2004. Photoreceptor regulation of CONSTANS protein in photoperiodic flowering. Science 303: 1003-1006.

van Dam JA, Aziz HA, Sierra MAA, Hilgen FJ, Ostende L, Lourens LJ, Mein P, van der Meulen AJ, Pelaez-Campomanes P. 2006. Long-period astronomical forcing of mammal turnover. Nature 443: 687-691.

van de Kam J, Ens B, Piersma T, Zwarts L. 1999. Ecologische Atlas van de Nederlandse Wadvogels. Haarlem: Schuyt.

van den Bosch F, Hengeveld R, Metz JAJ. 1992. Analysing the velocity of animal range expansion. J Biogeogr 19: 135-150.

van der Laan PA. 1959. Correlation between rainfall in the dry season and the occurrence of the white rice borer (*Scirpophaga innotata* Wlk.) in Java. Entomol Exp Appl 2: 12-20.

van Huizen THP. 1977. The significance of flight activity in the life cycle of *Amara plebeja* Gyll. (Coleoptera, Carabidae). Oecologia 29: 27-41.

van Hulst R. 1979. On the dynamics of vegetation: Markov chains as models of succession. Vegetatio 40: 3-14.

van Klinken RD, Walter GH. 1996. The ecology of organisms that breed in a divided and ephemeral habitat: insects of fallen rainforest fruit. Acta Oecol 17: 405-420.

van Oldenborgh GJ. 2005. Comments on "Predictability of winter climate over the North Atlantic European region during ENSO events". J Clim 18: 2773-2774.

Vaquero M, Vallverdu J, Rosell J, Pasto I, Allue E. 2001. Neandertal behavior at the Middle Palaeolithic site of Abric Romani, Capellades, Spain. J Field Archaeol 28: 93-114.

Varley GC, Gradwell GR, Hassell MP. 1973. Insect Population Ecology: an Analytical Approach. Oxford: Blackwell Scientific Publications.

Velasco LRI, Walter GH. 1992. Availability of different host plant species and changing abundance of the polyphagous bug *Nezara viridula* (Hemiptera: Pentatomidae). Environ Entomol 21: 751-759.

Verheyden C, Jouventin P. 1994. Olfactory behavior of foraging procellarii-forms. Auk 111: 285-291.

Vieira A, Miller DJ. 2006. Gamete interaction: Is it species-specific? Mol Reprod Dev 73: 1422-1429.

Vijaysegaran S, Walter GH, Drew RAI. 1997. Mouthpart structure, feeding mechanisms, and natural food sources of adult *Bactrocera* (Diptera: Tephritidae). Ann Entomol Soc Am 90: 184-201.

Vinson SB, Bin F, Vet LEM. 1998. Critical issues in host selection by insect parasitoids. Biol Control 11: 77-78.

Vitousek P. 2006. Ecosystem science and human-environment interactions in the Hawaiian archipelago. J Ecol 94: 510-521.

Vitousek PM, Walker LR, Whiteaker LD, Mueller-Dombois D, Matson PA. 1987. Biological invasion by *Myrica faya* alters ecosystem development in Hawaii. Science 238: 802-804.

Volis S, Mendlinger S, Ward D. 2002. Differentiation in populations of *Hordeum spontaneum* along a gradient of environmental productivity and predictability: life history and local adaptation. Biol J Linn Soc 77: 479-490.

Voous KH. 1962. Die Vogelwelt Europas und ihre Verbreitung. Ein tiergeographischer Atlas über die Lebensweise aller in Europa brütenden Vögel. Hamburg: Paul Parey.

Vors LS, Boyce MS. 2009. Global declines of caribou and reindeer. Global Change Biol 15: 2626-2633.

Vrba ES. 1980. Evolution, species and fossils: how does life evolve? S Afr J Sci 76: 61-84.

Vrba ES. 1984. Evolutionary pattern and process in the sister-group Alcelaphini—Aepycerotini (Mammalia: Bovidae). Pages 62-79 in Eldredge N, Stanley SM, eds. Living Fossils. New York: Springer Verlag.

Vrba ES. 1987. Ecology in relation to speciation rates: some case histories of Miocene-Recent mammal clades. Evol Ecol 1: 283-300.

Vrba ES, DeGusta D. 2004. Do species populations really start small? New perspectives from the Late Neogene fossil record of African mammals. Philos Trans R Soc Lond B Biol Sci 359: 285-292.

Waage JK, Hassell MP. 1982. Parasitoids as biological control agents—a fundamental approach. Pages 5-144 in Anderson RM, Canning EU, eds. Parasites as Biological Control Agents. Cambridge: Cambridge University Press.

Wada T, Seino H, Ogaea Y, Nakasuga T. 1987. Evidence of autumn overseas migration in the rice planthoppers, *Nilaparvata lugens* and *Sogatella furcifera*: analysis of light trap catches and associated weather patterns. Ecol Entomol 12: 321-330.

Wainwright PC, Reilly SM, (Eds). 1994. Ecological Morphology. Integrative Organismal Biology. Chicago: University of Chicago Press.

Walde SJ. 1994. Immigration and the dynamics of a predator-prey interaction in biological control. J Anim Ecol 63: 337-346.

Walker D. 1970. Direction and rate in some British post-glacial hydroseres. Pages 117-137 in Walker D, West R, eds. Studies in the Vegetational History of the British Isles. Cambridge: Cambridge University Press.

Wallace B. 1968. Topics in Population Genetics. New York: W.W. Norton.

Wallner WE. 1987. Factors affecting insect population dynamics: differences between outbreak and non-outbreak species. Annu Rev Entomol 32: 317-320.

Walls SC. 2009. The role of climate in the dynamics of a hybrid zone in Appalachian salamanders. Global Change Biol 15: 1903-1910.

Walsh RPD. 1996. Climate. Pages 159-205 in Richards PW, ed. The Tropical Rain Forest: An Ecological Study. Cambridge: Cambridge University Press.

Walsh RPD, Newbery DM. 1999. The ecoclimatology of Danum, Sabah, in the context of the world's rainforest regions, with particular reference to dry periods and their impact. Philos Trans R Soc Lond B Biol Sci 354: 1869-1883.

Walter GH. 1988a. Competitive exclusion, coexistence and community structure. Acta Biotheor 37: 281-313.

Walter GH. 1988b. Activity patterns and egg production in *Coccophagus bartletti*, an aphelinid parasitoid of scale insects. Ecol Entomol 13: 95-105.

Walter GH. 1991. What is resource partitioning? J Theor Biol 150: 137-143.

Walter GH. 1993a. The concept of interaction in ecological theory. Pages 131-148 in Kull K, Tiivel T, eds. Lectures in Theoretical Biology: The Second Stage. Tallinn: Estonian Academy of Sciences.

Walter GH. 1993b. Oviposition behaviour of diphagous parasitoids (Hymenoptera, Aphelinidae): a case of intersexual resource partitioning? Behaviour 124: 73-87.

Walter GH. 1995. Species concepts and the nature of ecological generalizations about diversity. Pages 191-224 in Lambert DM, Spencer HG, eds. Speciation and the Recognition Concept: Theory and Application. Baltimore: Johns Hopkins University Press.

Walter GH. 1999. Book review: Genetic structure and local adaptation in natural insect populations. Effects of ecology, life history, and behavior. Aust J Entomol 38: 45-46.

Walter GH. 2003. Insect Pest Management and Ecological Research. Cambridge: Cambridge University Press.

Walter GH. 2006. Book review: The Geographic Mosaic of Coevolution. Austral Ecol 31: 547-548.

Walter GH. 2008. Individuals, populations and the balance of nature: the question of persistence in ecology. Biol Philos 23: 417-438.

Walter GH. 2013. Autecology and the balance of nature—ecological laws and human-induced invasions. Pages 337-356 in Rohde K, ed. The Balance of Nature and Human Impact. Cambridge: Cambridge University Press.

Walter GH, Benfield MD. 1994. Temporal host plant use in three polyphagous Heliothinae, with special reference to *Helicoverpa punctigera* (Wallengren) (Noctuidae: Lepidoptera). Aust J Ecol 19: 458-465.

Walter GH, Donaldson JS. 1994. Heteronomous hyperparasitoids, sex ratios and adaptations. Ecol Entomol 19: 89-92.

Walter GH, Paterson HEH. 1994. The implications of palaeontological evidence for theories of ecological communities and species richness. Aust J Ecol 19: 241-250.

Walter GH, Paterson HEH. 1995. Levels of understanding in ecology: interspecific competition and community ecology. Aust J Ecol 20: 463-466.

Walter GH, Zalucki MP. 1999. Rare butterflies and theories of evolution and ecology. Pages 349-368 in Kitching RL, Scheermeyer E, Jones RE, Pierce NE, eds. Biology of Australian Butterflies. Melbourne: CSIRO Publishing.

Walter GH, Hengeveld R. 2000. The structure of the two ecological paradigms. Acta Biotheor 48: 15-46.

Walter GH, Hulley PE, Craig AJFK. 1984. Speciation, adaptation and interspecific competition. Oikos 43: 246-248.

Walter H, Walter E. 1953. Einige allgemeine Ergebnisse unserer Forschungsreise nach Südwestafrika 1952/53: Das Gesetz der relativen Standortskonstantz; das Wesen der Pflanzengesellschaften. Ber Deutsch Bot Ges 66: 228-236.

Wang JY. 1960. A critique of the heat unit approach to plant response studies. Ecology 41: 785-790.

Wang JY. 1962. The influence of seasonal temperature ranges on pea production. J Am Soc Hortic Sci 80: 436-448.

Wang JY. 1963. A graphical solution to temperature-moisture responses of tomato yield. J Am Soc Hortic Sci 82: 429-445.

Wang JY. 1972. Agricultural Meteorology. San Jose: Milieu Information Service.

Wanntorp H-E. 1983. Historical constraints in adaptation theory: traits and non-traits. Oikos 41: 157-160.

Waples RS, Gaggiotti O. 2006. What is a population? An empirical evaluation of some genetic methods for identifying the number of gene pools and their degree of connectivity. Mol Ecol 15: 1419-1439.

Wapshere AJ. 1989. A testing sequence for reducing rejection of potential biological control agents for weeds. Ann Appl Biol 114: 515-526.

Ward MP. 2005. The role of immigration in the decline of an isolated migratory bird population. Conserv Biol 19: 1528-1536.

Ward SA, Leather SR, Pickup J, Harrington R. 1998. Mortality during dispersal and the cost of host-specificity in parasites: how many aphids find a host? J Anim Ecol 67: 763-773.

Watson RT. 1989. Aggressive display and territoriality of the bateleur *Terathopius ecaudatus*. S Afr J Zool 24: 146-150.

Watson RT. 1990a. Population dynamics of the bateleur in the Kruger National Park. Ostrich 61: 5-12.

Watson RT. 1990b. Breeding biology of the bateleur. Ostrich 61: 13-23.

Webb JW, Moran VC. 1978. The influence of the host plant on the population dynamics of *Acizzia russellae* (Homoptera: Psyllidae). Ecol Entomol 3: 313-321.

Weber M. 1999. The aim and structure of ecological theory. Philos Sci 66: 71-93.

Weiner J. 1995. On the practice of ecology. J Ecol 83: 153-158.

Weiss H, Courty MA, Wetterstrom W, Guichard F, Senior L, Meadow R, Curnow A. 1993. The genesis and collapse of third millennium north Mesopotamian civilization. Science 261: 995-1003.

Weiss SB, White RR, Murphy DD, Ehrlich PR. 1987. Growth and dispersal of larvae of the checkerspot butterfly *Euphydryas editha*. Oikos 50: 161-166.

Weiss SB, Murphy DD, White RR. 1988. Sun, slope, and butterflies—topographic determinants of habitat quality for *Euphydryas editha*. Ecology 69: 1486-1496.

Wellington WG, Fettes JJ, Turner KB, Belyea RM. 1950. Physical and biological indicators of the development of outbreaks of the spruce budworm, *Choristoneura fumiferana* (Clem.) (Lepidoptera: Tortricidae). Can J Res D 28: 308-331.

Wellington WG, Johnson DL, Lactin DJ. 1999. Weather and insects. Pages 313-353 in Huffaker CB, Gutierrez AP, eds. Ecological Entomology. New York: John Wiley.

Werner T, Koshikawa S, Williams TM, Carroll SB. 2010. Generation of a novel wing colour pattern by the Wingless morphogen. Nature 464: 1143-U1157.

White TCR. 1969. An index to measure weather-induced stress of trees associated with outbreaks of psyllids in Australia. Ecology 50: 905-909.

White TCR. 1970a. The nymphal stage of *Cardiaspina densitexta* (Homoptera: Psyllidae) on leaves of *Eucalyptus fasciculosa*. Aust J Zool 18: 273-293.

White TCR. 1970b. Some aspects of the life history, host selection, dispersal, and oviposition of adult *Cardiaspina densitexta* (Homoptera: Psyllidae). Aust J Zool 18: 105-117.

White TCR. 1993. The Inadequate Environment. Nitrogen and the Abundance of Animals. Berlin: Springer-Verlag.

White TCR. 2001. Opposing paradigms: regulation or limitation of populations? Oikos 93: 148-152.

White TCR. 2008. The role of food, weather and climate in limiting the abundance of animals. Biol Rev 83: 227-248.

Whittaker RH. 1953. A consideration of climax theory. Ecol Monogr 23: 41-78.

Whittaker RH. 1967. Gradient analysis of vegetation. Biol Rev 42: 207-264.

Whittaker RH. (Ed.). 1978. Ordination of Plant Communities. The Hague: W. Junk.

Wiens JA. 1976. Population responses to patchy environments. Annu Rev Ecol Syst 7: 81-120.

Wiens JA, Moss MR, Turner MG, Mladenoff DJ. 2007a. The causes and consequences of spatial pattern. Introduction and review. Pages 125-130 in Wiens JA, Moss MR, Turner MG, Mladenoff DJ, eds. Foundation Papers in Landscape Ecology. New York: Columbia University Press.

Wiens JA, Moss MR, Turner MG, Mladenoff DJ, (Eds). 2007b. Foundation Papers in Landscape Ecology. New York: Columbia University Press.

Wilcox C, Cairns BJ, Possingham HP. 2006. The role of habitat disturbance and recovery in metapopulation persistence. Ecology 87: 855-863.

Williams CB. 1964. Patterns in the Balance of Nature and Related Problems in Quantitative Ecology. London: Academic Press.

Williams GC. 1966. Adaptation and Natural Selection. A Critique of Some Current Evolutionary Thought. Princeton: Princeton University Press.

Williams TD. 1995. Bird Families of the World. The Penguins. Spheniscidae. Oxford: Oxford University Press.

Williamson JE, Steinberg PD. 2002. Reproductive cycle of the sea urchin *Holopneustes purpurascens* (Temnopleuridae: Echinodermata). Mar Biol 140: 519-532.

Williamson JE, Steinberg PD. 2012. Fitness benefits of size-dependent diet switching in a marine herbivore. Mar Biol 159: 1001-1010.

Williamson JE, Carson DG, De Nys R, Steinberg PD. 2004. Demographic consequences of an ontogenetic shift by a sea urchin in response to host plant chemistry. Ecology 85: 1355-1371.

Willmer PG, Stone GN, Johnston IA. 2000. Environmental Physiology of Animals. Oxford: Blackwell Science.

Wilson EO. 1975. Sociobiology: The New Synthesis. Cambridge: Harvard University Press.

Wilson K, Gatehouse AG. 1993. Seasonal and geographical variation in the migratory potential of outbreak populations of the African armyworm moth, *Spodoptera exempta*. J Anim Ecol 62: 169-181.

Winkworth. 1973. Eco-physiology of mulga (*Acacia aneura*). Trop Grassl 7: 43-48.

Wolda H. 1978a. Fluctuations in abundance of tropical insects. Am Nat 112: 1017-1045.

Wolda H. 1978b. Seasonal fluctuations in rainfall, food and abundance of tropical insects. J Anim Ecol 47: 369-381.

Wolda H. 1983. "Long-term" stability of tropical insect populations. Res Popul Ecol 25, Issue 3 Supplement: 112-126.

Wolda H. 1989. The equilibrium concept and density dependence tests—what does it all mean? Oecologia 81: 430-432.

Wolda H, Dennis B, Taper ML. 1994. Density dependence tests, and largely futile comments: Answers to Holyoak and Lawton (1993) and Hanski, Woiwod and Perry (1993). Oecologia 98: 229-234.

Worster D. 1977. Nature's Economy. A History of Ecological Ideas. Cambridge: Cambridge University Press.

Wynn T, Coolidge FL. 2012. How to Think Like a Neandertal. Oxford: Oxford University Press.

Yaku A, Walter GH, Najar-Rodriguez AJ. 2007. Thrips see red—flower colour and the host relationships of a polyphagous anthophilic thrips. Ecol Entomol 32: 527-535.

Young AM. 1982. Population Biology of Tropical Insects. New York: Plenum.

Young CC. 2002. In the Absence of Predators. Conservation and Controversy on the Kaibab Plateau. Lincoln: Nebraska University Press.

Young JZ. 1975. The Life of Mammals. Their Anatomy and Physiology. 2nd ed. Oxford: Clarendon Press.

Yu X, McPhaden MJ. 1999a. Dynamical analysis of seasonal and interannual variability in the Equatorial Pacific. J Phys Oceanogr 29: 2350-2369.

Yu X, McPhaden MJ. 1999b. Seasonal variability in the equatorial Pacific. J Phys Oceanogr 29: 925-947.

Zaki SAH, Jordan WC, Reichard M, Przybylski M, Smith C. 2008. A morphological and genetic analysis of the European bitterling species complex. Biol J Linn Soc 95: 337-347.

Zalucki MP, Daglish G, Firempong S, Twine P. 1986. The biology and ecology of *Heliothis armigera* (Hubner) and *H. punctigera* Wallengren (Lepidoptera: Noctuidae) in Australia: What do we know? Aust J Zool 34: 779-844.

Zalucki MP, Clarke AR, Malcolm SB. 2002. Ecology and behavior of first instar larval Lepidoptera. Annu Rev Entomol 47: 361-393.

Zhang DY, Lin K, Hanski I. 2004. Coexistence of cryptic species. Ecol Lett 7: 165-169.

Zobel M. 1992. Plant species coexistence—the role of historical, evolutionary and ecological factors. Oikos 65: 314-320.

Index

A

[ ] Phrases in square brackets indicate complementary sources of information within index.

( ) Terms in round brackets are qualifiers or refer to an entry in the glossary. Not all glossary entries are listed here.

D

P

Q

R